Italian Manufacturing Association Conference
XVI AITeM

XVI AITeM (Italian Manufacturing Association) Conference, held in Naples, in the premises of the School of Engineering of the University of "Naples Federico II" on 13-15 September 2023

Editors
Luigi Maria Galantucci[1], Luca Settineri[2]

[1] Dipartimento di Meccanica Matematica e Management, Politecnico di Bari, Bari, Italy

[2] Dipartimento di Ingegneria Gestionale e della Produzione (DIGEP), Politecnico di Torino, Torino, Italy

Peer review statement

All papers published in this volume of "Materials Research Proceedings" have been peer reviewed. The process of peer review was initiated and overseen by the above proceedings editors. All reviews were conducted by expert referees in accordance to Materials Research Forum LLC high standards.

Copyright © 2023 by authors

Published under License by **Materials Research Forum LLC**
Millersville, PA 17551, USA

Published as part of the proceedings series
Materials Research Proceedings
Volume 35 (2023)

ISSN 2474-3941 (Print)
ISSN 2474-395X (Online)

ISBN 978-1-64490-270-7 (Print)
ISBN 978-1-64490-271-4 (eBook)

This book contains information obtained from authentic and highly regarded sources. Reasonable efforts have been made to publish reliable data and information, but the author and publisher cannot assume responsibility for the validity of all materials or the consequences of their use. The authors and publishers have attempted to trace the copyright holders of all material reproduced in this publication and apologize to copyright holders if permission to publish in this form has not been obtained. If any copyright material has not been acknowledged please write and let us know so we may rectify in any future reprint.

Distributed worldwide by

Materials Research Forum LLC
105 Springdale Lane
Millersville, PA 17551
USA
https://www.mrforum.com

Manufactured in the United State of America
10 9 8 7 6 5 4 3 2 1

Table of Contents

Assembly, disassembly and circular economy

Process and system simulation, optimization and digital manufacturing

Materials processing technology

Work in progress

Preface

This book presents selected peer reviewed papers presented at the XVI AITeM (Italian Manufacturing Association) Conference, held in Naples, in the premises of the School of Engineering of the University of "Naples Federico II" on 13-15 September 2023.

AITeM is an officially recognised association dedicated to the academic and industrial world, with a special focus on the young. AITeM is a meeting point for Research and Innovation players – manufacturing companies, universities, other public research institutions, as well as professionals and other experts – discussing the most urgent issues within our industry. It promotes the study, the development of knowledge and the application of manufacturing technologies and systems in all their aspects:

- transformation processes of products, from production to assembly, from testing to recycling;
- mechanical and technological characterisation of the transformed materials;
- methodologies and tools for designing transformation systems (instrumental goods), processes, and components;
- programming, management and control of production, assembly, testing, and recycling systems;
- quality management and environmental safeguard management for sustainable development.

AITeM promotes also events and actions for disseminating knowledge within the manufacturing sector, either autonomously or in collaboration with national and international partners.

Every two years, AITeM organises a national conference: the Association's key moment of vitality is an occasion for meeting and discussing the latest manufacturing topics.

This Conference proposes as a meeting place for the scientific community in the sector of technologies and processing systems. It is aimed at all researchers working on the issues of manufacturing technologies.

The scientific program of the conference is divided into sessions of different nature which can guarantee wide participation in the works of the conference both in the presentation and discussion phases.

The sessions of the XVI AITeM Conference are:

- Manufacturing processes design, optimization and engineering
- Additive manufacturing and reverse engineering
- Assembly, disassembly and circular economy
- Process and system simulation, optimization and digital manufacturing
- Materials processing technology
- Miscellaneous
- Work in progress

The sessions contain the contributions coming mainly from AITeM members. They are designed to enhance different types of papers: the traditional Scientific sessions contain contributions of experienced members who want to share the results achieved with colleagues, the Work in

Progress session contains interventions related to ongoing activities and laboratories that offer new ideas and propositions for collaboration and talk more about future research activities.

The selection of papers was made by the Editorial Committee on the basis of peer reviews. More than 150 reviewers were involved. The reviews considered the coherence of the topic with the AITeM topics, the novelty with respect to the state of the art, the overall scientific quality of the work.

Editorial Committee

Luigi Maria Galantucci - Politecnico di Bari (President)

Elena Bassoli - Università di Modena e Reggio Emilia

Luca Boccarusso -Università di Napoli Federico II

Davide Campanella - Università di Palermo

Gianni Campatelli - Università di Firenze

Antonio Del Prete – Università del Salento

Enrico Pisino - Competence Center +CIM 4.0 Torino

Loredana Santo – Università di Roma Tor Vergata

Enrico Savio - Università di Padova

Walter Terkaj - STIIMA CNR

Organizing Committee

Luigi Carrino – Università di Napoli Federico II - Chairman

Antonino Squillace – Università di Napoli Federico II - Coordinator

Giovanni Moroni – Politecnico di Milano - President of the AITEM 2021 Conference

Fabrizio Memola Capece Minutolo - Università di Napoli Federico II – Delegated by the Board

Other Components from the Dipartimento Di Ingegneria Chimica, dei Materiali e della Produzione Industriale (DICMAPI) – Università Di Napoli Federico II:

Antonello Astarita, Luca Boccarusso, Doriana D'Addona, Massimo Durante, Antonio Formisano, Antonio Langella, Valentina Lopresto, Luigi Nele, Ilaria Papa, Umberto Prisco, Fabio Scherillo, Antonio Viscusi

AITeM - Italian Manufacturing Association

c/o Dipartimento di Ingegneria Chimica, dei Materiali e della produzione industriale dell'Università Federico II di Napoli

Piazzale V. Tecchio, 80, 80125 – Napoli

Phone: +39 3441203243

Email: segreteria@aitem.org

Website: www.aitem.org

Conference web site: https://www.aitem.org/convegno2023/il-convegno

Manufacturing processes design, optimization and engineering

Italian Manufacturing Association Conference - XVI AITeM   Materials Research Forum LLC
Materials Research Proceedings 35 (2023) 2-9   https://doi.org/10.21741/9781644902714-1

A bio-inspired reinterpretation of symbiotic human-robot collaboration in assembly processes

Federico Barravecchia[a]*, Mirco Bartolomei[b], Luca Mastrogiacomo[c] and Fiorenzo Franceschini[d]

Politecnico di Torino, DIGEP (Department of Management and Production Engineering), Torino, Italy

[a]federico.barracchia@polito.it, [b]mirco.bartolomei@polito.it, [c]luca.mastrogiacomo@polito.it, [d]fiorenzo.franceschini@polito.it

Keywords: Human-Robot Collaboration, Robotics, Bio-Inspired

Abstract. The emergence of collaborative robotics allowed humans and robots to work closely together to perform manufacturing activities. By combining their distinctive strengths and abilities, humans and robots can support each other in completing complex tasks. The relationship between humans and robots is frequently described in the literature as symbiotic. However, the concept of symbiosis, originally conceived in natural science, is often oversimplified as the mere exchange of mutual benefits. In practice, the term 'symbiosis' encompasses a wide range of interactions, ranging from relationships with positive impacts to relationships with negative impacts. Understanding the foundation of Human-Robot Symbiosis is crucial for its management. Two are the primary aims of this paper: (i) reinterpreting the collaborative tasks in assembly processes according to the properties of symbiotic relationships; (ii) proposing a novel approach for evaluating assembly tasks based on the bio-inspired features of symbiotic Human-Robot collaborative systems.

Introduction

Collaborative robotics refers to the integration of human operators and robots working together to achieve a common goal in manufacturing processes [1]. Unlike traditional robotics, which typically involves robots working independently and autonomously, collaborative robots (cobots) facilitate the active participation of human operators in the process [2,3]. This allows for combining the strengths and abilities of human operators and cobots to achieve greater efficiency, precision, and safety in tasks [4].

In the literature, several studies refer to Human-Robot Symbiosis as a type of collaboration in which humans and robots work together in a mutually beneficial relationship where respective strengths are exploited to improve the overall performance and efficiency of the system [5]. It is important to acknowledge that although symbiosis is expected to result in a mutually beneficial relationship, it encompasses positive and negative relationships where both parties can be negatively impacted. In the context of human-robot symbiosis, this means that while the collaboration may improve overall performance and efficiency, it can also lead to negative effects. In this consideration, a more nuanced understanding of the dynamics involved in Human-Robot Symbiosis can help avoid potential adverse outcomes and optimise the benefits.

In this consideration, this paper aims to present a novel bio-inspired perspective on Human-Robot Collaboration (HRC). By drawing parallels to the relationships between organisms in natural ecosystems, the study seeks to deepen our understanding of human-robot symbiosis. To accomplish this, the paper proposes a categorisation of potential symbiotic relationships between humans and robots, examining them in detail and identifying the elements of exchange (symbiotic factors) that shape the relationship. Additionally, the research introduces a practical evaluation method, which can be used to discern the nature of the specific relationships established in collaborative assembly tasks, thus, identifying areas of strength and weakness and opportunities for improvement.

Human-Robot interactions

Based on literature, a summary of possible human-robot interactions is presented below, listed in order of complexity [5–7]:

- *Coexistence/Autarky*: refers to case in which human and robot performs different task with different work goals, but they share the physical space.
- *Supervising*: in this type of interaction, the robot has limited autonomy and requires constant input and direction from the human operator. The tasks are performed simultaneously and towards the same goals, but the robot has limited independence, and adaptability is not a requirement.
- *Cooperation*: refers to the coordinated effort between humans and robots to achieve a common goal, with each party working on a specific task or set of tasks. In this sense, cooperation can be defined as a structured way of working together, where roles and responsibilities are clearly defined and there is a clear division of activities.
- *Supportive*: robots or humans can act in a supportive way, i.e. in a master-slave relationship. Despite the sharing of the objective, resources and workspace there is no autonomy in the decision of the task for the supporter.
- *Collaboration*: refers to a process where robots and humans share tasks, information, and resources to achieve a common goal. Operations are carried out simultaneously and in direct contact, the autonomy in carrying out operations is divided equally between the agents.
- *Symbiotic Collaboration*: in this kind of interaction human and robot are mutually dependent on each other, the robot and human work together in a complementary way.

Reinterpreting Human-Robot Symbiosis

This section provides an overview of the various types of symbiotic relationships that can be outlined between humans and robots within the context of symbiotic collaboration. The taxonomy aims to support the analysis and design of human-robot symbiotic relationships. The same symbiotic relationships found in nature can be used to categorise potential symbiotic relationships between humans and robots. These include six typologies: mutualism, commensalism, parasitism, amensalism, incompatibility and neutralism. Following this scheme, Figure 1.A outlines the framework of the possible Human-Robot symbiotic relationships. In detail, the symbiotic relationships are the following:

- *Mutualism*, it refers to a symbiotic relationship where both the human and the robot benefit from working together towards a common goal. In HRC, this relationship can occur when the robot performs repetitive and physically demanding tasks while the human worker focuses on tasks requiring cognitive skills. An example is in an assembly process where the robot's precision and speed in completing repetitive tasks increases overall efficiency, and the human's cognitive skills enhance quality control, resulting in a mutually beneficial outcome.
- *Commensalism*, it is a relationship between humans and robots where one agent benefits, while the other is neither helped nor harmed. An example is using a robot to lift and move heavy finished products at the end of an assembly process. This benefits the human by reducing workload and risk of injury, while the robot is not directly impacted positively or negatively by the human's presence.
- *Parasitism*, it refers to a symbiotic relationship where one agent benefits at the expense of the other. In the context of HRC, an example of human-robot parasitism could be when the

robot is assigned a task that a human worker can complete faster. This negatively impacts the robot's efficiency, while the human worker benefits by saving physical effort.

- *Amensalism*, it is a symbiotic relationship where one agent has a negative effect on the other without any benefit to itself. An example in HRC is when a robot emits high levels of noise or vibrations, interfering with the human worker's ability to communicate and hear warning signals, leading to an increased risk of accidents. The human worker is negatively impacted by the robot's presence, while the robot doesn't benefit from the human's presence.
- *Incompatibility*, it occurs when the human and robot are unable to work together effectively or safely. An example is during a robotised welding task, where the robot can pose a risk to the worker's safety by exposing them to the welding flame, while the presence of the human worker can also hinder the robot's movement and speed.
- *Neutralism*, it is a symbiotic relationship where both the human and robot coexist without impacting or affecting each other. This can occur when they are working on different tasks or in different areas and do not interact with each other. In such cases, the mutual impacts are negligible, and neither agent benefits nor is harmed by the presence of the other.

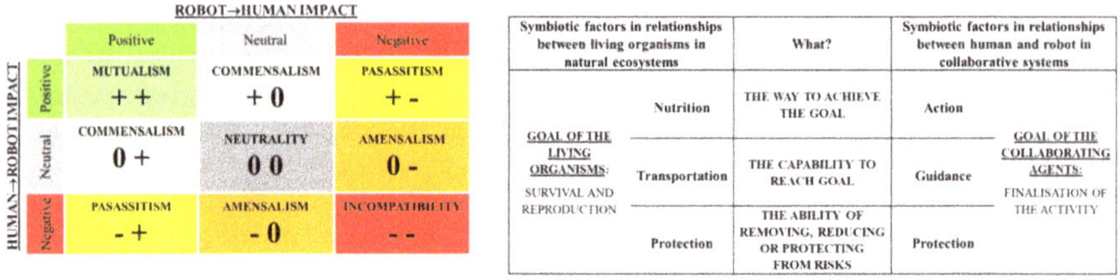

Fig. 1. (A) Classification of symbiotic human-robot relationships. Legenda: "+" positive impact of the relationship. "0" neutral impact of the relationship. "-" negative impact of the relationship. (B) Symbiotic factors in natural ecosystems and in collaborative systems.

Symbiotic relationships between living organisms are regulated by the exchange of symbiotic factors. To fully optimise human-robot symbiosis, it is necessary to identify the symbiotic factors exchanged between humans and robots and understand how they operate in the interaction. In order to identify human-robot symbiotic factors, we took a two-step approach. Firstly, we examined natural symbiotic relationships and then, through analogy, we identified the relevant symbiotic factors for HRC (see Figure 1.B).

Living organisms typically exchange nutrition, transportation and protection [8]. To find an analogy between natural symbiotic factors and human-robot symbiotic factors, we initially defined the objectives of the two types of symbioses. The symbiosis between living organisms aims to allow the survival and reproduction of natural organisms. On the other hand, the goal of the symbiotic relationship between collaborative agents (humans and robots) is to complete a task or an activity.

By analogy, considering the different objectives of the interaction, we identified the symbiotic factors between humans and robots as *action*, *guidance* and *protection*. Figure 1.B depicts the analogy process followed for the definition of the HRC symbiotic factors, which can be described as follows:

- *Action*, it refers to the process of doing or receiving the concrete actions that are necessary to complete a task. It encompasses the physical actions of the agents, such as grasping, moving, and manipulating objects.

- *Guidance*, it refers to the capability of an agent, whether human or robot, to lead the completion of an activity through understanding what needs to be done and sharing that knowledge with the other agent.
- *Protection*, it pertains to the ability of an agent to safeguard the other agent from any threats that may arise from the collaboration. This can include physical hazards, such as collision or malfunction, as well as ergonomic and psychological risks, such as repetitive stress injuries.

Evaluating symbiotic human-robot collaboration in assembly processes

This section introduces an evaluation tool designed to determine the nature of the relationship between humans and robots during collaborative processes. In detail, the proposed approach focuses on the analysis of existing collaborative processes.

The evaluation tool is based on the assessments of a team of experts who, after observing a collaborative task, assigns a rating to each symbiotic factor introduced in the previous sections. These factors (action, guidance and protection) are further detailed into specific dimensions to capture the distinguishing features of the symbiotic human-robot relationship.

In detail, the action factor is broken down into two dimensions:

- *Effort*: agents can provide the necessary effort to complete the task, or they can cause an increase in effort for the other agent.
- *Speed*: agents can speed up or slow down the execution of the task.

The guidance factor is divided into two specific aspects:

- *Knowledge*: agents can know and share the sequence of activities to be completed.
- *Decision-making*: agents can use their decision-making ability to choose which task to perform.

The protection factor is decomposed into the following dimensions:

- *Ergonomics*: the activity of one agent may affect the working conditions and ergonomics (physical and mental) of the other agent.
- *Safety*: agents can expose/protect the other agent from risks or threats

The evaluations focus on the individual elementary tasks of the assembly process. The team of experts uses the evaluation items listed in Table 1 to rate the mutual impact of the agents on each of these dimensions. The term impact is used here to refer to the effects or consequences the cobot has on human, and vice versa. The evaluations are expressed on a 7-level ordinal scale ranging from L1 (very negative impact) to L7 (very positive impact). The intermediate level (L4) represents the absence of impact on the dimension of analysis [9]

The combination of the partial impact ratings of the six dimensions allows for an assessment of the total impact of the relationship. The impact, whether it be from the robot to the human or vice versa, is determined by taking into account both the importance assigned to each dimension and the specific partial impacts within those dimensions.

To comprehensively evaluate the total impact of an agent on the other across all six dimensions, it is essential to adopt an effective aggregation method. One such approach may be the ME-MCDM (Multi Expert - Multi Criteria Decision Making) method [10–12]. The ME-MCDM method involves the use of max, min, and negation operators to combine linguistic information provided for non-equally important criteria [10,11]. According to the ME-MCDM method, the total impact (*TI*) can be calculated as follows [10,11]:

$$TI = \min_{k}[max(Neg(I_k), V_k)]. \tag{1}$$

Italian Manufacturing Association Conference - XVI AITeM   Materials Research Forum LLC
Materials Research Proceedings 35 (2023) 2-9   https://doi.org/10.21741/9781644902714-1

Being:

k the dimension of analysis, V_k the partial impact related to the k-th dimension, I_k the importance of the k-th dimension, $Neg(I_k)$ the negation of I_k. $Neg(L_i) = L_{q-i+1}$ where q is the number of rating level, for instance $Neg(L_7) = L_1$, $Neg(L_6) = L_2$ and $Neg(L_1) = L_7$.

The underlying logic of this method is that while low-importance criteria should have only a minimal impact on the overall aggregated value, highly important determinants should significantly contribute to the definition of the aggregated evaluation.

Table 2 illustrates a fictitious example of how the ME-MCDM method is applied in practice.

Case study

A simple case study is described to illustrate the application of the methodology in a real-world scenario. The case study concerns the collaborative assembly of a mechanical component, as shown in Figure 2.A. The assembly process was conducted within a collaborative environment with the involvement of a UR3-Universal Robot Cobot (see Figure 2.B).

The assembly process was decomposed into six elementary tasks (see the first column in Table 3), and through the rating of the 6 dimensions of analysis (see Table 1), the impacts of the agent's activity on the counterpart were evaluated. In the presented analyses, the weight of each sub-dimension was considered as follows: *effort* and *speed* were rated as very important (L6), while the other dimensions, including *guidance*, *decision-making*, *ergonomics*, and *safety*, were rated as slightly important (L3). The simplicity of the assembly operation and the absence of significant risks for the operator led the team of experts to assign greater importance to the sub-dimensions of the action compared to the other.

As an example, let us consider elementary task 5, which involves fixing an oval flange to the base. During this task, the cobot holds the flange in position while the human worker tightens the screws. In this case, the team of experts rated the impact of the cobot on the human worker's effort and speed as moderately positive (L6), as the cobot secures the workpiece, freeing the human worker's hands to tighten screws more easily and rapidly. Furthermore, the impact of the cobot on the human worker's knowledge was rated as slightly positive (L5), as the cobot's clamping of the oval flange indicates the manner in which the task is to be executed, thus providing guidance to the human worker. The impact on the other dimensions of analysis was rated as neutral (L4). On the other hand, the impact of the human on the robot has been rated as very positive (L7) for effort and speed, since the cobot would not be able to perform the task autonomously. The impact on the other dimensions of analysis was rated as neutral (L4).

By utilising the ME-MCDM aggregation technique, the outcome reveals in elementary task 5 a mutualistic relationship between the human and the cobot, as indicated by the positive total impact (L5) score for both.

The comprehensive outcomes of the analysis and the combined impact values for each elementary task are reported in Table 3. The relationship map depicted in Figure 2.C supports the identification of the resulting symbiotic relationship between humans and robots.

The analysis provides a preliminary foundation for optimising the collaborative assembly process. As an example, Task 4 was found to exhibit a parasitic relationship in which the robot gained an advantage at the expense of the human worker. Specifically, the cobot leaves the task of placing the oval flange in the correct position to be performed by the human worker. This has a negative impact on the human worker. After analysing the relationship, the need to redesign the task has emerged. This redesign involves assigning the responsibility of the task to the robot, thereby reducing the workload for the human worker.

Italian Manufacturing Association Conference - XVI AITeM   Materials Research Forum LLC
Materials Research Proceedings 35 (2023) 2-9   https://doi.org/10.21741/9781644902714-1

Tab. 1. Dimensions of analysis and rating scales.

Tab. 2. Application of the ME-MCDM method to a fictitious example (steps of the calculation).

Dimension (k)	Effort	Speed	Knowledge	Decision-making	Ergonomics	Safety
Importance (I_k)	L7	L4	L5	L5	L7	L7
Partial Impact (V_k)	L6	L2	L5	L4	L6	L4
$Neg(I_k)$	L1	L4	L3	L3	L1	L1
$max(Neg(I_k),V_k)$	L6	L4	L5	L4	L6	L4
Total Impact $\min_k[max(Neg(I_k),V_k)]$			L4			

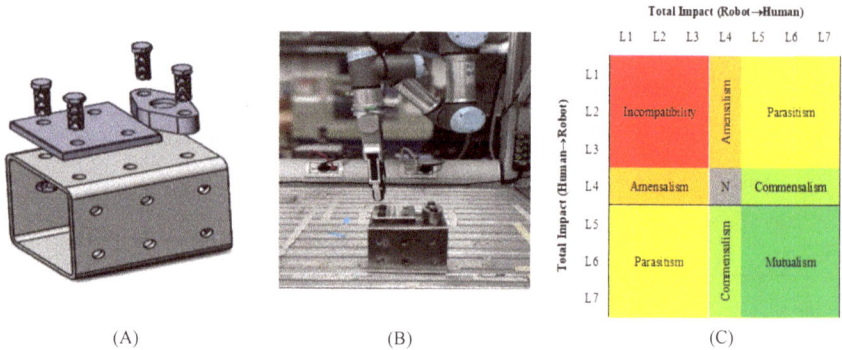

Fig. 2. (A) Scheme of the assembled mechanical equipment. (B) Snapshot of collaborative robot UR3e during the assembly process. (C) Relationship map. "N" refers to the relationship of neutralism.

List of elementary tasks	Allocation	Robot → Human						Human → Robot						Human → Robot Total Impact	Robot → Human Total impact	Relationship
		Effort	Speed	Knowledge	Decision making	Ergonomics	Safety	Effort	Speed	Knowledge	Decision making	Ergonomics	Safety			
1. Placement of the base in the working area.	R	L7	L5	L5	L4	L6	L4	L4	L4	L4	L4	L4	L4	L5	L4	C
2. Placement of the square flange on the base.	R	L6	L5	L5	L4	L6	L4	L4	L4	L4	L4	L4	L4	L5	L4	C
3. Fixing the square flange to the base with a pair of screws and nuts.	H	L6	L6	L5	L4	L4	L4	L7	L7	L4	L4	L4	L4	L5	L5	M
4. Placement of the oval flange on the base.	H	L3	L3	L4	L4	L4	L4	L5	L5	L5	L4	L4	L4	L3	L5	P
5. Fixing the oval flange to the base with a pair of screws and nuts.	H	L6	L6	L5	L4	L4	L4	L7	L7	L4	L4	L4	L4	L5	L5	M
6. Placement of the assembled component in another working area.	R	L7	L7	L4	L4	L7	L4	L4	L4	L4	L4	L4	L4	L5	L4	C

Tab. 3. List of elementary task and outcomes of the evaluation method. Allocation: H=human, C=cobot. Relationships: C=commensalism, M=Mutualism, P= Parasitism.

Conclusions

This article aims to provide a new perspective on Human-Robot Collaboration (HRC) by proposing a bio-inspired taxonomy of symbiotic relationships between humans and robots. The study identifies six different types of relationships depending on the type of impact generated by the robot on the human and vice versa. The proposed taxonomy can help to provide a comprehensive understanding of the nature of the interaction between humans and robots and provide a foundation for designing, evaluating, and improving HRC systems.

To apply the proposed perspective, an evaluation method to analyse the elementary tasks of an assembly process to identify relationships between humans and robots has been developed. The method enables the identification of potential areas for improvement, leading to optimised and enhanced HRC.

Italian Manufacturing Association Conference - XVI AITeM   Materials Research Forum LLC
Materials Research Proceedings 35 (2023) 2-9   https://doi.org/10.21741/9781644902714-1

The proposed framework presents some limitations, as it only considers direct interactions and overlooks the broader organizational context. Additionally, the evaluation tool provides a static representation of relationships without accounting for their evolution over time or potential skill loss.

Regarding the future, our aim is to further develop and refine our approach, with the goal of incorporating it into early design activities for HRC systems. The proposed perspective on Human-Robot Symbiosis could provide valuable insights for designers to develop effective and efficient HRC processes in manufacturing contexts.

References

[1] Z.M. Bi, M. Luo, Z. Miao, B. Zhang, W.J. Zhang, L. Wang, Safety assurance mechanisms of collaborative robotic systems in manufacturing, Robot Comput Integr Manuf. 67 (2021). https://doi.org/10.1016/j.rcim.2020.102022

[2] V. Villani, F. Pini, F. Leali, C. Secchi, Survey on human–robot collaboration in industrial settings: Safety, intuitive interfaces and applications, Mechatronics. 55 (2018). https://doi.org/10.1016/j.mechatronics.2018.02.009

[3] R. Gervasi, F. Barravecchia, L. Mastrogiacomo, F. Franceschini, Applications of affective computing in human-robot interaction: State-of-art and challenges for manufacturing, Proc Inst Mech Eng B J Eng Manuf. (2022). https://doi.org/10.1177/09544054221121888

[4] F. Barravecchia, L. Mastrogiacomo, F. Franceschini, A general cost model to assess the implementation of collaborative robots in assembly processes, International Journal of Advanced Manufacturing Technology. (2023). https://doi.org/10.1007/s00170-023-10942-z

[5] L. Wang, R. Gao, J. Váncza, J. Krüger, X. V. Wang, S. Makris, G. Chryssolouris, Symbiotic human-robot collaborative assembly, CIRP Annals. 68 (2019). https://doi.org/10.1016/j.cirp.2019.05.002

[6] S. El Zaatari, M. Marei, W. Li, Z. Usman, Cobot programming for collaborative industrial tasks: An overview, Rob Auton Syst. 116 (2019). https://doi.org/10.1016/j.robot.2019.03.003

[7] R. Müller, M. Vette, O. Mailahn, Process-oriented Task Assignment for Assembly Processes with Human-robot Interaction, in: Procedia CIRP, 2016. https://doi.org/10.1016/j.procir.2016.02.080

[8] M. Begon, C.R. Townsend, J.L. Harper, Ecology: From Individuals to Ecosystems, 4th Edition, Blackwell Publishing. (2005).

[9] F. Franceschini, M. Galetto, D. Maisano, Management for Professionals Designing Performance Measurement Systems, n.d. http://www.springer.com/series/10101

[10] R.R. Yager, Non-numeric multi-criteria multi-person decision making, Group Decis Negot. 2 (1993). https://doi.org/10.1007/BF01384404

[11] R.R. Yager, An approach to ordinal decision making, International Journal of Approximate Reasoning. 12 (1995). https://doi.org/10.1016/0888-613X(94)00035-2

[12] F. Barravecchia, L. Mastrogiacomo, F. Franceschini, The player-interface method: a structured approach to support product-service systems concept generation, Journal of Engineering Design. 31 (2020). https://doi.org/10.1080/09544828.2020.1743822

Italian Manufacturing Association Conference - XVI AIToM
Materials Research Proceedings 35 (2023) 10-18

Materials Research Forum LLC
https://doi.org/10.21741/9781644902714-2

Addressing idle and waiting time in short term production planning

Erica Pastore[1,a] *, Arianna Alfieri[1,b] and Claudio Castiglione[1,c]

[1]Department of Management and Industrial Engineering, Politecnico di Torino, Torino, Italy

[a]erica.pastore@polito.it, [b]arianna.alfieri@polito.it, [c]claudio.castiglione@polito.it

Keywords: Production Planning, Scheduling, Flowshop

Abstract. Production systems are facing the increase of economic and sustainability challenges in managing production resources, demand variability and variety, and the increasing shortage of materials. Thus, short-term production planning must include several aspects and consider multiple objective functions simultaneously. In this context, controlling and optimizing waiting and idle times might lead to various benefits, as they are among the main cost sources in production systems and can affect the feasibility of operations from a technological perspective. While waiting time is related to the work in process, idle time refers to a low utilization rate, and both may generate inefficiency and costs. This paper studies how different emphasis to waiting and/or idle time can affect the solution of short-term production planning with several industrially relevant objectives.

Introduction

Short-term production planning deals with scheduling jobs to be produced in the shop floor to optimize one or more criteria. These operations might consider technological constraints related to the jobs to be produced, the processes, and the production resources; optimization criteria are related to costs and challenges the company faces. Among these costs, those related to idle and waiting times are largely important [1]. Long waiting times usually transfer in high work in process (WIP) levels, with consequent high inventory costs and low service level. Instead, long idle times usually imply low utilization rate, possibly due to resource over-sizing, and related costs.

Apart from costs, some industries and technological processes might avoid idle and/or waiting times. For instance, temperature or other characteristics of the materials might require that each operation immediately follows the previous one, thus not allowing any waiting time (*no-wait*) [2]. Similarly, resources might use materials or consumables that become unusable if the machine stays idle for too long (e.g., paint can dry); in such cases, idle time is not allowed (*no-idle*). In addition, interrupting some processes may generate high costs, thus limiting the number of interruptions on some stages can be beneficial, i.e., limiting the occurrences for the machine to move from busy to idle and vice-versa [3]. This occurs, for instance, in casting processes where interrupting a continuous production flow implies maintenance and extensive cleaning, which in turn causes extra costs and delays in the entire production [4].

Although the literature on short-term production addressing idle and waiting time is vast, usually such problems are addressed from the algorithmic point of view. On the contrary, the objective of this paper is to study how different approaches (i.e., different ways of addressing idle and waiting time) can affect the solution of short-term production planning with several industrially relevant objectives. The industrial/managerial implications are the focus of the paper, while the algorithmic side is out of the scope of the paper. The paper studies a permutation flow shop (PFS) production system (i.e., jobs undertake a set of operations on a set of workstations with the same order, and the job sequence on each machine is the same).

Literature Review

The short-term planning of permutation flow shops has been widely studied in the literature. Many optimization criteria have been addressed, such as makespan, total flow time, total tardiness and

Italian Manufacturing Association Conference - XVI AITeM   Materials Research Forum LLC
Materials Research Proceedings 35 (2023) 10-18   https://doi.org/10.21741/9781644902714-2

so on [e.g. 5,6]. As idle and waiting times are the focus of the paper, in the following, only the literature addressing these two performance measures is reviewed.

Most of the literature on short-term production planning addressing idle and waiting time focuses on developing heuristic and meta-heuristic algorithms to solve various problem variants; however, as this paper aims at investigating idle time at the industrial level, the reader is referred to [1,7,8,9] for reviews of the solution methods available in the literature.

When idle time is considered, depending on the technological processes of the shop floor and on the criteria to optimize, it has been assumed either to be avoided (totally or in part) or to be minimized. The *no-idle* PFS scheduling problem is NP-hard, and due to its complexity, many authors developed heuristic and meta-heuristic algorithms to solve it. This problem arises in many real-life production systems such as in the production of integrated circuits, where the costs of steppers are so high that idle time is not desired [7]; also, in fiberglass production and in foundries, some machines (e.g., furnaces, casting machines, ceramic roller kilns) cannot be easily turned off and restarted due to the long machine setup times [10]. Sometimes, only some of the workstations composing the flow shop must respect the no-idle conditions. For instance, in ceramic frit production, only the central fusing kiln has the no-idle constraint; also, in steel production, only the final casting phase needs not to stop (i.e., to be no-idle) while the previous operations can admit idle time [11]. In other cases, idle time can be allowed, but it is linked to higher costs (i.e., there is no technological constraint to avoid idle time); in these cases, some authors have addressed the problem of minimizing the total amount of idle time in the production system [1], or the problem of minimizing the number of interruptions [3]. In such cases, obviously, machines can be idle.

When waiting time is considered, depending on the technological process, jobs can be required not to wait between two consecutive operations. For instance, in steel manufacturing, after being heated to a specific temperature, hot slabs cannot wait before rolling operations; otherwise, their temperature would significantly drop [12]. In general, all the manufacturing processes that require the WIP to be pre-heated to a high temperature may need no-wait conditions [13,14]. Also, robotic cells, which provide a highly coordinated manufacturing process, need to avoid waiting time between consecutive operations [15]. In some cases, avoiding waiting time is an efficient strategy to reduce WIP-related costs [1]. In such cases, the no-wait condition may be downgraded to the minimization of the total waiting time of jobs in the system [1,16], which implies the possibility that jobs wait between consecutive operations to optimize some other performance measure.

Problem formulation

This paper considers a flow shop production system. In this system, there is a set of J jobs to be processed, each requiring a set of M operations. Each operation is allocated to a single machine, and the order in which the operations have to be executed is the same for each job. Thus, the sequence of the operations is the same as the order of the machines, as depicted in Fig. 1. In the system, machines cannot perform more than one operation simultaneously, and jobs cannot be processed by more than one machine simultaneously. The operations, once started, cannot be interrupted (*non-preemption* assumption). For simplicity, a permutation flow shop is considered, in which the sequence of jobs is the same on each machine. Various problem variants are studied according to the assumptions made on idle and waiting time and on the performance measure.

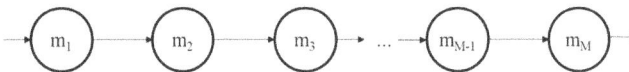

Fig. 1. Flow shop layout

Italian Manufacturing Association Conference - XVI AITeM
Materials Research Proceedings 35 (2023) 10-18

Materials Research Forum LLC
https://doi.org/10.21741/9781644902714-2

Specifically, the paper addresses the following systems:

- flow shop with allowed idle and waiting times (*general*),
- flow shop with no-idle constraint (*no-idle*),
- flow shop with no-wait constraint (*no-wait*).

For each system, four solutions are found, each minimizing one among the following objective functions (OF):

- *makespan* (*Cmax*), i.e., the completion time of the last job on the last machine;
- *total completion time* (*Ctot*), i.e., the sum of the completion times of all jobs in the sequence;
- *total core idle time* (*Cit*), i.e., the sum of idle times in all machines between the start of the first job of the sequence and the end of the last job;
- *total core waiting time* (*Cwt*), i.e., the sum of waiting times of all jobs between the first and the last operations.

In general, for any problem variant (i.e., for each combination system-OF), the aim is to decide how to process jobs (i.e., to find the job schedule) to optimize the selected OF. As the problem variants are mostly NP-hard, sub-optimal solutions are found in this paper by using a constructive heuristic algorithm. Specifically, the well-known NEH heuristic is adapted to the problem variants. The developed algorithm has the same general structure for all the problem variants, and only limited changes in the sorting rule (as explained in the following) are made to adapt it to each of them. The algorithm works as in the following.

1. Jobs are sorted according to a specific rule depending on which OF is minimized. For makespan and total completion time minimization (NEH_{Cmax}, NEH_{Ctot}), jobs are sorted according to the decreasing sum of processing times; for waiting time minimization (NEH_{Cwt}), jobs are sorted according to the index defined by [17] (that accounts for the variability of processing times); for idle time minimization (NEH_{Cit}), jobs are sorted according to the descending order of the index defined by [18] (that accounts for the variability, skewness and kurtosis of processing times).
2. Each job in the sorted list is inserted in the solution in the position that minimizes OF, thus originating the final schedule.

For each solution of each problem variant, all the OFs are evaluated and compared. The comparison aims at evaluating the differences among solutions (in terms of objective functions) found by modelling idle and waiting times in different ways.

Numerical results

The aim of the experiment is to assess the impact of the way idle and waiting times are modelled on the solution for short-term production planning. To this aim, the NEH is used to find solutions for all the problem variants previously described; the solutions are then compared with respect to their evaluated performance measures (*Cmax, Ctot, Cit, Cwt*). The experiment investigates the trade-off between imposing no-idle/no-wait conditions and paying idle/waiting time. This is particularly relevant for systems characterized by high idle and waiting time costs.

Design of experiment. The Taillard benchmark [19] is used to determine processing times of jobs on the machines. The number of jobs varies between 20 and 500, while the number of machines between 5 and 20. For each problem, 10 instances are available. For each instance, the three systems (general, no-idle, no-wait) are considered, and for each of them, four solutions are found by applying the NEH algorithm with different OFs (NEH_{Cmax}, NEH_{Ctot}, NEH_{Cit}, NEH_{Cwt}); note that in no-idle systems NEH_{Cit} cannot be used, as well as NEH_{Cwt} is not used in no-wait systems. Overall, 4800 experiments are run. For each solution, all the OF values are computed.

Results. Fig. 2 shows, for some of the considered problems, mean values and confidence intervals of some performance measures. Specifically, Fig.2 (a) shows the average values of *Cmax*,

Italian Manufacturing Association Conference - XVI AITeM   Materials Research Forum LLC
Materials Research Proceedings 35 (2023) 10-18   https://doi.org/10.21741/9781644902714-2

Ctot, Cit, Cwt of the solutions of the problems with 100 jobs. As an example, starting from the left part of the graph, the grey line shows the average completion time of solutions for no-idle systems found by minimizing *Cmax, Ctot* and *Cwt,* respectively; the second grey line shows the same for no-wait systems, and the third for general systems. Instead, Fig. 2 (b) shows the interval plot of makespan for all the problems with 500 jobs, grouped by system variant and NEH OF. As the figures show, for each system, minimizing different objective functions leads to difference system performance. For instance, in general systems, minimizing the waiting time leads to larger *Cmax* values, which, in turns, implies having low utilization levels of machines and lower production rates. Moreover, how idle and waiting times are modelled has an impact on the system performance measures. As Fig. 2 (a) shows, no-idle systems tend to have a larger waiting time than a general system in which idle time is minimized (i.e., general system – NEH_{Cit}). Obviously, if machines cannot be idle because of technological constraints, the consequence of increasing waiting times cannot be avoided. However, if machines can be idle but the idle time related cost is high, then minimizing idle time instead of imposing a no-idle condition can turn into lower waiting time (and, hence, WIP-related) costs. In this case, an economic trade-off should be evaluated.

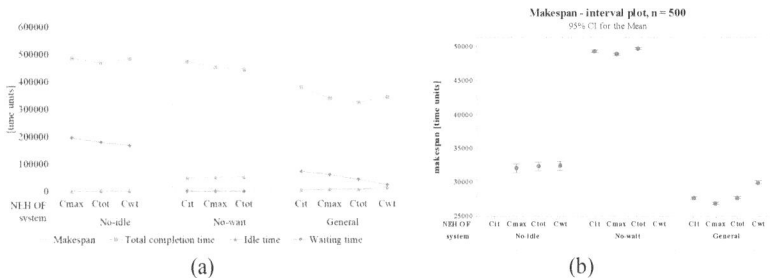

(a)   (b)

Fig. 2. (a) Average values of all performance measures for all problem variants with 100 jobs. (b) 95% confidence intervals of makespan in all problem variants with 500 jobs.

Table 1 shows the average values of all the performance measures, grouped by system, NEH OF, number of jobs and machines. The table shows that, as expected, the larger the instance, the larger the values of each measure. Obviously, for each system, each evaluated performance measure has its lowest value in the solution that minimizes it: for instance, for general systems, the makespan is lower in the solutions of NEH_{Cmax} than the one measured in the other solutions (the example refers to the rows with green borders of the table).

To assess the impact of the way idle and waiting times are modelled on the performance of the system, hypothesis tests are used for comparison. Starting from the *idle time*, it can be modelled either by imposing a no-idle condition, or by minimizing it as OF in a general system. The aim is to understand how the other performance measures (*Cmax, Ctot, Cwt*) change in these two cases. As an example, the blue rows of Table 1 display how makespan changes between a general system with NEH_{Cit} (denoted by $Cmax_{Gen,Cit}$) and a no-idle system with NEH_{Cmax} (denoted by $Cmax_{NI,Cmax}$). To evaluate the difference, the percentage difference Δ_{Cmax} is computed as:

$$\Delta_{Cmax} = \frac{Cmax_{Gen,Cit} - Cmax_{NI,Cmax}}{Cmax_{Gen,Cit}}. \qquad (1)$$

Table 1. Average performance measures for all problem variants, grouped by system, NEH OF, number of jobs and of machines

n. jobs		20			50			100			200		500
n. machines		5	10	20	5	10	20	5	10	20	10	20	20
Makespan [time units]													
System	**NEH OF**												
No-idle	C_{max}	1380	2007	3502	3015	3753	5877	5521	6887	9280	12174	15131	32032
	C_{tot}	1413	1995	3564	3068	3774	5935	5650	6980	9309	12428	15313	32326
	C_{wt}	1492	2061	3548	3124	3903	6069	5750	7073	9412	12597	15527	32433
No-wait	C_{it}	1494	2169	3302	3345	4578	6374	6381	8448	11361	16125	21225	49280
	C_{max}	1433	2006	3043	3312	4386	6090	6344	8291	11081	15810	20770	48835
	C_{tot}	1490	2092	3194	3440	4549	6263	6502	8585	11364	16114	21121	49696
General	C_{it}	1402	1802	2516	2913	3349	4274	5433	6038	7007	11133	12269	27650
	C_{max}	1261	1583	2319	2756	3135	3957	5272	5752	6634	10804	11753	26907
	C_{tot}	1343	1695	2463	2889	3351	4120	5470	6039	6973	11106	12217	27688
	C_{wt}	1439	1814	2561	3127	3640	4440	5846	6512	7547	12055	13396	29959
Total completion time [time units]													
No-idle	C_{max}	18308	31712	60716	91810	130469	235129	310747	449960	697524	1446917	2089681	9931072
	C_{tot}	17937	30567	61106	87618	126870	233150	293060	435216	680733	1409108	2045126	9695973
	C_{wt}	19455	31690	61326	92247	133463	240541	312290	449817	692068	1458294	2107398	9848739
No-wait	C_{it}	17998	26441	42571	95621	129115	184689	342539	452853	625838	1698083	2235262	12631925
	C_{max}	16602	24704	40486	87432	120392	176732	327895	430746	600214	1610098	2163203	12368030
	C_{tot}	15670	24252	39944	82172	118434	174832	302911	430587	599178	1586040	2179182	12578382
General	C_{it}	18768	25484	38795	90943	108210	147843	331945	365163	443562	1294577	1462361	7843045
	C_{max}	15530	22015	35586	77701	97664	134507	273004	328974	417138	1171820	1385930	7478628
	C_{tot}	14607	20945	34252	72579	92739	129430	264163	313461	398242	1129317	1314814	7173633
	C_{wt}	15472	21687	34054	76418	97054	132443	284449	331726	417910	1202852	1403763	7603568
Total core idle time [time units]													
No-wait	C_{it}	1456	7258	25707	3362	16172	57887	6118	29888	109181	55958	203359	465365
	C_{max}	1569	7503	26913	3702	16619	59145	6439	30527	109980	56117	204157	466044
	C_{tot}	1665	7710	27848	4025	17508	60405	6829	32241	113159	58204	209402	481218
General	C_{it}	140	953	4169	433	1312	6468	319	2377	8964	2707	11195	15356
	C_{max}	420	2012	7134	660	3044	10961	707	3726	15108	4648	18806	22500
	C_{tot}	625	2384	8401	970	3853	12511	1102	4978	17444	5874	22108	29321
	C_{wt}	1056	3488	10629	1894	6602	18320	2666	9374	28849	14480	44620	73747
Total core waiting time [time units]													
No-idle	C_{max}	4682	13360	32180	24955	49243	128670	62588	167571	357830	395643	920298	3420189
	C_{tot}	4602	12842	32520	20391	46117	126908	45026	153468	343209	358420	872828	3155334
	C_{wt}	4154	12960	31769	15171	45069	128630	32280	136670	337244	305421	865828	2923396
General	C_{it}	3880	7016	9717	13437	23579	40602	43084	67241	104676	175804	291344	1087660
	C_{max}	2329	4421	6684	13030	17163	30105	39687	55440	86463	168936	262840	1177607
	C_{tot}	1714	3556	5772	7460	12927	25641	19296	41601	67913	101469	182405	713631
	C_{wt}	899	2295	4195	2926	7309	16820	8203	19182	42068	45134	100114	340582

The tested hypothesis is $\Delta_{Cmax} = 0$. Over all the experiments, the average Δ_{Cmax} is -0.1663, and the T-test of the tested hypothesis has a p-value equal to zero, thus the mean percentage difference cannot be considered equal to zero. In practice, the makespan of solutions of no-idle systems in which the NEH_{Cmax} is used ($Cmax_{NI,Cmax}$) is larger than that of a general system that minimizes the idle time ($Cmax_{Gen,Cit}$). This means that imposing a no-idle condition on one hand avoids idle time related costs but, on the other hand, it increases the utilization related costs. Fig. 3 graphically displays the trade-off between these two performance measures. If the no-idle systems (red diamonds) do not have any cost related to idle times, the costs related to the makespan are larger than the general systems (blue circles).

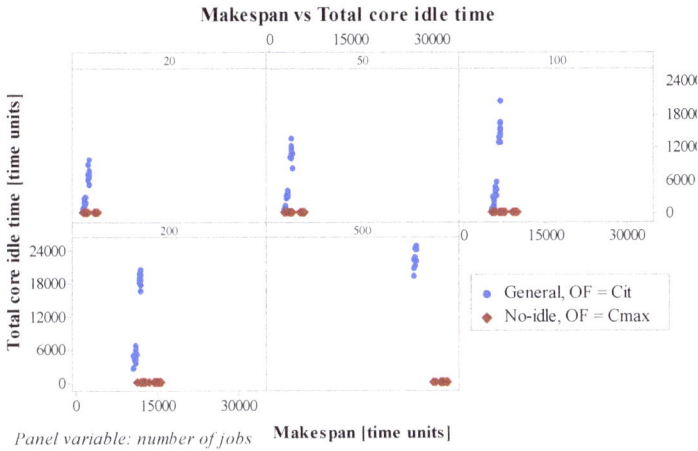

Fig. 3. Dispersion of cmax (x-axis) and cit (y-axis), grouped by number of jobs. Compared systems: general with NEH$_{Cit}$ (blue circles), no-idle with NEH$_{Cmax}$ (red diamonds).

The same analysis has been performed for total completion times and waiting times. In both cases, the difference between the two systems when such performance measures are considered is statistically significant.

The same comparison has been made to address the *waiting time* modelling. The considered systems are: general system with waiting time minimization (*Gen, Cwt*), no-wait systems with their OFs (*NW, Cmax – NW, Ctot – NW, Cit,* alternatively). As an example, let consider the total completion time as the performance measure to be evaluated; then, the compared systems are the ones written with the purple color in Table 1. The percentage difference is computed as:

$$\Delta_{Ctot} = \frac{Ctot_{Gen,Cwt} - Ctot_{NW,Ctot}}{Ctot_{Gen,Cwt}}. \tag{2}$$

Over all the experiments, the average Δ_{Ctot} is equal to -0.2708, and the T-test for the hypothesis $\Delta_{Ctot} = 0$ has a p-value equal to zero, thus the mean percentage difference differs from zero. In practice, minimizing the total completion time in a no-wait system leads to larger completion times than minimizing the total core waiting time in a general system. Fig. 4 graphically shows how the two measures are distributed.

Italian Manufacturing Association Conference - XVI AITeM
Materials Research Proceedings 35 (2023) 10-18

Materials Research Forum LLC
https://doi.org/10.21741/9781644902714-2

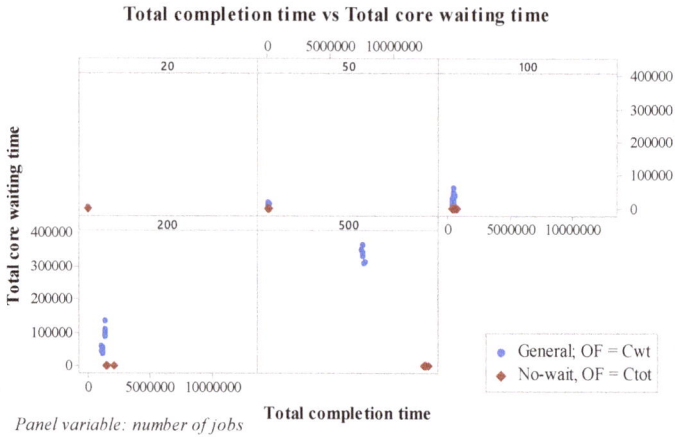

Total completion time vs Total core waiting time

Panel variable: number of jobs

Fig. 4. Dispersion of Ctot (x-axis) and Cwt (y-axis), grouped by number of jobs. Compared systems: general with NEH$_{Cwt}$ (blue circles), no-wait with NEH$_{Ctot}$ (red diamonds).

As the figure shows, imposing the no-wait condition increases the total completion time, leading to larger costs related to flow time, WIP and service level reduction.

Conclusions

Idle and waiting times are very relevant performance measures in production systems, as they are critical in some technological processes, and they both generate costs. In short-term production planning, they can be avoided by imposing no-idle and no-wait conditions, or they can be optimized to reduce them. This paper studies how different ways to model them can generate different schedules with different total costs, thus affecting the solution of short-term production planning. The addressed systems are permutation flow shops with and without no-idle/no-wait conditions, in which several performance measures are optimized. Numerical results on benchmark problems available in the literature show that the way idle and waiting times are modelled significantly impacts on other performance measures such as makespan and total completion time. These are in turns related to utilization, WIP, and flow time costs.

If the technological characteristics of the process or the materials impose no-idle/no-wait conditions, the other performance measures will suffer from these constraints, but no actions can be implemented to improve them. For all the other cases in which idle/waiting times can occur but with high costs, the economic trade-off of allowing some idle/waiting time but reducing makespan and/or total completion time related costs should be considered. The numerical results of the paper specifically show that sometimes allowing (and thus paying) idle/waiting times is beneficial to reduce other cost sources such as utilization and/or flow time, WIP, etc.

Finally, as the used heuristic algorithm, developed for different problems (even though adapted for the considered problem), could have had some effect on the performed comparisons, future research will be devoted to developing more sophisticated ad hoc solution algorithms that include the economic trade-off in finding the optimal solution.

Italian Manufacturing Association Conference - XVI AITeM
Materials Research Proceedings 35 (2023) 10-18

Materials Research Forum LLC
https://doi.org/10.21741/9781644902714-2

References

[1] A. Alfieri, M. Garraffa, E. Pastore, F. Salassa, Permutation flowshop problems minimizing core waiting time and core idle time. Comp. & Ind. Eng., 108983. (2023) https://doi.org/10.1016/j.cie.2023.108983

[2] H. Aydilek, A. Aydilek, M. Allahverdi, A. Allahverdi, More effective heuristics for a two-machine no-wait flowshop to minimize maximum lateness. Inter. J. of Ind. Eng. Comp., 13(4), (2022) 543-556. https://doi.org/10.5267/j.ijiec.2022.7.002

[3] H. Öztop, M.F. Tasgetiren, L. Kandiller, Q.K. Pan, Metaheuristics with restart and learning mechanisms for the no-idle flowshop scheduling problem with makespan criterion. Comp. & Oper. Res., 138, 105616. (2022) https://doi.org/10.1016/j.cor.2021.105616

[4] W. Höhn, T. Jacobs, N. Megow, On Eulerian extensions and their application to no-wait flowshop scheduling. J. of Sched., 15(3) (2012) 295-309. https://doi.org/10.1007/s10951-011-0241-1

[5] M. de Fátima Morais, M.H.D.M. Ribeiro, R.G. da Silva, V.C. Mariani, L. dos Santos Coelho, Discrete differential evolution metaheuristics for permutation flow shop scheduling problems. Comp. & Ind. Eng., 166, 107956. (2022) https://doi.org/10.1016/j.cie.2022.107956

[6] R.g. Saber, M. Ranjbar, Minimizing the total tardiness and the total carbon emissions in the permutation flow shop scheduling problem. Comp. & Oper. Res., 138, 105604. (2022) https://doi.org/10.1016/j.cor.2021.105604

[7] R. Ruiz, E. Vallada, C. Fernández-Martínez, Scheduling in flowshops with no-idle machines. in: Computational intelligence in flow shop and job shop scheduling, (2009) 21-51. https://doi.org/10.1007/978-3-642-02836-6_2

[10] P.J. Kalczynski, J. Kamburowski, A heuristic for minimizing the makespan in no-idle permutation flow shops. Comp. & Ind. Eng., 49(1) (2005) 146-154. https://doi.org/10.1016/j.cie.2005.05.002

[1] Q.K. Pan, R. Ruiz, An effective iterated greedy algorithm for the mixed no-idle permutation flowshop scheduling problem. Omega, 44 (2014) 41-50. https://doi.org/10.1016/j.omega.2013.10.002

[12] H. Yuan, Y. Jing, J. Huang, T. Ren, Optimal research and numerical simulation for scheduling no-wait flow shop in steel production. J. of Appl. Math. (2013) https://doi.org/10.1155/2013/498282

[13] R. Alvarez-Valdés, A. Fuertes, J.M. Tamarit, g. Giménez, R. Ramos, A heuristic to schedule flexible job-shop in a glass factory. Eur. J. of Oper. Res., 165(2) (2005) 525-534. https://doi.org/10.1016/j.ejor.2004.04.020

[14] B. Na, S. Ahmed, G. Nemhauser, J. Sokol, A cutting and scheduling problem in float glass manufacturing. J. of Sched., 17 (2014) 95-107. https://doi.org/10.1007/s10951-013-0335-z

[15] A. Agnetis, D. Pacciarelli, Part sequencing in three-machine no-wait robotic cells. Oper. Res. Letters, 27(4), (2000) 185-192. https://doi.org/10.1016/S0167-6377(00)00046-8

[16] A.P. De Abreu, H.Y. Fuchigami, An efficiency and robustness analysis of warm-start mathematical models for idle and waiting times optimization in the flow shop. Comp. & Ind. Eng., 166, (2022) 107976. https://doi.org/10.1016/j.cie.2022.107976

[8] H. Singh, J.S. Oberoi, D. Singh, Multi-objective permutation and non-permutation flow shop scheduling problems with no-wait: a systematic literature review. RAIRO-Oper. Res., 55(1), (2021) 27-50. https://doi.org/10.1051/ro/2020055

[9] A. Allahverdi, A survey of scheduling problems with no-wait in process. Eur. J. of Oper. Res., 255(3), (2016) 665-686. https://doi.org/10.1016/j.ejor.2016.05.036

[17] K. Maassen, A. Hipp, P. Perez-Gonzalez, Constructive heuristics for the minimization of core waiting time in permutation flow shop problems. In 2019 International Conference on Industrial Engineering and Systems Management (IESM) (pp. 1-6). IEEE. (2019) https://doi.org/10.1109/IESM45758.2019.8948147

[18] W. Liu, Y. Jin, M. & Price, A new Nawaz-Enscore-Ham-based heuristic for permutation flow-shop problems with bicriteria of makespan and machine idle time. Eng. Opt., 48(10), (2019) 1808-1822. https://doi.org/10.1080/0305215X.2016.1141202

[19] E. Taillard, Benchmarks for basic scheduling problems. Eur. J. of Oper. Res., 64(2), (1993) 278-285. https://doi.org/10.1016/0377-2217(93)90182-M

Italian Manufacturing Association Conference - XVI AITeM
Materials Research Proceedings 35 (2023) 19-27

Materials Research Forum LLC
https://doi.org/10.21741/9781644902714-3

A numerical methodology for improving the thermoforming process of complex thermoplastic composite components

Antonios G. Stamopoulos[1, *], Francesco Lambiase[1] and Alfonso Paoletti[1]

[1] University of L'Aquila, Department of Industrial and Information Engineering and Economy, Piazzale Pontieri 1, Roio, L'Aquila, Italy

Keywords: Thermoforming, Composites, Automotive

Abstract. In recent years, there has been an increasing demand for lightweight composite structures. Thermoplastic composite materials appear to be a very promising solution to this direction considering their unique aspects and their capability to be heated and stamped. Nevertheless, the cost of the development of the dies that are necessary to fulfill the requirements of the process was based, until recently, on trial-and-error tests. In the present work, an automotive component is considered, and the corresponding dies are developed using a fast and efficient process simulation software. The process parameters as well as the characteristics of the cavity are defined aiming to minimize the process induced defects.

Introduction

Since the introduction of new mobility sollutions based on partially or fully electric vehicles, overall vehicle weight reduction is essential for augmenting their efficiency. Needless to mention that the overall structural performance should remain at least the same or even ameliorated. To this end, thermoplastic composite materials are considered as an appealing solution that could bring many advantages such as superior mechanical performance with lower weight, long term conservation without requirements and the ease to produce and among them a combination of superior mechanical performance and weight saving. For the aeronautical industry, both the non-payload and the payload are factors that influence the fuel consumption [1-2] as well as for the automotive industry in which is widely accepted that a weight saving of 100 kg could potentially reduce the vehicle consumption significantly [3].

Among the existing variety of composite materials in terms of both fibre and matrix systems, the thermoplastic-based materials are very appealing, especially to the automotive industry, due to their unique features such as their recyclability and their low requirements of energy spent for both storage and production [4]. Most of these components are fabricated using the traditional injection molding process [4-5] where also short fiber composites may be used. Whenever the quality and structural performance of the produced parts is important, continuous fibers are implemented, usually in the form of textiles (woven or braided). Among the existing techniques to this direction is the thermoforming process [6] that is based on the hot stamping of a semi-finished composite product (plate) by also applying temperature cycle that brings the composite plate to its melting point or even above it. After the stamping and the cooling down of the component, the final product may be extracted.

Even though significant research has been performed in the field of the formability of composite materials, there is a significant number of parameters that have a strong effect on the output. Firstly, the material type in terms of textile type reinforcement and the polymeric matrix system that defines the temperature window in which the composite plate can be thermoformed [5-7]. Secondly the relative crosshead speed between the stamping and mold tools. Nevertheless, one important parameter is the geometry of the component that needs to be produced that defines the cavity of the dies. In research level, the majority of works are referring to simple geometries such

Italian Manufacturing Association Conference - XVI AIIeM   Materials Research Forum LLC
Materials Research Proceedings 35 (2023) 19-27   https://doi.org/10.21741/9781644902714-3

as hemispheres [8-11] or double-dome geometries [12-13] that are considered as the benchmark ones.

Among the most important aspects is the development of the dies, especially considering the increasing costs of both the materials and the energy required. Therefore, a lot of attention is paid on reducing the design to production time, an important part of which is the accurate design of the molds for hosting the composite semi-finished product. Most of this part was, until recently, conducted applying a trial-and-error experimental procedure where the defects (wrinkles, textile shearing, polymer mitigation, undesired thickness variation) and product quality was observed and adjustments were made to the molds design to minimize them. However, in the recent years there have been observed FE thermoforming process simulations of hemispheres [8-9,11] or double domes [13] and more recently case studies of significantly more complex geometries such as battery trays [14] or other automotive components [15]. It should be noted that when it comes to the study of more complex, non-symmetric components production, there is no protocol to follow and, therefore, the procedure is based mostly on empirical observations.

In the present work, a non-symmetric triangular profile automotive component (corner bracket) is considered. Numerical analyses of the thermoforming process using beyond stat-of-the-art FE models are conducted, based on the outcome of an experimental campaign for characterizing the visco-elastic behaviour of the composite material. Several adjustments are imposed in the geometry of the dies for assessing their effect on the formation of wrinkles, textile shearing and residual stresses. The results were analysed using indicators for understanding the provenance with respect to the material forming behaviour. This way, a methodology is proposed that inter-connects the dies geometry design with the production of the component.

Methodology Outline

As previously mentioned, the component of interest is a 3-way corner bracket that is used as the basis for welding the laminates of the rear part of the cabin of a commercial vehicle. The methodology followed is depicted in Fig.1. It starts with the definition of the input, in terms of the component to produce, and the composite material forming behavior.

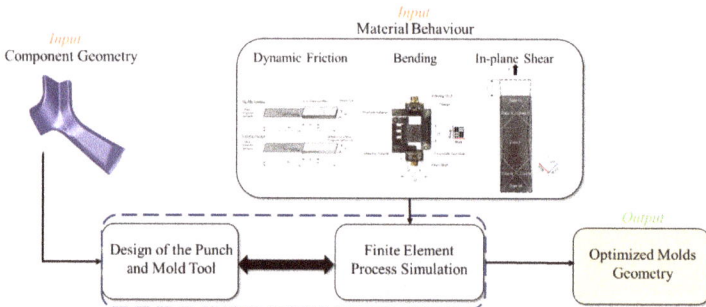

Fig. 1: Outline of the numerical methodology adopted by the present work.

Since the geometry is very complex, a first step is the definition of an initial design of the molds (punch and mold tool) that complies with the characteristics regarding the alignment of the component and the principal directions of the reinforcement. Subsequently, a first finite element simulation of the thermoforming process takes place for identifying the critical zones of high probability for wrinkles and the residual stresses as well as the fiber shearing (the in-plane deformation of the textile fabric). The ideal conditions in terms of stamping speed and material temperature are defined. Considering the fact that the component is highly non-symmetric, the cavity characteristics in terms of orientation and inclination insider the molds assessed as a method

Italian Manufacturing Association Conference - XVI AITeM   Materials Research Forum LLC
Materials Research Proceedings 35 (2023) 19-27   https://doi.org/10.21741/9781644902714-3

for relieving the final product from residual stresses and for eliminating the wrinkles. Therefore, after having conducted a limited number of finite element simulations, the optimized molds may be obtained, saving a lot of time and reducing the cost of a potential trial-and-error experimental investigation.

Materials

The composite material utilized is the TEPEX 104 RG600(x)/47% (Bond Laminates GmbH, Brilon, Germany). It consists of a polypropylene matrix system reinforced by 47% with an E-glass 2-2 twill weave fabric. The fabric is balanced (50/50) having equal number of fibers in the warp and weft directions. Its nominal ply thickness is roughly 0.5 mm. This composite semi-finished plate is produced in the European Union using a continuous press process where 2 films of polypropylene and the textile are pressed under a well-defined pressure and temperature cycle to achieve a complete impregnation. According to past works [14], the composite material's melting point is roughly 164 °C while the solidification one at 121 °C. A very interesting part is the temperature opening windows that define the onset of both processes. For the melting process, the onset temperature is 127 °C while the extrapolated end is 172 °C. For the crystallization onset, the temperature is at 130 °C while the material crystallization ends at 110 °C. To this end, the temperatures 160 °C and 190 °C are considered, since at 140 °C the material is fully (when heated up) or partially (when firstly melted and then cooled down) solidified. The data regarding its forming behaviour in the 2 temperatures of interest and in various stamping speeds were adopted from past research [14].

Thermoforming Simulation

Starting with the necessary assumptions, the component studied in the present work should be made of 3 layers of the TEPEX 104 RG600(x)/47% composite material. In addition, the fabrication process should be horizontal and not vertical for mainly two reasons; the first one is related to the gravity which can cause a shaggy deformation of the heated composite plate while the second one is related to the absence of auxiliary equipment such as tensioners/springs. Therefore, responding to the industrial requirements, no additional supporting fixture is foreseen. Furthermore, the crosshead speed of the stamping tools, and thus the relative closing speed of the dies, should be 5 mm/s, as required by the manufacturer of the component. Finally, the composite plate imposed has dimensions of 688 mm length and 434 mm width.

The assumptions of the second category are related to the restrictions of the simulation strategy and the FE package used. In the present work, Aniform™ FE code [14,16] is utilized which assumes an isothermal process. Thus, during the process, it is assumed a constant material temperature without any thermal interaction between the stamping tools and the composite plate. Moreover, the cool-down phase of the thermoforming is not taken into consideration in the analysis. Finally, it is assumed that the punch and mold tools are closing completely, leaving a cavity of a thickness equal to the laminate nominal thickness. The most interesting fact of this software package though, as seen in previous works [13], is the ability of utilizing a novel and sophisticated approach to derive accurate predictions with a relatively lower time consumption compared to traditional FE approaches. The Aniform™ software incorporates the innovative LTR3D membrane elements alongside with a DKT (Direct Kirchhoff Triangle) shell elements that enclose the in-plane shear behaviour of the material and the bending behaviour. The dynamic friction behaviour is assigned to the contact elements. To this end, there can be placed contact elements incorporating the dynamic friction between the dies and the composite plate while, for adding information regarding the interlayer behaviour, contact elements incorporating the results of the experimental campaign for the identification of these properties are inserted. To this end, the 3-layer configuration used in the present work is schematically described in Fig.2(a).

Italian Manufacturing Association Conference - XVI AITeM   Materials Research Forum LLC
Materials Research Proceedings 35 (2023) 19-27   https://doi.org/10.21741/9781644902714-3

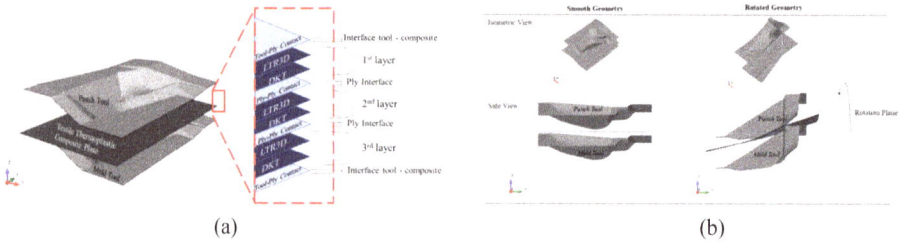

(a)   (b)

Fig. 2: The stacking sequence of the elements of the FE model (a) and the characteristics of the smoothed and inclined geometrical configurations (b).

Consequently, the following simulations are performed in the Initial geometry design and the Smoothed one as seen in Table 1 for capturing the optimal design and process parameters. The Initial geometry refers to the first design of the stamps and the component. After the execution of the first loop of simulations, the critical zones are identified, and the stamps are smoothed, eliminating potential zones of geometrical instability.

Table 1: The process parameters imposed for performing the thermoforming of the composite component.

Geometry	Stamping Speed [mm/min]	Stamping Temperature [°C]	Additional Rotation [degrees]	Simulations
Initial	50, 100, 200	160, 190	-	6
Smoothed	50,100, 200	160, 190	-	6
Rotated	100	190	42	1

In addition to that, FE simulation is conducted introducing an out-of-plane rotation of the cavity formatted by the dies for assessing a potential amelioration in terms of reduction of the defects introduced by the process. In the Rotated geometry configuration, the dies (punch and mold tool) are aligned in a way that a major part of the component to produce is fully aligned to the y-axis (xy plane) as seen in Fig.2(b). This way, to the heated composite plate different stresses are developed locally during the application of pressure by the punch tool and different stress states and deformations are expected.

Results and Discussion

The Effect of Stamping Speed and Material Temperature

In almost symmetrical, box shaped geometries, the synergistic effect between the stamping speed and the material temperature has been well addressed [14]. The FE software output can either be the ply or the laminate characteristics in terms of textile deformation, normal or shear stresses and the distribution of the thickness as seen in Fig.3 in the case of the Initial configuration, 160 °C material temperature and 50 mm/min stamping speed.

Italian Manufacturing Association Conference - XVI AITeM   Materials Research Forum LLC
Materials Research Proceedings 35 (2023) 19-27   https://doi.org/10.21741/9781644902714-3

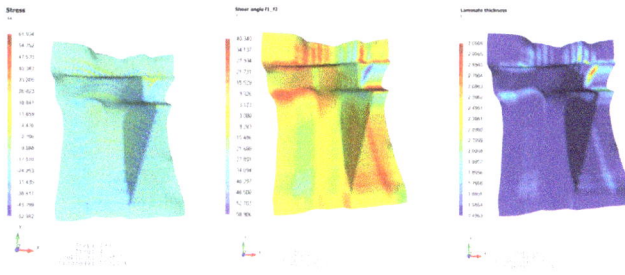

Fig. 3: The normal stress x, the shear angle (textile shearing) deformation and the laminate thickness distribution of the Initial configuration.

It is obvious a significant increase of the laminate thickness in a particular zone inside the product form where a wrinkle is expected. The laminate thickness prior to the hot stamping was 1.5 mm and, at some points, the final thickness values were up to roughly 3 mm which is a strong indication of the presence of folds/wrinkles. On the other hand, there can be noticed a severe deformation of the textile (Textile Shear Deformation) that is attributed to the complex form of the component and the aggressive angles of the cavity. The intense residual stresses in the x-axis are also attributed to the cavity geometrical characteristics. After having captured the maximum values of each of the abovementioned characteristics of the same product in different stamping speeds and material temperatures, the synergistic effect of these two manufacturing parameters are addressed in the comparative histograms of the maximum values of Fig.4. The maximum normal stresses x and y were tensile. Moreover, increased compressive stresses were observed that are a pre-requisite for the formation of wrinkles. In addition, the textile shear angular deformation was negative that practically indicates that the warp and weft bundles angle was reduced. Finally, by increasing the material temperature, a stress relief was observed with less textile shearing and lower deviation between the final and the nominal laminate thickness.

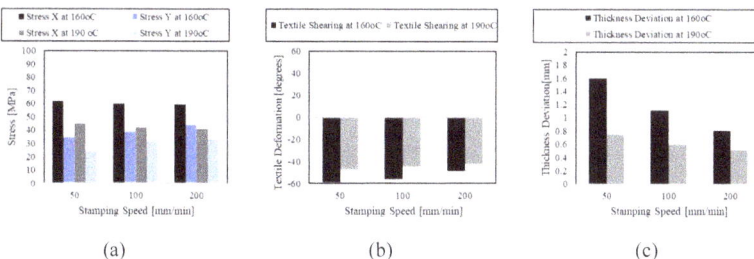

(a)   (b)   (c)

Fig. 4: Results regarding the maximum residual stresses in x and y direction (a), the textile shearing (b) and the thickness deviation (c) from the nominal value for the Initial geometry configuration.

Amelioration Through Edge Smoothening
As previously mentioned, a trial-and-error procedure was adopted for ameliorating the characteristics of the cavity for relieving the final product from defects. The procedure was explained in section 4. A comparison between the values of stresses, laminate thickness and textile shearing is presented in Fig.5. There can be noticed a net reduction of the residual stresses at both 160 °C and 190 °C in all the stamping speeds of interest. In fact, by introducing smoother stamping tools, the maximum stress x became compressive with significantly lower value compared to the

Initial configuration for the case of 160 °C. At 190 °C and stamping speeds between 50 and 100 mm/min the maximum normal stress is not compressive but tensile, a fact explained by the increased formability of the composite material in higher temperatures. In addition, the textile shearing became positive, therefore the textile angle tends to open which is another indication of less folding and, therefore, less possibility of the development of wrinkles. The increase of both the stamping speed and the material temperature introduced a significant decrease of the textile deformation and the thickness deviation from the nominal value. However, there should be noticed a tendency for the residual stresses to increase. Especially in the case of the stresses towards the y axis with a stamping speed of 200 mm/min, a fact that may introduce a negative effect on the structural performance of the component.

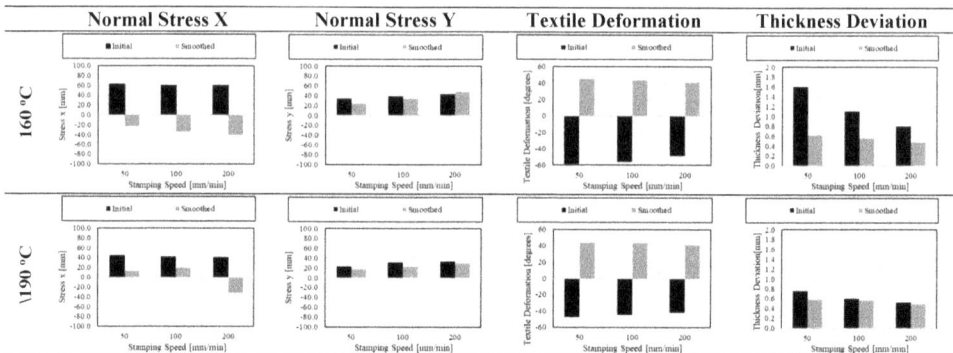

Fig. 5: The output of the simulations for the 2 geometries at the 2 temperatures and various stamping speeds.

Experimental Validation

After completing the tasks of the proposed methodology, the stamping tools developed using the iterative procedure explained in the previous sections were produced and placed in a 4-axis universal industrial press as seen in Fig.6, at the premises of the Crossfire srl (Faenza, Italy). The validation of the methodology was conducted by replicating the exact conditions of the FE simulation in which the material is heated up to 160 °C and stamped imposing a speed of 50 mm/min using the same composite material. In the same figure, presented is a comparison between the output of the FE analysis and the actual production of the composite component. The Aniform software incorporates an AI add-on that calculates the probability of the introduction of some defects, based on the information gained regarding the residual stresses, thickness variation and textile shearing, delivering values from 0 (absence) to 1 (certain) in the zones of the thermoformed composite plate. From this comparison, it is evident the accuracy of the simulation regarding the critical zones (highlighted within the red circles and formed out of the form of the component) and the good quality of the thermoformed component. In fact, by observing the thermoformed composite laminate, the useful zone of the component is free of defects such as wrinkles (folding) while the textile deformation was found to be in accordance with the predictions of the FE analysis.

Italian Manufacturing Association Conference - XVI AITeM
Materials Research Proceedings 35 (2023) 19-27

Materials Research Forum LLC
https://doi.org/10.21741/9781644902714-3

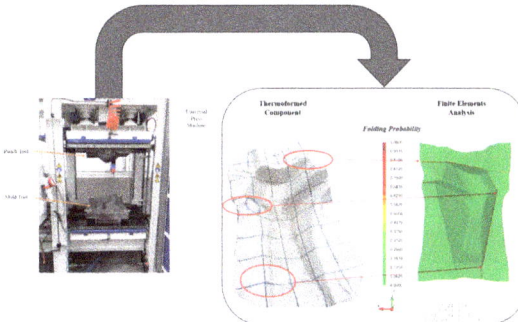

Fig. 6:Comparison between the results of the conducted experimental and numerical analyses.

The Effect of the Stamping Plane Rotation

The results in terms of total laminate thickness, textile shearing angle and stresses x are presented in the following Fig.7. Starting from the laminate thickness, the inclination of the cavity inside the dies appears to derive a bracket uniform enough with values near the nominal thickness while the critical zones of folding appeared to be reduced. However, the maximum value obtained was found to be 22% higher compared to the normal Smoothed configuration. What is worse though, is the prevision about the shearing angle that appears to be distributed not uniformly, introducing an unbalanced textile deformation. This deformation of the fibers direction (textile shearing) is the cause of the unbalanced distribution of the residual stresses. In fact, the deformation of the textile introduced an in-plane waviness that increases the residual stresses locally, leading to a net worsening of the component from both the quality and structural point of view.

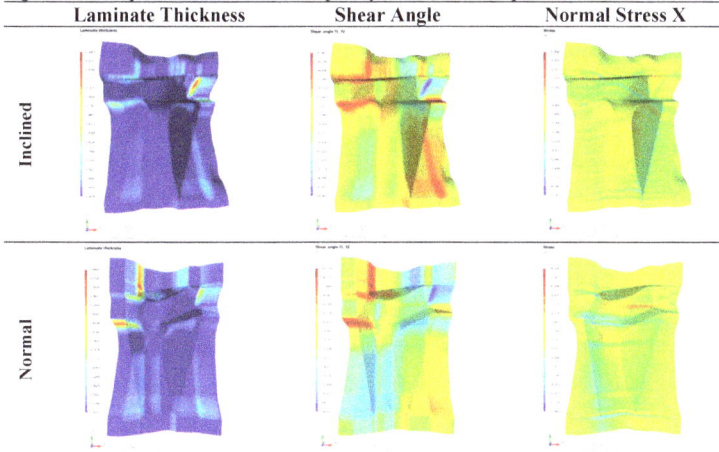

Fig. 7: Comparison between the Normal and the Inclined smoothened molds geometry effect on various outputs of the final component

Conclusions

In the present work, there was presented a numerical methodology for ameliorating the thermoforming process applied to thermoplastic textile composites. The increase of the material temperature in general appears to have a positive effect on some of the characteristics while the

Italian Manufacturing Association Conference   XVI AITeM   Materials Research Forum LLC
Materials Research Proceedings 35 (2023) 19-27   https://doi.org/10.21741/9781644902714-3

increase of the stamping speed appears to ameliorate the uniformity of the thickness and, as a result of the increased formability, the decrease of the deformation of the textile. However, at stamping speeds of 200 mm/min there was observed the introduction of more residual stresses to the component. By the implementation of the artificial intelligence module of the software the critical zones of the initial and the smoothed geometry were identified, leading to the conclusion about the effectiveness of the edges smoothening strategy that potentially contributes to the stresses redistribution during the stamping procedure. The inclination of the stamping plane though, even if it contributes to the uniformity of the overall product thickness, appears to introduce a severe textile shearing and more residual stresses with an unbalanced stress field. Finally, this procedure appears to be an effective solution for developing the dies of processes such as the present one, since it contributes to the virtualization of the process and the reduction of the process development time and cost. It should be noted that similar approaches are considered as pillars of the transition to the modern industry 4.0 years and to more competitive process lines, also considering the cost of the dies for producing components of high geometrical complexity such as the one treated in the present paper.

References

[1] Turgut E.T: An Analysis of the Effect of the Non-Payload Weight on Fuel Consumption for a Wide-Bodied Aircraft. Anadolu University Journal of Science and Technology 18(1), 59-68 (2017). https://doi.org/10.18038/aubtda.300429

[2] Lathasree P, Sheethal RM.: Estimation of aircraft fuel consumption for a mission profile using neural networks. In: Proceedings of the International Conference on Aerospace Science and Technology INCAST 2008, 26 - 28 June 2008, Bangalore, India.

[3] Patel M., Pardhi B., Chopara S., Pal M.: Lightweight Composite Materials for Automotive - A Review. International Research Journal of Engineering and Technology 5 (11), 41-47 (2018).

[4] Campbell F.C.: Manufacturing processes for advanced composites". Elsevier Science, ISBN 9781856174152, (2004).

[5] Stamopoulos A.G., Gazza F., Neirotti G. Assessment of the compressive mechanical behavior of injection molded E-glass/polypropylene by mechanical testing and X-ray computed tomography. The International Journal of Advanced Manufacturing Technology, 1-15 (2023). https://doi.org/10.21203/rs.3.rs-2298336/v1

[6] Ashter S.A.: Thermoforming of single and multilayer laminates. William Andrew Publishing, ISBN 9781455731725, (2014).

[7] De Luca P., Lefebvre P., Pickett A.K.: Numerical and experimental investigation of some press forming parameters of two fibre reinforced thermoplastics: APC2-AS4 and PEI-CETEX. Composites Part A: Applied Science and Manufacturing 29A: 101-110, (1998). https://doi.org/10.1016/S1359-835X(97)00060-2

[8] Guzman-Maldonado E., Hamila N., Boisse P., Bikard J.: Thermomechanical analysis, modelling and simulation of the forming of pre-impregnated thermoplastic composites. Composites Part A: Applied Science and Manufacturing 78: 211-222, (2015). https://doi.org/10.1016/j.compositesa.2015.08.017

[9] Haanappel S.N., Ten Thije R.H.W., Sachs U., Rietman B., Akkerman R.: Formability analyses of uni-directional and textile reinforced thermoplastics. Composites: Part A 56: 80-92, (2014). https://doi.org/10.1016/j.compositesa.2013.09.009

Italian Manufacturing Association Conference - XVI AITeM
Materials Research Proceedings 35 (2023) 19-27

Materials Research Forum LLC
https://doi.org/10.21741/9781644902714-3

[10] Haanappel S.P., Sachs U., ten Thije R.H.W., Rietman B., Akkerman R.: Forming of Thermoplastic Composites. Key Engineering Materials 504 (506), 237-242 (2012). https://doi.org/10.4028/www.scientific.net/KEM.504-506.237

[11] D'Emilia G., Gaspari A., Natale E., Stamopoulos A.G., Di Ilio A. Experimental and numerical analysis of the defects induced by the thermoforming process on woven textile thermoplastic composites. Engineering Failure Analysis 135, 106093 (2022). https://doi.org/10.1016/j.engfailanal.2022.106093

[12] Harisson P., Gomes R., Curado-Correia N. Press forming a 0/90 cross-ply advanced thermoplastic composite using the double-dome benchmark geometry. Composites Part A: Applied Science and Manufacturing 54, 56-69 (2013). https://doi.org/10.1016/j.compositesa.2013.06.014

[13] Sargent J., Chen J., Sherwood J., Cao J., Boisse P., Willem A., Vanclooster K., Lomov S.V., Khan M., Fetfatsidis K., Jauffres D. Benchmark study of Finite Element Models for simulating the thermostamping of woven-fabric reinforced composites. International Journal of Material Forming 3, 683-686 (2010). https://doi.org/10.1007/s12289-010-0862-5

[14] Stamopoulos A.G., Di Genova L.G., Di Ilio A. Simulation of the thermoforming process of glass-fiber reinforced polymeric components: investigation of the combined effect of the crosshead speed and the material temperature. The International Journal of Advanced Manufacturing Technology 117, 2987-3009 (2021). https://doi.org/10.1007/s00170-021-07845-2

[15] Guzman-Maldonado E., Hamila N., Naouar N., Moulin G., Boisse P. Simulation of thermoplastic prepreg thermoforming based on a visco-hyperelastic model and a thermal homogenization. Materials and Design 93, 431-442 (2016). https://doi.org/10.1016/j.matdes.2015.12.166

[16] Haanappel S.N., Ten Thije R.H.W., Sachs U., Rietman B., Akkerman R.: Formability analyses of uni-directional and textile reinforced thermoplastics. Composites: Part A 56: 80-92, (2014). https://doi.org/10.1016/j.compositesa.2013.09.009

Italian Manufacturing Association Conference - XVI AITeM
Materials Research Proceedings 35 (2023) 28-36

Materials Research Forum LLC
https://doi.org/10.21741/9781644902714-4

Study of compostable materials for the production of transparent food containers

Giulia Cappiello[1,a] *, Daniele Rocco[2,b], Clizia Aversa[1,c], and Massimiliano Barletta[1,d]

[1]Università degli Studi Roma Tre, Dipartimento di Ingegneria Industriale, Elettronica e Meccanica, Via della Vasca Navale 79, 00146 Roma (Italy)

[2]Sapienza Università degli Studi di Roma, Dipartimento di Ingegneria Meccanica e Aerospaziale, Via Eudossiana 18, 00184 Roma (Italy)

[a]giulia.cappiello@uniroma3.it, [b]daniele.rocco@uniroma1.it, [c]clizia.aversa@uniroma3.it [d]massimiliano.barletta@uniroma3.it

Keywords: Biobased materials, Extrusion, Thermoforming

Abstract. The aim of this work is to develop a new class of transparent containers for food packaging relying on bioderived polyesters and additives. Usually, transparent containers are manufactured using different plastic materials to achieve mechanical strength, thermal resistance, printability, good visual appearance. Yet, these containers cannot be recycled, being made by multiple polymers. This makes the management of their end-of-life troublesome. In contrast, bioderived polyester can be composted or recycled. In the present study blends of polylactic acid PLA with N,N'-ethylene(bis-stearamide) were processed by a co-rotating twin-screw extruder. The compounds were reprocessed by cast extrusion to make transparent films. Fine dispersion of ethylene(bis-stearamide), EBS, in PLA resulted in increased crystallinity, mechanical strength and thermal resistance, without compromising the transparency of the extruded films. The films were thermoformed to get the containers, whose thermo-mechanical performance were assessed.

Introduction

Currently, there is growing interest in technological solutions that can reduce the environmental footprint resulting from the use of fossil-sourced plastics. This interest has increased significantly since 2019, thanks to EU Communication COM 2019/904, in which a ban on the use of fossil-sourced plastics for single-use products was established. [1] Alternative strategies to the use of fossil-derived plastics are therefore necessary to enable the packaging sector to exceed the regulatory limits. One such strategy of particular interest is biobased and compostable bioplastics. There are already compostable materials on the market that are used in the food packaging sector, but it is currently particularly difficult to make compostable and transparent products [2]. Biopolymers have received considerable research attention due to their biodegradable nature and at the same time less environmental impact than their non-biodegradable petroleum-based counterparts. [3] Among the most promising biopolymers is polylactic acid PLA, which can only be composted in an industrial context. The behavior of compostable materials is defined by the EN 13432 standard, when disposed in an industrial composer [4] PLA is a material derived from renewable sources and represents a viable solution to the problem of food waste disposal [5]. PLA has mechanical properties and physical properties comparable to some petroleum-based polymers, such as polystyrene (PS) and poly (ethylene terephthalate) (PET), due to its high elastic modulus. However, further improvements are needed to achieve the accuracy and repeatability required for industrial applications. To improve the properties of bioplastic materials, it is necessary to mix them with other biopolymers or process additives and fillers that opacify or colour the final formulation [6].

Italian Manufacturing Association Conference - XVI AITeM   Materials Research Forum LLC
Materials Research Proceedings 35 (2023) 28-36   https://doi.org/10.21741/9781644902714-4

For industrial applications, biopolymers such as PLA are often blended with mineral fillers, talc, to increase the overall mechanical properties of the blend. This totally decreases the transparency of the film produced [7]. To obtain a transparent, compostable film with high mechanical properties at the same time, it is necessary to choose additives that do not compromise its transparency. The present article is part of this contest and aims to study possible compostable, transparent formulations that can be used in the food packaging sector. PLA applications are limited by several factors, such as low glass transition temperature, low thermal resistance and brittleness. In the literature, there are several studies where attempts are made to improve these properties with additives, but the transparency is compromised [8] [9] [10]. The present study is devoted to the engineering of PLA blends with N,N'-ethylene(bis-stearamide and the thermal and physical properties associated with these formulations were evaluated. Blends of PLA polylactic acid with N,N'-ethylene (bis-stearamide) were processed by a co-rotating twin-screw extruder to make pellets and subsequently reprocessed by cast extrusion to obtain transparent films. The films were thermoformed to obtain containers, whose thermo-mechanical performance was evaluated.

Materials and methods

Materials. Luminy poly(lactic acid) (PLA) grades L175 and LX175 (Total Corbion PLA _Stadhuisplein, NS Gorinchem, The Netherlands) were chosen as the base polymers for the preparation of all formulations. PLA L175 and LX175 have a high molecular weight with an MFI of 8 g/10 min, suitable for the thermoforming process. Two PLA grades have a different -D isomer content, the LX175 grade has a higher -D isomer content of approximately 3%. This parameter affects the crystallization kinetics of the material and consequently the degree of transparency. As a clarifying nucleating agent, EBS EVIWAX 140 (Eigenmann & Veronelli S.p.A., Milan, Italy) is included in the formulations. EBS EVIWAX 140 is an active organic nucleating agent for PLA especially during heating heat treatments, based on N,N'-ethylenebis(stearamide) and is also used for its action as a process release agent [11].The table shows the formulations designed for this study.

Table 1 Composition of the blends

Materials	L [%wt.]	LX [%wt.]	LE [%wt.]	LXE [%wt.]
PLA L175	99		99	
PLALX175		99		99
EBS			1	1

EBS has low molecular weight and flexible hydrocarbon segment with the existence of polar amide groups (–CONH–) in the molecular structure of EBS make it compatible with some of the polar polymer such as PLA [12].All the materials involved in this study are safe for direct food contact and have the relevant manufacturer's certification, although the food contact compliance of the final formulations developed needs to be reassessed.

Processing. The LE and LXE materials, without any drying treatment, were extruded through the Leistrizt ZSE 27 IMaxx 27 twin-screw corotating extruder used to mix the formulations. This machine consists of 10 thermoregulated zones where the temperature can be customised for each formulation. The temperature profile is set to be parabolic within the extrusion barrel (up to T9), as shown in Table 2. The head temperature is increased by 10-15 degrees. The head temperature is increased by 15°C, compared to T9, to prevent obstructions from passing through the die. The temperature profiles adopted for each material are shown below and are based on the melting temperature of the high melting material present in the formulation. The temperature profile used must be higher than the melting temperature of the materials in order to obtain a polymeric melt. [6] [7].

Table 2 Profile temperature for reative extrusion of LE, LXE blend

Materials	T_{S1} [C°]	T_{S2} [C°]	T_{S3} [C°]	T_{S4} [C°]	T_{S5} [C°]	T_{S6} [C°]	T_{S7} [C°]	T_{S8} [C°]	T_{S9} [C°]	T_{S10} [C°]
LE	160	175	180	185	185	190	190	185	190	190
LXE	140	155	165	175	175	170	170	170	175	175

The maximum temperature of the thermal profile is set 10-15 °C higher than the melt temperature in the vicinity of the powder dosing zone (zone 4) to facilitate the incorporation process of the powder additive and the melt polymer. The operating parameters were set respectively at 230 rpm for the screw speed and a melting pressure at 33 bar. This process resulted in top-quality pellets that were dried for 4 hours at 65°C. Films were then made using a cast Minicast Plus extruder, manufactured by EUR.EX.MA. XTR 20. The operating parameters used for sheet extrusion are listed in the following tables.

Table 3 Operating parameters for cast extrusion of the films

Materials	Speed [rpm]	Thickness [mm]	T_1 [°C]	T_2 [°C]	T_3 [°C]	T_4 [°C]	T_{sx} [°C]	T_c [°C]	T_{dx} [°C]
L	135	400	195	195	195	195	197	195	197
LE	140	400	195	195	195	195	197	195	197
LX	155	400	190	190	190	190	192	190	192
LXE	170	400	190	190	190	190	192	190	192

The films were thermoformed using a machine called Formech 450DT. A ribbed rectangular tray suitable for food packaging and preservation was chosen as the mould. The operating parameters for the process are given in Table 4.

Table 4 Operating parameters for thermoforming process

Materials	Residence time [s]	Film temperature [°C]
L	25	160
LE	28	167
LX	22	138
LXE	32	138

The final manufactured products are shown in the following images.

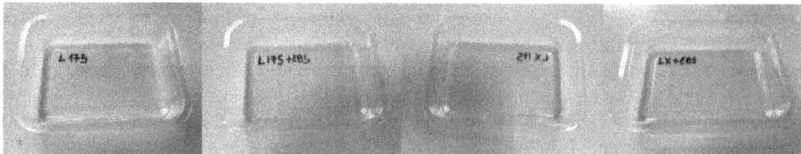

Figure 1 Final transparent products

Thermal characterization. The thermo-rheological properties of the manufactured compounds were characterized through DSC analysis, using a calorimeter DSC3 (from Mettler Toledo). Measurement was conducted under a nitrogen atmosphere, according to ASTM D-3418-15. The crystallization behavior of the solid samples was explored by heating/cooling the samples between 20 °C and 190 °C, at heating/cooling rates of 10 °C/min. Glass transition, cold crystallization and melting temperature (T_g, T_{cc} and T_m) such as ΔH_m and ΔH_{cc} were determined from second heating scan. The crystallinity degree (X_c) was calculated from the second scan as reported in Eq (1):

$$X_c = \frac{\Delta H_m - \Delta H_{cc}}{\Delta H_{mo} * (1 - m_f)} \times 100. \qquad (1)$$

Where ΔH_m and ΔH_{cc} are the enthalpies of melting and cold crystallization, respectively. ΔH_{mo} is enthalpy of melting for a 100% crystalline PLA sample.

Haze test. This test method involves the evaluation of specific light transmission and light scattering properties of planar sections of translucent or transparent plastic materials. The test was performed in accordance with ASTM D 1003 - 00. A UV-Vis-NIR Spectrophotometer V-670 equipped with a single-beam integrating sphere ISN-723 (Jasco, Inc., Easton, MD, USA) was used. The spectral bandwidth is 5 nm. To proceed with the Haze calculation, it is necessary to take several measurements in different configurations to calculate transmittance values. Haze value can be calculated using the following equation:

$$\text{Haze } [\%] = (T_d / T_t) * 100. \qquad (2)$$

where T_d is the diffuse transmittance and T_t is the total transmittance.

Thermal stability test. Thermal stability is evalueted through tests were carried out on the prototype trays to assess their thermo-mechanical resistance to contact with hot liquids. Specifically, 200 ml of a hot liquid (water) was poured into the thermoformed trays at a temperature of 65°C to assess any deformation. The test was repeated three times for each sample and the observation time was set at 5 minutes. This test is considered passed if the product does not show any deformation in its geometry after the observation period. In order to evaluate the dimensional parameters at the end of each test, the treated tray was inserted into a new one to see if they matched.

Mechanical characterization. The Izod impact test is a standard method approved by ASTM to determine the notch toughness of a material. Toughness is the ability of the material to absorb energy and deform plastically before fracture. Toughness, which exactly represents the energy absorbed by the specimen during impact, measured in Joules, is:

$$K = F_p(H-h). \qquad (3)$$

Where K is the hardness of the notch, H and h are the final and initial heights of the pendulum respectively, and F_p is the force-weight of the pendulum. The test is performed in accordance with ISO 180. The specimens used for this type of test are rectangular and were moulded with an injection moulding machine RAY-RAN, equipped with a heatable cylinder with a capacity of 57 cm^3 at a temperature of 190°C.

Italian Manufacturing Association Conference - XVI AITeM
Materials Research Proceedings 35 (2023) 28-36

Materials Research Forum LLC
https://doi.org/10.21741/9781644902714-4

Results and discussion

From the DSC thermal characterization, the graphs corresponding to the second heating scan for the four formulations are shown. From the curves, it was possible to visualize and calculate the change-of-state temperatures and crystalline fraction index of PLA present in each formulation.

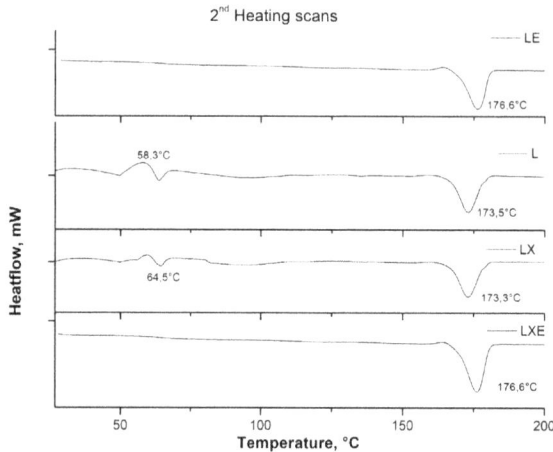

Figure 2 DSC scans for the transparent films

From the three scans performed, the parameters shown in Table 5.

Table 5 DSC results

Samples	L	LX	LE	LXE
Tg [°C]	58,3	64,5	-	-
Tc (PLA) [°C]	156,31	154,51	158,4	160,2
Tm (PLA) [°C]	173,5	173,3	176,6	176,6
Xc PLA [%]	14,3	27,1	28,9	41,2

EBS with both PLA L175 and LX175 increased the crystalline fraction. as evidenced by the decreased degree of disorder in the amorphous region of the crystalline fraction, due to the presence of EBS acting as a nucleant and promoter of PLA crystallization [13].

The crystalline fraction is an important parameter affecting the physical properties of a PLA film. According to a study by Mohanty et al., [11], increasing the crystalline fraction in a PLA film can improve its tensile strength, but at the same time reduce its deformability. However, the addition of a compatibilizer such as EBS could have a significant effect on the crystalline fraction of the film. According to research by Li et al., [9], EBS can promote the crystallization of PLA and increase its crystalline fraction, leading to higher mechanical strength of the film. Furthermore, EBS can also affect the morphology of the crystalline fraction of PLA, as observed by Dong et al., [14], who reported that the addition of EBS can increase the crystal size of PLA. In conclusion, the addition of EBS can positively influence the crystalline fraction of a PLA film, improving its

Italian Manufacturing Association Conference - XVI AITeM   Materials Research Forum LLC
Materials Research Proceedings 35 (2023) 28-36   https://doi.org/10.21741/9781644902714-4

mechanical properties. However, it is also important to consider the side effects that may result from changing the morphology of PLA crystals, e.g. in terms of transparency or permeability of the film. For this reason, the transparency of extruded films has been evaluated with the Haze test. The haze test showed slightly less transparency in the formulations with EBS, but overall, the values were all below 10%.

Table 6 Haze results

Materials	T_t [%]	T_d [%]	Haze [%]
L	92.8	2.9	3.1
LE	92.6	5.2	5.6
LX	92.0	4.2	4.6
LXE	92.0	5.9	6.4

This result agrees with studies of Liu et al., [12], in which the addition of EBS to PLA led to an increase in the Haze value, due to the formation of compatible microphases between PLA and EBS. Specifically, the Haze value increased from 5.5% of the pure PLA film to 15.8% of the PLA-EBS film, with 10% by weight of EBS. Similarly, Wu et al., [15], also reported that the addition of EBS to PLA led to an increase in the Haze value of the film. Specifically, the Haze value increased from 1.7% in the pure PLA film to 4.4% in the PLA-EBS film to 5% by weight of EBS. The formation of compatible microphases between PLA and EBS refers to the creation of localized regions within the material, the two polymer phases tend to form homogenous microspheres or microstructures that can help improve certain properties of the composite material, such as mechanical strength or transparency. The compatibility between polymers depends on their chemical properties, such as polarity, molecular weight and molecular morphology, and the mixing conditions used to prepare the composite material. When polymers are compatible, they tend to mix homogeneously, forming a uniform polymer blend with improved physical and mechanical properties compared to individual polymers [16]. To assess the thermal stability of the formulations, a test at 70°C and 100°C was carried out on the thermoformed products. In particular, the deformation of the trays was assessed, using hot water, after an observation time of 5 minutes. After the test at 70°C, the trays in L and LX appeared visibly deformed, while those in LXE and LE showed no changes from their initial shape. The test repeated at 100°C also resulted in the deformation of all trays. PLA is known to have good thermal resistance, with a glass transition temperature (T_g) of approximately 60-65 °C. However, the addition of EBS to PLA increases the thermal resistance of the composite material. For example, Wu et al. showed that adding 10 wt% EBS to PLA can increase the T_g of the composite material up to 80 °C, [15]. Adding EBS to PLA can also improve the thermal stability of the material by reducing the thermal degradation rate of PLA [9]. However, the effect of EBS on thermal resistance also depends on the amount of EBS present in the formulation. Some studies have shown that a high amount of EBS can reduce the thermal resistance of the composite material. For example, another study reported that the addition of 30 wt% EBS to PLA can reduce the T_g of the composite material to about 57 °C [12]. In general, the addition of EBS to PLA can improve the thermal resistance of the composite material at low percentages but can reduce the thermal resistance at high percentages of EBS. Therefore, LXE and LE formulations compared to L and LX are optimal from a transparency-thermal resistance perspective. Despite the various transformation processes, the thermal properties of the final product remain optimum. The addition of EBS at 1 %wt. creates the right balance between transparency, thermal stability and mechanical properties of the formulation. From a mechanical point of view, an Izod characterization test was performed. The addition of EBS to PLA can affect the impact strength of the composite material.

Italian Manufacturing Association Conference - XVI AITeM   Materials Research Forum LLC
Materials Research Proceedings 35 (2023) 28-36   https://doi.org/10.21741/9781644902714-4

Table 7 Izod results

Samples	Impact energy accumulated [J]	Impact Strenght [kJ/m²]
L	2,89	70,6
LX	2,94	73,5
LE	4,75	103,32
LXE	5,84	114,50

In this case, an increase in the impact strength of PLA after the addition of EBS of approximately 60 % is observed. The LXE formulation is the most impact resistant as it contains a higher concentration of D isomer than the respective LE. These values agree with other studies. For example, Wang et al. showed that the addition of 5% by weight of EBS to PLA can increase the impact strength of the composite material by 60% compared to pure PLA [10]. Furthermore, Chen et al. reported an increase in the impact strength of PLA with the addition of EBS at 10 wt% [8]. However, when the percentage of EBS is more than 10%wt., it can reduce the impact strength of the composite material at high EBS percentages. For example, Liu et al. reported that the addition of 30% by weight of EBS to PLA can reduce the impact strength of the composite material by 50% compared to pure PLA [12]. In general, the effect of EBS on impact strength depends on the amount of EBS present in the formulation. The formation of compatible microphases between the PLA and EBS, which can promote the dissipation of impact energy and improve the toughness of the composite material [17]. The formation of a fibrillar structure of EBS in the PLA matrix can act as a bridge between the phases and improve the toughness of the composite material [8] [18].

Conclusion
In this study, four compostable formulations were proposed as suitable for interesting applications in transparent food packaging. Formulations engineered with EBS were found to be suitable for thermoforming processes, which is the most adopted production technique. Experimental results showed that the inclusion of EBS within the formulation allowed for a combination of transparency, thermal and mechanical resistance of the bioplastic blends used. The experimental results showed that the correct concentration of EBS in the evaluated formulations leads to the production of an excellent quality compound through the reactive extrusion process using a co-rotating twin-screw extruder. The good quality of the plastic films made the thermoforming process possible to produce products (food trays) with an excellent finish. The PLA-based compounds LX175 and EBS showed both good thermal and mechanical resistance, not deforming plastically under the action of a hot liquid (water) at 70°C. The L175 and EBS system showed higher transparency but lower impact strength. This behaviour is dictated by the different structure of the two formulations the compatibility between the chosen grade of PLA and EBS can improve the interaction between the two phases and increase the toughness of the composite material. The experimental results obtained are also influenced by the type of processing and the processing parameters used. The result obtained derives from a rational design of the formulation aimed at achieving thermal characteristics comparable to those of the fossil materials currently used. As future studies are planned to further improve the thermal performance of the product (>100°C) to extend its applicability.

References

[1] DIRETTIVA (UE) 2019/904 sulla riduzione dell'incidenza di determinati prodotti di plastica sull'ambiente, 2019.

[2] M. K. H. W. Y. F. D. Y. G. K. K. Hamad, «Properties and medical applications of polylactic acid: A review,» eXPRESS Polymer Letters, vol. 9, n. 5, p. 435 - 455, 2015. https://doi.org/10.3144/expresspolymlett.2015.42

[3] S. B. Palai*, «Synergistic effect of polylactic acid(PLA) and Poly(butylene succinate-co-adipate) (PBSA) based sustainable, reactive, super toughened eco-composite blown films for flexible packaging applications,» Polymer Testing, vol. 83, n. 106130, 2020. https://doi.org/10.1016/j.polymertesting.2019.106130

[4] UNI EN 13432-Commissione Europea, 2002.

[5] P. D. M. Murariu, « PLA composites: From production to properties,» Elsevier, 2016. https://doi.org/10.1016/j.addr.2016.04.003

[6] K. A. C. N. M. B. M. G. A. S. R. G. Y. A. ,. Eraslan, «Poly(3-hydroxybutyrate-co-3-hydroxyhexanoate) (PHBH): Synthesis, properties, and applications - A review,» European Polymer Journal,, vol. 167, 2022. https://doi.org/10.1016/j.eurpolymj.2022.111044

[7] L. L. O. G. M. B. S. V. M. Castillo, «Crystalline morphology of thermoplastic starch/talc nanocomposites induced by thermal processing.,» Heliyon, Vol. %1 di %2 5, e01877, 2019. https://doi.org/10.1016/j.heliyon.2019.e01877

[8] J. Z. X. &. Y. D. Chen, «Preparation and properties of poly(lactic acid)/ethylene-butylene-styrene copolymer (PLA/EBS) blends,» ournal of Materials Science, vol. 54, n. 13, pp. 9893-9903, 2019.

[9] Z. L. L. Z. Y. L. C. L. Y. L. S. &. Q. Y. Li, «Improving the mechanical properties and biodegradability of polylactic acid/ethylene-butylene-styrene copolymer blends by adding chain extenders.,» Polymer Testing, vol. 76, pp. 128-136, 2019.

[10] H. Z. X. Z. Y. &. Y. D. Wang, «Preparation and properties of poly(lactic acid)/ethylene-butylene-styrene copolymer (PLA/EBS) blends,» Journal of Applied Polymer Science, vol. 136, n. 33, p. 47854, 2019.

[11] W. T. W. P. B. e. a. Chow, «Mechanical and Thermal Oxidation Behavior of Poly(Lactic Acid)/Halloysite Nanotube Nanocomposites Containing N,N′-Ethylenebis(Stearamide) and SEBS-g-MA,» J Polym Environ, vol. 26, p. 2973-2982, 2018. https://doi.org/10.1007/s10924-018-1186-7

[12] X. C. M. Z. W. Y. H. &. W. L. Liu, «he effect of ethylene-butylene-styrene (EBS) on the crystallization, rheology, and mechanical properties of poly(lactic acid) (PLA),» Polymer Engineering & Science, vol. 59, n. 4, pp. 799-809, 2019.

[13] S. G. K. e. al., «Effect of ethylene-butylene-styrene copolymer on the crystallinity and thermal properties of polylactide,» Journal of Applied Polymer Science, vol. 117, n. 1, pp. 342-351, 2010.

[14] W. Q. B. W. W. Z. L. Y. M. C. Y. &. L. Y. Dong, «Effect of ethylene-butylene-styrene triblock copolymer on the crystallization and rheology behaviors of polylactide,» Journal of Polymer Research, vol. 22, n. 11, p. 204, 2015.

[15] C. X. J. X. X. &. Z. S. Wu, «Effects of ethylene-butylene-styrene content on the properties of poly(lactic acid)/ethylene-butylene-styrene blend films,» Journal of Applied Polymer Science, vol. 136, n. 47, p. 48113, 2019.

[16] L. S. W. Z. Z. W. X. &. L. J. Wang, «Ethylene-butylene-styrene (EBS) modified poly(lactic acid) (PLA) with improved properties for packaging applications.,» Journal of Applied Polymer Science, vol. 137, n. 22, p. 48771, 2020.

[17] H. M. X. &. G. S. Zhou, «Mechanical and thermal properties of poly(lactic acid)/ethylene-butylene-styrene copolymer blends,» ournal of Applied Polymer Science, vol. 133, n. 28, p. 43710.., 2016.

[18] B. R. V. d. S. E. e. a. Matos, «Evaluation of commercially available polylactic acid (PLA) filaments for 3D printing applications,» J Therm Anal Calorim, vol. 137, p. 555-562, 2019 https://doi.org/10.1007/s10973-018-7967-3

Italian Manufacturing Association Conference - XVI AITeM
Materials Research Proceedings 35 (2023) 37-44

Materials Research Forum LLC
https://doi.org/10.21741/9781644902714-5

Manufacturing of a hybrid component in Ti6Al4V-ELI alloy by combining diffusion bonding and superplastic forming

Pasquale Guglielmi[1,a] *, Antonio Piccininni[1,b], Angela Cusanno[1,c] and Gianfranco Paumbo[1,d]

[1]Department of Mechanics, Mathematics and Management, Politecnico di Bari, via Orabona 4, 70125 Bari, Italy

[a]pasquale.guglielmi@poliba.it, [b]piccininni.antonio@poliba.it, [c]angela.cusanno@poliba.it, [d]gianfranco.palumbo@poliba.it

Keywords: Bonding, Sheet Forming, Titanium Alloys

Abstract. Through Diffusion Bonding (DB) large surfaces can be joined with low distortions and localized microstructural changes. Heterogeneous Titanium (Ti) components, composed, for example, by a porous layer over a bulk one, which play a key role for biomedical applications, can be obtained. In the present work, the possibility to combine the SuperPlastic Forming (SPF) process (for creating the part's shape) and the DB process (for joining two different layers) is investigated. In particular, the porous layer was obtained by solid state foaming, made possible thanks to a heat treatment set on a slice cut from a billet produced by compacting Ti powders via Hot Isostatic Pressing (HIP). SPF was used to shape both the Ti sheet and the slice cut from the HIPed billet. Since the SPF is conducted at high temperature, a porous structure could be obtained in the HIPed material (solid state foaming occurred); setting the proper pressure and time, the two layers could be successfully joined by DB. All the investigated pressure levels revealed to be able to produce a complete solid state joint, without any discontinuity; in addition, the final hybrid component could be manufactured according to the desired geometry.

Introduction

Diffusion bonding is a solid state joining process. The joint made using such a process are affected exclusively by a microscopic deformation and characterized by high homogeneity, without secondary materials or liquid phases. During the process, two surfaces are joined at elevated temperature (between 50 and 80 percent of the melting point) by means of a pressure applied to the interface. The pressure must be sufficiently reduced so as to avoid obvious deformations of the parts to be joined; this usually determines that the characteristic times of the process are around two hours and more depending on the geometry of the part. Since diffusion bonding is caused by atomic migration across a solid-state interface, there are theoretically no microstructural discontinuities at the interface region and, consequently, the mechanical properties and microstructure in the bonded region are no different from those of the base metal. In addition, when differences exist between the two materials to be joined, the interface zone would be able to mediate their properties [1]. DB can be considered a valid solution in obtaining joints in materials, such as Ti alloys, usually difficult by conventional techniques [2]. Among others, aspects of significant importance in the adoption of such a technology are attributable to (i) reduced distortions, (ii) applicability to large areas, as well as the aforementioned (iii) limited microstructural distortions. [3]. A further interesting aspect is related to the possibility of combining the DB and SPF processes for the fabrication of highly resistant joints with complex geometries that would otherwise be hardly produced. In fact, it has been demonstrated that this strategy can make it possible to obtain high-performance and sustainable final components from an economic point of view [4]. Typical applications involve sandwich structures with cost and weight reductions of between 30 and 50 percent compared to what is possible with conventional

Italian Manufacturing Association Conference - XVI AITeM   Materials Research Forum LLC
Materials Research Proceedings 35 (2023) 37-44   https://doi.org/10.21741/9781644902714-5

techniques [5]. The diffusion bonding process must be optimized according to three parameters: time, applied pressure and temperature; the choice of suitable values for such parameters is essential to promote the diffusion which is necessary for the joining. In addition, specific geometric parameters such as (i) the final shape and (ii) the number of layers shall be taken into account at the design stage [6]. Furthermore, the microstructure of the starting material should be considered [6]. Finally, two necessary conditions that must be satisfied during the process are represented by (i) the intimate contact between the two surfaces and (ii) the absence of surface contaminating substances that could interfere with the bonding.

In the present work the DB process has been investigated for the fabrication of a joint characterized by two layers in Ti6Al4V-ELI alloy; specifically, the two layers were produced using two different processes: the monolithic layer was obtained through the rolling process, while the other one was obtained from through the Hot Isostatic Pressing (HIP) process [7]. The porous Ti alloys, due to their high mechanical performances and good biocompatibility, have found wide use in the biomedical sector for a wide range of bone implants [8,9]. In fact, the porous structure makes the mechanical characteristics (Young's modulus) more similar to the human bone's ones, thus reducing the stress shielding phenomena. Such a reduction of the Young's modulus is intimately linked to the level of porosity inherent in the material. In addition, a porous bone-like prosthesis topography may promote bone growth [10,11]. The present study is aimed to evaluate the feasibility of the DB process, both in the presence and absence of a superplastic deformation, focusing the attention on both the bonding capability of the two layers and the possibility of increasing the porosity of the layer produced by HIP.

Material and Methodology

Investigated material

The experimental activity discussed in the present work was conducted on Ti6Al4V-ELI alloy circular samples (disks with a diameter of 75mm). More specifically, the DB process was performed by joining one disk extracted from a tolled sheet (1mm thick) and one disk extracted by Electrical Discharge Machining from a billet produced by HIP (1.5mm thick). As concerns the HIP process, the following parameters were used: (i) Ti6Al4V-ELI powders with max diameter equal to 50μm, (ii) Argon pressure of 0.2 MPa and (iii) HIP pressure equal to 80MPa.

In order to distinguish the two types of layers involved in the process, in this paper the rolled material and that deriving from the HIP process will be called "bulk" and "HIPed" respectively. The chemical composition of the investigated material is reported in Table 1.

Table 1 Chemical composition (weight %) of the investigated Ti6Al4V-ELI alloys.

Al%	V%	Fe%	C%	N%	H%	O%	Ti
6.15	3.87	0.15	0.008	0.006	0.001	0.08	Bal.

Adopted methodology

All the tests performed were conducted in atmospheric conditions, although it is known that the presence of oxygen could hinder the diffusion bonding process. In fact, there are several testimonies in the literature that focus the attention on a possible effect of the presence of oxygen during the diffusion bonding process. However, in this regard, Lee et Al [12] focused the attention on this aspect. The evidence that emerges shows the capability of the DB process to take place effectively even in the presence of oxygen. In fact, a suitable preparation of the surfaces in contact and to be welded together coupled with the capability of self-evacuation from the interstices between the two plates in contact following the pressure applied by the gas on the entire sandwich would make the aspect relating to the presence of oxygen less pressing than that it might be thought. In the light of this, it is anticipated that the presence of oxygen during the process did not

Italian Manufacturing Association Conference - XVI AITeM   Materials Research Forum LLC
Materials Research Proceedings 35 (2023) 37-44   https://doi.org/10.21741/9781644902714-5

represent an obstacle to obtaining the complete welding of the various interfaced plates. An overview of the equipment used to perform the experimental activities is reported in Figure 1.

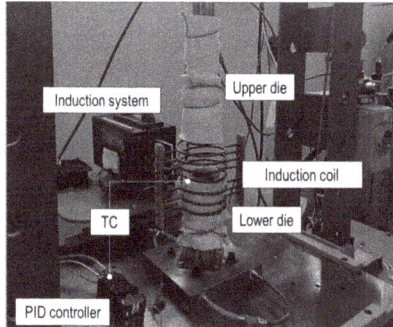

Figure 1 Overview of the experimental equipment used for DB experiments.

From a methodological point of view, the DB tests were carried out by exploiting the action of a pressurized gas (Argon) acting on a central area of the circular sample.

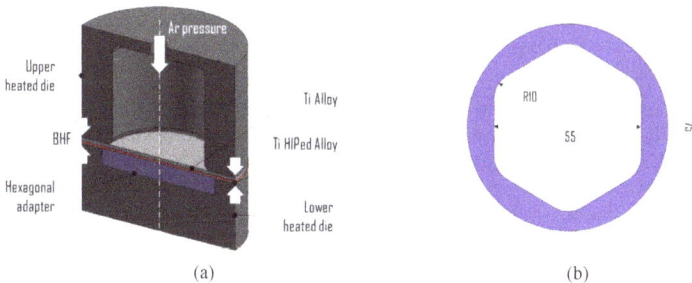

(a)   (b)

Figure 2 Scheme of the experimental DB setup (a) and the adapter for conducting tests with plastic deformation (b).

In addition, in order to avoid gas leakages during the test, a blank holder was used to apply a load on the peripheral area of the sample (see scheme in Figure 2); in the present work a constant value of the Blank Holder Force (BHF) was used (equal to 8.85MPa). All tests were conducted at a temperature of 850°C, since it is well known in literature that it is the optimal condition for the superplastic behavior of the investigated Ti alloy [13].

During each test, a constant and homogeneous temperature in the test area was ensured by the adoption of an induction heating system. The temperature was controlled during each test by a type K thermocouple (TC) welded to the interface of the two dies and directly connected to a PID system, as shown in Figure 1.

The experimental activity was initially focused on the feasibility of the DB process by investigating different Argon pressures; subsequently, for one of the previously analyzed conditions, the combined effect of the superplastic deformation was also considered. For all performed tests a fixed duration of 240 minutes was considered. Finally, regarding the test with superplastic deformation a hexagonal adapter was used (Figure 2b) with a depth of 7.5mm directly integrated into the interface between the two visible die in the experimental setup (purple block in Figure 2a). Table 2 summarizes the experimental conditions investigated in the present work. The same table shows the nomenclature used to distinguish the investigated test conditions.

Italian Manufacturing Association Conference - XVI AITeM   Materials Research Forum LLC
Materials Research Proceedings 35 (2023) 37-44   https://doi.org/10.21741/9781644902714-5

Table 2 Investigated experimental conditions.

Nomenclature	Method	Argon pressure, MPa
0.75DB	DB	0.75
1.00DB	DB	1.00
1.40DB	DB	1.40
1.40DBF	DB + SPF	1.40

Before each test (3 replications), the surfaces to be joined were properly prepared by pre-polishing to eliminate residues of the previous processes and reduce the roughness (less than 1 µm). Processed samples were analyses by means of light microscopy and Vickers microhardness; finally, the porosity level was measured to evaluate the effect of both the pressure and the deformation on the foaming capability of the HIPed material. The level of porosity (quantified in terms of percentage area and average diameter) was assessed by investigating samples extracted from the component after the DB process and subjected to a proper metallographic preparation.

Metallographic analyses aimed at determining both the joined zone and changes in terms of porosities. For this purpose, the Nikon MA200 microscope and the ImageJ software for digital image processing were used. For this latter purpose each analysis was carried out considering 3 different micrographs. Finally, the mechanical properties of the hybrid component were evaluated through 10 different Vickers microhardness measurements for each zone (HIPed and bulk layers, as well as the interface) using an automatic Qness Q10 Microhardness tester with a load of 500g.

Results
The micrographs obtained for the evaluation of the hybrid structures obtained are shown from Figure 2 to Figure 5. For all tested conditions a perfect adhesion between the two different layers was recorded, supporting the effectiveness of the proposed approach. Although micrographs have indicated the two different layers of the joint (HIPed and Bulk), it is possible to distinguish them easily because of the different type of manufacturing process from which they derive: in fact, noticeable black circular areas associated with the registered porosity are visible in the layer resulting from HIP process. Red arrows indicate the interface zone: no discontinuities or clusters of porosity were found in this area. Furthermore, in accordance with the literature [12], an aspect of not negligible importance is that relating to the absence of inclusions (oxygen) in the area where the two layers are joined. This consideration is even more supported when high magnifications are considered.

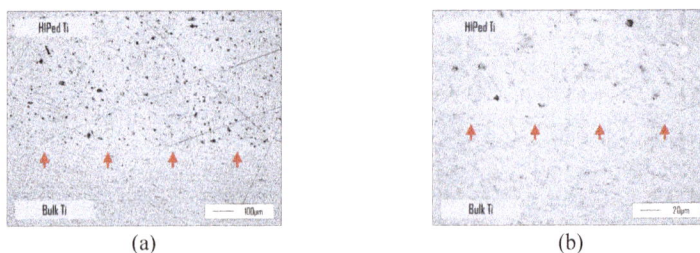

(a)   (b)

Figure 3 DB test conducted with Argon pressure of 0.75MPa at two different magnifications: (a) 70X (b) 350X.

Italian Manufacturing Association Conference - XVI AITeM   Materials Research Forum LLC
Materials Research Proceedings 35 (2023) 37-44   https://doi.org/10.21741/9781644902714-5

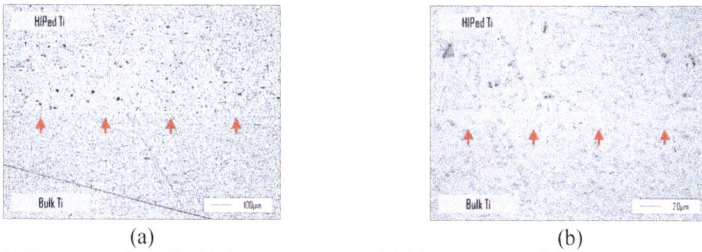

(a)   (b)

Figure 4 DB test conducted with Argon pressure of 1.00MPa at two different magnifications: (a) 70X (b) 350X.

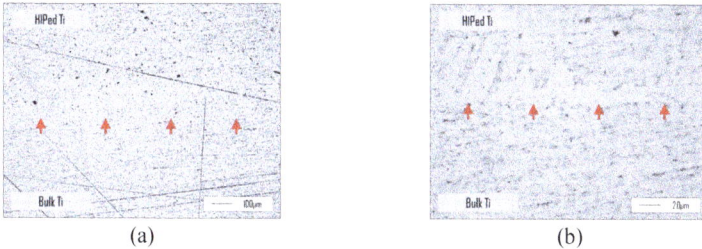

(a)   (b)

Figure 5 DB test conducted with Argon pressure of 1.40MPa at two different magnifications: (a) 70X (b) 350X.

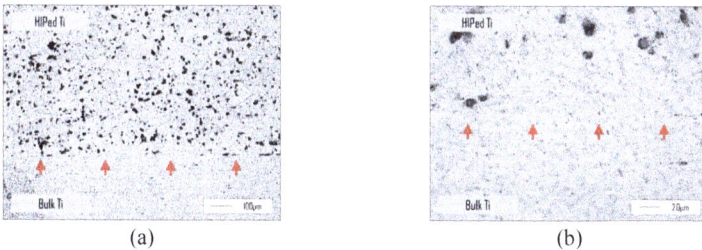

(a)   (b)

Figure 6 DB test coupled with SPF conducted with Argon pressure of 1.40MPa at two different magnifications: (a) 70X (b) 350X.

In addition, although after the diffusion bonding process there are no particular propensities of the material to foam (this is due to the low process temperature, i.e. 850°C), the contribution of the deformation seems to stimulate the porosity growth mechanism. In this regard, analyses of the porosities, both in terms of percentage area and average size are reported respectively in Figure 6a and Figure 6b. In the same figures the values referred to the "As Received" (AR) condition, deriving from the HIP process and not yet subjected to any DB process, are reported to better understand how the Argon pressure can influence the foaming behavior of the Ti alloy. For both output variables considered, the DB process allows to increase the typical values that can be associated with the AR condition. Regarding the percentage of porous areas, a certain effect of the Argon pressure applied during the DB process was recorded; on the other hand, when no deformations are provided during the process, the average diameter of these porosities does not undergo statistically significant variations. Finally, by combining the DB process with the SPF process, in addition to obtaining a further increase in the percentage of porosity (Figure 6b), there is an increase - this time significant and equal to about 50% - in terms of the average diameter of the porosities (Figure 6b).

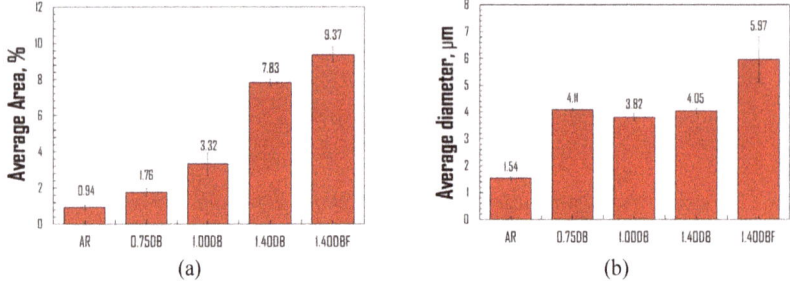

(a)   (b)

Figure 7 Analysis of the porosities for the different conditions used for the diffusion bonding process: (a) Average percentage area and (b) Average diameter of the porosities.

For the test performed by combining DB and SPF processes the attention was focused on the three different areas shown in Figure 7 (the wall, the corner and the flat area). The correspondent micrographs are shown in Figure 8.

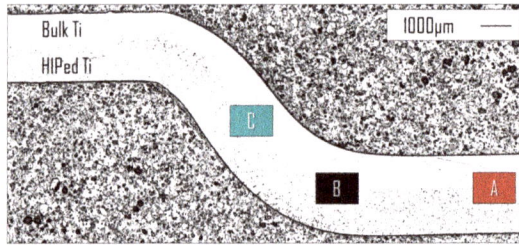

Figure 8 Macrography of the three investigated areas: (A) Flat, (B) Corner and (C) Wall.

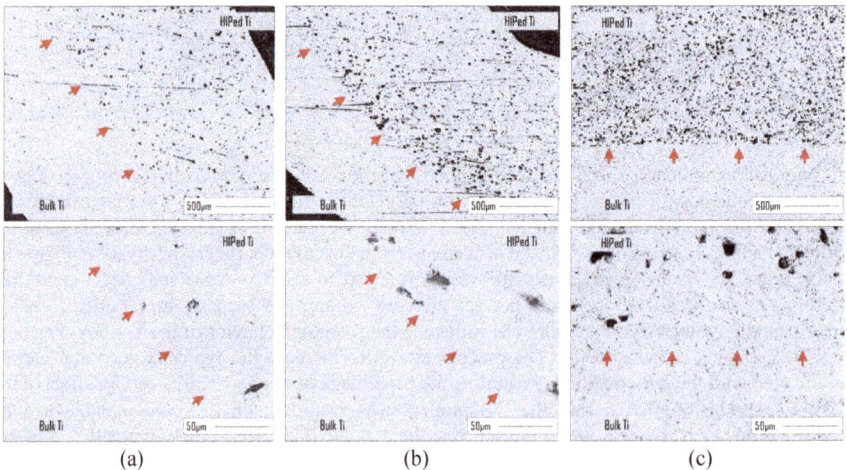

(a)   (b)   (c)

Figure 9 metallographic analysis at different magnifications referred to (a) the wall, (b) the corner and (c) the flat area of the hexagonal component.

For all the investigated areas it is possible to note that a complete adhesion of the two layers occurred; furthermore, the flat (Figure 8c) area and the one in the corner (Figure 8b) are more porous than the wall (Figure 8a), since they are the most deformed ones. Therefore, the energy supplied to the system during the combined process (DB and SPF) allowed not only to ensure the bonding of the two layers, but also to increase the level of porosity in the HIPed layer.

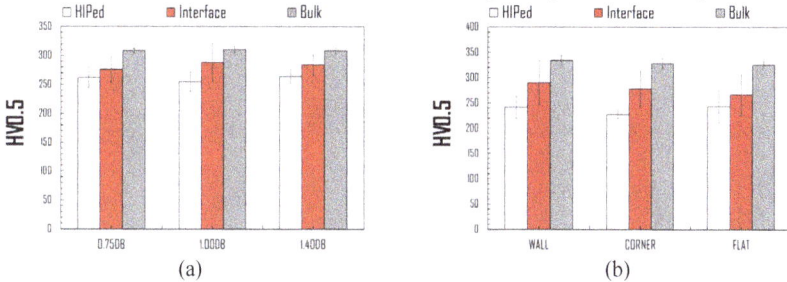

Figure 10 Microhardness analysis: (a) comparison between the different DB conditions and (b) comparison between the different areas referring to the combined process of DB and SPF.

From a mechanical point of view, Vickers microhardness analyses (Figure 9) confirm the effectiveness of the processes investigated since for all the cases analyzed (for DB tests) and the areas investigated (for DBF tests) the microhardness at the interface is intermediate between the one of the bulk layer and the one of the HIPed layer, guaranteeing a perfect merge of the mechanical properties of the two materials.

Conclusions

In the present work the diffusion bonding process aimed at the fabrication of a hybrid component characterized by two Titanium layers (one produced by rolling, one obtained through the HIP process) was investigated. In particular, the pressure levels applied during the solid state diffusion process (0.75, 1.00 and 1.40 MPa) for 240 minutes were able to produce the complete adhesion of the surfaces. With reference to the highest Argon pressure level (1.40 MPa) and by combining the DB process with the SPF process, it was possible to obtain a perfectly sound and continuous component at the interface. Furthermore, the effect of both temperature and pressure during the DB allowed to emphasize the foaming behavior of the HIPed Ti alloy; this is particularly evident when the deformation occurs (the growth of the porosity level is statistically significant). This aspect could be crucial in the design of hybrid structures for which specific mechanical properties influenced by the density of one of the two layers involved in the process are required. Finally, in order to fully investigate the obtained joints, future developments include in-depth analysis of the interface layer and any imperfections by means of more targeted analyses such as SEM and/or XRD.

References

[1] B. Hamilton, S. Oppenheimer, D.C. Dunand, D. Lewis, Diffusion Bonding of Ti-6Al-4V Sheet with Ti-6Al-4V Foam for Biomedical Implant Applications, Metall. Mater. Trans. B. 44 (2013) 1554–1559. https://doi.org/10.1007/s11663-013-9942-5

[2] B. Derby, E.R. Wallach, Theoretical model for diffusion bonding, Met. Sci. 16 (1982) 49–56. https://doi.org/10.1179/030634582790427028

[3] O.D. Sherby, J. Wadsworth, Superplasticity-Recent advances and future directions, 1989. https://doi.org/10.1016/0079-6425(89)90004-2

[4] Y.H. Kim, J.M. Lee, S.S. Hong, Optimal design of superplastic forming processes, J. Mater. Process. Technol. 112 (2001) 166–173. https://doi.org/10.1016/S0924-0136(00)00880-3

[5] W. Han, K. Zhang, G. Wang, Superplastic forming and diffusion bonding for honeycomb structure of Ti-6Al-4V alloy, J. Mater. Process. Technol. 183 (2007) 450–454. https://doi.org/10.1016/j.jmatprotec.2006.10.041

[6] M. Dewidar, Mechanical and Microstructure Properies of High Porosity Sintered Ti- 6Al-4V Powder for Biomedical Applications, JES. J. Eng. Sci. 34 (2006) 1929–1940. https://doi.org/10.21608/jesaun.2006.111343

[7] F.X. Zimmerman, Hot isostatic pressing: today and tomorrow, J. Mater. Sci. Technol. 1 (2008) 1–11.

[8] A.T. Sidambe, Biocompatibility of advanced manufactured titanium implants-A review, Materials (Basel). 7 (2014) 8168–8188. https://doi.org/10.3390/ma7128168

[9] M. Sarraf, E. Rezvani Ghomi, S. Alipour, S. Ramakrishna, N. Liana Sukiman, A state-of-the-art review of the fabrication and characteristics of titanium and its alloys for biomedical applications, Bio-Design Manuf. 5 (2022) 371–395. https://doi.org/10.1007/s42242-021-00170-3

[10] F. Li, J. Li, H. Kou, L. Zhou, Porous Ti6Al4V alloys with enhanced normalized fatigue strength for biomedical applications, Mater. Sci. Eng. C. 60 (2016) 485–488. https://doi.org/10.1016/j.msec.2015.11.074

[11] P. Guglielmi, A. Piccininni, A. Cusanno, A.A. Kaya, G. Palumbo, Mechanical and microstructural evaluation of solid-state foamed Ti6Al4V-ELI alloy, Procedia CIRP. 110 (2022) 105–110. https://doi.org/10.1016/j.procir.2022.06.021

[12] H.S. Lee, J.H. Yoon, C.H. Park, Y.G. Ko, D.H. Shin, C.S. Lee, A study on diffusion bonding of superplastic Ti-6Al-4V ELI grade, J. Mater. Process. Technol. 187–188 (2007) 526–529. https://doi.org/10.1016/j.jmatprotec.2006.11.215

[13] D. Sorgente, G. Palumbo, A. Piccininni, P. Guglielmi, L. Tricarico, Modelling the superplastic behaviour of the Ti6Al4V-ELI by means of a numerical/experimental approach, Int. J. Adv. Manuf. Technol. 90 (2017) 1–10. https://doi.org/10.1007/s00170-016-9235-7

Italian Manufacturing Association Conference - XVI AITeM
Materials Research Proceedings 35 (2023) 45-52

Materials Research Forum LLC
https://doi.org/10.21741/9781644902714-6

Comparison between two tailored press hardening technologies by means of physical and numerical simulation

Maria Emanuela Palmieri[1,a] *, Luigi Tricarico[1,b]

[1]Department of Mechanical Engineering, Mathematics & Management Engineering, Politecnico of Bari, Via Orabona, 4, 70125 Bari, Italy

[a]mariaemanuela.palmieri@poliba.it, [b]luigi.tricarico@poliba.it

Keywords: Sheet Forming, Automotive, Tailored Properties

Abstract. Tailored Tool Tempering (TTT) and Intermediate Pre-Cooling techniques are two tailored press hardening technologies studied for automotive applications to obtain structural components with good energy absorption characteristics and high strength. The aim of this work is the comparison between these two tailored technologies in terms of mechanical properties on the part. An automotive B-Pillar in 22MnB5 steel was considered as case study. Two Finite Element (FE) models were developed for simulating both technologies. FE thermal cycles were experimentally reproduced on specimens using a Gleeble physical simulator. After physical simulation, metallographic, tensile and hardness tests were carried out to evaluate the mechanical properties. Optimal values of process parameters that guarantee ductile and resistant zones on the same component were detected. In these optimal conditions, the TTT technology guarantees greater fracture deformability in the component zone which is to absorb energy.

Introduction

The reduction of fuel consumption and the improvement of safety for passengers are the main objectives for transport industries [1]. To fulfill these objectives, for several decades, the focus is on reducing the vehicle structures mass and optimizing production processes [2]. The trend to reduce vehicle weight was further increased due to the production of electric vehicles, where the battery mass should be compensated [3]. Ultra-strength materials such as advanced high-strength boron steels and the press hardening process promote the reduction of vehicle body structures mass [1]. The press hardening process is a combination of two operations, namely, the hot forming of a fully austenitized blank in a furnace at about 900 °C and the quenching heat treatment in the forming tools that leads to a martensitic microstructure on the component [1]. This technology is well suited for the production of components with diversified mechanical properties (tailored components) [4], in order to simultaneously guarantee high impact resistance and good energy absorption capacity. These components can be produced either by tailored blank technologies or by tailored process technologies [1]. The tailored blank technologies include: Tailored Rolled Blanks (TRB) [5], Tailored Welded Blanks (TWB) [6] and Patchwork blanks [7]. Tailored process technologies, on the other hand, allow to obtain components with customized mechanical properties by modifying microstructural properties of the part through heat treatments after the press hardening process or through the variation of thermal cycles (modification of the press hardening process). Some examples of tailored process technologies are detailed below: laser partial annealing [8], Tailored Blank Heating (TBH), known also as partial heating [4], Tailored Tool Tempering (TTT) and the relatively recent Intermediate Pre-Cooling (IPC). In this work the attention is focused on the TTT and the IPC technologies. The first one involves the use of tools with heated segments (by means of cartridge heaters) in areas where a ductile area should be generated and cooled segments (by means of cooling channels) where resistant regions are desired [9]. This technology is also known in the literature as differential cooling [4]. The second one uses, downstream to the conventional austenitizing furnace, an additional furnace (tempering station),

Italian Manufacturing Association Conference - XVI AITeM
Materials Research Proceedings 35 (2023) 45-52

Materials Research Forum LLC
https://doi.org/10.21741/9781644902714-6

where selected blank regions are cooled to guarantee the required part ductility while the other blank areas are maintained at the austenitization temperature [10]. This new technology differs from the conventional partial heating technology since the temperature partitioning of the blank takes place after the complete autenitization phase. The TTT technology is better described in Fig. 1, where the scheme of the process (Fig. 1a) and of the thermal cycles (Fig. 1b) are shown in correspondence of resistant and ductile regions. In Fig. 1b, the solid orange curve represents the thermal cycle of the ductile region, while the dashed blue curve represents the thermal cycle of the resistant region. During the quenching phase, the high cooling rates in the resistant regions of the component lead to a martensitic microstructure at the end of the process. Conversely, lower cooling rates lead to more ductile microstructures, e.g., bainitic microstructure. Several scientific works showed that the microstructure and mechanical properties of the part in the ductile region is mainly influenced by the temperature of heated tools and the quenching time in the tools [11]. The IPC technology and resulting thermal cycles in ductile and resistant regions are schematized in Fig. 2. Specifically, from Fig. 2a it can be seen that the blank is first austenitized in the furnace and then it is moved into a second furnace where the intermediate pre-cooling is obtained by masking the blank regions which define the ductile area of the part. The masked areas are cooled below approximately 700 °C, while the unmasked areas are reheated to about 950 °C (austenitization temperature). This solution was patented by the AP&T Company with the TemperBox® tempering furnace [12]. In Fig. 2b, the thermal cycle of the ductile region is represented with the solid orange line, while the thermal cycle of the resistant region is represented with the dashed blue line. The slow cooling in the tempering station in correspondence of the masked areas leads to a predominantly ferritic-pearlitic microstructure that confers ductility properties to these areas. This latter technology is quite recent, therefore still few works studied it. Moreover, for the best authors knowledge, nobody compared yet these two investigated technologies which lead to different microstructures in the soft region. The objective of this work is to compare the TTT technology with the new IPC technology in terms of mechanical properties of the part at the end of the process, after evaluating the influence of the process parameters for both technologies. For the TTT technology, the process parameters evaluated were the temperature of the heated tools (T_q) and the quenching time in the tools (t_q). Instead, for the IPC technology, the parameters considered were the time taken in the tempering station ($t_{precooling}$) and the temperature to which the blank drops in correspondence with the ductile regions ($T_{precooling}$).

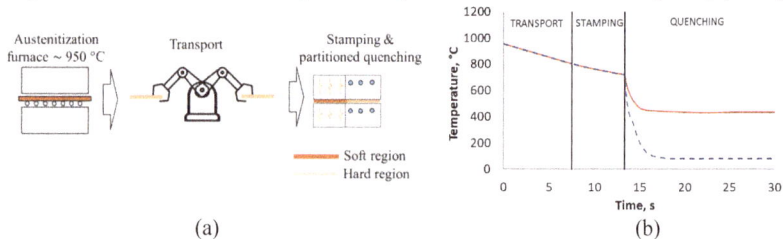

(a)   (b)

Fig. 1. Press hardening with tailored tool tempering approach (a) scheme of the process and (b) scheme of thermal cycles in ductile and resistant regions.

Italian Manufacturing Association Conference - XVI AITeM   Materials Research Forum LLC
Materials Research Proceedings 35 (2023) 45-52   https://doi.org/10.21741/9781644902714-6

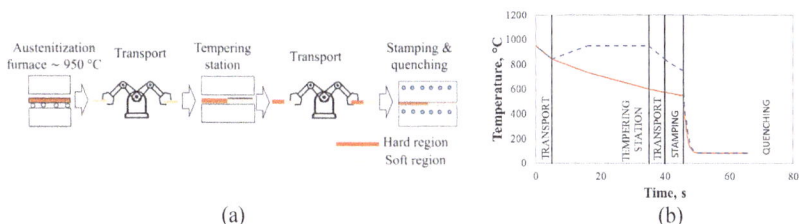

(a)

(b)

Fig. 2. Press hardening with intermediate pre-cooling approach (a) scheme of the process and (b) scheme of thermal cycles in ductile and resistant regions.

Materials and Method

In order to compare the two investigated tailored technologies, an automotive structural component was chosen as a case study, i.e., the B-Pillar shown in Fig. 3, manufactured starting from a 1.3 mm thick blank in 22MnB5 steel (C=0.217 %, Mn=1.16 % and B=0.0029 %). As can be seen in Fig. 3, a more strength central area and two more ductile areas (lateral ones) are required for this component. The study was carried out by means of physical and numerical simulations. First, using the Finite Element (FE) commercial software, AutoForm®R10, two FE models were developed; one allows the numerical simulation of the press hardening process with the tailored tool tempering technology, the other, the numerical simulation of the intermediate pre-cooling technology. The numerical simulations were performed varying the process parameters (t_q and T_q for the TTT technology and $t_{precooling}$ and $T_{precooling}$ for the IPC technology). The numerical simulations were set as in Tab. 1 for the TTT technology and as in Tab. 2 for the IPC technology. The ranges of process parameters were identified based on previous studies [9, 13].

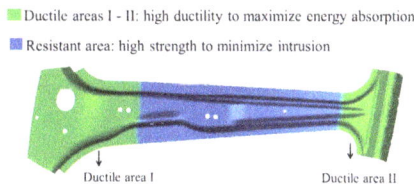

Fig. 3. Case study: B-Pillar with one central resistant area and two lateral ductile areas

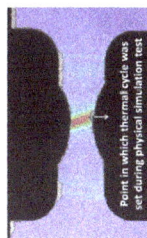

Fig. 4. Notched specimen for tensile testing.

Tab. 1. FE simulations plan for TTT technology.

T_q, °C	t_q, s
430	5-20-35-50
465	5-20-35-50
500	5-20-35-50

Tab. 2. FE simulations plan for IPC technology.

$T_{precooling}$, °C	$t_{precooling}$, s
600	30-90-150-210
650	30-90-150-210
700	30-90-150-210

Both FE models take into account the influence of the temperature and the strain rate on the flow stress and phase transformations. Specifically, the flow curves of the 22MnB5 steel are defined for different microstructural phases (austenite, ferrite-pearlite, bainite and martensite), for a temperature between 20 °C and 850 °C and a strain rate between 0.01 s^{-1} and 1 s^{-1}. For estimating microstructural phases and hardness on the component, in both FE models the continuous cooling transformation phase diagram (CCT) and the steel composition were defined. The heat transfer coefficient (HTC) between blank and ambient was set equal to 0.02 mW/(mm^{2}K) for a temperature of 20 °C and equal to 0.075 mW/(mm^{2}K) for a temperature of 950 °C. Instead, the HTC between blank and tools was defined as a function of the contact pressure and the gap between blank and tools. For the lubrication conditions the Coulomb model was used and a friction coefficient equal to 0.4 was set. Finally, Elastic Plastic Shell (EPS) elements were adopted for the numerical simulations.

At the end of numerical simulations, for both technologies, the numerical hardness values and the predicted microstructure were evaluated. Moreover, the thermal cycles in the resistant area and in one of the two ductile areas (area II) were extracted. The numerical thermal cycles were then physically simulated on 1.3 mm thick 22MnB5 steel specimens, using the Gleeble®3180 system. The physically simulated specimens were then subjected to Vickers hardness tests (load 10 kg and dwell time 10 s), metallographic analyses (after etching with 2 % nital solution), and tensile tests. The tensile tests were assisted by a Digital Image Correlation (DIC) system in order to acquire the mechanical behaviour in the specimen centre where the thermal cycle was set during physical simulation test. To allow the localization of the deformation at this point, as can be seen from Fig. 4, the specimens were notched.

Results and discussion

The numerical simulations results show that in the resistant area, regardless of the process parameters values, a completely martensitic microstructure is estimated with an average hardness of 490 HV10. This is valid for both investigated technologies. In the ductile areas, on the other hand, the microstructure and the hardness are significantly influenced by the process parameters. Some FE results in terms of microstructural phases at the end of the process are reported in Fig. 5 for the TTT technology and in Fig. 6 for the IPC technology.

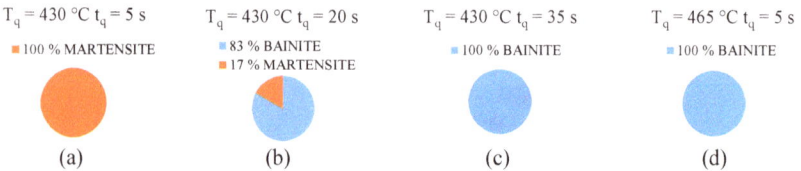

Fig. 5. Pie chart of the microstructure in the ductile area of the component at the end of the press hardening process with TTT technology as the parameters T_q and t_q vary: (a) $T_q = 430$ °C – $t_q = 5$ s; (b) $T_q = 430$ °C – $t_q = 20$ s; (c) $T_q = 430$ °C – $t_q = 35$ s; (d) $T_q = 465$ °C – $t_q = 5$ s.

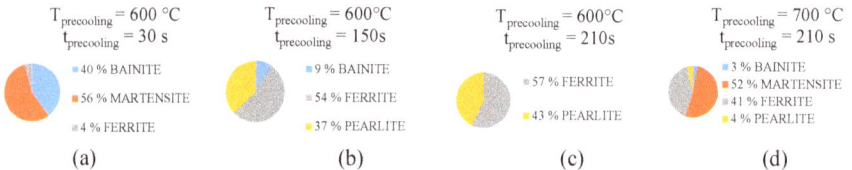

Fig. 6. Pie chart of the microstructure in the ductile area of the component at the end of the press hardening process with IPC technology as the parameters $T_{precooling}$ and $t_{precooling}$ vary: (a) $T_{precooling} = 600$ °C – $t_{precooling} = 30$ s; (b) $T_{precooling} = 600$ °C – $t_{precooling} = 150$ s; (c) $T_{precooling} = 600$ °C – $t_{precooling} = 210$ s; (d) $T_{precooling} = 700$ °C – $t_{precooling} = 210$ s.

Fig. 5 shows that at low temperatures ($T_q = 430$ °C) as the quenching time increases, the fraction of bainite (more ductile microstructural phase respect to the martensite microstructure) increases. For T_q equal to 430 °C, the complete bainitic transformation is reached for t_q equal to 35 s. At high temperatures ($T_q = 465$ °C - 500 °C), instead, a quenching time of 5 s is sufficient to complete the

Italian Manufacturing Association Conference - XVI AITeM   Materials Research Forum LLC
Materials Research Proceedings 35 (2023) 45-52   https://doi.org/10.21741/9781644902714-6

bainitic transformation. Fig. 6, on the other hand, shows that as $t_{precooling}$ increases, the percentage of more ductile microstructures (ferrite and pearlite) increases. However, at high temperatures ($T_{precooling}$ = 700 °C), even for the greater $t_{precooling}$ a certain percentage of the harder martensite microstructure is always estimated.

Consistent with the microstructural results, the FE hardness results in the ductile area show that in the TTT technology an increase in the heated tools temperature and in the quenching time lead to a reduction in hardness. For a fixed value of the heated tool temperature, there is a threshold value of the quenching time such that the complete bainitic transformation is obtained and the hardness value remains constant. This threshold value is equal to 35 s for T_q = 430 °C.

Meanwhile, FE hardness results in the ductile area of the component stamped with the IPC technology show that the hardness decreases as the $T_{precooling}$ decreases and $t_{precooling}$ increases.

Such FE hardness results are shown in Fig. 7 (Fig. 7a for TTT technology and Fig. 7b for IPC technology). Fig. 7a shows that the minimum hardness condition for TTT technology is reached for T_q = 500 °C already for t_q = 5 s. Instead, in the IPC technology the minimum hardness condition is obtained for $T_{precooling}$ = 600 °C and $t_{precooling}$ = 210 s. Under these conditions of maximum softening (i.e., minimum hardness), comparing the two technologies, it can be derived that for the IPC approach, in the ductile region, the hardness is approximately 40 % lower than that achieved in the TTT approach.

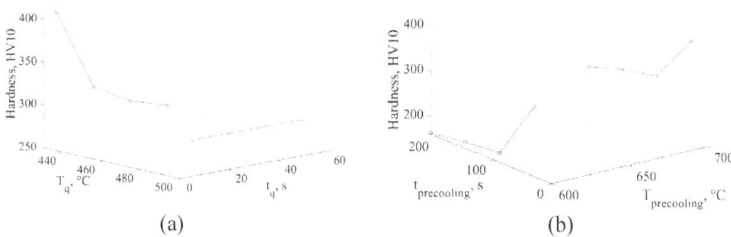

Fig. 7. FE hardness results as a function of process parameters for (a) the TTT technology and (b) the IPC technology.

As described in the methodology section, the numerical results were then validated through experimental tests. From the hardness and metallographic analyses on specimens subjected to the thermal cycles of the TTT technology, a good agreement with the numerical results was found. Specifically, the experimental hardness values differ of about 2 % from the FE data. Furthermore, metallographic analyses confirm a completely martensitic microstructure in the resistant area for each value of the T_q and t_q, and a predominantly bainitic microstructure in the ductile area already for T_q = 465 °C and t_q = 5s (Fig. 8).

The results of hardness tests on specimens subjected to the thermal cycles of the IPC technology show a good agreement with the numerical predictions only in the ductile area (error percentage of about 5 %). In the resistant area, instead, the experimental hardness shows a decreasing trend as the time taken in the tempering station increases. As an example, Fig. 9 shows the results for $T_{precooling}$ = 600°C. This reduction in hardness is justified by the grain growth that occurs when the material is heated at high temperature for long time. The grain growth is confirmed by comparing the microstructure obtained for a $t_{precooling}$ = 30 s (image on the left in Fig. 9) with the one obtained for $t_{precooling}$ = 210 s (image on the right in Fig. 9). A martensitic microstrucutre is observed in both figures, however in the microstructure for the lower value of $t_{precooling}$ the size of the lath martensite is smaller than that in the microstructure observed for the greater value of $t_{precooling}$.

Italian Manufacturing Association Conference - XVI AITeM
Materials Research Proceedings 35 (2023) 45-52

Materials Research Forum LLC
https://doi.org/10.21741/9781644902714-6

(a)   (b)

Fig. 8. Micrographs (1000X) corresponding to the thermal cycles of the resistant area (a) and ductile area (b) for T_q equal to 465 °C and t_q equal to 5 s.

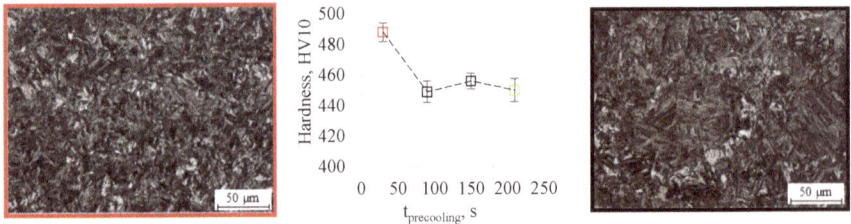

Fig. 9. Experimental hardness values as the $t_{precooling}$ varies referred to the resistant area obtained with IPC technology and micrographs (500 X) corresponding to the thermal cycles for $t_{precooling}$ equal to 30 s (image on the left) and $t_{precooling}$ equal to 210 s (image on the right).

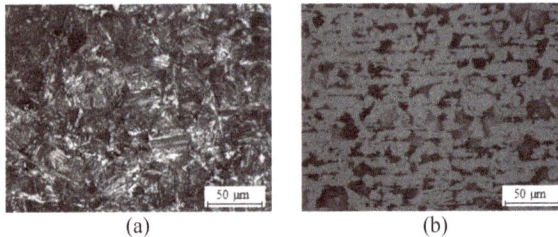

(a)   (b)

Fig. 10. Micrographs (500X) corresponding to the thermal cycles of the ductile area for (a) $t_{precooling}$ equal to 30 s and (b) $t_{precooling}$ equal to 210 s.

In the ductile area obtained with the IPC technology, the metallographic analyses confirm the predictions of FE simulations. As an example, Fig. 10 shows the microstructure obtained for $T_{preccoling}$ = 600 °C both for $t_{precooling}$ = 30 s (Fig. 10a) and $t_{precooling}$ = 210 s (Fig. 10b). A mixed microstructure is observed in Fig. 10a, whereas a ferritic-pearlitic microstructure is observed in Fig. 10b, as predicted from numerical results summarised in Fig. 6.

With the aim of optimizing the TTT technology to obtain components with tailored properties, the results described so far allow to state that the optimal condition can be achieved by imposing a temperature of the heated tools of 465 °C and a quenching time equal to 5 s. These values guarantee at the same time low energy consumption, short process cycle time and completely bainitic microstructure in ductile areas and martensitic microstructure in resistant areas.

For the IPC technology, the optimal condition could be achieved by cooling down to a temperature of 600 °C in a time of 30 s the areas that are desired to be ductile. These process parameter values are optimal because a $T_{precooling}$ equal to 600 °C guarantees the maximum softening in ductile areas and a $t_{precooling}$ equal to 30 s avoids grain growth phenomenon in the resistant area.

After physical simulation tests, the specimens were subjected to tensile tests with the aim of comparing the ultimate tensile strength and the fracture deformability (strain at break) between the ductile zone obtained with the TTT technology and the one obtained with the IPC technology. Fig. 11 compares the engineering stress-strain curve of the TTT technology in the optimal condition ($T_q465°C-t_q5s$) with the IPC technology curves both in the optimal condition ($T_{precooling}600°C-t_{precooling}30s$) and in the maximum softening ($T_{precooling}600°C-t_{precooling}210s$). The deformations shown in Fig. 11 were obtained locally (at the breaking point) by means of the DIC system.

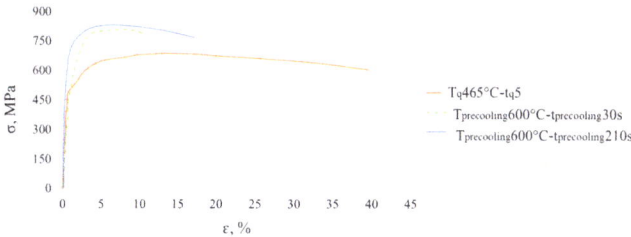

Fig. 11. Engineering stress-strain curves of (i) the TTT technology in the optimal condition, (ii) the IPC technology in the optimal condition and (iii) the IPC technology in the maximum softening condition.

From the comparison of the tensile testing curves in optimal conditions it can be observed that the TTT technology guarantees a greater fracture deformability in ductile regions, although the ultimate tensile strength is comparable. However, the greatest fracture deformability is reached during IPC process in the condition of maximum softening.

Conclusions

This work compared two tailored technologies for press hardening process, namely TTT and IPC for guaranteeing on the same component a resistant area with high strength and a ductile area with high fracture deformability. In TTT technology the ductile region (mainly bainitic) is obtained by increasing the heated tool temperature and the quenching time. However, to optimize the process, values of process parameters were chosen to ensure at the same time a good level of softening, low cycle times (low t_q) and low energy consumption (low T_q). In IPC technology, on the other hand, the ductile region (mainly ferritic-pearlitic) is obtained by increasing the time taken in the tempering station and lowering the temperature to which the blank drops in correspondence with the ductile regions. However, with this study was observed that an increase in $t_{precooling}$ leads to the problem of martensitic grain growth in the resistant region. Therefore, the optimal values of the process parameters were chosen in order to avoid this problem and ensure a good softening on the ductile part.

Comparing the conditions of maximum softening, the IPC technology guarantees a 40% lower hardness and a fracture deformability about double compared to the TTT technology. In terms of optimal conditions, the experimental plan shows that greater fracture deformability is reached with TTT technology. However, future work intends to investigate precooling times between 30 s and 90 s, which could guarantees higher softening in the soft zone and no grain growth problems in the hard zone.

Acknowledgments

The authors are grateful to PNRR - Spoke 11 (Innovative Materials and Lightweighting) project for funding this research and the Italian Ministry of University and Research that partly supported this research under the Programme "Department of Excellence" Legge 232/2016 (Grant No. CUP - D93C23000100001)".

References

[1] Billur, Eren. "Hot stamping of ultra high-strength steels." From a technological and business perspective. Cham (2019). https://doi.org/10.1007/978-3-319-98870-2

[2] Neugebauer, R., et al. "Press hardening-An innovative and challenging technology." Archives of civil and mechanical engineering 12 (2012): 113-118. https://doi.org/10.1016/j.acme.2012.04.013

[3] Boretti, Alberto. "Plug-in hybrid electric vehicles are better than battery electric vehicles to reduce CO2 emissions until 2030." International Journal of Energy Research 46.14 (2022): 20136-20145. https://doi.org/10.1002/er.8313

[4] Merklein, Marion, et al. "Hot stamping of boron steel sheets with tailored properties: A review." Journal of materials processing technology 228 (2016): 11-24. https://doi.org/10.1016/j.jmatprotec.2015.09.023

[5] Cheng, Wei, et al. "A process-performance coupled design method for hot-stamped tailor rolled blank structure." Thin-Walled Structures 140 (2019): 132-143. https://doi.org/10.1016/j.tws.2019.03.037

[6] Samadian, Pedram, et al. "Fracture response in hot-stamped tailor-welded blanks of Ductibor® 500-AS and Usibor® 1500-AS: experiments and modelling." Engineering Fracture Mechanics 253 (2021): 107864. https://doi.org/10.1016/j.engfracmech.2021.107864

[7] Mori, Ken-ichiro, et al. "Combined process of hot stamping and mechanical joining for producing ultra-high strength steel patchwork components." Journal of Manufacturing Processes 59 (2020): 444-455. https://doi.org/10.1016/j.jmapro.2020.10.025

[8] Harrer, T., and M. Schäfer. "Laser softening of press hardened steel for novel automotive parts." Advanced High Strength Steel and Press Hardening: Proceedings of the 4th International Conference on Advanced High Strength Steel and Press Hardening (ICHSU2018). 2019. https://doi.org/10.1142/9789813277984_0079

[9] Palmieri, Maria Emanuela, Francesco Rocco Galetta, and Luigi Tricarico. "Study of Tailored Hot Stamping Process on Advanced High-Strength Steels." Journal of Manufacturing and Materials Processing 6.1 (2022): 11. https://doi.org/10.3390/jmmp6010011

[10] Information on https://formingworld.com/hotforming-tailored-temporing-methodologies/, accessed October 2022.

[11] Palmieri, Maria Emanuela, et al. "Analysis of transition zone on a hot-stamped part with tailored tool tempering approach by numerical and physical simulation." steel research international https://doi.org/10.1002/srin.202200665. https://doi.org/10.1002/srin.202200665

[12] Information on https://www.aptgroup.com/company/news/apt-launching-temperbox%C2%AE-new-cycle-time-neutral-production-solution-enables-tailored, accessed February 2023.

[13] Palmieri, Maria Emanuela, and Luigi Tricarico. "Numerical-experimental study of a tailored press-hardening technology with intermediate pre-cooling to manufacture an automotive component in advanced high strength steel." Sheet Metal 2023 25 (2023): 447. https://doi.org/10.21741/9781644902417-55

Italian Manufacturing Association Conference - XVI AITeM
Materials Research Proceedings 35 (2023) 53-61

Materials Research Forum LLC
https://doi.org/10.21741/9781644902714-7

Workload and stress evaluation in advanced manufacturing systems

Graziana Blandino[1,a] *, Francesca Montagna[1,b] and Marco Cantamessa[1,c]

[1]Department of Management and Production Engineering, Politecnico di Torino,10129 Torino, Italy

[a]graziana.blandino@polito.it, [b]francesca.montagna@polito.it, [c]marco.cantamessa@polito.it

Keywords: Ergonomics, Smart Manufacturing, Industry 4.0/5.0

Abstract. Industry 5.0 emphasizes the development of human-centred work environments, shifting the focus from technologies embedded in manufacturing systems to workers. Efforts in the literature focus on operators' well-being for workstation configuration or on stress in collaborative environments, but few papers consider stress induced by management practices in advanced manufacturing contexts, although "lean" or "agile" for instance could in principle lead to more stressful workplaces. This paper reviews the literature, evaluating the mental and physical workload of production line operators who perform mentally demanding tasks and experience stress in advanced manufacturing systems. The goal is to design and to perform a pilot test on an innovative and rigorous research protocol, to be adopted in 'non-fictional' experiments, and able to compare push vs pull settings and their effects on workers' workload and stress (WLS). The results will highlight new sources of stress, contributing to the development of human-centred and socially sustainable manufacturing systems.

Introduction

The fifth industrial revolution, Industry 5.0, defines a new paradigm of efficient and productive cooperation between autonomous machines, robots and human workers in advanced manufacturing systems. Automation and digitalization of processes have ensured efficiency and system optimization, but have had several consequences on workers that have acquired new roles [1,2] and had to develop new competencies, with possible impact on their safety and health [3]. Learning new skills, as well as changing roles, have been found to contribute to workers' discomfort, stress and fatigue [3]. In view of that, detecting signs of workload and stress (WLS) experienced by workers while executing work tasks at these contemporary manufacturing workplaces can contribute to developing new strategies aimed at preserving their well-being [4].

The literature on advanced manufacturing systems presents some limits. On the one hand, WLS phenomena are sometimes confused or considered as synonyms, and only few contributions [5,6] attempt to define them properly. On the other hand, contributions tend to explore only physical factors such as workload, fatigue and ergonomics and neglect the stress phenomenon. Some current contributions focus on human factors, such as cognitive load in manufacturing [7–9] and especially in assembly tasks [10,11] or push/pull tasks [12,13], though without assessing the phenomenon of stress.

The present work reviews the human factors measurement methods and the experimental protocols used in the literature, focusing on the analysis of WLS assessment. It then proposes a protocol to be adopted in 'non-fictional' experiments, along with a pilot study conducted to validate it and to assess if the measurements are suitable to compare effectively push vs pull settings.

The first section of the paper reviews the literature on human factors in advanced manufacturing systems and on experimental protocols. The following sections illustrate the aim of the paper, the research methodology and the designed protocol, followed by the discussion. Finally, the conclusions and limitations are shown.

Literature review

Workload is defined as the cost of performing a task and depends on several factors, such as the requirements of the task, the context in which the task is performed, and the skills of the worker [6]. Stress at work, on the other hand, is the phenomenon that occurs when the demand for work exceeds the worker's ability to perform it [14].

Three main categories of measurements, as in Table 1, exist in the literature: physiological, physical and psychological. The physiological and physical categories are defined as objective since they measure data that are not influenced by the perceptions of participants to experiments. Psychological measurements are subjective, and focus on workers' emotional state of WLS.

The physical methods measure stress and physical workload through the analysis of body posture and the ergonomics of workstations. In this regard, studies focus on postural measurements [15], body motion indicators [9,15,16], body language [9] or workers' performance [11,13,17].

Physiological measurements record the unconscious physiological and cognitive processes of workers while executing tasks. A number of studies focus on heart rate [18], or on associated composite indicators [15,17,19,20], measuring the distance between two heartbeats on the cardiac signal [6,17,21]. Besides, electrodermal activity (EDA) is investigated [5] by examining skin conductance (SC), in reaction to external stimuli e.g. [17,20,21]. Finally, other measures are possible: the breath rate [22], even if not so diffusely [15,19]; the electroencephalographic (EEG) signal [23], especially for stress measurements; and face temperature [24]. Finally, workload can be evaluated through the blink rate and the pupil size indicators of the ocular activity [25].

Psychological measurements examine the subjective perceptions of workers through the submission of questionnaires and tests, as in Table 1. The main advantage of these measurements relies on the fact that questionnaires and tests can be proposed during different stages of the experimental activities, or some days after, allowing different comparisons e.g. [6,20]. Moreover, these methods enable a better calibration of measurement methods, reducing the misinterpretation of data [26], but they require large samples of respondents to provide reliable results [27].

Table 1. Physical, Physiological and Psychological measurements and indicators

Physical measurement	Indicators	Workload	Stress	Reference
Postural	Ovako Working Posture Analysis System (OWAS)		X	[15]
	Rapid Entire Body Assessment (REBA)		X	[15]
	Rapid Upper Limb Assessment (RULA)		X	[15]
Body motion	Assembly line speed		X	[16]
	Occupational Repetitive Action (OCRA)	X		[8]
	Vector Magnitude Units (VMU)		X	[15]
	Hyperactivity		X	[9]
	Self-touching		X	
Body language/Behavioural	Reaction time	X		[11]
	Error rate	X		[17]
	Completion time	X		[13]
	Accuracy	X		
Physiological measurement	**Indicator**	**Workload**	**Stress**	**Reference**
Cardiac activity	Inter-beat intervals	X	X	e.g.[17,21]
	Heart Rate (HR), HR Variability (HRV)	X	X	e.g.[15,28]
EDA	Skin Conductance (SC)	X	X	e.g.[20,21]
Breathing activity	Breathing rate	X	X	[15,19]
Facial activity	Face temperature	X		[24]
Cerebral activity	High Beta frequency	X	X	[23,29]
Ocular activity	Pupil size changes	X		[25]
	Blink rate	X		[25]
	Eyes movement	X		[29]

Psychological data collection Method	Workload	Stress	Reference
Cognitive Load Assessment for Manufacturing (CLAM)	X		[7]
National Aeronautics Space Administration-Task Load Index (NASA-TLX)	X	X	e.g.[19,30]
Rating Scale Mental Effort (RSME)	X		[30]
Subjective workload assessment technique (SWAT)	X	X	[24,31]
Workload profile (WP)	X		[31]
Instantaneous Self-Assessment (ISA)	X		[17]
Modified Cooper-Harper Scale (MCH)	X		[24]
State-Trait Anxiety Inventory (STAI)		X	[20]
Depression Anxiety Stress Scales (DASS)		X	[12]
Valence-Arousal Test		X	[23]
Numeric Analog Scale (NAS)		X	[6]
Body Part Discomfort (BDP) scale		X	[32]
Perceived Stress scale (PSS)		X	[33]
Short Stress State Questionnaire (SSSQ)		X	[27]

The comparison of methods shows an overlap of measurements and indicators for WLS and suggests combining multiple types of measurements to avoid incomplete analyses. The experimental tasks usually include motor and cognitive activities. As in Table 2, motor tasks consist of manual assembly activities [9,15,32] that, in some studies, are executed by using collaborative robots [21,23] or augmented reality glasses [10], or working in different operational (e.g., push vs. pull) contexts [12,19]. In other cases, the task type involves other activities, such as crimping [8] or replacement of spare parts [6]. On the other hand, cognitive tasks aim to replicate high levels of attention and mental concentration in real manufacturing contexts. Examples are the N-back task [13,19], [28,29], the auditory stimulus detection task [11], or the visual search task [17]. Motor tasks are instead proposed to measure physical workload, while cognitive tasks are usually related to mental workload; however, standard tasks have still not emerged. Experimental activities can be carried out in laboratories and in real industrial environments. In laboratory experiments, augmented and virtual reality are adopted to simulate manufacturing contexts while executing tasks, and sensitive and fragile measurement equipment [15], difficult to integrate in real manufacturing environments, is used. In a real context [11, 12, 21], such as an automotive assembling line [32], wearable technologies for measurements aim to not interfere with the tasks.

Table 2. Experimental tasks and contexts

References	Experimental task				Experimental environment	
	Motor task			Cognitive		
	Assembly	Push vs Pull	Other	task	Laboratory	In-field
[10,32]	X					X
[11]	X			X		X
[6,8]			X		X	
[28]			X	X	X	
[9,15,21,23]	X				X	
[13,29]	X			X	X	
[12]		X				X
[19]		X		X	X	
[17]				X	X	

As a general comment about human factors measurements and these experimental protocols, one can observe the little attention paid to demographic variables, such as workers' age and gender [34] or to environmental factors, such as noise and temperature levels, which instead have been demonstrated to influence humans' stress and workload [35].

Italian Manufacturing Association Conference - XVI AITeM   Materials Research Forum LLC
Materials Research Proceedings 35 (2023) 53-61   https://doi.org/10.21741/9781644902714-7

Aim of the paper and research questions

This paper describes the protocol design and validation phase of a research aimed at comparing push vs pull settings and their effects on workers' WLS indicators. The first research question to be answered is: what are the most appropriate WLS measurement methods to be employed in the case of in-field experiments? The ensuing research question is: how do WLS indicators change when working in push vs. pull operational settings?

The literature review suggests adopting a combination of methods for measuring WLS. In line with this, the research methodology implements cardiac, electrodermal and breathing activity data collection to calculate physiological indicators, and NASA-TLX and PSS-10 questionnaires for psychological indicators.

The originality of the research is due to the possibility of:
- conducting these complementary measures through professional biomedical devices recording the integral physiological signal in a real industrial environment, with levels of accuracy and detail in the parameters that would be impossible to achieve by using widely-diffused consumer-grade wearable devices;
- conducting the measures in a real industrial environment, to consider environmental factors.

The validation process of the experimental protocol consisted of physiological and psychological signals analyses firstly to prove their suitability for in-field WLS measurements and, subsequently, to verify changes in the associated indicators according the experimental conditions.

The designed protocol and the adopted methodology of analysis

The protocol designed for the pilot experiment consists of three consecutive phases (i.e. before, during and after the task execution), as in Fig. 1. Before the task execution, the consent form and the sociodemographic questionnaire are filled; health status is certified, and gender and age data are collected to allow further investigations that are now neglected in the literature. Moreover, the participant wears the non-invasive biomedical devices to test functionality and for calibration; physiological data are collected for 5 minutes in rest condition and the subject is then instructed on the procedure details. Each data collection session consists of almost 15 minutes of physiological signals recording while the participant performs assembly tasks. Then, the biomedical devices are removed and a rest pause is ensured for the subject. Finally, WLS questionnaires are anonymously filled in by the participant referring to the task executed.

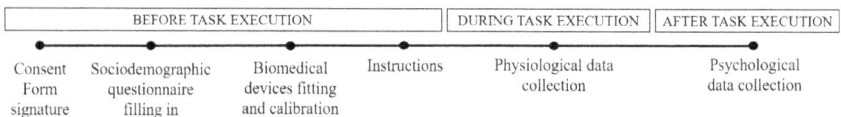

BEFORE TASK EXECUTION				DURING TASK EXECUTION	AFTER TASK EXECUTION
Consent Form signature	Sociodemographic questionnaire filling in	Biomedical devices fitting and calibration	Instructions	Physiological data collection	Psychological data collection

Figure 1. Experimental protocol

The device to be used for data collection are Encephalan Mini ABP-10, connected to ECG, EDA and thorax/abdominal breath sensor, in combination with Encephalan-EEGR Elite software. The ECG is recorded through three electrodes (Fig. 2.a), placed one for each wrist and the third on the left forearm. The sensor for respiratory activity is a band positioned on the participant's abdomen (Fig. 2.b), while EDA device comprehends ring sensors in the index and middle fingers of the left hand (Fig. 2.c). Electrodes and sensors are connected to the central data collection unit (Fig. 2.d), which in turn is connected to the SW for data storage. HR for the ECG technique, SC for EDA, and finally breathing rate for respiratory activity are the chosen WLS indicators.

Italian Manufacturing Association Conference - XVI AITeM   Materials Research Forum LLC
Materials Research Proceedings 35 (2023) 53-61   https://doi.org/10.21741/9781644902714-7

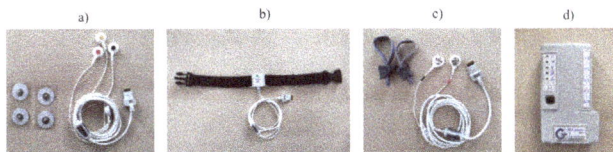

Figure 2. a) ECG device. b) Respiratory sensor. c) EDA device. d) Central data collection unit

The experimental protocol is aimed at tasks typical of an assembly environment and includes picking components, carrying out pre-assembly processes, assembly and storing the finished good. In the "push" setting, the order follows the production cycle, without restrictions on working times and unlimited buffers between stations. In the pull setting, production follows a Kanban system that defines limited buffers, and tack time defines the production pace. All physiological signals are analysed by MATLAB. The power spectral density of the noise is plotted, as well as the appropriate filters are applied to clean the signal. The physiological WLS indicators are calculated for each session, both in task execution and rest condition, in order to validate the difference among the experimental conditions. Then, statistical analyses of the WLS questionnaires for the assessment of correlation with the physiological results are conducted on MINITAB.

Validation of the protocol
The validation experiment was conducted in cooperation with a company operating in the plastic components industry, which has both "push" and "pull" operations. The company is a small-to-medium-sized enterprise with high Technology Readiness Level and medium-high maturity and leadership levels, according to ISO 9004:2018 standard. Since time constraints have been identified as a source of stress [36], one would expect to find physical and psychological stress levels to be higher under pull than push conditions. Therefore, in view of the validation, if the protocol works in push settings, it could work in the pull condition as well. Among the push sections of the company's operations, we have chosen the assembly process (Fig 3.a) for customized heating radiator covers (Fig. 3.b).

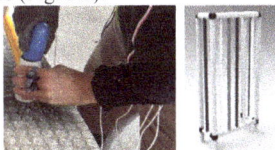

Figure 3. a) Assembly process. b) Heating radiator cover

The process consists in analysing the specific customer order and picking the required plastic tubes, cutting the tubes to measure, drilling holes in tubes, assembling and delivering to outgoing port. These steps are carried out by a single operator, who is free to organize his work, since the process is not rigid, and who moves along the workstations dedicated to each step. Cycle time depends on the customer order and averages 5 minutes per part. After set-up and calibration, physiological data were collected for 5 minutes in rest condition, followed by a working session of 15 minutes. A total of 4 sessions were carried out at different times. The breathing rate indicator was calculated for each step and for the entire session and was compared between the rest and the working conditions, as in Fig 4.a. By using research-grade equipment, the indicator recorded changes significantly for each session between the rest and the push task execution, coherently with the literature [17], and showed differences in WLS between the process steps, as in Fig 4.b, indicating the effectiveness of the protocol in detecting WLS differences. After each session, the NASA-TLX and PSS-10 questionnaires were compiled; the responses were examined, but not statistically analysed due to the limited sample of responses.

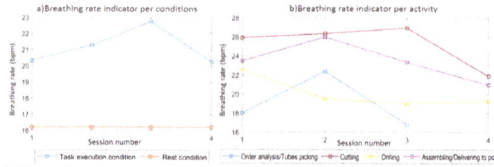

Figure 4. a) Breathing rate vs task conditions. b) Breathing rate vs task activities

The NASA-TLX responses for the 4 sessions (S1-S4) in Fig. 5.a, revealed that the mental demand (item 1) is higher than physical and temporal demand (items 2, 3). In general, the task was successfully performed (item 4) in each session at an intermediate level of effort and frustration (items 5, 6). Finally, from the PSS-10 assessments, in Fig. 5.b, intermediate scores were registered and negative emotions (items 1,2,3,6,9,10) were never perceived, while positive ones (items 4,5,7,8) occurred at times. The capability to appreciate differences in mental, physical and temporal demand, and also to capture the intensity and the nature of the perceived emotions of the workers, allowed the validation of both the two proposed questionnaires.

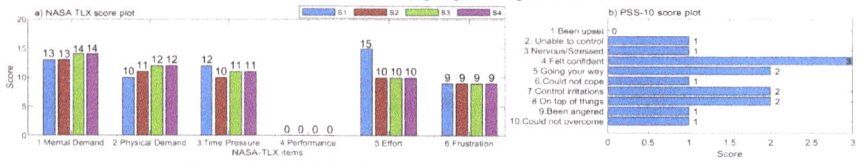

Figure 5. a) NASA-TLX responses. b) PSS-10 responses

Conclusion and limitations

The paper describes the design and validation of an experimental protocol for the evaluation of WLS indicators in push vs pull conditions in real industrial environments, using research-grade physiological equipment. The experiment was conducted in a plastics component company in a push condition. Cardiac, breathing, electrodermal activities data and NASA-TLX and PSS-10 answers have been collected, even if by way of example only the breathing rate data is reported. The indicators, among which here the respiration rate is discussed, have proved to be appropriate for WLS detection, and the two questionnaires confirmed that the protocol is suitable to study WLS in real push/pull industrial contexts. Furthermore, the results suggest that the analysed push task is demanding. The task variety and schedule flexibility balanced the stress due to medium-high time pressure, effort, mental and physical demand, according to the International Labour Organisation [37]. Consequently, it will be interesting to compare with pull situation. A larger number of participants and longer sessions will characterise the next experiments to monitor the evolution of WLS indicators and establish robust evidence under various work conditions.

References

[1] J. Alves, T. M. Lima, P. D. Gaspar, Is Industry 5.0 a Human-Centred Approach? A Systematic Review, Processes, 11.1 (2023) 193. https://doi.org/10.3390/pr11010193

[2] M. C. Zizic, M. Mladineo, N. Gjeldum, and L. Celent, From Industry 4.0 towards Industry 5.0: A Review and Analysis of Paradigm Shift for the People, Organization and Technology, Energies, 15.14 (2022) 5221. https://doi.org/10.3390/en15145221

[3] J. Leng, W. Sha, B. Wang, P. Zheng, C. Zhuang, Q. Liu, L. Wang, Industry 5.0: Prospect and retrospect, J Manuf Syst, 65 (2022) 279-295. https://doi.org/10.1016/j.jmsy.2022.09.017

[4] V. Villani, M. Gabbi, L. Sabattini, Promoting operator's wellbeing in Industry 5.0: detecting mental and physical fatigue, in Conference Proceedings - IEEE International Conference on

Systems, Man and Cybernetics, 2022 (2022), 2030-2036.
https://doi.org/10.1109/SMC53654.2022.9945324

[5] C. Setz, B. Arnrich, J. Schumm, R. la Marca, G. Tröster, U. Ehlert, Discriminating stress from cognitive load using a wearable eda device, IEEE Transactions on Information Technology in Biomedicine, 14.2 (2010) 410-417. https://doi.org/10.1109/TITB.2009.2036164

[6] A. Brunzini, M. Peruzzini, F. Grandi, R. K. Khamaisi, M. Pellicciari, A preliminary experimental study on the workers' workload assessment to design industrial products and processes, Applied Sciences, 11.24 (2021) 12066. https://doi.org/10.3390/app112412066

[7] P. Thorvald, J. Lindblom, R. Andreasson, On the development of a method for cognitive load assessment in manufacturing, Robot Comput Integr Manuf, 59 (2019) 252-266. https://doi.org/10.1016/j.rcim.2019.04.012

[8] E. Giagloglou, P. Mijovic, S. Brankovic, P. Antoniou, I. Macuzic, Cognitive status and repetitive working tasks of low risk, Saf Sci, 119 (2019) 292-299. https://doi.org/10.1016/j.ssci.2017.10.004

[9] M. Lagomarsino, M. Lorenzini, E. de Momi, A. Ajoudani, An Online Framework for Cognitive Load Assessment in Industrial Tasks, Robot Comput Integr Manuf, 78 (2022). https://doi.org/10.1016/j.rcim.2022.102380

[10] H. Atici-Ulusu, Y. D. Ikiz, O. Taskapilioglu, T. Gunduz, Effects of augmented reality glasses on the cognitive load of assembly operators in the automotive industry, Int J Comput Integr Manuf, 34.5 (2021) 487-499. https://doi.org/10.1080/0951192X.2021.1901314

[11] M. Drouot, N. le Bigot, E. Bricard, J. L. de Bougrenet, V. Nourrit, Augmented reality on industrial assembly line: Impact on effectiveness and mental workload, Appl Ergon, 103 (2022) 103793. https://doi.org/10.1016/j.apergo.2022.103793

[12] M. Petrovic, A. M. Vukicevic, M. Djapan, A. Peulic, M. Jovicic, N. Mijailovic, K. Jovanovic, Experimental Analysis of Handcart Pushing and Pulling Safety in an Industrial Environment by Using IoT Force and EMG Sensors: Relationship with Operators' Psychological Status and Pain Syndromes, Sensors, 22.19 (2022) 7467. https://doi.org/10.3390/s22197467

[13] F. N. Biondi, A. Cacanindin, C. Douglas, J. Cort, Overloaded and at Work: Investigating the Effect of Cognitive Workload on Assembly Task Performance, 63.5 (2021) 813-820. https://doi.org/10.1177/0018720820929928

[14] D. M. Wegner, Stress and Mental Control'. Stress and mental control. Handbook of life stress, cognition and health, In S. Fisher & J. Reason (Eds.), 1988, 683-697.

[15] M. Peruzzini, F. Grandi, M. Pellicciari, Exploring the potential of Operator 4.0 interface and monitoring, Comput Ind Eng, 139 (2020) 105600. https://doi.org/10.1016/j.cie.2018.12.047

[16] V. K. Rao Pabolu, D. Shrivastava, M. S. Kulkarni, A Dynamic System to Predict an Assembly Line Worker's Comfortable Work-Duration Time by Using the Machine Learning Technique, in Procedia CIRP, 106 (2022) 270-275. https://doi.org/10.1016/j.procir.2022.02.190

[17] E. M. Argyle, A. Marinescu, M. L. Wilson, G. Lawson, S. Sharples, Physiological indicators of task demand, fatigue, and cognition in future digital manufacturing environments, International Journal of Human Computer Studies, 145 (2021) 102522. https://doi.org/10.1016/j.ijhcs.2020.102522

[18] R. Castaldo, P. Melillo, U. Bracale, M. Caserta, M. Triassi, L. Pecchia, Acute mental stress assessment via short term HRV analysis in healthy adults: A systematic review with meta-

Italian Manufacturing Association Conference - XVI AITeM   Materials Research Forum LLC
Materials Research Proceedings 35 (2023) 53-61   https://doi.org/10.21741/9781644902714-7

analysis, Biomedical Signal Processing and Control, 18 (2015) 370-377. https://doi.org/10.1016/j.bspc.2015.02.012

[19] F. N. Biondi, B. Saberi, F. Graf, J. Cort, P. Pillai, B. Balasingam, Distracted worker: Using pupil size and blink rate to detect cognitive load during manufacturing tasks, Appl Ergon, 106 (2023) 103867. https://doi.org/10.1016/j.apergo.2022.103867

[20] M. Ciccarelli, A. Papetti, M. Germani, A. Leone, G. Rescio, Human work sustainability tool, J Manuf Syst, 62 (2022) 76-86. https://doi.org/10.1016/j.jmsy.2021.11.011

[21] R. Gervasi, K. Aliev, L. Mastrogiacomo, F. Franceschini, User Experience and Physiological Response in Human-Robot Collaboration: A Preliminary Investigation, Journal of Intelligent and Robotic Systems: Theory and Applications, 106.2 (2022) 36. https://doi.org/10.1007/s10846-022-01744-8

[22] A. Nicolò, C. Massaroni, E. Schena, M. Sacchetti, The importance of respiratory rate monitoring: From healthcare to sport and exercise, Sensors, 20.21 (2020) 1-45. https://doi.org/10.3390/s20216396

[23] A. T. Eyam, W. M. Mohammed, J. L. Martinez Lastra, Emotion-driven analysis and control of human-robot interactions in collaborative applications, Sensors, 21(2021) 4626. https://doi.org/10.3390/s21144626

[24] J. Kang, K. Babski-Reeves, (2009). Evaluation of methods for determining optimal mental workload levels. In IIE Annual Conference. Proceedings, Institute of Industrial and Systems Engineers (IISE) (2009) 913.

[25] S. Chen, J. Epps, Using task-induced pupil diameter and blink rate to infer cognitive load, Hum Comput Interact, 29. 4 (2014) 390-413. https://doi.org/10.1080/07370024.2014.892428

[26] Y. Z. Abd Elgawad, M. I. Youssef, T. M. Nasser, New methodology to detect the effects of emotions on different biometrics in real time, International Journal of Electrical and Computer Engineering, 13.2 (2023) 1358-1366. https://doi.org/10.11591/ijece.v13i2.pp1358-1366

[27] L. Gualtieri, F. Fraboni, M. de Marchi, E. Rauch, Development and evaluation of design guidelines for cognitive ergonomics in human-robot collaborative assembly systems, Appl Ergon, 104 (2022) 103807. https://doi.org/10.1016/j.apergo.2022.103807

[28] D. Cavallo, F. Facchini, G. Mossa, Information-based processing time affected by human age: An objective parameters-based model, in IFAC-PapersOnLine, 54.1 (2021) 7-12. https://doi.org/10.1016/j.ifacol.2021.08.001

[29] J. Morton, A. Zheleva, B.B. Van Acker, W. Durnez, P. Vanneste, C. Larmuseau, K. Bombeke, 'Danger, high voltage! Using EEG and EOG measurements for cognitive overload detection in a simulated industrial context', Appl Ergon, 102 (2022) 103763. https://doi.org/10.1016/j.apergo.2022.103763

[30] A. Widyanti, W. Larutama, The relation between performance of lean Manufacturing and employee' mental workload, in IEEE International Conference on Industrial Engineering and Engineering Management, 2016 (2016) 252-256. https://doi.org/10.1109/IEEM.2016.7797875

[31] S. Rubio, E. Díaz, J. Martín, J. M. Puente, Evaluation of Subjective Mental Workload: A Comparison of SWAT, NASA-TLX, and Workload Profile Methods, Applied Psychology, 53.1 (2004) 61-86. https://doi.org/10.1111/j.1464-0597.2004.00161.x

[32] V. Kopp, M. Holl, M. Schalk, U. Daub, E. Bances, B. Garcia, U. Schneider, Exoworkathlon: A prospective study approach for the evaluation of industrial exoskeletons, Wearable Technologies, 3 (2022) e22. https://doi.org/10.1017/wtc.2022.17

[33] M. Mailliez, S. Hosseini, O. Battaïa, R. N. Roy, Decision Support System-like Task to Investigate Operators' Performance in Manufacturing Environments, in IFAC-PapersOnLine, 53 (2020) 324-329. https://doi.org/10.1016/j.ifacol.2021.04.110

[34] V. di Pasquale, S. Miranda, W. P. Neumann, Ageing and human-system errors in manufacturing: a scoping review, International Journal of Production Research, 58.15 (2020) 4716-4740. https://doi.org/10.1080/00207543.2020.1773561

[35] A. M. Abbasi, M. Motamedzade, M. Aliabadi, R. Golmohammadi, L. Tapak, Combined effects of noise and air temperature on human neurophysiological responses in a simulated indoor environment, Appl Ergon, 88 (2020) 103189. https://doi.org/10.1016/j.apergo.2020.103189

[36] J. R. Kelly, J. E. Mcgrath, Effects of Time Limits and Task Types on Task Performance and Interaction of Four-Person Groups, 49.2 (1985) 395. https://doi.org/10.1037/0022-3514.49.2.395

[37] International Labour Organization. Workplace Stress: A Collective Challenge. International Labour Office: Geneva, Switzerland (2016).

Italian Manufacturing Association Conference - XVI AITeM   Materials Research Forum LLC
Materials Research Proceedings 35 (2023) 62-69   https://doi.org/10.21741/9781644902714-8

Modelling the laser overageing treatment of a 6xxx Al alloy by means of physical simulation tests

Antonio Piccininni[1,a*], Pasquale Guglielmi[1,b], Angela Cusanno[1,c] and Gianfranco Palumbo[1,d]

[1] Department of Mechanical Engineering, Mathematics and Management, Politecnico di Bari, 70126, Italy

[a]antonio.piccininni@poliba.it, [b]pasquale.guglielmi@poliba.it, [c]angela.cusanno@poliba.it, [d]gianfranco.palumbo@poliba.it

Keywords: Laser Heat Treatment, Aluminium Alloys, Hardness

Abstract. The local modification of the material properties is a promising strategy to broaden the range of applications for Aluminum (Al) alloys, since excessive thinning are avoided, and sound parts obtained. For example, by means of laser heating, the strain behaviour of age hardenable Al alloys can be locally affected to enhance the formability at room temperature. But predicting the properties modification (occurrence of overageing/solutioning) due to the local heating still needs investigations. In this work, short-term heat treatments on AA6063-T6 samples were conducted using the Gleeble system (able to subject the material to high heating rates combined with large temperature gradients). Wide ranges of temperature and time were thus explored, and the change of mechanical properties assessed by hardness tests. Experimental data were used to create a model and thus define the heating parameters able to bring the material to the overaged state or, alternatively, to the fully solutioned one.

Introduction

The reduction of fuel consumption in the transportation sector is becoming one of the most urgent issues to be tackled [1]. The reduction of sprung/unsprung masses in vehicles [2] is one of the most direct solutions to reduce pollution trying, at the same time, to meet the standards of passengers' comfort and safety [3]. In light of this, the scientific research has focused the attention on the definition of new products able to combine limited weight, good level of strength, and high standard of safety. Aluminum (Al) and its alloys are regarded as ideal candidates to match all the mentioned requirements: in particular, especially for the automotive sector, the adoption of tubular structural parts has shown several advantages since they can combine the necessary stiffness without excessively increasing the vehicle's weight. Despite all the advantages in terms of limited density and high strength-to-weight ratio, the adoption of Al alloys for structural and complex components is partially hindered by the low formability at room temperature [4]. Therefore, innovative manufacturing solutions have been investigated. Among the others, the approach based on the local modification of the material properties by means of a short-term heat treatment has demonstrated huge potentialities [5]: in such a way, the properly optimized alternation of strength and ductility improves the alloy formability at room temperature. Moreover, being the forming operations carried out at room temperature, costs related to the equipment are sensibly lowered. Literature confirms the effectiveness of the mentioned approach especially when applied to sheet metal parts [6,7], whereas only few examples deal with the possibility to apply a local modification to tubular components [8]. Nevertheless, it is reported that laser treatments, when applied to AA6063-T6 tubular samples, influences the deformation behaviour when the tube is deformed by the action of an elastomer [9]. Nevertheless, the question is not trivial: according to the precipitation kinetics in Al-Mg-Si alloys [10], if subjecting the material in the T6 "peak hardening" state to a heat treatment, two different outcomes can be obtained: (i) a temporary alteration of the

Italian Manufacturing Association Conference - XVI AITeM   Materials Research Forum LLC
Materials Research Proceedings 35 (2023) 62-69   https://doi.org/10.21741/9781644902714-8

properties brining the alloy to the fully solutioned condition which is characterized by higher formability but highly metastable (if kept at room temperature, it evolves due to the triggering of the natural ageing) or (ii) a permanent alteration, i.e. the overaged condition, which is characterized by an improvement of the material formability that is stable with time [11]. Given that, the gap still to be bridged is the definition of an accurate methodology that is able to predict the effect of short-term heating on the properties distribution.

Therefore, in the present work, a methodology based on the physical simulation is proposed to numerically predict the occurrence of the overaged state in a AA6063-T6 tubular component locally heated by means of a laser heat treatment. Thanks to a reduced number of tests using the Gleeble system, the combination of temperature and time able to determine the overaged or the fully solutioned state could be determined by hardness measurements along the heat-treated samples. In particular, hardness values were then fitted by a logistic function that was subsequently implemented in a python script. Laser heating of the tubular part (initially supposed in the T6 condition) was simulated by means of the Finite Element (FE) code Abaqus and results post-processed by means of the above-mentioned script to numerically predict the final distribution of hardness over the treated area. Numerical predictions were eventually validated by experimental laser heating tests. The proposed results can be regarded as preliminary steps toward a flexible and on-demand manufacturing approach of tubular components (starting from high-strength but difficult-to-form materials of automotive interest) with tailored behaviour coming from local features.

Calibration of the overageing heat treatment

Specimens were extracted from AA6063 tubular samples (200 mm long) in the T6 condition and heat treated using the physical simulator Gleeble 3180. Figure 1a shows a detailed view of the specimen positioned in the test chamber and clamped, at its end, by means of two cooled grips. The specimens is heated by Joule effect due to the current flow which is continuously modulated according to the temperature data acquired by a thermocouple (TC2) welded in the middle part of the specimen (black spot in Figure 1a). The performed heat treatment was composed of: (i) a heating step with a rate of 400°C/s (thus approximately reproducing the passage of a laser), (iii) a soaking step during with the test temperature was kept in the region of the TC2 and (iii) a final cooling step by blowing air inside the test chamber. Being the grips cooled, the sample experienced a temperature decrease moving from the middle position (where TC2 was positioned) toward the two ends: a parabolic distribution was supposed, which determined several and simultaneous heat treatments within the same tests. Thanks to the welding of the four thermocouples and assuming that the temperature distribution was symmetric with respect to the TC2 position, the parabolic trend could be reconstructed based on 7 points as shown in Figure 1b (plotted curve refers to the test temperature at 350°C for 5 seconds; filled markers refer to the acquisition from the TC).R

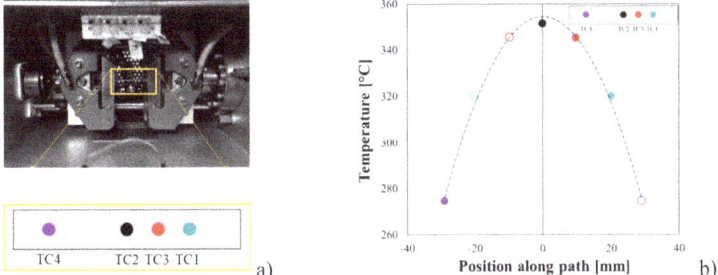

a)   b)

Figure 1 Set-up for the preliminary characterization: (a) detail of the Gleeble test chamber, (b) parabolic distribution of properties for the test temperature of 350°C

Italian Manufacturing Association Conference - XVI AITeM   Materials Research Forum LLC
Materials Research Proceedings 35 (2023) 62-69   https://doi.org/10.21741/9781644902714-8

Once cooled down to room temperature, specimens were subjected to hardness measurements (using the Qness micro hardness tester Q10 A+, setting a load of 200 g and a dwell time of 15 s) along the longitudinal direction and at regular interval of time to evaluate the achieved overaged condition or, alternatively, the fully solutioned one.

Experimental laser heating tests

Laser heating on tubular components were carried out using a 2.5 kW CO_2 laser head (A in Figure 2a) equipped with a Diffractive Optical Element (DOE) to obtain a top-hat energy distribution. A preliminary test was carried on a 200 mm long AA6063 tube (external diameter of 40 mm, average thickness of 1.9 mm) initially in the T6 condition: the tube (B in Figure 2a) was preliminarily sprayed with a black paint for high temperature (to increase the absorption of the laser radiation), positioned on a designed fixture (C in Figure 2a) and laser heated under a constant power of 600 W (square laser spot of 10 mm) for 5 seconds at a distance of 63 mm from the one of its ends. The temperature evolution was monitored by a wire thermocouple (D in Figure 2a) welded on the inner surface of the tube in correspondence of the laser spot's center (the correspondent temperature-time curve is plotted in Figure 2b). Additional laser heating tests were finally carried out to validate the numerical predictions.

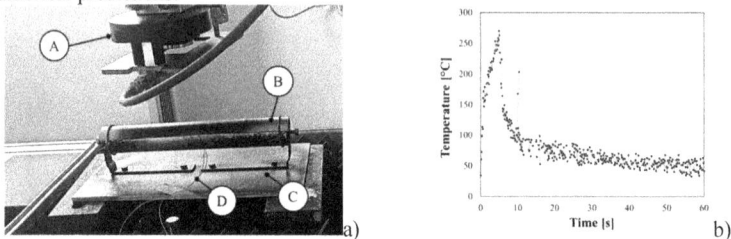

Figure 2 Laser heating tests: (a) experimental set-up; (b) temperature acquired from the thermocouple

Numerical simulation of the laser heating

The tube laser heating on a tubular component was simulated with the FE commercial code Abaqus/CAE. The tube was modelled as a deformable body and meshed with 46400 DC3D8 8-node solid elements (average size of 1 mm). The heat treatment of the tubular specimens was simulated solving the pure thermal transient problem (using the implicit solver): preliminary simulations were run to calibrate the unknown thermal boundary conditions – i.e. the quantity of absorbed radiation by the material and the heat transfer coefficient between the tube and the surrounding environment – by minimizing the difference between the experimental temperature acquisition from the thermocouple (as shown in Figure 2b, it was welded to the underside of the sample) and the numerical temperature curve extracted from the node at the same location. Once calibrated the thermal model, subsequent numerical simulations, keeping the same heating strategy (laser spot at 63 mm from the tube's end), were run changing the laser power and the heating time over three levels (600 W, 800 W, 1000 W for the laser power and 5 s, 10 s and 15 s for the heating time). Results from the numerical simulations were post-processed by means of a python script to predict the final hardness distribution.

Results from the calibration of the overageing heat treatment

The first heating tests were carried out setting the temperature to 450°C and the soaking time to 2 s. Local temperature evolutions could be extracted (see Figure 3a): during the soaking time the parabolic distribution embraced temperatures between 450°C (at the TC2 location) and 400°C (at the TC4 location). The results of the hardness measurement, shown in Figure 3b, demonstrate that all the specimen's points heated at temperatures included in the parabolic distribution were brought

Italian Manufacturing Association Conference - XVI AITeM   Materials Research Forum LLC
Materials Research Proceedings 35 (2023) 62-69   https://doi.org/10.21741/9781644902714-8

to the fully solutioned state: the hardness, in fact, evolved (natural ageing) from an average value of 50 HV (1 h after the heat treatment) up to around 80 HV (one week after).

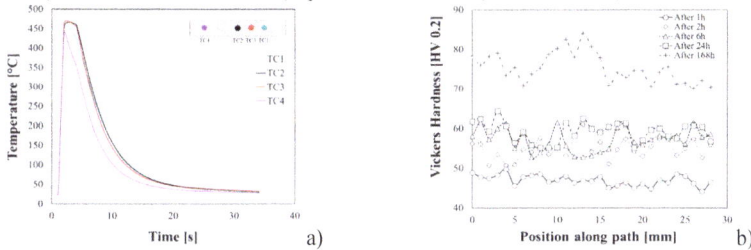

Figure 3 Results from the first Gleeble test (450°C for 2 s): (a) local temperature curves, (b) time evolution of the hardness values along the longitudinal path.

According to the results of the first tests, the subsequent ones were carried out at a lower temperature (350°C) and varying the soaking time over three levels (2. 5 and 10 seconds). As an example, the evolution of the hardness values of the specimen heat-treated setting the temperature to 350°C and the time to 5 s (see Figure 4a) suggested that the drop in the material hardness was remarkable (the reference value in the T6 condition is around 88 HV) and, above all, constant over time (up to one week), thus confirming the occurrence of the overaging.

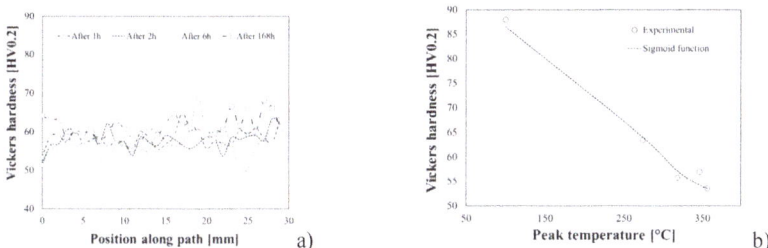

Figure 4 Results from Gleeble test at 350°C for 5 s: (a) time evolution of hardness along the measuring path, (b) the constructed sigmoid function

Similar results were obtained also in the other two tests setting different values of the holding time. The hardness measured close to the thermocouples could be then univocally related to the correspondent temperature. Measured values could be plotted as a function of the peak temperature reached during the heating test (corresponding to the temperature during the soaking step). Hardness values, as shown in Figure 4b, were fitted by a logistic function [12] whose analytical formulation is expressed by Equation 1.

$$HV = 88 - \frac{A}{1+\exp[\lambda(T_0-T_{max})]} \tag{1}$$

being A the maximum drop in the hardness (from the initial T6 state), T_0 the function's midpoint and λ the steepness of the curve. Nevertheless, an important aspect had to be assessed: the hardness measured at a specific location had not to be taken as the result of the only soaking step but had to be related to the whole thermal history (heating, soaking and cooling). Therefore, the integral below each extracted temperature/time curve (considered as an accurate indicator of the thermal history) was calculated and divided by the peak temperature reached during the test (i.e. the soaking temperature), thus obtaining the equivalent time (t^*). In such a way, the hardness value was considered to be determined by an equivalent heat treatment characterized by a constant

Italian Manufacturing Association Conference - XVI AITeM   Materials Research Forum LLC
Materials Research Proceedings 35 (2023) 62-69   https://doi.org/10.21741/9781644902714-8

temperature (equal to the peak temperature) and a duration equal to t^*. Table 1 reports the calculated t^* values for all temperature curves; moreover, values referring to curves from the same Gleeble test (i.e. same soaking time) were then averaged (t^*_{avg}).

Table 1 List of extracted heating conditions along with the correspondent equivalent time

ID test	Peak Temp. [°C]	Soaking time [s]	t^* [s]	t^*_{avg} [s]
T353-t2	353	2	7.26	
T352-t2	352	2	7.09	6.99
T329-t2	329	2	7.21	
T300-t2	300	2	6.39	
T351-t5	351	5	10.16	
T345-t5	345	5	9.86	10.05
T320-t5	320	5	10.34	
T274-t5	274	5	9.84	
T352-t10	352	10	14.69	
T314-t10	314	10	15.51	15.50
T256-t10	256	10	16.30	

Eventually, the hardness values from each Gleeble test were fitted by the sigmoid function, each related to a certain value of the average equivalent time. Therefore, the constants of the sigmoid (T_0 and λ) could be expressed as a linear function of the average equivalent time, as shown in Figure 5 (a and b).

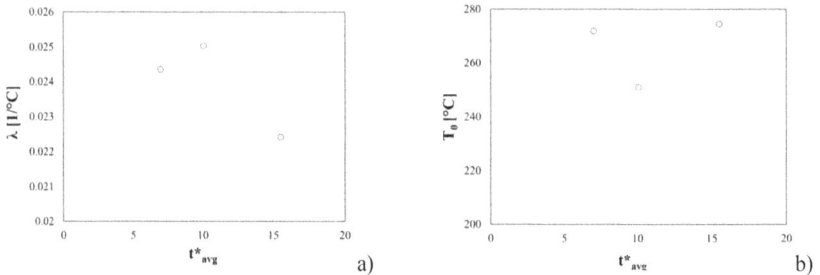

Figure 5 Constants of the logistic formulation as a function of t^*_{avg}: (a) λ and (b) T_0

Numerical results.
Preliminary FE simulations were run to calibrate the unknown thermal boundary conditions. Temperature acquired by the thermocouple (see Figure 2b) were used as target for tuning the FE model: differences between experimental data and the nodal temperature from the same location was minimized by setting the absorption coefficient to 0.49 and the convective heat transfer coefficient to 1 W/m²K. The model could be thus used to simulate the effect of different values of the laser power and the holding time. Results in terms of temperature evolutions of the node positioned at the center of the laser spot have been plotted in Figure 6.

Italian Manufacturing Association Conference - XVI AITeM

Materials Research Proceedings 35 (2023) 62-69

Materials Research Forum LLC

https://doi.org/10.21741/9781644902714-8

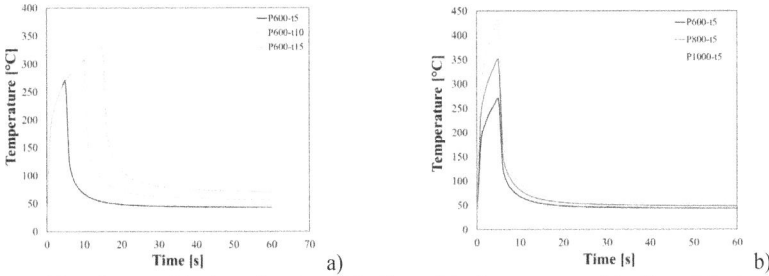

Figure 6 Simulation of the laser heating: (a) effect of the beam power, (b) effect of the heating time

Numerical results revealed a good accordance with the theory, since a remarkable increase in the peak temperature was reached when increasing the heating time (Figure 6a) or, alternatively, increasing the laser power (Figure 6b). For all the simulated conditions, the integral bounded by the temperature time curve extracted from the node located at the center of the spot was calculated and, once known the maximum temperature, the equivalent time was calculated. The equivalent time for all the simulated conditions have been reported in Table 2.

Table 2 Determination of the equivalent time for the investigated heating conditions

ID simulation	Laser power [W]	Heating time [s]	Max. Temp [°C]	Eq. time [s]
P600-t5	600	5	269.11	14.66
P600-t10	600	10	307.74	19.60
P600-t15	600	15	330.55	24.30
P800-t5	800	5	349.67	13.35
P800-t10	800	10	399.594	18.61
P800-t15	800	15	430.788	23.50
P1000-t5	1000	5	429.62	12.55
P1000-t10	1000	10	492.00	17.99
P1000-t15	1000	15	530.985	22.98

Post-processing of the numerical results. Starting from the numerical results of the laser heating, the final distribution of hardness was calculated by means of a Python script: the nodal temperature of each node was extracted by the script for all time instants, thus allowing to identify the peak temperature and to calculate the integral below the temperature-time curve. From the ratio between the integral and the peak temperature, the equivalent time could be calculated and the final hardness estimated using Equation 1. According to the results from the Gleeble tests at 450°C, heating simulations characterized by a nodal peak temperature close or higher then 400°C were not post-processed (presumably the treated region would have been brought to the fully solutioned state). Figure 7 shows the predicted final hardness distribution for the other post-processed conditions: only when setting the laser power at 800 W for 5 s, the drop in the material properties was much more pronounced and the predicted hardness in the heated zone below 60 HV.

Italian Manufacturing Association Conference - XVI AITeM
Materials Research Proceedings 35 (2023) 62-69

Materials Research Forum LLC
https://doi.org/10.21741/9781644902714-8

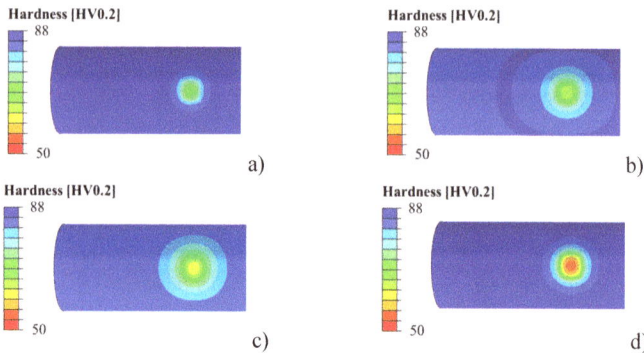

Figure 7 Numerical prediction of the final hardness distribution: (a) P600-t5, (b) P600-t10, (c) P600-t15, (d) P800-t5

Validation of the numerical predictions. According to the numerical results, the laser heating test conducted setting the laser power to 800 W and the heating time to 5 s was replicated. Once the tube cooled down to room temperature, the portion subjected to the laser heating was extracted and the hardness (in the thickness direction) monitored at different times. As reported in Figure 8, the hardness did not varied, thus revealing that overaged condition was reached. In addition, the measured values were in good accordance with the numerical prediction (continuous green line).

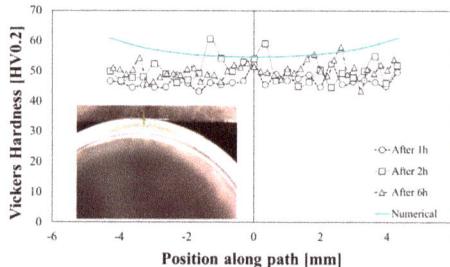

Figure 8 Experimental measured hardness vs. predicted values from the numerical simulations (P800-t5)

Conclusions

In the present work, an original methodology to predict the change of material properties determined by a laser heating on AA6063-T6 tubular parts has been proposed. The Gleeble physical simulator allowed to gather a large quantity of data by means of a limited number of tests. It was demonstrated that all the temperature close or above 400°C, even for very small holding time (few seconds), brought the material to the fully solutioned state; temperature of 350°C allowed to reach a more stable overaged condition. The preliminary characterization provided enough data to construct an analytical model: a sigmoid function was adopted to correlate the measured local hardness value to the peak temperature reached during the heating test. Moreover, in order to precisely correlate the final properties to the whole heat treatment (heating, soaking and cooling), measured hardness were also expressed as a function of the equivalent time t^*. The FE simulation of the tube laser heating, after having inversely calibrated the unknown thermal boundary conditions, allowed to discard all the combination of laser power and heating time that

Italian Manufacturing Association Conference - XVI AITeM   Materials Research Forum LLC
Materials Research Proceedings 35 (2023) 62-69   https://doi.org/10.21741/9781644902714-8

led to a maximum temperature in the spot higher than 400°C in order to avoid the fully solutioning of the material (according to the results from the preliminary Gleeble tests at 450°C). Results from the FE simulations were post-processed by means of a Python script implementing the fitted sigmoid function: in such a way, it was possible to numerically predict the final hardness distribution. The experimental test reproducing the chosen heating condition (laser power equal to 800 W and heating time to 5 s) allowed to determine a good accordance between the numerical predicted hardness values and the experimental ones, thus confirming the effectiveness of the proposed approach. At the same time, future steps will be aimed at improving the model capability by enriching the characterization step using the Gleeble system: in particular, focusing the attention on a narrower range of temperatures around 400°C, it would be possible to evaluate the threshold condition discerning the overaged condition from the fully solutioned one. In such a way, any laser heating would ideally be designed according to the type of local modification to be achieved (fully solutioning rather than overageing).

References

[1] Mayer RM, Poulikakos LD, Lees AR, Heutschi K, Kalivoda MT, Soltic P. Reducing the environmental impact of road and rail vehicles. Environ Impact Assess Rev 2012;32:25-32. https://doi.org/10.1016/j.eiar.2011.02.001

[2] Hirsch J. Recent development in aluminium for automotive applications. Trans Nonferrous Met Soc China (English Ed 2014;24:1995-2002. https://doi.org/10.1016/S1003-6326(14)63305-7

[3] DIN EN 45545-2. Railway applications - Fire protection on railway vehicles - Part 2: Requirements for fire behaviour of materials and components 2016.

[4] Kalpakjian S, Schmid SR. Manufacturing processes for engineering materials. Singapore; London: Pearson Education; 2017.

[5] Geiger M, Merklein M, Vogt U. Aluminum tailored heat treated blanks. Prod Eng 2009;3:401-10. https://doi.org/10.1007/s11740-009-0179-8

[6] Piccininni A, Palumbo G. Design and optimization of the local laser treatment to improve the formability of age hardenable aluminium alloys. Materials (Basel) 2020;13. https://doi.org/10.3390/ma13071576

[7] Kahrimanidis A, Lechner M, Degner J, Wortberg D, Merklein M. Process design of aluminum tailor heat treated blanks. Materials (Basel) 2015;8:8524-38. https://doi.org/10.3390/ma8125476

[8] Peixinho N, Soares D, Vilarinho C, Pereira P, Dimas D. Experimental study of impact energy absorption in aluminium square tubes with thermal triggers. Mater Res 2012;15:323-32. https://doi.org/10.1590/S1516-14392012005000011

[9] Piccininni A, Magrinho JP, Silva MB, Palumbo G. Formability Analysis of a Local Heat-Treated Aluminium Alloy Thin-Walled Tube. Springer International Publishing; 2021. https://doi.org/10.1007/978-3-030-75381-8_230

[10] Buchanan K, Colas K, Ribis J, Lopez A, Garnier J. Analysis of the metastable precipitates in peak-hardness aged Al-Mg-Si(-Cu) alloys with differing Si contents. Acta Mater 2017;132:209-21. https://doi.org/10.1016/j.actamat.2017.04.037

[11] Sekhar AP, Nandy S, Ray KK, Das D. Prediction of Aging Kinetics and Yield Strength of 6063 Alloy. J Mater Eng Perform 2019;28:2764-78. https://doi.org/10.1007/s11665-019-04086-z

[12] Maalouf M. Logistic regression in data analysis: An overview. Int J Data Anal Tech Strateg 2011;3:281-99. https://doi.org/10.1504/IJDATS.2011.041335

Italian Manufacturing Association Conference - XVI AITeM
Materials Research Proceedings 35 (2023) 70-77

Materials Research Forum LLC
https://doi.org/10.21741/9781644902714-9

An insight into friction stir consolidation process mechanics through advanced numerical model development

Abdul Latif[1,a*], Riccardo Puleo[1,b], Giuseppe Ingarao[1,c], Livan Fratini[1,d]

[1]Department of Engineering, University of Palermo, Viale delle Scienze, Palermo, 90128, Italy

[a]abdul.latif@community.unipa.it, [b]riccardo.puleo01@unipa.it, [c]giuseppe.ingarao@unipa.it, [d]livan.fratini@unipa.it

Keywords: Circular Economy, Aluminum Alloys, Numerical Modelling

Abstract. Friction stir consolidation (FSC) is a solid-state process adopted to recycle machining scraps with aim to reduce the adverse impact of obtaining metals from their primary source. FSC was also applied to offer plausible new routes for alloying and upcycling from powder and scrap metal and thus drew the attention of many researchers. During FSC process, a rotating tool with a certain force is applied to a given chips batch enclosed in a die chamber turning it into a consolidated billet. It is assumed that favorable process conditions for chips bonding are acquired by the combined effect of friction, stirring action, and pressure of the tool. However, the real process is quite complex, and it can be understood only by developing proper solid bonding criteria through numerical modeling that can forecast the consolidation process. Therefore, in this research, an attempt was made to implement different existing bonding criteria. Some of these were good enough to predict favorable conditions for sound bonding of particular case studies, however a uniform criteria with a single threshold value that is applicable to all case studies could not be achieved. Therefore, this study suggests for a new approach to accurately predict the bonding integrity of the FSC process.

Introduction

Aluminum consumption is rapidly increasing due to its growing demand for lightweight applications particularly in the transport, packaging, construction, and electronic industries. The accelerating demand is putting immense pressure on industries to increase the production rate [1]. Roughly 100 million metric tons of aluminum are currently produced per year. But for each 1 ton of aluminum production from the primary source, 12-16 tons of greenhouse gas (GHG) are produced and thus obtaining aluminum from its primary source is also one of the greatest causes of greenhouse gas (GHG) emission. Further, aluminum production is an energy-intensive process causing 13 Exajoules of energy consumption that accounts for almost 1% of total global energy consumption [2].

Interestingly, aluminum is an infinitely recyclable material [3]. Almost 35 % of the aluminum demand is met by recycling aluminum scraps [4]. Recycling is a highly energy efficient process, and it requires 5% of the energy compared to obtaining aluminum from bauxite ore (primary source) [1]. Nevertheless, the conventional recycling method has significant limitations, especially during recycling aluminum machining chips. Due to their high surface-to-volume ratio, these machining chips are prone to oxidation, causing permanent material loss during the melting process. They cause adverse environmental impact, high cost, and significant permanent material loss. Therefore, the researchers turned to Solid-State Recycling (SSR) techniques. During SSR methods, metal scraps transform into finished or semi-finished billet through mechanical means such as high pressure, friction, rotational speed, and force [5]. Recently, SSR processes have been analyzed as a potential alternative for recycling machining scraps.

The complex interactions between the process parameters make the SSR process difficult to be designed and controlled. Accurate predictions rely on the formulation of comprehensive weld

Italian Manufacturing Association Conference - XVI AITeM
Materials Research Proceedings 35 (2023) 70-77

Materials Research Forum LLC
https://doi.org/10.21741/9781644902714-9

models or criteria taking into account all physical aspects. Güley et al. [6] reported that the main challenges to predict the chips welding are connected to the complex stress states, and random position and orientation of contacting surfaces.

Akeret et .al [7] proposed a maximum pressure criterion where the contact pressure between the surfaces to be bonded is greater than the flow stress of the material at that point, however the model did not provide information effect of process parameters on weld quality. This aspect was studied by Plata and Piwnik [8], that compared the deformation energy as the sum of energies in the shear and compressive directions with the threshold value of the adhesion energy at the given point. The authors assumed that the energy in shear is much smaller compared with that in compressive strain. It was proved that Plata-Piwnik approach overestimated the weld quality at dead zone of extrusion process [9]. In this respect, Donati and Tomesani [9] introduced the flow speed as an additional correction factor to the Plata and Piwnik criterion. Then several numerical models were proposed for solid state process.

Ceretti et .al [10] presented a new approach in determining the critical value of the bonding criterion for flat rolling. They reported that solid bonding was dependent not only on the interface pressure, but also on the temperature at which the contact takes place. Schulze et al. [11] developed a numerical model to predict individual chips weld quality during hot extrusion. They found that local chips weld quality depended on the die type and hot extrusion parameters for different profile geometries. However, in Friction Stir Consolidation (FSC) process a rotating tool is applied to a desired chips mass enclosed inside a die chamber. The chips welding is due to the combined effect of pressure, force, friction and other factors characterizing the process mechanics [12]. Baffari et al. [13] developed a preliminary numerical model for bonding integrity of FSC process based on temperature and density. Although researchers have attempted numerical modeling of solid-state processing [11-13], proper bonding criteria are still under investigation to accurately predict the chips welding during FSC process. The current study proceeds with a detailed experimental campaign and numerical simulation for FSC process. To understand the bonding integrity, four different numerical criteria: Plata-Piwnik, Donati-Tomesani, Akeret and Plata-Piwnik-Zener were implemented. These criteria are widely used to quantify the bonding quality during solid-state processing. The goal was to develop a more accurate bonding criterion that can forecast the quality of FSCed recycled billet at given process conditions. Implementing such criterion as a quality indicator might be one of the key interests that can pave the path of FSC applicability at the industrial scale.

Materials and Methods

Aluminum alloy AA7075 was considered in the current study due to its popularity in aerospace manufacturing industries, where 90% of the input material turns into machining chips and is available as pre-consumer scraps [14]. Further, AA7075 is the hardest among wrought aluminum alloys with relatively low ductility. Therefore, analyzing the quality of AA7075 recycled billet can provide a broad understanding to forecast the quality of relatively soft and ductile alloys. In the ongoing investigation, the as-received material had an average Vickers hardness value of HV 150. As-received material was a 3 mm thick rolled sheet of AA7075-T6 that was turned into chips by milling operations. The chips were cleaned by submerging them in acetone for 30 minutes. First, 15 g chips were loaded in a cylindrical die with a nominal diameter of 25.4 mm and then compacted at 5 kN force by an H13 steel cylindrical tool with a 25 mm diameter. The die and pressing tool system were integrated with ESAB-LEGIO (Fig. 1a and Fig. 1b), a dedicated friction stir welding machine. Finally, a consolidated billet was manufactured after applying the tool at 20 kN with a tool rotational speed of 1500 rpm. It is important to note that due to machine limitation, it was not possible to instantly increase load from 5 kN compaction force to 20 kN consolidation force, but rather it was gradually increased with a step increment of 0.5 kN/s. Upon reaching the desired load, finally, the whole charge was consolidated at 20 kN constant force for the processing time of

Italian Manufacturing Association Conference - XVI AITeM   Materials Research Forum LLC
Materials Research Proceedings 35 (2023) 70-77   https://doi.org/10.21741/9781644902714-9

40 and 60 seconds. In short, the whole process was completed in two steps: the transition phase, and the consolidation phase. Two billets were manufactured at processing time 40 and 60 seconds. The process parameters were selected based on the previous studies [15].

Measured outputs
The billet was sectioned, mounted, and polished with a series of abrasive papers assisted by distilled water and alumina. The hardness and microstructure were analyzed throughout the section at 119 points equally spaced at a distance of 1.5 mm, as shown in Fig. 1d. The hardness was measured through the Vickers hardness test by applying a load of 49 N (5 kg) for 15 seconds. Keller's reagent was used to reveal the microstructure.

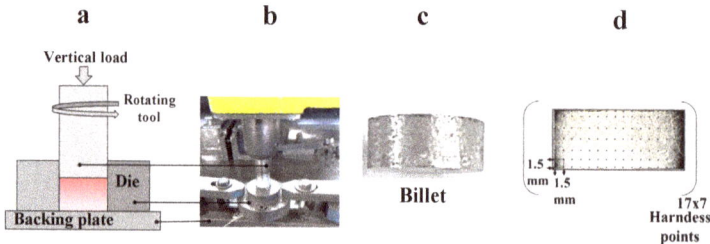

Fig.1: (a) Schematic diagram, (b) Friction stir consolidation (c) billet and (d) Analyzed loci on billet section.

Analysis of solid-state welding of the chips
The numerical model was implemented using the commercial FEA software SFTC DEFORM. The chips were modeled as a single block porous material part, using a Shime-Oyane model [16], with a relative density of 0.44 and a mesh of 60000 elements. Particular attention was paid to the tool-material contact zone using a mesh refinement. The relative density was found experimentally through a known volume of compacted chips at 5 kN and the volume of the consolidated billet at 20 kN force. The other three objects: the tool, die, and backing plate, were modeled as a rigid body with 15000 mesh elements (Fig. 2a). The simulation was a force-based numerical model in which the input was the force registered during the experimental process. The numerical model was validated through vertical tool velocity-time and stroke-time data that were obtained experimentally from the machine database, as shown in Fig. 2b.

To understand the bonding integrity, four different solid bonding criteria: Akeret [7], Plata-Piwnik [8], Donati-Tomesani [9], and Plata-Piwnik-Zener, were implemented. These criteria are widely used to quantify the quality of bonding during solid-state processing. For example, Plata-Piwnik uses a threshold value (W_{lim}) that indicates the ratio between the normal pressure (p) and the actual material effective stress (σ_{eff}) with the time integral. The Plata-Piwnik criterion is reported below in Eq. (1); once the ratio reaches this threshold value, then the solid bonding is assumed to be occurred.

$$W = \int \frac{p}{\sigma_{eff}}\, dt \geq W_{lim}. \tag{1}$$

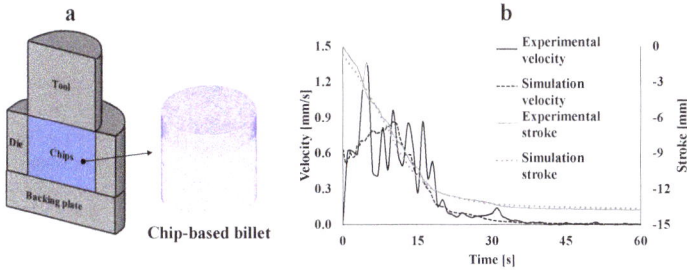

Fig. 2: (a) Numerical model for FSC, and (b) Tool experimental and simulated velocity-time and stroke-time graphs for model validation.

Results and Discussion

The numerical model was developed from two different data sets that were obtained through experimental measurements and numerical simulation for billets of processing time 40s and 60s. The experimental data collected were Vickers hardness values and grain size measured at 119 observation loci across the section, as shown in Fig 3 and Fig. 4. In order to interpolate the experimental values for hardness and grain size for the whole section, heatmap MATLAB in-built function was used. The hardness values were ranging 80-140 HV, while the grain's size was around 1.5-7 μm. The top zone of the billet was characterized by a high hardness value around HV 140 with a bigger grain size of 5-7μm compared to the bottom portion, which had a hardness value and grain size below HV 80 and 2 μm, respectively. It is assumed that the top portion of the billet is subject to high strain, strain rate, pressure, and temperature compared to the bottom portion. Furthermore, increasing processing time led to expanding the consolidated zone, and therefore, higher hardness and grain size can be noticed in the consolidated area of the 60s billet, compared to the section of the 40s billet.

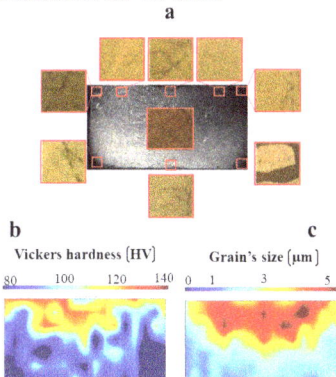

Fig. 3: FSC billet 40 seconds sample (a) microstructure (b) hardness (c) grain's size distribution.

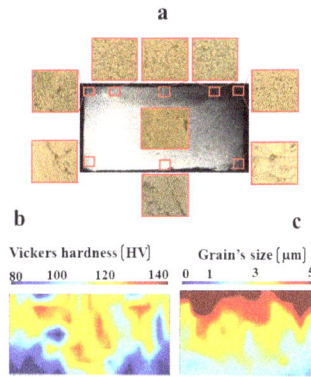

Fig. 4: FSC billet 60 seconds sample (a) microstructure (b) hardness (c) grain size distribution.

Based on hardness value and grain size, an experimental criterion was designed by setting a specific threshold value that indicated the fully consolidated zone. This study considers both

Italian Manufacturing Association Conference - XVI AITeM   Materials Research Forum LLC
Materials Research Proceedings 35 (2023) 70-77   https://doi.org/10.21741/9781644902714-9

hardness and grain size values for setting up the experimental criterion. After detailed analysis, the threshold for full consolidation was set to a minimum HV 90 for hardness and 2 μm for grain size. Regions of FSCed billet characterized by values exceeding these threshold values represent a fully consolidated zone; if either of these values was satisfied, that means partial consolidation occurred; if neither of these two values was satisfied no consolidation occurred at all. For numerical solid bonding criteria: Plata-Piwnik, Donati-Tomesani, Akeret, and Plata-Piwnik-Zener were implemented. The input data such as temperature, stress, strain, strain rate, pressure, etc. were obtained from numerical simulation. For a given criterion, the threshold was identified for both of the analyzed processing times (40s, 60s), this threshold will be referred to as local thresholds and all these numerical criteria were plotted using the local threshold as the upper bound of the scale. All the numerical criteria were plotted using an image comparison procedure by using MATLAB. The bonding map area was compared with the experimental criteria map. The local threshold of bonding criteria was varied until similarity between numerical bonding criteria and experimental bonding one reached the best value. For the sake of simplicity, the similar index was defined only for the consolidated zones. The schematic of the implemented procedure is presented in Fig. 5.

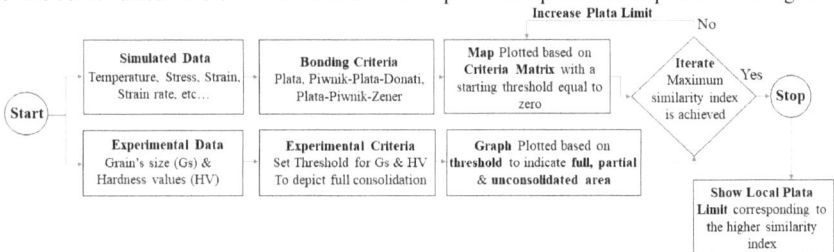

Fig. 5: Flow diagram for developing numerical model and their comparison with experimental criteria to forecast bonding integrity.

The comparison of experimental criteria and four different bonding criteria are presented in Fig. 6. The consolidated zone of the experimental criteria was considered as a reference, and all four numerical criteria were compared. No bonding criteria were found to accurately predict the experimental bonding occurrence. Although, Plata-Piwnik-Zener criterion seems to provide a better prediction of the shape of the bonding area, actually the best performance is provided by the Donati and Tomesani one as reported in Table 1. All the obtained results are summarized in Fig. 6 and Table 1. It is important to note that the error and the local limit value for each numerical bonding criteria are different and were found to be dependent on the processing time of the two analyzed billets. Further, the change in processing time also led to a variation in the local bonding limit.

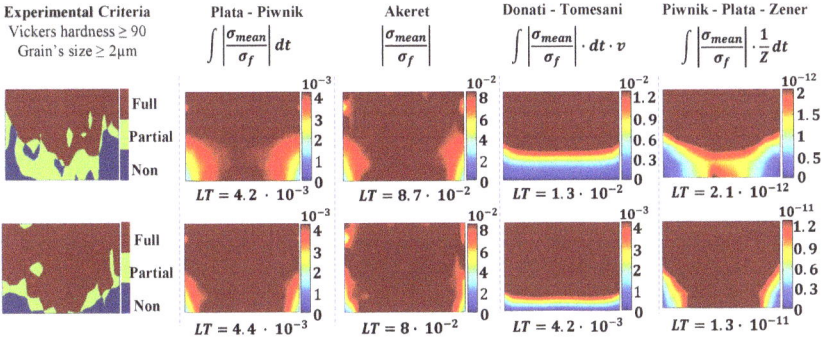

Fig. 6: Comparison of numerical local thresholds (LT) bonding value map with experimental criteria map for 40s and 60s.

Table 1. Comparison of bonding criteria in terms of numerical local thresholds (LT) and similarity index.

Bonding criteria	LT (40 s)	% Similarity index (40 s)	LT (60 s)	% Similarity index (60 s)
Plata-Piwnik	0.0042	58	0.0044	39
Akeret	0.087	64	0.08	40
Donati-Tomesani	0.013	80	0.0042	87
Plata-Piwnik-Zener	2.1E-12	84	1.3E-11	84

Due to the variations of the local bonding criteria limit with processing time, an average value of a criterion was utilized to analyze the effect of a unique value on estimating the bonding quality. The Plata-Piwnik-Zener was used for this purpose and the obtained maps are reported in Fig. 7, with details mentioned in Table 2. The results show that the average Plata-Piwnik-Zener threshold cannot satisfactorily predict the bonding quality compared to the maps of experimental criteria (Fig. 7a and b). Either underestimation or overestimation are clearly visible for the consolidated zone of the 40s and 60s (Fig. 7c and Fig. 7d), respectively. Then setting a limit suitable to 60s on a given local Plata-Piwnik-Zener provided underestimated consolidated zone for 40s (Fig. 7e). On the other hand, setting a local bonding limit favorable to 40s overestimated consolidated zone for 60s (Fig. 7f). These results reveal that no single unique threshold was achieved that can properly represent the consolidated zone for both case studies. Such kind of misleading results occurred for all the criteria analyzed in the present paper.

Italian Manufacturing Association Conference - XVI AITeM   Materials Research Forum LLC
Materials Research Proceedings 35 (2023) 70-77   https://doi.org/10.21741/9781644902714-9

Table 2. Details of thresholds value for Plata-Piwnik-Zener.

Bonding criteria	LT (40 s)	LT (60 s)	Average LT
Plata-Piwnik- Zener	2.1E-12	1.3E-11	7.5E-12

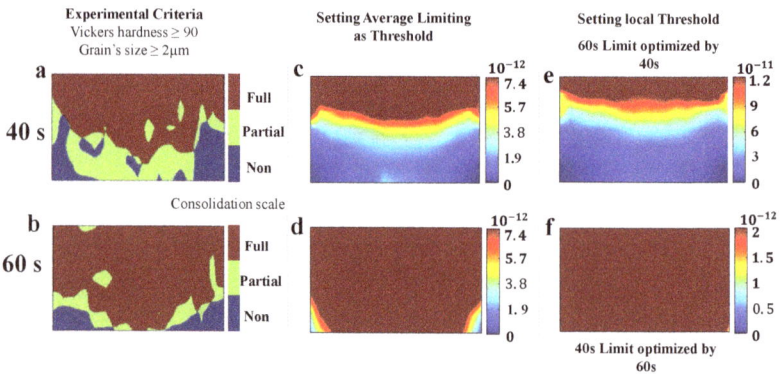

Fig. 7: Estimation of Plata-Piwnik-Zener maps based on different single average and local thresholds values.

Conclusions

In this paper a numerical-experimental procedure was proposed to set-up solid bonding criteria in Friction Stir Consolidation processes. The adopted procedure was implemented on different bonding criteria and the results were compared one another. The selected criteria can qualitatively predict the consolidation trends of the FSC processes; despite that some research is still needed to obtain a robust and reliable criterion for predicting the consolidation level of FSC samples; the following conclusions can be drawn.

1. No single bonding threshold value exists that can be applied to predict the bonding integrity for FSC billet.

2. The bonding thresholds were dependent on processing time, and therefore each case study has its own local threshold value.

3. If a single bonding threshold value e.g., average limiting value, was applied for all case studies, then it could not lead to satisfactory results, either causing underestimating or overestimating consolidated zones.

The peculiarity of FSC process demands the development of comprehensive bonding criteria that can predict the quality of chips bonding during the FSC process. New variables and methodologies should be considered in the analyses in order to find the correct metric that actually affect the chips bonding in FSC processes.

References

[1] T.G. Gutowski, S. Sahni, J.M. Allwood, M.F. Ashby, E. Worrell, The energy required to produce materials: constraints on energy-intensity improvements, parameters of demand, Philosophical Transactions of the Royal Society A: Mathematical, Physical and Engineering Sciences 371.1986 (2013): 20120003. https://doi.org/10.1098/rsta.2012.0003

[2] D. Raabe, C.C. Tasan, E.A. Olivetti, Strategies for improving the sustainability of structural metals, Nature. 575.7781 (2019): 64-74 https://doi.org/10.1038/s41586-019-1702-5

[3] K.R. Barbara, T. E. Graedel, Challenges in metal recycling, Science 337.6095 (2012): 700. https://doi.org/10.1126/science.1221806

[4] Information on https://www.iea.org/reports/material-efficiency-in-clean-energy-transitions

[5] B. Wan, W. Chen, T. Lu, F. Liu, Z. Jiang, M. Mao, Review of solid state recycling of aluminum chips, Resources, Conservation and Recycling.125 (2017):37-47. https://doi.org/10.1016/j.resconrec.2017.06.004

[6] V. Güley, A. Güzel, A. Jäger, N.B. Khalifa, A.E. Tekkaya, W.Z. Misiolek, Effect of die design on the welding quality during solid state recycling of AA6060 chips by hot extrusion. Materials Science and Engineering: (2013) A, 574, 163-175. https://doi.org/10.1016/j.msea.2013.03.010

[7] R. Akeret, Extrusion welds-quality aspects are now center stage. In Proceedings of the 5th International Aluminium Extrusion Technology Seminar, 1992.

[8] M. Plata, J. Piwnik, J, Theoretical and experimental analysis of seam weld formation in hot extrusion of aluminum alloys. In Proceedings of International Aluminum Extrusion Technology Seminar (Vol. 1, pp. 205-212).

[9] L. Donati, and L. Tomesani, The prediction of seam welds quality in aluminum extrusion, Journal of Materials Processing Technology 153 (2004): 366-373. https://doi.org/10.1016/j.jmatprotec.2004.04.215

[10] E. Ceretti, L. Fratini, F. Gagliardi, C. Giardini, A new approach to study material bonding in extrusion porthole dies. CIRP annals, 2009, 58.1: 259-262. https://doi.org/10.1016/j.cirp.2009.03.010

[11] A. Schulze, O. Hering, A.E. Tekkaya, Welding of Aluminium in Chip Extrusion. In: Forming the Future: Proceedings of the 13th International Conference on the Technology of Plasticity. Springer International Publishing, 2021. p. 139-147. https://doi.org/10.1007/978-3-030-75381-8_11

[12] A. Latif, G. Ingarao, M. Gucciardi, L. Fratini, A novel approach to enhance mechanical properties during recycling of aluminum alloy scrap through friction stir consolidation. The International Journal of Advanced Manufacturing Technology, 2022, 119.3-4: 1989-2005 https://doi.org/10.1007/s00170-021-08346-y

[13] D. Baffari, A.P. Reynolds, X. Li, L. Fratini, Bonding prediction in friction stir consolidation of aluminum alloys: A preliminary study. In: AIP Conference Proceedings. AIP Publishing LLC, 2018. p. 050002. https://doi.org/10.1063/1.5034875

[14] J.M. Allwood, J.M. Cullen, M.A. Carruth, D.R. Cooper, M. McBrien, R.L. Milford, M.C. Moynihan, A.C. Patel, Sustainable materials: with both eyes open, Vol. 2012, Cambridge, UK: UIT Cambridge Limited, 2012

[15] A. Latif, G. Ingarao, L. Fratini, Multi-material based functionally graded billets manufacturing through friction stir consolidation of aluminium alloys chips. CIRP Annals, 2022, 71.1: 261-264. https://doi.org/10.1016/j.cirp.2022.03.035

[16] S. Shima, M. Oyane. Int. J. Mech. Sci., 18(6), 285-291.(1976) Retrieved May 2, 2017, from https://doi.org/10.1016/0020-7403(76)90030-8

Italian Manufacturing Association Conference - XVI AITeM   Materials Research Forum LLC
Materials Research Proceedings 35 (2023) 78-85   https://doi.org/10.21741/9781644902714-10

On the role of intermetallic and interlayer in the dissimilar material welding of Ti6Al4V and SS 316L by friction stir welding

Harikrishna Rana[1,a] *, Gianluca Buffa[1,b] and Livan Fratini[1,c]

[1]Department of Engineering, University of Palermo, Viale Delle Scienze, 90128 Palermo, Italy

[a]harikrishnasinh.rana@unipa.it, [b]gianluca.buffa@unipa.it, [c]livan.fratini@unipa.it

Keywords: Friction Stir Welding, Titanium Alloy, Stainless Steel

Abstract. Joining titanium with stainless steel can lighten the structure of numerous industrial applications. However, a vast disparity of thermal, physical, and chemical properties between these alloys leads to defects in conventional arc welding techniques, viz., brittle intermetallic compounds, pores, cracks, etc. Friction stir welding (FSW) is a renowned solid-state joining technology for creating dissimilar material joints producing visco-plastic material flow at the interface. The present investigation compares the intermetallic layer thickness and properties as a function of the thickness of the Cu interlayer sandwiched in lap joints. Macrostructural and microstructural characterizations were carried out to understand the localized microstructural evolution comprising intermetallic, grain refinement, defects, etc. Mechanical properties were also evaluated for prepared lap joints.

Introduction

The idea of using the appropriate material in the correct location has sparked a rise in the use of multi-material components. This concept has led to the development of several cutting-edge technologies, such as additive manufacturing, solid-state joining and processing, and selective laser melting, which have numerous applications in industries such as space, chemical, shipping, automotive, and transportation[1, 2]. These technological advances offer vast potential for creating customized multi-material structures and components. It is difficult to fathom the production of a multi-material component without dissimilar material joining, yet despite the significant interest in these industries, many unanswered questions remain. Weight reduction translates directly to cost savings in these industries. Such multi-material components make the structure lighter and more cost-effective at the same time. However, these benefits do not come without challenges. The mismatch in coefficients of thermal expansion between Ti_6Al_4V and SS 316L, which are 8.6 $\times$ 10^{-6} /°C and 17.2 $\times$ 10^{-6} /°C respectively, leads to a joint with significant residual stress [3]. Moreover, the differences in properties such as density, melting point, and chemical affinity also create negative effects during joining. The formation of brittle intermetallic compounds (IMCs), such as $TiFe$, $Ti_5Fe_{17}Cr_5$, and $TiFe_2$, is inevitable due to extreme heat input during fusion welding [4, 5]. TIG, MIG, braze welding, diffusion welding, and laser welding technologies have been widely reported to fabricate dissimilar material joints [4, 6-8]. However, these joints often fail at the interface due to the presence of IMCs.

Efforts have been made to reduce heat input during joining and eliminate IMCs. One feasible solution is to use an interlayer that acts as a diffusion barrier to suppress the reaction between Fe and Ti [4]. Various interlayers have been reported for SS/Ti joints, including V and Ta in laser welding, Ni, Cu, Al, and Ag in diffusion bonding[9-14], and Ag-Cu, Cu-Ti, Ag-Cu-Zn, Ag, and Ti-based alloys in brazing[15-17]. However, none of these investigations have reported a bonding interface free from IMCs or defects. In a nutshell, solid-state welding is more suitable than fusion-related methods since the major problems associated with melting can be eliminated. Nonetheless,

Italian Manufacturing Association Conference - XVI AITeM   Materials Research Forum LLC
Materials Research Proceedings 35 (2023) 78-85   https://doi.org/10.21741/9781644902714-10

it remains challenging to create high-quality dissimilar material joints, and further research is needed to develop effective solutions.

Over the last couple of decades, Friction Stir Welding (FSW) has emerged as a highly effective solid-state technique for producing joints that are free from defects, even when joining materials that have significant differences in properties and compositions [18-20]. This is achieved by employing low heat input mechanisms, which sets FSW apart from both conventional and modern fusion-based processes [21-23]. In a study conducted by Fazel et al.[24], a commercially pure Ti/SS 304 lap joint was produced using various processing conditions, ultimately resulting in a joint efficiency of 73%. Although the amount of IMC present (TiFe) was reduced when compared to other solid-state techniques (such as diffusion bonding), it was still not possible to eliminate it entirely. Additionally, it was reported that at elevated temperatures, the thickness of the IMC layer increased, and the formation of a Ti oxide layer contributed to a decline in joint properties. Therefore, it is highly desirable to identify a suitable interlayer and determine the lowest possible heat input parameters in order to achieve a successful Ti/SS FSW joint.

Based on the information and studies elaborated it is evident that FSW has the potential to produce efficient Ti/SS joints accompanied by limited literature on producing defect-free joints. Therefore, the main aim of this study is to explore the possibility of creating a defect-free Ti_6Al_4V/SS 316L FSW joint through experimentation with the introduction of different Cu interlayer thicknesses. In addition, the study aims to analyze joint strength, microhardness, and microstructural features such as grain morphology and intermetallic compounds.

Materials & Methods

Dissimilar lap joints were created between 2 mm thick austenitic stainless steel AISI 316L and 2 mm thick titanium alloy Ti_6Al_4V employing pure Cu interlayer with 3 different thicknesses viz. 0.05, 0.1-, and 0.2-mm. The chemical compositions of the as-received material are mentioned in Table 1. Both the skin and stringer plate was maintained at an identical size of 140 × 90 mm. When employing FSW on high-strength alloys such as titanium alloys, selecting the appropriate tool material is crucial. The author has illustrated the effectiveness of W25Re compared to conventional tungsten carbides alloys like K10 and K10-K30 in this regard.[25]. The tool design is represented schematically in Fig. 1. The experiments were conducted at VR of 16, wherein the TR was 600 rota. min^{-1} and 37 mm min^{-1}, with varying the cu interlayer thickness from 0.05 – 0.2.

Table 1 Chemical compositions for the substrates

WT %	Ti	Al	V	Fe	C	Cr	Mo	Mn	Ni	Si	Other
Ti-6Al-4V	87.7÷91	5.5÷6.75	3.5÷4.5	<0.4	<0.08	-	-	-	-	-	<2.1
SS 316L	-	-	-	61.9÷72	<0.03	16÷18	2÷3	<2	10÷14	<1	<9.1

The microstructure specimens were prepared as per the standard metallographic specimen preparation standard. Kroll's reagent (HF 2 ml + HNO$_3$ 6 ml + H$_2$O 92 ml) was used for etching Ti_6Al_4V sections, while Carpenters reagent (FeCl$_3$ 8.5gm + CuCl$_2$ 2.4 gm, HCL 122 ml+ HNO$_3$ 6ml + C$_2$H$_6$O 122 ml) for AISI 316L section. Various microstructural features such as grain morphologies, material flow patterns, intermetallic particles, fractured particles, and other compositional elements were examined through the use of optical microscopy (OM) (OLYMPUS, Model-Inverted Metallurgical Microscope GX51), scanning electron microscopy (SEM), and energy dispersive spectroscopy (Zeiss, Model: Ultra-55 SEM).

Italian Manufacturing Association Conference - XVI AITeM   Materials Research Forum LLC
Materials Research Proceedings 35 (2023) 78-85   https://doi.org/10.21741/9781644902714-10

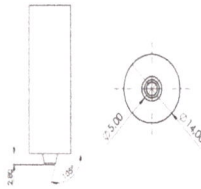

Fig. 1 *W25Re tool geometry for FSW*

To investigate microhardness, an Eseway 4302 Vickers hardness tester was used, following the ASTM E-384 standard. A square-based pyramid diamond indenter (136° intersects) was utilized with a 5 kg load and 15 seconds dwell time. For tensile tests, specimens with a width of 10 mm were employed in a conventional tensile testing machine with a velocity of 2 mm/minute.

Result & Discussion:

Microstructural Characterization

The macrostructures of the prepared Ti/SS lap joints are displayed in Fig. 2. It is evident from the micrographs that as the thickness of the interlayer increases, there is a marked increase in the penetration of SS to Ti in the form of a hook. Such hooks are formed due to distinct material flow patterns ensuing from distinctive mechanical properties. The profile of the hook has a significant impact on the resulting properties because of the interlocking between the skin and stringer materials. It is interesting to note that the phenomenon is prominent on the retreating side (RS). Two factors contribute to the formation of "hook" (i) material flow and (ii) heat flow. During FSW, the rotating tool pin pushes the stringer material underneath it, which extrudes in the opposite direction. Eventually, as the stringer material flows further upward being constrained and surrounded by the skin material, an extruded hook is engendered close to the thermo-mechanically affected zone (TMAZ). An identical phenomenon ensuing "extruded hook", has been reported during FSW of Al-Ti lap joint configurations[26].

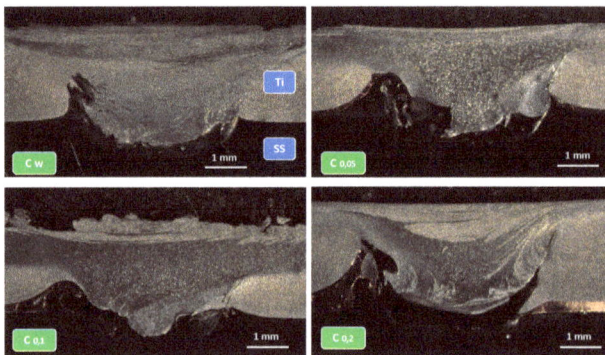

Fig. 2. *Macrostructures of the prepared Ti/SS lap joints*

Italian Manufacturing Association Conference - XVI AITeM   Materials Research Forum LLC
Materials Research Proceedings 35 (2023) 78-85   https://doi.org/10.21741/9781644902714-10

On the one hand, the joints prepared without interlayer exhibited a limited SS material penetration to Ti skin accompanied by minimal defect generation. On the other hand, the joints prepared with the Cu interlayer exhibited comparatively larger SS material penetration to the Ti skin in the form of a hook. Simultaneously, the larger inflow of the SS material engenders a larger void into the stringer material and leaves behind a large cavity or a wormhole. Moreover, with the increase in the thickness of Cu interlayer, the size of the void enlarges owing to the mechanical deformation superimposed upon supposed material mixing at elevated temperature. Such a phenomenon can be justified by the heat transfer incurring for the two cases, one without interlayer and with interlayer as schematically represented by Fig. 3 (a) and (b) respectively.

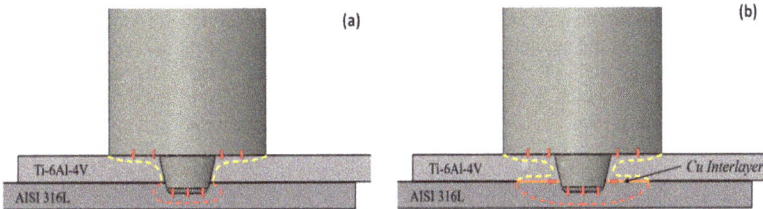

Fig. 3. Proposed heat flow during Ti/SS FSW for specimens prepared (a) Without Cu interlayer (b) With Cu interlayer.

The thermal conductivity of the titanium alloy is almost half that of stainless steel (k_Ti=6.70 W/(m K), k_SS=15 W/(m K)). This means that heat conducts more easily in SS substrate than Ti, resulting in a larger thermally-influenced layer compared to titanium when no interlayer is present. Whereas, for the Ti skin the heat is concentrated at the center only with a very limited heat span. In the absence of an interlayer, the thermal layer in titanium remains near the interface, while in steel, there is a slight expansion of the thermal layer at the sides and under the pin. Introducing copper interlayer having very high thermal conductivity than the other two substrates the rate of heat dissipation increases and conducts the frictional heat away from the stirring zone resulting in mechanical deformation instead of supposed material mixing at elevated temperatures.

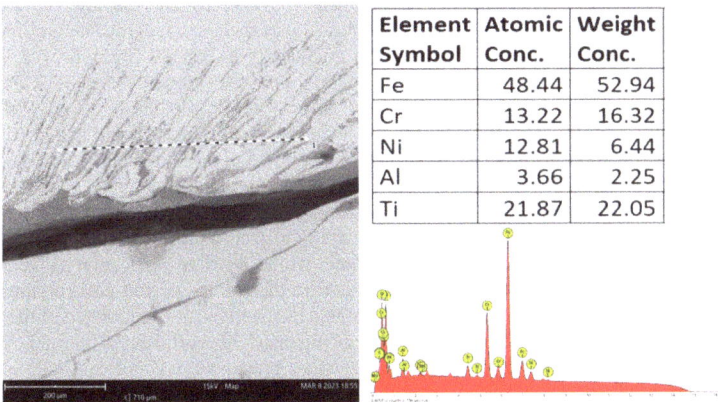

Element Symbol	Atomic Conc.	Weight Conc.
Fe	48.44	52.94
Cr	13.22	16.32
Ni	12.81	6.44
Al	3.66	2.25
Ti	21.87	22.05

Fig. 4. Line scan near Ti/SS interface displaying the $Ti_5Fe_{17}Cr_5$ intermetallic.

Italian Manufacturing Association Conference - XVI AITeM   Materials Research Forum LLC
Materials Research Proceedings 35 (2023) 78-85   https://doi.org/10.21741/9781644902714-10

After selecting the most promising sample, it was subjected to SEM and EDS mapping. The resulting SEM-EDS analysis uncovered the existence of a $Ti_5Fe_{17}Cr_5$ IMC layer that was extremely thin and sporadic, found close to the interface of Ti/SS (See Fig. 4). Remarkably, the thickness of the IMC layer may be considered relatively thin compared to the observed thickness in conventional arc welded joints [26, 27]. However, a thicker layer of intermetallic was observed in the vicinity of the extruded SS hook. On the contrary, when the extruded SS hook penetrates the Ti, the temperatures of both substrates increase, resulting in Ti-Fe diffusion and the development of a thick IMC layer. $Ti_5Fe_{17}Cr_5$ was the predominant IMC phase detected in NZ. The mechanical characteristics of the FSW joints produced are significantly impacted by the location, size, and quantity of these IMCs. However, the specimens prepared with Cu interlayer did not exhibit such IMCs along the interface.

Apart from that, highly diverse grain morphologies were observed in the different zones of the FSW joints. Surprisingly, the nugget zone (NZ) majorly occupied by the Ti skin was characterized by larger grains as compared to the parent material (PM). This can be attributed to the lower heat transfer coefficient of Ti which does not allow the heat to dissipate fast ensuing the larger grains led by the slow cooling rate. The TMAZ was characterized by distorted and elongated grains, whereas heat affected zone (HAZ) with the finer grains as compared to the NZ. Several defects such as porosity, tunnels, and voids were evident in the NZ owing to a greater mismatch of the properties for skin and stringer materials.

Mechanical Properties

The tensile test results for all specimens are in Fig. 5a. On the one hand, specimens prepared without Cu interlayer exhibited the highest shear strength of 421 N/mm. On the other hand, the moderate shear strengths of 150, 370, and 250 N/mm were obtained for the specimens prepared with Cu interlayer thicknesses of 0.05, 0.1, and 0.2 mm respectively. The lower strengths recorded in those specimens can be attributed to defects such as porosity, cavities, wormholes, and recesses. However, the dimensions of the hook in those specimens have greatly influenced the shear strength results owing to the mechanical interlocking achieved. The specimens prepared with the 0.1 mm interlayer thickness offered the highest strength among the Cu interlayer specimens. Such an increase may have resulted from the two predominant mechanisms: (i) Longer hook (ii) Reduction in the amount of Fe_3Ti brittle IMC.

As the hardness results are concerned, Ti_6Al_4V and SS 316L substrates displayed a hardness of ~275 HV and ~180 HV, respectively. The NZ of the Ti skin specimen recorded a marginal rise of ~ 15% as compared to the substrate as indicated graphically in Fig 5b. The improved hardness values can be attributed to the Ti-SS composite structure, and IMCs present in the NZ of the welded samples. However, several peaks in the hardness values were recorded in the vicinity of TMAZ for the specimen prepared without Cu interlayer, owing to the presence of hard and brittle IMCs. While the sudden fall in the hardness values can be attributed to the softer SS 316L hook formed near the TMAZ region. The observed increase in hardness values could be attributed to the effectiveness of shear lag and dislocation strengthening mechanisms. On the one side, the shear lag mechanism involves the transfer of load from the Ti matrix to the hard $TiFe3$ IMC present in the NZ. This generates shear stress at the interface, which restricts dislocation movement and improves material properties. On the other side, during the cooling period post-thermos-plastic deformation caused by FSW, geometrically inexorable dislocations are formulated in the vicinity of the Ti/SS interface and IMCs due to the significant mismatch of coefficient of thermal expansion between two substrates. These dislocations impede crack propagation, which results in increased shear strength and hardness. In summary, the observed "dual" metallurgical-mechanical bonding mechanism can be effective in producing dissimilar Ti/SS lap joints via FSW, making it a promising method for producing lap joints using dissimilar materials.

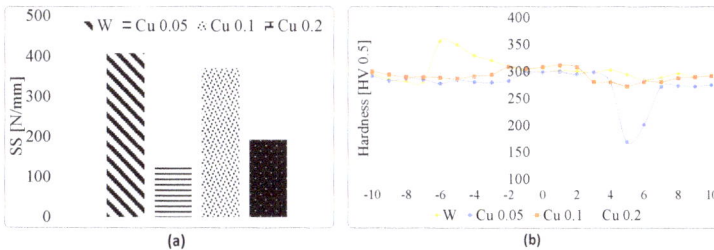

Fig. 5. *Specimen comparison for (a) Shear strength (b) Hardness Distribution*

Conclusions

This study documents the successful fabrication of dissimilar lap joints utilizing Ti_6Al_4V and SS 316L substrates. The effects of the thickness of the Cu interlayer on the joints' macrostructural, microstructural, and mechanical properties were thoroughly explored. The following conclusions were reached:

- The joints prepared without Cu interlayers exhibited superior properties owing to almost defect-free NZ.
- The dissimilar material Ti/SS joint displayed interesting material flow patterns. The development and geometry of the extruded hook were found to be critical to mechanical interlocking, resulting in improved joint strength. However, Cu interlayer specimens were characterized by multiple defects such as cavities and recesses owing to heat dissipation by the highly conductive Cu interlayer to the surrounding zones.
- The highest shear strength of 421 N/mm was recorded with the specimen prepared without Cu interlayer owing to minimal defect rate. The prepared Ti/SS lap joint hardness values were marginally increased by 15% attributable to the strengthening mechanisms led by IMCs.

References

[1] H. Gugel, A. Schuermann, W. Theisen: Laser welding of NiTi wires. Materials Science and Engineering: A 481, 668-671 (2008). https://doi.org/10.1016/j.msea.2006.11.179

[2] G. Kale, R. Patil, P. Gawade: Interdiffusion studies in titanium-304 stainless steel system. Journal of nuclear materials 257(1), 44-50 (1998). https://doi.org/10.1016/S0022-3115(98)00179-2

[3] C.Y. Ho, R.E. Taylor, Thermal expansion of solids, ASM international1998.

[4] S. Chen, M. Zhang, J. Huang, C. Cui, H. Zhang, X. Zhao: Microstructures and mechanical property of laser butt welding of titanium alloy to stainless steel. Mater. Des 53, 504-511 (2014). https://doi.org/10.1016/j.matdes.2013.07.044

[5] B. Shanmugarajan, G. Padmanabham: Fusion welding studies using laser on Ti-SS dissimilar combination. Optics and Lasers in Engineering 50(11), 1621-1627 (2012). https://doi.org/10.1016/j.optlaseng.2012.05.008

[6] X. Yue, P. He, J. Feng, J. Zhang, F. Zhu: Microstructure and interfacial reactions of vacuum brazing titanium alloy to stainless steel using an AgCuTi filler metal. Mater. Charact. 59(12), 1721-1727 (2008). https://doi.org/10.1016/j.matchar.2008.03.014

[7] Z. Cheng, J. Huang, Z. Ye, H. Liu, J. Yang, S. Chen, X. Zhao: Interfacial microstructure evolution and mechanical properties of TC4 alloy/304 stainless steel joints with different joining modes. Journal of Manufacturing Processes 36, 115-125 (2018). https://doi.org/10.1016/j.jmapro.2018.09.027

[8] R. Shiue, S. Wu, J. Shiue: Infrared brazing of Ti-6Al-4V and 17-4 PH stainless steel with (Ni)/Cr barrier layer (s). Materials Science and Engineering: A 488(1-2), 186-194 (2008). https://doi.org/10.1016/j.msea.2007.10.075

[9] S. Kundu, B. Mishra, D. Olson, S. Chatterjee: Interfacial reactions and strength properties of diffusion bonded joints of Ti64 alloy and 17-4PH stainless steel using nickel alloy interlayer. Mater. Des 51, 714-722 (2013). https://doi.org/10.1016/j.matdes.2013.04.088

[10] S. Kundu, S. Chatterjee: Characterization of diffusion bonded joint between titanium and 304 stainless steel using a Ni interlayer. Mater. Charact. 59(5), 631-637 (2008). https://doi.org/10.1016/j.matchar.2007.05.015

[11] S. Kundu, M. Ghosh, A. Laik, K. Bhanumurthy, G. Kale, S. Chatterjee: Diffusion bonding of commercially pure titanium to 304 stainless steel using copper interlayer. Materials Science and Engineering: A 407(1-2), 154-160 (2005). https://doi.org/10.1016/j.msea.2005.07.010

[12] A. Elrefaey, W. Tillmann: Solid state diffusion bonding of titanium to steel using a copper base alloy as interlayer. J. Mater. Process. Technol. 209(5), 2746-2752 (2009). https://doi.org/10.1016/j.jmatprotec.2008.06.014

[13] P. He, X. Yue, J. Zhang: Hot pressing diffusion bonding of a titanium alloy to a stainless steel with an aluminum alloy interlayer. Materials Science and Engineering: A 486(1-2), 171-176 (2008). https://doi.org/10.1016/j.msea.2007.08.076

[14] E. Atasoy, N. Kahraman: Diffusion bonding of commercially pure titanium to low carbon steel using a silver interlayer. Mater. Charact. 59(10), 1481-1490 (2008). https://doi.org/10.1016/j.matchar.2008.01.015

[15] C. Liu, C. Ou, R. Shiue: The microstructural observation and wettability study of brazing Ti-6Al-4V and 304 stainless steel using three braze alloys. Journal of materials science 37(11), 2225-2235 (2002). https://doi.org/10.1023/A:1015356930476

[16] H. Dong, Z. Yang, Z. Wang, D. Deng, C. Dong: CuTiNiZrV amorphous alloy foils for vacuum brazing of TiAl alloy to 40Cr steel. J. mater. sci. technol. 31(2), 217-222 (2015). https://doi.org/10.1016/j.jmst.2014.04.003

[17] T. Chung, K. Jungsoo, B. Jeongseok, R. Byoungho, N. Daegeun: Microstructures of brazing zone between titanium alloy and stainless steel using various filler metals. Transactions of Nonferrous Metals Society of China 22, s639-s644 (2012). https://doi.org/10.1016/S1003-6326(12)61778-6

[18] P. Patel, H. Rana, V. Badheka, V. Patel, W. Li: Effect of active heating and cooling on microstructure and mechanical properties of friction stir-welded dissimilar aluminium alloy and titanium butt joints. Weld. World. 64(2), 365-378 (2020). https://doi.org/10.1007/s40194-019-00838-6

[19] Y. Su, W. Li, F. Gao, A. Vairis: Effect of FSW process on anisotropic of titanium alloy T-joint. Mater. Manuf. Process 37(1), 25-33 (2022). https://doi.org/10.1080/10426914.2021.1942911

[20] G. Buffa, M. De Lisi, E. Sciortino, L. Fratini: Dissimilar titanium/aluminum friction stir welding lap joints by experiments and numerical simulation. Adv. Manuf. 4(4), 287-295 (2016). https://doi.org/10.1007/s40436-016-0157-2

[21] M.P. Mubiayi, E.T. Akinlabi, Friction stir welding of dissimilar materials: an overview, Proceedings of World Academy of Science, Engineering and Technology, World Academy of Science, Engineering and Technology (WASET), 2013, pp. 65-69.

[22] P. Goel, N.Z. Khan, Z.A. Khan, A. Ahmari, N. Gangil, M.H. Abidi, A.N. Siddiquee: Investigation on material mixing during FSW of AA7475 to AISI304. Mater. Manuf. Process 34(2), 192-200 (2019). https://doi.org/10.1080/10426914.2018.1544717

[23] J. Verma, R.V. Taiwade, C. Reddy, R.K. Khatirkar: Effect of friction stir welding process parameters on Mg-AZ31B/Al-AA6061 joints. Mater. Manuf. Process 33(3), 308-314 (2018). https://doi.org/10.1080/10426914.2017.1291957

[24] M. Fazel-Najafabadi, S. Kashani-Bozorg, A. Zarei-Hanzaki: Joining of CP-Ti to 304 stainless steel using friction stir welding technique. Mater. Des 31(10), 4800-4807 (2010). https://doi.org/10.1016/j.matdes.2010.05.003

[25] G. Buffa, L. Fratini, F. Micari, L. Settineri: On the choice of tool material in friction stir welding of titanium alloys. Proceedings of NAMRI/SME 40, (2012).

[26] Y. Chen, C. Liu, G. Liu: Study on the joining of titanium and aluminum dissimilar alloys by friction stir welding. Open Mater. Sci. 5(1), 6-10 (2011). https://doi.org/10.2174/1874088X01105010256

[27] Z. Ma, X. Sun, S. Ji, Y. Wang, Y. Yue: Influences of ultrasonic on friction stir welding of Al/Ti dissimilar alloys under different welding conditions. Int. J. Adv. Manuf. Technol. 112(9), 2573-2582 (2021). https://doi.org/10.1007/s00170-020-06481-6

Italian Manufacturing Association Conference - XVI AITeM
Materials Research Proceedings 35 (2023) 86-93

Materials Research Forum LLC
https://doi.org/10.21741/9781644902714-11

Validation of charge welds and skin contamination FEM predictions in the extrusion of a AA6082 aluminum alloy

Marco Negozio[1,a*], Riccardo Pelaccia[2,b], Lorenzo Donati[1,c], Barbara Reggiani[2,d], Sara Di Donato[1,e]

[1] DIN Department of Industrial Engineering - University of Bologna, Viale Risorgimento 2, 40136, Bologna, Italy

[2] DISMI Department of Sciences and Methods for Engineering - University of Modena and Reggio Emilia, Via Amendola 2, 42122, Reggio Emilia, Italy

[a]marco.negozio2@unibo.it, [b]riccardo.pelaccia@unimore.it, [c]l.donati@unibo.it, [d]barbara.reggiani@unimore.it, [e]sara.didonato2@unibo.it

Keywords: Extrusion, Aluminum Alloy, FEM Simulation, Charge Welds, Skin Contamination

Abstract. The reduction of scraps related to Charge Welds and Skin Contamination defects is getting an increased industrial attention in order to improve the extrusion process overall efficiency. Recently, FEM simulations allowed the prediction of these defects under different die designs or processing conditions without performing time-consuming and expensive experimental analysis. However, the validation of the FEM codes has not been fully experimentally assessed. In this paper, Charge Welds and Skin Contamination defects were experimentally analysed in the extrusion of a AA6082 aluminum alloy profile produced under strictly monitored processing conditions. The collected data were used to assess the accuracy of the predictions made by using two commercial FEM codes Qform Extrusion UK and Altair HyperXtrude. The final aim of this work is to discuss the reliability of the FEM simulations and to validate their applicability in the industrial field.

Introduction

The extrusion of lightweight alloys represents a widely used forming process to produce profiles with constant cross-section, high geometry complexity and excellent mechanical properties. During an extrusion cycle, at the beginning and the end of each extruded profile, defects may occur affecting the properties of the components and leading to the scrap of material [1, 2].

"Charge Welds" (fig. 1) are an intrinsic defect caused by the continuous extrusion of consecutive billets [3]. At the end of each ram stroke, after the removal of the billet rest, the die remains completely filled with the already extruded material. When the next billet is loaded into the container and the extrusion starts, the new material interacts with the old one present in the die thus generating a transition zone where the profile contains a mixture of new and old billet material. This transition zone is defined as Charge Welds extent. The part of the extruded profile in which this interaction is present must be discarded because it is usually contaminated by oxides, dust, or lubricant collected during the loading into the press, thus resulting in the lowering of the mechanical properties. The scrap of material due to the Charge Welds defect usually starts from the stop mark, which is a visible surface defect generated during the billet change due to the contact between the material and the die bearings, and ends when the cross-section of the profiles contains only the new billet material (Fig. 1c).

Italian Manufacturing Association Conference - XVI AITeM   Materials Research Forum LLC
Materials Research Proceedings 35 (2023) 86-93   https://doi.org/10.21741/9781644902714-11

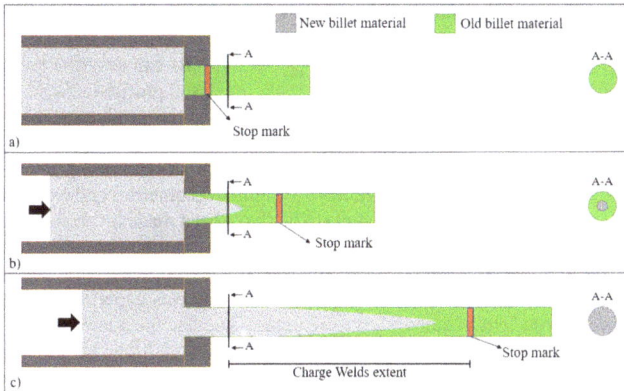

Figure 1: Schematization of the Charge Welds generation. a) Start of the extrusion process, b) interaction between new and old billet material in the profile, c) Charge Welds extent.

The second type of defect investigated in this work is known as "Billet Skin Contamination" as depicted in fig. 2. It is related to the outer layer (i.e. 'skin') of the billets which has a different chemical composition and microstructure with respect to the inner billet material due to the DC-casting process conditions and, if not specifically removed, it may also contain contaminations as oxides, dust or impurities collected in the billet pre-heating and handling phases [4]. During the ram stroke, if the billet rest is too short, the billet skin can flow inside the die until it reaches the extruded profile, thus generating a decrease of the mechanical proprieties and, consequently, the scrap of material on the right side of the stop mark. Indeed, by analysing the Fig. 2a, it is possible to notice that the skin contamination may occur before the stop mark at the end of the ram stroke thus increasing the length of the profile to be discarded. If the billet rest is too long, there is no contamination on the profile but uncontaminated billet material is discarded in the billet rest (Fig. 2c). Fig. 2b represents the optimized condition with no billet sin contamination in the profile before the stop mark and also a minimal thickness of the billet rest.

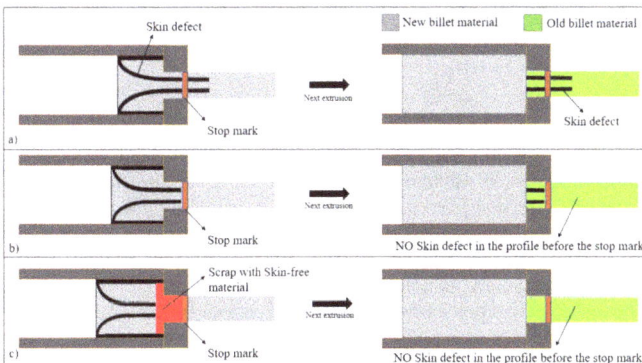

Figure 2: Schematization of the Skin Contamination generation. a,c) Unoptimized billet rest, b) optimized billet rest.

Italian Manufacturing Association Conference - XVI AITeM

Materials Research Forum LLC

Materials Research Proceedings 35 (2023) 86-93

https://doi.org/10.21741/9781644902714-11

In order to experimentally analyse the evolution of these defects and determine the exact amount of contaminated profile length to be discharged, time-consuming activities such as profile cutting, grinding and etching must be performed [5]. The results can describe the Charge Welds and Skin Contamination trends only for the specifically analysed profile-tools and, in addition to that, the experimental analysis can be carried out only after the production phase, leading to additional costs that could be avoided by optimizing the process parameters and tools geometries at a die design stage. For these reasons, numerical simulations became an interesting tool for extrusion die manufacturers due to the possibility to optimize geometries and extrusion parameters before the production stage. Although different studies have been made in the simulation of Charge Welds and Skin Contamination behaviours [6-8], further investigations are still needed in order to assess the accuracy and the reliability of FEM codes.

In this work, the evolution of the two investigated defects was experimentally analysed in a AA6082 extruded profile. The acquired data were used to validate the numerical predictions made by using two commercial FEM codes, Qform Extrusion UK and Altair HyperXtrude. The final aim of this investigation was to assess the reliability of the FEM codes for the extrusion process optimization thus proving their applicability in the industrial field.

Experimental Investigation

The geometry of the investigated extruded profile is shown in Fig. 3: the AA6082 solid profile was produced by Indinvest LT plant of Latina (Italy) on an industrial 35 MN press by means of a flat die with a single opening. The extrusion cycle involved the processing of fifty billets under strictly monitored conditions: the profile exit temperature was acquired by means of a pyrometer pointed in the center of the top surface of the profile, the ram force behaviour was also collected by the press. The investigated profile cross-sections were extracted in the transition between the 6th and the 7th billet in order to analyse samples from a steady-state process condition. All the process parameters are reported in Tab. 1.

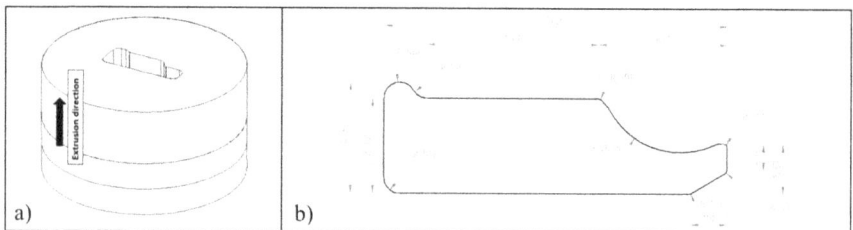

Figure 3: Geometries of the a) die and b) profile under investigation.

The industrial scrap made by the company due to Charge Welds and Skin Contamination was collected and further analysed: 1500 mm in the "front part" of the profile (starting from the stop mark, in the opposite direction from the extrusion one) and 4500 mm in the "back part" of the profile (starting from the stop mark, in the extrusion direction). The whole scrap length was initially cut into samples of 100 mm. After that, in the extent of the profiles in which the defect was expected to increase rapidly, samples were further sectioned for an in-depth investigation. All the samples were grinded and polished with abrasive papers and subsequently etched for about 90 sec in a sodium hydroxide solution (30% in H2O at 60°C). The results of the etching are shown in Fig. 4. These images were elaborated in order to recreate the evolution of the Charge Welds and Skin Contamination along the extruded profile using the stop mark as a reference point.

Italian Manufacturing Association Conference - XVI AITeM
Materials Research Proceedings 35 (2023) 86-93

Materials Research Forum LLC
https://doi.org/10.21741/9781644902714-11

Table 1: *Process parameters and geometry tolerances.*

Process Parameters and geometry tolerances	Profile
Aluminum alloy	AA6082
Extrusion ratio	20
Ram speed [mm/s]	7.64
Container temperature [°C]	440
Billet initial temperature [°C]	530
Die initial temperature [°C]	450
Ram acceleration time [s]	5
Billet length [mm]	990
Billet diameter [mm]	254
Container diameter [mm]	264
Billet Rest length [mm]	15

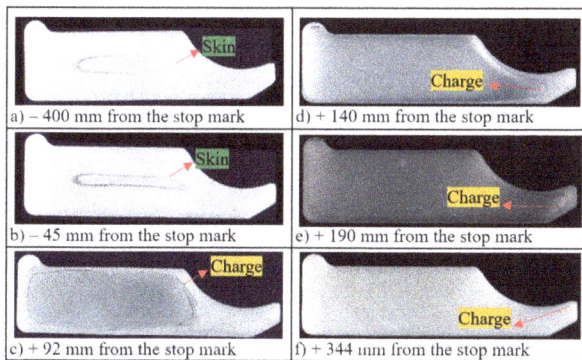

Figure 4: Cross-section of the profiles after etching: evolution of Skin Contamination and Charge Welds defects.

Fig. 5 reports the behaviours of the defects within the profile section at different distances from the stop mark: negative values on the x-axis represent samples extracted from the end of the billet 6th (billet skin defect) while positive ones represent samples extracted from the transition from billet 6th to 7th (charge weld defect). In the y-axis the percentage of profile section contaminated area by the defects is reported while the vertical lines represent the extremes of the scrap made by the company based on the experience of the technicians. The figure shows a clear unoptimized amount of profile length scrap since both the Charge Welds and the Skin Contamination extents are lower than the scrap made by the company. The Charge defect appears in the center of the cross-section of the profile and starts at +70 mm from the stop mark, increasing rapidly until it reaches the 90% of contamination at +190 mm and up to 95% at +400 mm, which is considered the defect extinction. The Skin Contamination also appears in the center of the profile at -3700 mm from the stop mark. The main difference between the two defects is that the Skin remained nearly constant in terms of contaminated area (12-15%) until it flattened out towards the shorter sides (4.5% at +68 mm), thus suggesting the upcoming of the Charge Welds defect.

To evaluate the Skin Contamination behaviour during the extrusion, a slice of the billet taken from the experimental batch was also analysed: a specimen of the billet surface was extracted, polished, and etched to assess the initial billet skin thickness. A value of 250 μm was found as skin depth in the billet.

Italian Manufacturing Association Conference - XVI AITeM   Materials Research Forum LLC
Materials Research Proceedings 35 (2023) 86-93   https://doi.org/10.21741/9781644902714-11

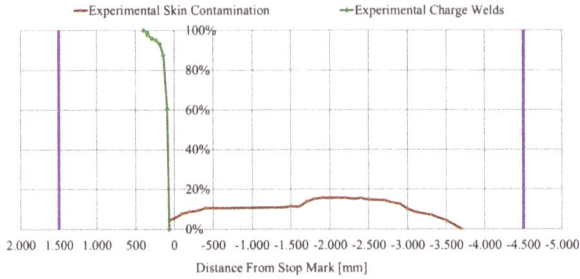

Figure 5: Experimental evolutions of the Charge Welds and Skin Contamination compared to the industrial scrap (purple lines in the graph).

Numerical and Analytical Investigation

The investigated case study was simulated using two commercial FEM codes: Qform Extrusion UK (Fig. 6a) and Altair HyperXtrude (Fig. 6b). The first FEM code is a Qform tool tailored for the simulation of the extrusion process. Through the use of this tool, it is possible to automatically generate the mesh and perform thermomechanical simulations using an ALE (Arbitrarian-Lagrangian-Eulerian) approach. The same numerical approach is used by HyperXtrude, which is an Altair product also focused on the extrusion simulation. Both codes are capable of simulating the material flow, the thermal field, the extrusion load and several extrusion defects, including Charge Welds and Skin Contamination starting from the CAD geometries of workpiece-tools and the definition of the process parameters.

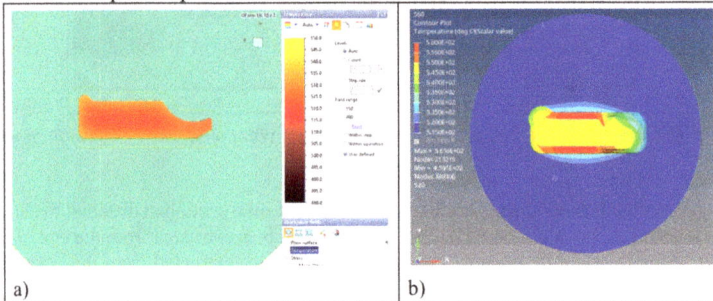

Figure 6: Temperature simulation using: a) Qform Extrusion UK, b) Altair HyperXtrude.

Both simulations were made using the same flow stress law: the following Hensel-Spittel equation was selected because it considers the effect of strain, strain rate and temperature [9]:

$$\bar{\sigma} = A \cdot e^{m_1 T} \cdot \bar{\varepsilon}^{-m_2} \cdot \dot{\bar{\varepsilon}}^{-m_3} \cdot e^{\frac{m_4}{\bar{\varepsilon}}} \cdot (1 + \bar{\varepsilon})^{m_5 T} \cdot e^{m_7 \bar{\varepsilon}} \cdot \dot{\bar{\varepsilon}}^{m_8 T} \cdot T^{m_9} \tag{1}$$

where $\bar{\sigma}$ is the flow stress, $\bar{\varepsilon}$ the strain, $\dot{\bar{\varepsilon}}$ the strain rate, T the temperature (K) and A, m_1-m_9 the material parameters to be regressed over experimental trials. These material parameters were taken from [10] and are summarized in Tab. 2.

Italian Manufacturing Association Conference - XVI AITeM
Materials Research Proceedings 35 (2023) 86-93

Materials Research Forum LLC
https://doi.org/10.21741/9781644902714-11

Table 2: AA6082 Hensel-Spittel parameters [12].

Hensel-Spittel Parameters	AA6082
A	568000 [MPa]
m1	-0.002117 [K^{-1}]
m2	0.1059
m3	0.098
m4	0.0009266 [K^{-1}]
m5	-0.00065
m7	0.02343
m8	0.00006471[K^{-1}]
m9	-1.208

Table 3: Friction conditions [10].

Surface	Friction condition
Billet-Container	Sticking condition
Billet-Ram	Sticking condition
Billet-Die	Sticking condition
Bearings	Levanov model (m = 0.3, n = 1.25)

The friction conditions between workpiece and tools were taken from [10]. The optimized default values for extrusion are reported in Tab. 3 while the AA6082 physical properties used in the simulations are reported in Tab. 4.

Table 4: Material parameters for the AA6082 aluminum alloy [10].

Material Properties	AA6082
Density [Kg/m3]	2690
Specific heat [J/kg K]	900
Thermal conductivity [W/m K]	200
Thermal expansivity [m/K]	2.34*10-5
Young's modulus [GPa]	68.9
Poisson's ratio	0.33

Results and Discussion

At first, the results of the simulations were validated by comparing the experimental and numerical exit temperature. The experimental data were acquired during multiple billets extrusion, although only the transition between the 6th and 7th was etched and analysed. Table 5 reports the average profile exit temperature including its standard deviation during the steady-state phase of three consecutive billets extruded in the same process conditions. The temperatures were acquired by the use of a pyrometer pointed in the top part of the profile immediately after the press exit. Table 5 also reports the average peak extrusion load for the same three investigated billets (5th, 6th and 7th). The simulation results show an error in the prediction of the two parameters always below the 5% using both FEM codes.

Table 5: Comparison of profile exit temperature and maximum extrusion load.

	Altair HyperXtrude	Qform Extrusion UK	Experimental
Profile Exit Temperature [°C]	542 (1.2% error)	529 (1.1% error)	535 ±5
Max Extrusion Load [MN]	17.5 (2.3% error)	17.7 (3.5% error)	17.1 ±0.1

After validating the simulations, the numerical predictions of the Charge Welds and Skin Contamination made by using Qform Extrusion UK and Altair HyperXtrude were carried out and the results are reported in Fig. 6 and Fig. 7. Both data are compared to the experimentally found values of the defects' evolutions.

According to the Qform Extrusion UK simulation, the Charge Welds onset was found at +85 mm and its extinction at +280 mm, showing a slight overestimation of 15 mm for onset and an underestimation of 120 mm for the defect extent. The Skin Contamination extent was predicted at -3500 mm while it was experimentally found at -3700 mm, with an underestimation of 200 mm.

Moreover, the numerical predictions show that the percentage evolution is accurate both in the case of Charge Welds and Skin Contamination (Fig. 7).

According to the Altair HyperXtrude simulation, the Charge Welds onset was found at +92 mm and its extinction at +406 mm, showing an overestimation of 22 mm for onset and of 6 mm for the defect extent. The Skin Contamination extent was predicted at -4500 mm while it was experimentally found at -3700 mm, with an overestimation of 800 mm. The numerical prediction of the Charge Welds shows a great accuracy in the percentage evolution simulation, while the Skin Contamination results were less accurate since the defect does not enlarge till the 100% of the sample area but remains almost constant at around the 14% (Fig. 8). This significant difference was caused by the increase in the Skin volume calculated by the code during the ram stroke, while it is supposed to remain constant during the entire process.

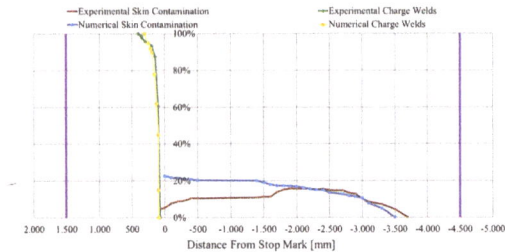

Figure 7: Qform Extrusion UK

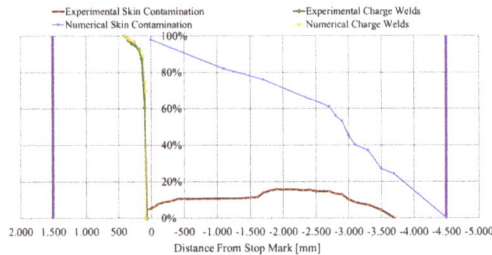

Figure 8: Altair HyperXtrude

The numerical simulations show an accuracy considerably higher if compared to the scrap made by technician's experience. The industrial error on the Charge Welds extent is 1100 mm while the Qform and HyperXtrude predictions errors are 120 mm and 6 mm, respectively. Moreover, the industrial error on the Skin Contamination extent is 800 mm while the Qform and HyperXtrude predictions errors are 200 mm and 800 mm, respectively. In summary, the error in the simulation is always lower than the one made by the industrial scrap except in the prediction made by Altair HyperXtrude on the Skin Contamination prediction, which resulted the same as the industrial one.

Conclusions

In the present work, experimental and numerical investigations were carried out for evaluating the accuracy of the Qform Extrusion UK and Altair HyperXtrude FEM codes in the prediction of Charge Welds and Skin Contamination evolutions on a solid extruded profile made by AA6082 aluminum alloys. The main outcomes of this work can be summarized as follows:

- A very good correlation between experimental and numerical data on Charge Welds predictions, both in terms of extent and of percentage evolution, was confirmed using the two tested FEM codes. 120 mm of error was found in the defect extent prediction using Qform Extrusion UK while 6 mm using Altair HyperXtrude. Both discrepancies with the experimental defect extent were extremely lower than the one found considering the industrial scrap (1100 mm).

- A good numerical-experimental matching was found for the Skin Contamination defect: Qform Extrusion UK accurately predicted both the extent of the defect (with an error of 200 mm) and its percentage evolution. However, Altair HyperXtrude prediction was less accurate in the defect extent (with error of 800 mm, the same as the industrial scrap) and in its percentage evolution.

- The numerical investigations proved the reliability of the FEM codes for the extrusion process optimization thus proving their applicability in the industrial field. Further experimental and numerical analyses are still required to assess the accuracy of the FEM simulation in different 6XXX aluminum alloys extrusions.

References

[1] N. Hashimoto, Application of Aluminum Extrusions to Automotive Parts, Kobelco Technology Review 35 (2017) 69-75.

[2] J. Hirsch, Automotive trends in aluminium - The European perspective, Materials Forum 28(3) (2004) 15-23.

[3] A.J. Den Bakker, L. Katgerman, S. Van Der Zwaag, Analysis of the structure and resulting mechanical properties of aluminium extrusions containing a charge weld interface, Journal of Material Processing Technology 229 (2016) 9-21. https://doi.org/10.1016/j.jmatprotec.2015.09.013

[4] Y.T. Kim, K. Ikeda, Flow behavior of billet surface layer in porthole die extrusion of aluminum. Metall Mater Trans A 31(6) (2000) 1635-1643. https://doi.org/10.1007/s11661-000-0173-4

[5] B. Reggiani, L. Donati, Experimental, numerical, and analytical investigations on the charge weld evolution in extruded profiles. International Journal of Advanced Manufacturing Technology 99 (2018) 1379-1387. https://doi.org/10.1007/s00170-018-2595-4

[6] A.A. Ershov, V.V. Kotov, YuN. Loginov. Capabilities of QForm-extrusion based on an example of the extrusion of complex shapes, Metallurgist, 2012, 55(9-10), p 695-701. https://doi.org/10.1007/s11015-012-9489-8

[7] P. Chathuranga. Case study of extrusion die design optimization using innovative cartridge type die, Light Metal Age, 2014, 77(5), p 20-27.

[8] N. Biba, S. Stebunov, A. Lishny, Simulation of material flow coupled with die analysis in complex shape extrusion, Key Engineering Materials, 2014, 585, p 85-92. https://doi.org/10.4028/www.scientific.net/KEM.585.85

[9] A. Hensel, T. Spittel, Kraft und Arbeitsbedarf bildsamer Formgeburgsverfahren, 1. Auflage, Leipzig: VEB Deutscher Verlag fur Grundstoffindustrie, 1978.

[10] M. Negozio, R. Pelaccia, L. Donati, B. Reggiani, T. Pinter, L. Tomesani, Finite Element Model Prediction of Charge Weld Behaviour in AA6082 and AA6063 Extruded Profiles, Journal of Material Engineering and Performance 30 (2021) 4691-4699. https://doi.org/10.1007/s11665-021-05752-x

Additive manufacturing and reverse engineering

Italian Manufacturing Association Conference - XVI AITeM
Materials Research Proceedings 35 (2023) 95-102

Materials Research Forum LLC
https://doi.org/10.21741/9781644902714-12

Comparison of specific cutting energy in dry and wet post- process turning of Ti6Al4V EBM parts

Ersilia Cozzolino[1,a] *, Stefania Franchitti[2,b], Rosario Borrelli[2,c], Antonello Astarita[1,d]

[1]Dipartimento di Ingegneria Chimica, dei Materiali e della Produzione Industriale, Università degli Studi di Napoli Federico II, Piazzale Tecchio 80, 80125 Napoli, Italy

[2]CIRA (Italian Aerospace Research Center), 81043 Capua CE, Italy

[a]ersilia.cozzolino@unina.it, [b]s.franchitti@cira.it, [c]r.borrelli@cira.it, [d]antonello.astarita@unina.it

Keywords: Sustainable Manufacturing, Electron Beam Melting, Machining

Abstract. Additive Manufacturing (AM) is accelerating more and more today. Among its advantages, AM is claimed to be green technology. However, AM parts usually require postprocessing to improve their surface finishing and to be assembled. In this study, some Ti6Al4V cylindrical samples have been manufactured by Electron Beam Melting and then post- processed by turning. Both dry and wet turning has been performed under the same process parameters. Surface roughness has been measured both before and after each turning pass along the parallel and perpendicular direction to the cylindrical axis and energy consumption has been recorded during each turning pass. Specific Cutting Energy (SCE) has been calculated to evaluate the energy efficiency of the turning process. The results of this study demonstrate that dry turning is more energy efficient than wet turning by selecting the same machining process parameters while obtaining a comparable surface roughness.

Introduction

Sustainable Manufacturing (SM) is a crucial topic today in both the research community and industry. It consists of the creation of manufactured products through processes that minimize the negative environmental impacts derived from energy consumption, material waste and inefficient use of natural resources. There is also an excellent diversity of interpretations and concepts related to SM [1].

Concerning the materials investigated, the titanium alloy Ti6Al4V was noted as the most popular material utilized in the latest SM technology research due to its superior material properties, such as corrosion resistance, high strength, low density, high fracture toughness, biocompatibility, high industrial demand, including its well-established history in the aerospace sector [2], and its suitability for various manufacturing technologies, including additive manufacturing. Among the nickel alloys, Inconel 718 was observed to be the foremost popular workpiece material due to the fact that it is a high-performance superalloy that provides corrosion resistance along with strength at both atmospheric and high-temperature ranges [3].

Additive Manufacturing (AM) consists of producing three-dimensional objects by adding layers of material based on a three-dimensional computer model. AM is currently becoming a key technology in industries such as aerospace, biomedicine, and manufacturing as it allows fabricating customized products thanks to its ability to create complex objects with advanced attributes. AM is claimed to be a green technology because it holds great potential in improving materials efficiency, reducing life-cycle impacts, and enabling greater engineering functionality compared to conventional technologies. Nevertheless, the literature lacks guidelines for obtaining AM having good mechanical properties while saving energy and reducing environmental impacts. Also, surface finishing of AM parts hardly ever is acceptable for the industrial quality standards. For this reason, postprocessing is always required. Very few studies exist in the literature having

Italian Manufacturing Association Conference - XVI AITeM   Materials Research Forum LLC
Materials Research Proceedings 35 (2023) 95-102   https://doi.org/10.21741/9781644902714-12

as objective the minimization of both Ra roughness parameter and power and energy consumption in post-process machining of Ti6Al4V EBM parts. This study aims to fill this gap of knowledge. In particular, in this study a cylindrical sample has been post-processed by both dry and wet turning by selecting the same process parameters. Power consumption has been acquired in both situations and surface roughness has been measured before and after the turning processes. The sustainability of the turning processes has been evaluated by means of the sustainability index called specific energy consumption (SEC), calculated by means the power measurements and process parameters adopted in this study.

Materials and Methods

A cylindrical sample has been manufactured by EBM with the axis parallel to the building one. It has a height of 103.38 mm and a diameter of 30 mm. The sample has been built by means of the Arcam A2X machine by using Ti6Al4V powders having the 93.7% of particle size between 45 μm to 106 μm. As usual, EBM has been performed under a vacuum to reduce contamination and minimize electron collisions with air molecules. Also, to prevent electrostatic charging and smoke events, a helium pressure of 10^{-3} mbar has been applied. The standard Ti6Al4V build time, the layer thickness equal to 50 μm and the line offset of 0.1 mm have been fixed to build the sample in the EBM machine, which worked in automatic mode [4].

The sample has been then post-processed by machining, by means of the FEL-660HG lathe, to improve its surface finishing and to meet industrial quality requirements. Dry turning and wet turning have been performed on two sides of the sample, both for a length equal to 50 mm, under the same process conditions. The lubricant oil Siroil Emulg has been employed for the wet turning process. New inserts, Sandvik CNMG 12 04 08-SM H13A, have been used to perform both dry and wet turning by adopting a spindle speed of 300 rev/min. Table 1 contains the process parameters adopted in this study, according to the best results obtained in the literature in terms of surface roughness by turning [5].

Table 1. Process parameters in post-process machining of the cylindrical sample

Feed rate (mm/rev)	Depth of cut (mm)	Number of cutting passes	MRR (mm³/s)
0.28	0.8	3	105.5

Surface roughness has been investigated before and after the turning processes by means of the confocal microscope 3D Optical Surface Metrology System Leica DCM3D. After the surface acquisition, the Ra roughness parameter has been measured by selecting three profiles on both the parallel and perpendicular directions to the build one.

Power consumption has been recorded over time by means of the power quality analyser CA8331. It measures current and tension by means of specific sensors, whose number depends on the electrical connection of the machine to be analysed. The lathe adopted in this study has a three-phase without neutral connection 32 A 380 V. Thus, three tension cables, three crocodile clips and three current sensors were employed to measure power over time, by selecting a sampling period of 1 second.

To evaluate the energy efficiency of the turning processes, the specific cutting energy (SCE) has been calculated as follows [6]:

$SCE = P_{cut}/MRR$ (1)

Where $Pcut$ is the cut power, that is calculate as:

$P_{cut} = P_{actual} - P_{air}$ (2)

Where $Pactual$ is the actual power, that is the average power measured by means of the power device during the material removal, and $Pair$ is the air power, which is the power measured while the lathe is energized but there is no contact between cutting tool and the workpiece. In other words, air power is the power consumption measured when the lathe is working, under the process conditions fixed, but the material is not removed. The air power was measured both during wet and dry turning as it is affected by the pump energy contribution. Material removal rate (MRR) is calculated as follows:

$MRR = V_c * a * f$ (3)

Where a is the depth of cut, f is the feed rate, and V_c is the cutting speed calculated as:

$V_c = \frac{\pi D n}{1000}$ (4)

Where D is the diameter of the sample to be cut and n is the spindle speed.

Results and Discussion

A result of our study is the feasibility of the post-process machining by means of the parallel lathe on the EBM Ti6Al4V cylindrical sample either with and without the lubricant usage. Fig. 1 shows the machining process performed with the lubricant.

As previously described, surface roughness has been investigated before and after the turning process with and without the lubricant along both the parallel and perpendicular directions to the cylindrical axis. Table 1 contains all the numerical results in this regard. Each Ra measurement is the average of three profiles taken on both the principal directions of the cylinder.

Figure 1. Post-process turning of the EBM Ti6Al4V sample with the lubricant

It can be observed that the roughness parameter Ra along the parallel direction is higher than Ra measured along the perpendicular directions. In particular, before the machining, Ra along the

Italian Manufacturing Association Conference - XVI AITeM   Materials Research Forum LLC
Materials Research Proceedings 35 (2023) 95-102   https://doi.org/10.21741/9781644902714-12

parallel direction to the building one is found equal to 35.81 μm whereas Ra along the perpendicular direction is 28.79 μm, as a result of the staircase effect. That is due to the fact that EBM technology consists in melting the material layer by layer. Thus, the cylindrical sample has been post-processed to improve its surface roughness to meet the industrial quality requirements. As previously mentioned, two lengths have been machined on the same cylindrical sample with and without lubricant by using the same process parameters. The results displayed in Table 2 show that a higher percentage reduction has been obtained after three passes without lubricant along both the parallel and perpendicular directions.

Moreover, it can be observed that the reduction percentage of Ra in the perpendicular direction is always higher than that obtained in the parallel direction, with and without the lubricant. This is due both to the higher roughness to be smoothened, according to the EBM staircase effect, and the technological signature of the turning process [7][8].

Table 2. Surface roughness along the principal directions of the cylindrical sample

Ra// [μm] as built	Ra// [μm] with lubricant	Ra// [μm] without lubricant	Reduction % Ra// with lubricant	Reduction % Ra// without lubricant
35.81	2.20	1.77	93.86	95.05
Ra⊥ [μm] as built	Ra⊥ [μm] with lubricant	Ra⊥ [μm] without lubricant	Reduction % Ra⊥ with lubricant	Reduction % Ra⊥ without lubricant
28.79	0.34	0.42	98.81	98.17

As previously described, power consumption has been measured by means of a power device that gives as output of current and tension measurements the power recorded over time. Fig. 2 shows the power trend during the three turning passes to remove a depth of cut equal to 2.4 mm with and without lubricant.

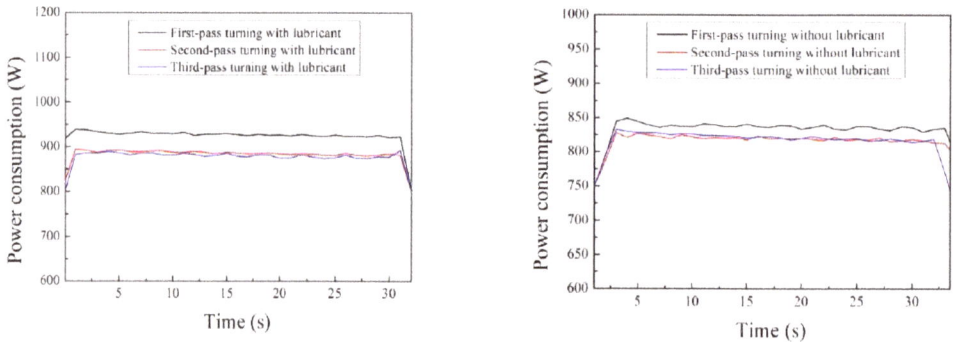

Figure 2. Power consumption over time of all the turning passes with and without the lubricant

The curves on the two diagrams show the same trend, just the curves representing the power consumption of all the turning passes with lubricant are translated upwards because of the contribution of the lubricant pump. All of them show an increase in power consumption at the beginning, when the cutting tool become in contact with the workpiece and, in opposite, power decreases at the end of the machining process, when the cutting tool turns away from the sample.

Also, it can be seen that in both situations the power curve related to the first-pass turning is the highest one, as at the beginning surface roughness is the highest one and cutting forces involved are higher. Therefore, power consumption to remove the material is higher at the beginning of the process. In fact, second-pass turning and third-pass turning are represented by lower curves as the cutting forces and then the power consumption required is lower.

Moreover, the first-pass turning without lubricant shows some oscillations over time: this is due to the fact that at the beginning the surface roughness is very high as well as the cutting forces involved are higher, as a result of the absence of the lubricant oil, which helps to reduce wear and tear.

In general, higher power consumption is required during the first pass turning since the roughness is higher. However, by comparing dry and wet first-pass turning, such difference is lower in the "without lubricant" phase. This is due to the fact that, even the surface roughness is lower, after the first pass turning, higher cutting forces are involved in dry turning than wet turning, so higher power consumption is required. We can say that, after the first-pass turning, cutting forces effect prevails on the roughness effect.

Table 3 contains the results derived from the power measurements (actual power and air power) and calculations (cut power, material removal rate and specific cutting energy), according to the equations described in the previous sections. Air power is the power consumption of the lathe when it is energized but the tool is not in contact with the material whereas the actual power is the power consumption during the material removal. It can be observed that the air power is higher than the air power without the lubricant usage, as expected, because of the energy consumption of the lubricant pump.

Table 3. Results of the power analysis in the post-process machining

Number of the turning pass	Pactual (W) with lubricant	Pactual (W) without lubricant	P_{air} (W) with lubricant	P_{air} (W) without lubricant	Pcut (W) with lubricant	Pcut (W) without lubricant	MRR (mm³/s)	SCE (J/mm3) with lubricant	SCE (J/mm3) without lubricant
1	917	835	800	750	117	85	105.5	1.1	0.8
2	886	818	800	750	86	68	102.6	0.8	0.7
3	878	815	800	750	78	65	99.9	0.8	0.7

The material removal rate has been calculated according to Eq. 3 by considering the current diameter after each turning pass. Obviously, after each turning pass, the diameter to be cut decreases, so cutting speed decreases and then material removal rate decreases. Also, cut power decreases after each turning pass according to the fact that the surface roughness decreases. Fig. 3 depicts the relationship between cut power and material removal rate. It can be noted that cut power increases by increasing the material removal rate both with and without lubricant usage. This is probably the effect of the changing diameter between the turning passes.

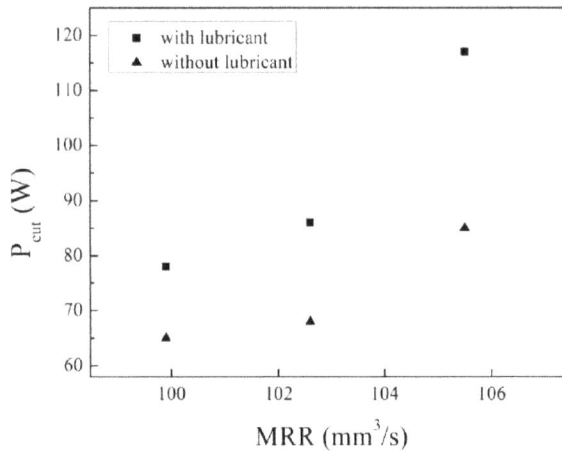

Figure 3. Cut power for each material removal rate with and without the lubricant

Specific cutting energy (SCE) has been calculated according to Eq. 1 by means of the power measurements. Numerical results ranging from 0.7 and 1.1 J/mm^3 have been obtained (Table 3). Fig. 4 displays specific cutting energy results for each turning pass with and without lubricant. It represents the energy spent to cut a material volume equal to 1 mm^3. Both with and without the lubricant, SCE decreases with the increase of the turning passes. Also, SCE is higher for turning with the lubricant as the cut power is higher.

These results underline that dry turning is more energy efficient than wet turning by selecting the same machining process parameters. Also, comparable results have been obtained in terms of surface roughness (Table 2) with and without lubricant along both the parallel and perpendicular direction tot the building one. For these reasons, according to our results, it is worthwhile to select the dry turning rather than the wet turning as it results in similar results in terms of surface roughness by minimizing the specific cutting energy.

Italian Manufacturing Association Conference - XVI AITeM   Materials Research Forum LLC
Materials Research Proceedings 35 (2023) 95-102   https://doi.org/10.21741/9781644902714-12

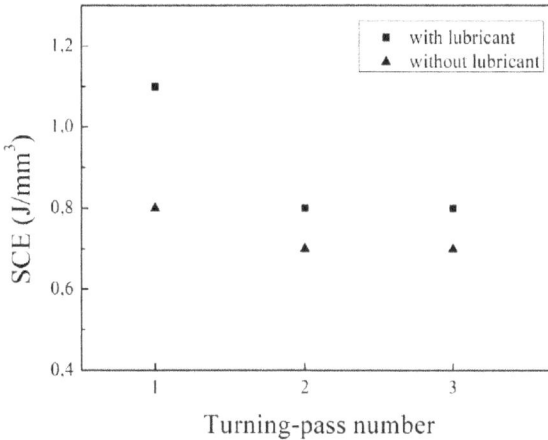

Figure 4. Specific cutting energy for each turning pass with and without the lubricant

Conclusions

This experimental study demonstrates the feasibility of performing dry turning as postprocessing of an EBM Ti6Al4V cylindrical sample. Roughness has been measured before and after the turning process with and without lubricant along the parallel and perpendicular directions to the build one. Power consumption has been recorded to investigate the energy efficiency of the process.
The main conclusions of this study are:

- Roughness parameter Ra along the parallel direction to building one was found equal to 35.81 μm whereas Ra along the perpendicular direction was 28.79 μm. Thus, in the parallel direction the roughness is higher than that in the perpendicular direction, as a result of the staircase effect.
- A higher power consumption is required to remove the first layer of material as higher are the cutting forces involved, as a consequence of the higher surface roughness.
- Power curves show the same trend both with and without the lubricant. The first layer is always the most energy consuming.
- Air power is 800 W with the lubricant, and it is 750 W without the lubricant. Thus, the power consumption of the lubricant pump is about 50 W.
- After each turning pass, cut power decreases according to a reduction in surface roughness.
- Cut power increases by increasing material removal rate both in dry and wet post-process turning.
- Specific cutting energy values ranging from 0.7 and 1.1 J/mm^3 have been obtained in our experimental study. It decreases with increasing the turning pass number.
- As comparable Ra numerical measurements have been obtained with and without lubricant and dry turning is more energy efficient, it is good to choose dry postprocess turning whenever possible.

This is just a preliminary study on the feasibility of performing dry turning as postprocess machining of an EBM Ti6Al4V part. Further investigation will include additional tests with different parameters to deeply investigate the two strategies, with and without lubricant oil during machining of AM parts, to find the best scenario for minimizing specific energy consumption. Also, a comprehensive sustainability assessment will be carried out to consider the impact of both the lubricant and tool wear on the environment.

References

[1] A. Sartal, R. Bellas, A.M. Mejías, A. García-Collado, The sustainable manufacturing concept, evolution and opportunities within Industry 40: A literature review, Adv. Mech. Eng. 12 (2020). https://doi.org/10.1177/1687814020925232

[2] S. Liu, Y.C. Shin, Additive manufacturing of Ti6Al4V alloy: A review, Mater. Des. 164 (2019) 107552. https://doi.org/10.1016/j.matdes.2018.107552

[3] R. Vaßen, J. Fiebig, T. Kalfhaus, J. Gibmeier, A. Kostka, S. Schrüfer, Correlation of Microstructure and Properties of Cold Gas Sprayed INCONEL 718 Coatings, J. Therm. Spray Technol. 29 (2020) 1455-1465. https://doi.org/10.1007/s11666-020-00988-w

[4] M. Koike, K. Martinez, L. Guo, G. Chahine, R. Kovacevic, T. Okabe, Evaluation of titanium alloy fabricated using electron beam melting system for dental applications, J. Mater. Process. Technol. 211 (2011) 1400-1408. https://doi.org/10.1016/j.jmatprotec.2011.03.013

[5] V. Upadhyay, P.K. Jain, N.K. Mehta, In-process prediction of surface roughness in turning of Ti-6Al-4V alloy using cutting parameters and vibration signals, Meas. J. Int. Meas. Confed. 46 (2013) 154-160. https://doi.org/10.1016/j.measurement.2012.06.002

[6] F. Hojati, A. Daneshi, B. Soltani, B. Azarhoushang, D. Biermann, Study on machinability of additively manufactured and conventional titanium alloys in micro-milling process, Precis. Eng. 62 (2020) 1-9. https://doi.org/10.1016/j.precisioneng.2019.11.002

[7] E. Cozzolino, A. Astarita, R. Borrelli, S. Franchitti, V. Lopresto, C. Pirozzi, A Preliminary Investigation of Energy Consumption for Turning Ti6Al4V EBM Cylindrical Parts, 926 (2022) 2355-2362. https://doi.org/10.4028/p-vm4f1y

[8] E. Cozzolino, S. Franchitti, R. Borrelli, C. Pirozzi, A. Astarita, Energy consumption assessment in manufacturing Ti6Al4V electron beam melted parts post processed by machining, Int. J. Adv. Manuf. Technol. (2023) 1289-1303. https://doi.org/10.1007/s00170-022-10794-z

Italian Manufacturing Association Conference - XVI AITeM
Materials Research Proceedings 35 (2023) 103-110

Materials Research Forum LLC
https://doi.org/10.21741/9781644902714-13

Preliminary evaluation of an additive manufacturing procedure for producing patient-specific upper-limb orthotic devices

Francesca Sala[1,a] *, Mariangela Quarto[1,b], Gianluca D'Urso[1,c], and Claudio Giardini[1,d]

[1]University of Bergamo - Department of Management, Information and Production Engineering, via Pasubio 7/b, Dalmine (BG), 24044, Italy

[a]francesca.sala@unibg.it, [b]mariangela.quarto@unibg.it, [c]gianluca.d-urso@unibg.it, [d]claudio.giardini@unibg.it

Keywords: Material Extrusion, Reverse Engineering, Health Care

Abstract. In the orthopedic field, the need for patient-specific devices is crucial to ensure a rapid and successful care treatment. The traditional techniques for manufacturing customized orthopedic systems, specifically orthoses, are laborious and present multiple and time-consuming steps. The present research analyzed the possibility of optimizing the conventional process for manufacturing personalized orthoses by leveraging the principles of Reverse Engineering (RE) and Additive Manufacturing (AM). Digital orthotic models of different anatomical regions were obtained using 3D laser scanning and semi-automated CAD processing, whilst the prototypes were produced using a Fused Deposition Modelling (FDM) printer and polymeric filaments suitable for the intended use. Furthermore, topological optimization was employed to improve the shape and the weight of the different medical devices. Potential advantages and drawbacks of the discussed procedure were evaluated through a preliminary indication of production times and costs.

Introduction

Additive Manufacturing (AM) techniques widely spread in the orthopedic sector [1,2]. Nowadays, the capability of AM technologies to fabricate articulated geometries with a variety of biocompatible materials represents a potential and valuable solution to the disadvantages related to the traditional manufacturing of patient-specific orthoses [3]. Indeed, it is common knowledge that the traditional practice of orthotics customization is not always effective: the skills and experience of the medical operator have a great impact over the quality of the final product [4] and, if the outcome is not good enough, the likelihood of the patient rejecting the rehabilitation therapy increases [5].

In the present article, an alternative methodology to the traditional fabrication of orthoses dedicated to the treatment of upper limb regions was studied in an effort to bring more comfortable and high-performance medical solutions. The approach involved three main steps: acquiring the anatomical region of interest as a 3D scanning, shaping the orthotic model on the basis of the individual diagnostic case and fabricating the final medical device. Three anatomical regions of different size and level of detail were used as test objects for the evaluation of the validity and repeatability of the manufacturing process under investigation. The anatomical regions involved the upper extremities of the limbs, particularly the wrist and finger joints.

Pathological conditions affecting the wrist joint may require the use of orthoses that immobilize the wrist joint, while enabling full mobility of the metacarpophalangeal (MP) joint and thumb. Clinicians might rely on a variety of wrist orthotic patterns depending on the anatomical region along which they extend: volar, dorsal, ulnar and circumferential. As the name suggests, circumferential devices completely envelop the wrist joint, bringing greater stability, and are especially indicated in fractures and complex regional pain syndrome.

Italian Manufacturing Association Conference - XVI AITeM   Materials Research Forum LLC
Materials Research Proceedings 35 (2023) 103-110   https://doi.org/10.21741/9781644902714-13

On the basis of the medical consultation, pathological conditions affecting the fingers may need a medical device involving only the targeted finger or may necessitate a device involving the whole hand. Diagnoses such as, finger sprains, mallet finger, boutonniere and swan-neck deformities, may require finger-based orthoses to constrain only the movement of the proximal interphalangeal (PIP) and/or distal interphalangeal (DIP) joint under investigation, thus leaving the metacarpophalangeal (MP) joint unrestricted and allowing the mobility of the other digits. On the other side, other clinical conditions, like osteoarthritis and traumatic injuries of the thumb, may require more structured orthopedic devices, that, as opposed to finger-based splints, cross the entire palm and dorsum of the hand. This is the case of hand-based thumb orthoses, which constrain both interphalangeal (IP) and metacarpophalangeal (MP) thumb joints, allowing for wrist and other digits mobility.

Hence, from the present research, three medical devices, characterized by a different degree of complexity, were developed. At a later stage, different orthotics designs for each application were designed exploiting topologically optimization too. Their feasibility, along with the entire process validity, was assessed through a preliminary evaluation on 3D printing costs and times.

Methodology

Acquisition of the anatomical regions. The production of patient-specific orthoses started with the activity of acquiring the surface geometry information about the anatomical area to be healed. The acquisition was carried out by means of a handheld 3D scanner, in particular, the technology was Hexagon Absolute Arm 7-Axis equipped with an RS6 Laser Scanner. The instrumentation provides contactless, rapid (max 1.2 million points/s point acquisition rate, max. 300 Hz line rate) and detailed (0.026 mm precision) scans, in accordance with the current clinical requirements [6].

The three anatomical regions were acquired through a single scan in the form of a point cloud, directly exportable to stl-format. The scanning parameters remained unchanged for each application, while the acquisition set-up was in accordance with the dimension and level of detail of the specific anatomical region and, most importantly, the therapeutic goal. This implies that, in the presence of injuries that prevent active movement of the joint under consideration, it is essential for the physician to passively relocate the patient's joint in the correct alignment in order to scan a physiological joint and prevent the unsuccessful development of a pathological scan. This consideration safely applies to the pathologies outlined in the introductory chapter. Key information concerning the 3D scanning set-up of the different anatomical regions of the upper extremities is presented in Table 1.

Table 1. 3D scanning set-up of the three applications.

Upper limb region	Joints	Number of stabilizers	Position
Wrist	Radiocarpal, ulnocarpal and distal radioulnar joints	3	The arm is positioned horizontally and stabilized by three supports at: - radius head; - index, middle, and ring fingers middle phalanges; - thumb distal phalanx.
Thumb	Interphalangeal (IP) and/or metacarpophalangeal (MP) joints	3	The hand is positioned horizontally and stabilized by three supports at: - ulna head; - index, middle, and ring fingers middle phalanges; - thumb distal phalanx.

Finger (Index)	Proximal interphalangeal (PIP) and distal interphalangeal (DIP) joints	2	The (index) finger is positioned horizontally and stabilized by two supports at: - metacarpals; - (index) finger distal phalanx.

Elaboration of the customized anatomical regions into orthotic models. The 3D scans were imported into a software developed with the purpose of transforming the surface anatomical information into a solid model of the orthosis. The program was initially designed for a specified application, namely arm modelling [7], and partially automates its manual modelling in a series of simple steps (described in Table 2), based on the Python language. In the present research, the use of this software was extended to other atomic regions belonging to the upper-limb extremities (fingers) in an effort to test its adequacy.

Table 2. Operations of the modelling software.

Function	Description
Centering	Translation of the stl scan from its original reference system to a new reference system with (0;0;0) origin.
Fixing	Removal of duplicated/isolated vertices/faces, edge repair, hole closure, normal correction, mesh reconstruction and simplification.
Smoothing	Homogenization of the surface texture.
Expanding	Shift of the faces along their normal and towards the outside to arbitrarily expand the mesh.
Solid creation	Converting the surface into a solid through the generation of a thickness of arbitrary size.
Cutting	Removal of the element extremities through cuts perpendicular to Cartesian axes. Cuts in different directions are executed by orienting the orthotic element accordingly.
Lightening	Creation of holes (of arbitrary shape and dimension) along the solid structure to reduce the weight and volume of the orthotic element.
Dividing	Splitting the model into two halves.
Combining	Creation of connection points on the external surfaces of the two halves and implementation of elements that prevent slippage of the union surfaces during the assembly of the two halves.

From the software, two orthotic designs per anatomical application were retrieved: a full model and a lightened model. The latter design presented a pattern of rhomboid-shaped holes, realized through the lightening operation. An additional orthotic design was developed from the full model: simulations were performed using the optimization module of the finite-element analysis (FEA) software Abaqus in order to lighten the structure by considering reasonable forces acting on the orthotic element during the rehabilitation pathway.

Static simulations simplified the patient-device system by considering only the orthotic element. Load conditions were conceived to recreate the flexion of the joints considered in each application and, in the current optimization study, were represented as followed:

- Wrist – Application of a torque (with $F = 200$ N and $d = 100$ mm) on the distal perpendicular surface of the orthosis (in the proximity of the fingers), forcing the proximal perpendicular surface of the orthosis with an encastre constraint.
- Thumb – Application of a torque (with $F = 30$ N and $d = 100$ mm) on the perpendicular thumb surface of the orthosis, forcing with an encastre constraint the perpendicular surface of the orthosis near the wrist.
- Finger (Index) - Application of a torque (with $F = 30$ N and $d = 60$ mm) on the distal perpendicular surface of the index finger, forcing with an encastre constraint the proximal perpendicular surface of the index finger.

Italian Manufacturing Association Conference - XVI AITeM   Materials Research Forum LLC
Materials Research Proceedings 35 (2023) 103-110   https://doi.org/10.21741/9781644902714-13

The optimization process determined a new material distribution based on the minimization of the strain energy, while satisfying a volume (below the 60% of the original volume) and geometric (preservation of the load and boundary condition areas) constraint.

Additive manufacturing of the orthotic models. The orthotic models were manufactured by Ultimaker S5 desktop Fused Deposition Modelling (FDM) machine, an additive manufacturing (AM) technology based on heating and extrusion of a thermoplastic filament from a nozzle to a building platform. The polymeric filament was polylactic acid (PLA) and, besides being one the most commonly used material in 3D printing, it combined the needs for cost-effectiveness [8]. The mechanical properties (Table 3) and printing settings (extra fast modality, Table 4) of Ultimaker PLA were retrieved from the technical data sheets. Besides, the use of the manufacturing material, the use of support material was kept to a minimum: only orthotic models characterized by large overhangs and hanging features necessitated the use of Ultimaker Breakaway support material (the minimum overhang value for which the support was printed was 65°).

Table 3. Mechanical properties of Ultimaker PLA.

	Density	Elastic modulus	Ultimate tensile strength	Poisson ratio
PLA	1.24 g/cm^3	2347 MPa	46 MPa	0.33

Table 4. Printing settings of Ultimaker PLA (extra fast modality).

	Layer height	Infill % , pattern	Printing temperature	Printing speed
PLA	0.6 mm	100%, line	210°C	70 mm/s

Results

Process evaluation. The process of manufacturing customized orthoses for upper limb rehabilitation progressed according to the three stages stated in the methodology.

Scans of the three anatomical regions under consideration were obtained in multiple attempts owing to the high resolution of the scanning medium (RS6 Laser Scanner), which inevitably detected the presence of unintentional muscle contractions inducing misalignment issues in the mesh. The resulting scans were of high quality and the meshes were extremely dense (i.e., the arm scan featured over 150,000 nodes).

Subsequently, the scans were entered into the modeling software. By subjecting the arm scan to the modeling commands, orthotic models (full and lightened designs) for wrist rehabilitation were obtained flawlessly. Also, the software responded positively to the generation of orthotic models of a different shape and geometry, indeed, orthotic models for rehabilitation of the thumb finger and index finger were successfully obtained. Nevertheless, two remarks concerning the development of the latter two orthotic models should be made; these scans were not subjected to the fixing command, as this algorithm (combined with the smoothing function) was prone to excessive shrink the original mesh in an attempt to optimize the number of nodes and triangles (as shown in Figure 1, representing the index finger application). One further remark concerned the orthotic model for the treatment of the index finger, dividing and combining operations were not necessary since the orthosis could be modeled as a single part.

Different designs for the three applications were developed (as presented in the following paragraph, in Table 5-7):

- Model 1 – full model.
- Model 2 – lightened model obtained with the lightening function.
- Model 3 – lightened model obtained with the optimization module of Abaqus.

Regarding model 3, at the completion of the topological optimization process, the resulting geometry was definitely rough such that extra edge refinements are necessary. Particularly, the finishes would be functional to facilitate the 3D printing process, besides providing practical wearability. Depending on the extension and medical function of the specific anatomical region,

Italian Manufacturing Association Conference - XVI AITeM   Materials Research Forum LLC
Materials Research Proceedings 35 (2023) 103-110   https://doi.org/10.21741/9781644902714-13

different orthotic thicknesses were made: the wrist and thumb orthoses were built with a thickness of 2 mm, whilst the index finger orthosis of 1 mm. The thicknesses conformed to medical indications and are slightly thinner than traditional devices.

The medical prototypes were printed in PLA using FDM technology (Ultimaker S5). In almost every instance the use of support material (Ultimaker Breakaway) was necessary, with the only exception of models 1 and 2 of the index finger orthosis. Models derived from topological optimization (model 3) maximized the use of support material.

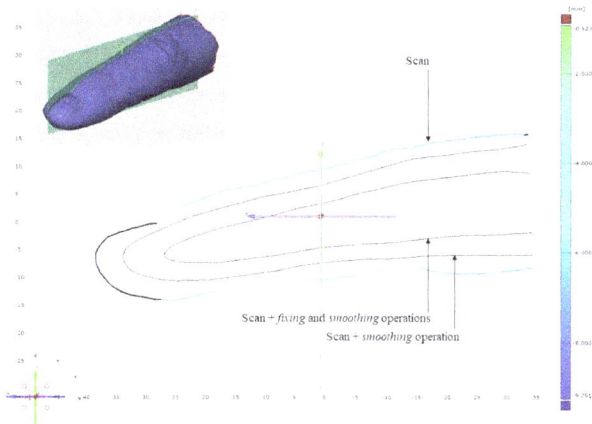

Figure 1. Graphication of the error (mm) between the finger original scan and the finger model which was developed using the fixing command. In the upper left quadrant measurement section was reported.

Product evaluation. The current procedure was complemented with a time-cost valuation of the medical devices. Time analysis comprised the estimation of the production time (material deposition time) through the use of the Ultimaker Cura slicing program. Cost analysis involved the cost of the materials used to produce the prototypes, thus the manufacturing material (PLA, about 0.57 €/m) and the support material (Breakaway, about 0.92 €/m). Cost analysis was based on the length of deposited filament, captured through the Ultimaker Cura slicing software. Also, the analysis was supplemented with the information on prototype weight and mechanical behavior under reasonable working condition to assess the feasibility of the two strategies of volume reduction. The mentioned information is discussed below and reported in Table 5-7.

In terms of time, in each medical application, the progressive lightening of the device showed an increase in printing time prompted by the need of employing support material for the manufacturing of geometries characterized by large overhangs and thin-walled structures. The effect was particularly emphasized in topologically optimized models (model 3), whose designs were characterized by printing times at least twice as long as the full models (model 1). In percentage terms, the worst-case scenario was the orthosis dedicated to the rehabilitation of the index finger conditions: the printing time of the orthotic model 3 was 2.6 times longer than the printing time of the orthotic model 1.

Table 5. Assessment of the wrist orthotic models.

	Wrist – Model 1	Wrist – Model 2	Wrist - Model 3
Printing time	122 min	176 min	294 min
Material cost	7.07 €	8.27 €	11.66 €
Weight	95 g	92 g	64 g
Max. stress	30 MPa	31 MPa	32 MPa
Max. displacement	2.7 mm	3.2 mm	4.0 mm

Table 6. Assessment of the thumb orthotic models.

	Thumb – Model 1	Thumb – Model 2	Thumb - Model 3
Printing time	74 min	95 min	151 min
Material cost	4.95 €	5.43 €	6.71 €
Weight	64 g	62 g	42 g
Max. stress	11 MPa	11 MPa	11 MPa
Max. Displacement	0.5 mm	0.5 mm	0.5 mm

Table 7. Assessment of the (index) finger orthotic models.

	Finger – Model 1	Finger – Model 2	Finger - Model 3
Printing time	12 min	14 min	31 min
Material cost	0.40 €	0.35 €	0.80 €
Weight	6 g	5 g	3 g
Max. stress	14 MPa	24 MPa	25 MPa
Max. Displacement	0.5 mm	0.6 mm	0.5 mm

A similar trend is shown in the cost analysis too. Indeed, in each medical application, the progressive lightening of the device led to a rise in material cost. The implication of the above consideration was that the manufacturing material cost saved through the volume reduction strategy (particularly that referring to topological optimization) was totally recovered by the use of support material (which for the same quantity has a higher purchase price than PLA). The worst-case scenario was, once again, the orthosis dedicated to the treatment of the index finger: cost of model 3 was twice as much as model 1, however, in absolute terms, the cost was moving from € 0.40 to € 0.80.

From a mechanical behavioral perspective, during the conducted FEA simulations, all the medical applications and each different design of medical application were considered compliant in terms of tensile strength and deformation. Regarding the maximum stress values recorded, the stress was always found to be less than the UTS of the manufacturing material, thus preventing failure during service. In terms of deformations, the highest displacement values were recorded in the case of the orthosis dedicated to the wrist joint care: these values were slightly more than 3 mm, but were still considered acceptable as they occur near the distal end of the device, thus, leaving the wrist area almost unaffected; displacements of the wrist region varied from 0.5 mm to

1.1 mm. Regarding the thumb and the index finger orthoses, the registered maximum displacements were definitely contained, thus making the proposed solutions viable.

Considering all the factors mentioned above, some observations emerged. First, production time was a significant variable in the evaluation process of the analyzed methodology. Production times were meaningful (especially for large volume applications) and prolonged by additional time variables not taken into account in the current study (e.g. acquisition and modelling times). Given the circumstances, it is unlikely to guarantee ready-to-use devices and, therefore, it would be necessary to adopt some measures to streamline the overall process time. While no particular advantage emerged from the time analysis, sources of competitiveness did emerge from the cost evaluation. Indeed, the estimated material cost for 3D production was significantly lower than the cost of material used in the production of conventional orthoses (LTT materials).

The strategy of reducing the volume of orthotic models with the aim of optimizing the production process did not achieve the desired results in terms of time and cost (except for the finger index orthosis model 2, whose output might be a considerable alternative). Even more dubious was the role of volume reduction implemented through topological optimization. Indeed, there was also evidence of an even greater increase in material cost and production time: the irregular shapes derived from the topological optimization simulation required an immoderate amount of support material, which drastically affected the material deposition process. In contrast, from the mechanical performance perspective, satisfactory stress and displacement values were ensured for moderate volume (and weight) reduction in both the strategies of volume reduction. Of course, topological optimization, whose inherent purpose is to define the best material distribution while satisfying a peculiar operating condition, allowed for even lighter solutions.

Certainly, through the illustrated methodology, it was possible to make extremely customized and lightweight geometries able, at the same time, to ensure adequate mechanical strength, which up to date it is not possible with traditional splinting techniques.

Conclusions

The present research evaluated the process of customized splinting for the treatment of upper limb pathologies through the utilization of advanced techniques, specifically Reverse Engineering (RE) and Additive Manufacturing (AM).

The manufacturing process of patient-specific orthoses, analyzed in the current study, was considered functional and straightforward. Some general observations may be drawn. Regarding the acquisition activity, although the scanning laser under consideration was highly performant, it is advisable to evaluate technologies that facilitate the scanning operation for the medical worker, even in the face of a reduction of the image quality. Inherent to the modelling activity, the software developed with the purpose of transforming the arm surface into an arm orthosis successfully designed also medical devices of various kinds, demonstrating the possibility of using a single application for modelling diverse anatomical areas. Optimization of the software algorithms would allow even more agile handling of meshes of different sizes. On the 3D printing side, there was an emerging need to optimize printing time, for instance using higher-performance 3D printers. In addition, post-printing surface finishing treatments of the devices should be provided too.

Two volume-reduction approaches were investigated by developing lightened orthotic models. Nevertheless, from the current product evaluation, the most competitive design in terms of cost, time and mechanical behavior resulted to be the standard full prototype (model 1). Thus, the role of the two emptying strategies found to be still uncertain: the costs and production times associated with the deployment of these approaches must be justified by the value and specificity of the considered application. Surely, the use of the topological optimization technique is a powerful tool, able to greatly lighten medical devices while ensuring good mechanical performances. However, in some cases the role of topological optimization compared to the volume reduction performed with the lightening command was questionable (e.g., in the case of model 3 of the wrist and thumb orthoses improved lightness was achieved and mechanical performance simulations suggested that lighter structures can be further realized), in some others it was clearly the option to be discarded (e.g., in the case of model 3 of the index finger orthosis, whose efforts would not bring additional value compared to model 2).

To conclude, the present study consisted in a preliminary analysis, evaluating few factors (such as production cost, time and mechanical behavior); future research needs to be conducted to provide a more exhaustive assessment.

References

[1] M. Javaid, A. Haleem, Additive manufacturing applications in orthopaedics: A review, J. Clin. Orthop. Trauma. 9 (2018) 202–206. https://doi.org/10.1016/J.JCOT.2018.04.008

[2] A. Abdudeen, J.E. Abu Qudeiri, A. Kareem, A.K. Valappil, Latest Developments and Insights of Orthopedic Implants in Biomaterials Using Additive Manufacturing Technologies, J. Manuf. Mater. Process. 2022, Vol. 6, Page 162. 6 (2022) 162. https://doi.org/10.3390/JMMP6060162

[3] Y.A. Jin, J. Plott, R. Chen, J. Wensman, A. Shih, Additive manufacturing of custom orthoses and prostheses - A review, Procedia CIRP. 36 (2015) 199–204. https://doi.org/10.1016/J.PROCIR.2015.02.125

[4] Y. Wang, Q. Tan, F. Pu, D. Boone, M. Zhang, A Review of the Application of Additive Manufacturing in Prosthetic and Orthotic Clinics from a Biomechanical Perspective, Engineering. 6 (2020) 1258–1266. https://doi.org/10.1016/J.ENG.2020.07.019

[5] I. Safaz, H. Türk, E. Yaşar, R. Alaca, F. Tok, I. Tuğcu, USE AND ABANDONMENT RATES OF ASSISTIVE DEVICES/ORTHOSES IN PATIENTS WITH STROKE -, Gulhane Med. J. 57 (2015) 142–144. https://doi.org/10.5455/GULHANE.152325

[6] A. Paoli, P. Neri, A. V. Razionale, F. Tamburrino, S. Barone, Sensor architectures and technologies for upper limb 3d surface reconstruction: A review, Sensors (Switzerland). 20 (2020) 1–33. https://doi.org/10.3390/S20226584

[7] F. Sala, M. Carminati, G. D'Urso, C. Giardini, A feasibility analysis of a 3D customized upper limb orthosis, Procedia CIRP. 110 (2022) 207–212. https://doi.org/10.1016/J.PROCIR.2022.06.038

[8] V. DeStefano, S. Khan, A. Tabada, Applications of PLA in modern medicine, Eng. Regen. 1 (2020) 76–87. https://doi.org/10.1016/J.ENGREG.2020.08.002

Italian Manufacturing Association Conference - XVI AITeM   Materials Research Forum LLC
Materials Research Proceedings 35 (2023) 111-117   https://doi.org/10.21741/9781644902714-14

Additive foam manufacturing

Luca Landolfi[1,a], Andrea Lorenzo Henri Sergio Detry[1,b], Daniele Tammaro[1,c],
Massimiliano Maria Villone[1,d], Pier Luca Maffettone[1,e] and Antonino Squillace[1,f]*

[1]University of Naples Federico II, Dept. of Chemical, Materials and Industrial Production
Engineering, Piazzale V. Tecchio 80, 80125 Naples (Italy)

[a]luca.landolfi@unibg.it, [b]andrea.detry@unina.it, [c]daniele.tammaro@unina.it,
[d]massimilianomaria.villone@unina.it, [e]pierluca.maffettone@unina.it, [f]squillac@unina.it

Keywords: Material Extrusion, Polymers, Design Optimization

Abstract. Traditional foams manufacturing processes are used to create quickly and cost-effectively high-strength and low-weight structures; lately there have been efforts to produce foams in Additive Manufacturing (AM), making it possible to produce free-form foamed structures. One of the biggest advantages of the additive foam manufacturing (AFM) technique proposed in this paper is the use of a physical blowing agent (PBA), which, as opposed to a chemical blowing agents (CBA), can be used to foam almost any thermoplastic polymer without modifying its chemical properties, resulting in a polymer life-cycle advantage, especially from the recycling point of view. The research being presented aims to investigate the effect of process parameters on the microstructure and properties of 3D printed physical foams, in terms of density, dimensions and mechanical properties.

Introduction

Polymer cellular materials are used in several applications and technological fields (e.g., biomedical, engineering, aerospace, nautical, sport and leisure), offering distinctive characteristics that derive from their cellular morphology and pores structures (i.e., dimension, orientation, density) [1]. Many examples are also present in nature where high performances are reached at minimum material cost using foams [2]. As a matter of fact, Nature has often chosen optimized cellular structures to shape life on our planet. The pores' characteristic size, shape, and organization are important factors in determining these materials' structure–property relationship. Natural cellular materials, such as echinoid and beeswax honeycombs [3, 4], are usually complex foamed structures, designed to carry out a specific task or optimize a specific property. The development of AM made it feasible to increase the design potentials of a component: generative design and lattice structures are only few of the instruments which allow to improve the performances of highly engineered parts for specific mechanical properties, with high strength-to-relative-density ratio. However, incorporating lattice structures into the design of an object can be challenging, as it requires careful consideration of multiple design parameters and introduces complexities in predicting the resulting mechanical behavior. [5]. Nowadays foams could be easily and cost-effectively produced by AM technology with control over micro and macro morphology thanks to an innovative foam AM (FAM) that has been developed by Tammaro et al. [6], by pre-treating the filament with a physical blowing agent (PBA).

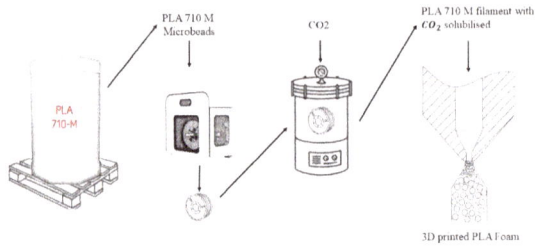

Figure 1 Schematic representation of the steps in a FAM process

During the fused filament deposition (FFM) the PBA expands creating bubbles in the polymer. In contrast with CBA, it can be used to foam almost any thermoplastic polymer without altering the chemical composition. This turns out in a polymer life-cycle advantage, especially from the point of view of its recycling. Nofar et al. [7] reviewed the influences of the main FAM parameters on the foam properties. FAM process consists of two phases: solubilization and extrusion. In the first phase, the solubilization, the polymeric filament is kept in a autoclave in which the PBA is insufflated by controlling the pressure and the temperature of the vessel, and the time of absorption. This allows the PBA to be solubilized within the polymer filament. After this phase, if the polymer is simply stored at room condition a partial desorption of the PBA can be induced: the longer is the desorption time the higher is the percentage of lost PBA. By carefully managing the two previous phases it is possible to have a designable gradient of PBA along the cross section of the filament. The second phase consists of the extrusion through a capillary (nozzle) with the desired exit diameter, controlling the temperature and the speed of the extrusion. [8]. During the extrusion phase, the polymer experiences a rapid pressure drop and a temperature rise from the inside to the outside of the nozzle. Due to this effect, the PBA expands and causes the polymer foaming. The rapid temperature rise and expansion allow the foamed polymer to crystallize.

Process Parameters

Each of the process parameters, reported in Table 1, should be investigated to map its importance on the microstructure and the mechanical behavior of the thermoplastic polymer foam produced.

Table 1 Process parameters and their unit

Variable	Description	Unit
P_a	Pressure of absorption	bar
t_a	Time of absorption	h
t_d	Time of desorption	h
T_e	Temperature of extrusion	°C
S_e	Speed of extrusion	mm/min
N_d	Nozzle diameter	mm

The percentage of PBA solubilized in the polymer could be easily set up by changing the P_a value: the higher the P_a and the higher the PBA in the polymer. t_a and t_d values control the gradient of the concentration of the PBA in the cross-section of the foamed strand. The larger the t_a value, the deeper the PBA solubilizes in the polymer, the lower the foam density and the more the bubbles will show up toward the core. On the other hand, as the t_d value increases, the PBA tends to desorb from the polymer and the number of bubbles on the surface reduces. In the extrusion phase the polymer filament, with the PBA solubilized inside, is pushed by a gear system in the extruder. As

Italian Manufacturing Association Conference - XVI AITeM   Materials Research Forum LLC
Materials Research Proceedings 35 (2023) 111-117   https://doi.org/10.21741/9781644902714-14

is represented in Figure 2 the extruder consists of two zones, a heated zone, referred to as the hot end, and a cooled area, referred to as the cold end.

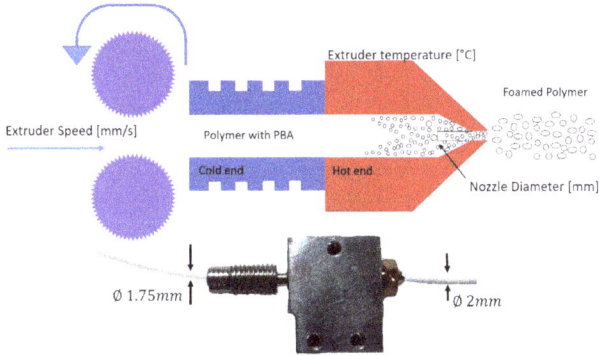

Figure 2 Additive Manufacturing of foams extruder: sketch (top) and real picture (bottom).

The hot end is a heated convergent conduit with a final capillary (nozzle) in which the polymer is melted. The cold end is the connection conduit from the gear system to the hot end, it is cooled usually with a fan system, in order to keep the temperature of the pushed filament constant and lower than the melting temperature of the polymer. In the present research, the influence of the temperature of the extruder T_e, the pushing speed of the gear system S_e, and the nozzle diameters N_d on the foam morphology and properties have been investigated. The foaming phenomenon is given to the equilibrium of two forces at the interface between the polymer and the PBA. First the PBA, previously solubilized in the polymer, tends to expand during heating. On the other hand, the polymer pushed in the nozzle from the gear system results in pressure. Inside the nozzle, the polymer pressure is predominant, and the PBA cannot expand. At the exit of the nozzle, the pressure of the polymer drops and the over-pressurized PBA quickly expands promoting the polymer foaming. The T_e performs a critical role in the viscosity of the polymer and affects the flow behaviour, as showed in [11]. Experimental studies have shown that the foam density decreases with an increase in temperature during extrusion. These relationships have been widely studied and reported in the literature on polymer extrusion and foaming. For example, [9] provide a comprehensive overview of the complex relationship between foam density and temperature. An increase in Se leads to a higher pressure of the polymer melt in the nozzle due to the increased shear rate and shear stress, consequently, the viscosity of the polymer melts decreases [10]. The geometry of the conduit N_d affects the speed, the pressure of the polymer, and the heat transfer from the nozzle to the polymer. Because the FAM is a multiphase extrusion of polymer and PBA, all the extrusion parameters have an influence on the expansion force of the PBA and on the interaction between the two. Since FAM involves the multiphase extrusion of polymer and PBA, all extrusion parameters can influence the expansion force of the PBA and the interaction between the two phases. Changes in the extrusion temperature, pressure, speed, and die geometry can alter the rheological properties of the polymer melt and the PBA, affecting the viscosities, and interfacial tensions this led to variations in the foam morphology, cell size, and density [12]. The process results are repeatable and controllable through a statistical experimental campaign, and mutual influences of process parameters on foam morphology can be studied.

Materials and methods

In this research it was chosen to validate the AM of thermoplastic polymer foams with a custom-made polylactic acid filament, to have control of the thermal history and the rheology of the filament. Each foamed strand was then characterized in terms of microstructure with SEM images, mechanical properties via mini-tensile tests and finally in terms of density and cross section size measurements. The polylactic acid (PLA) used in this research is the PLA 710 grade M of the Bewi Synbra bought as microbeads, whose characteristics are reported from the manufacturer [13]. The carbon dioxide pure at 99,95% from the Sol Group S.p.a was used as a blowing agent.

Table 2 Polymer properties

Description	Value	Unit	Note	Standard
Specific gravity	1,24	g/cc		D792
Flux index	6	g/10min	T=210°C; 2.16kg	D1238
Relative viscosity	1	g/10min	T=210°C; 2.16kg	D1238
Color			Transparent	
Melting Temperature	145-160	°C		D3418
Glass Transition Temperature	55-60	°C		D3418
Ultimate Strenght	60	MPa		D882
Young Modulus	3,6	MPa		D882

The strand densities were measured using the Gibertini Eternity Balance, which employs Archimedes' principle by comparing the weight in air and in water. Foamed strands were examined using a Hitachi High-Technologies Corporation TM3000 electronic microscope (SEM) to investigate the micromorphology of the foam. The specimens were prepared by cutting and flash-freezing with liquid nitrogen using an Astra Platinum blade. Metallization of the specimens was achieved using a K650X Sputter Coater from Quarum Technologies with gold as the filler material for surface conductivity. The mechanical behavior of each foamed strand was analyzed using micro-tensile tests conducted with a Deben Microtest 200N instrument. Custom jaws were designed and fabricated to securely grip the strand with epoxy resin during testing. In Figure 3, it is possible to see 3 cross section images made by scanning electron microscope of the strands realized at different extruder temperatures keeping constant the others process parameters. It can be noticed how, as the temperature increases, the bubbles tend to coalescence and escape as gas from the polymer. The strand realized with the lowest temperature has a final diameter of 2.6 mm and the one realized with the highest temperature has a diameter of 1.2mm. This put in evidence the need to accurately control the temperature during the printing to obtain a product with the desired microstructure.

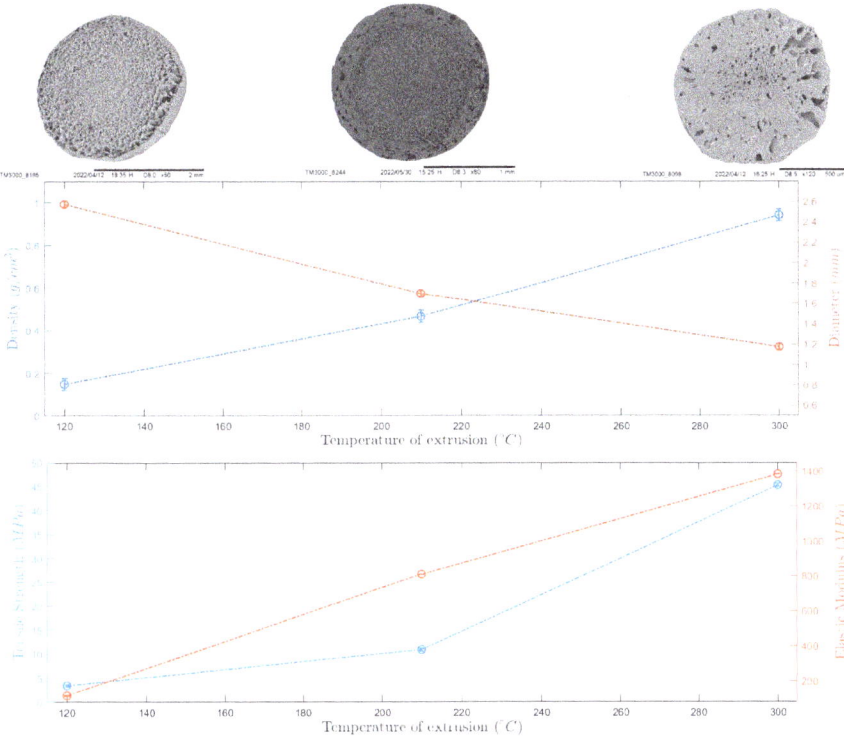

*Figure 3 Influence of the temperature on the foam properties. P_a=32.5bar; T_a=33h;
T_d=138h; N_d=0.7mm; S_e=675mm/min*

The final density and diameter of the foamed strand strictly depends on bubbles size and location, as reported in figure 3. In the same figure, tensile strength and elastic modulus of foamed filaments are reported versus extrusion temperature. To establish the degree of robustness of the process and obtain a response surface methodology, we conducted a design of experiment with replication of the tests at the central point. While we will not report all of the test replications in this paper, it is important to note that our work with the response surface methodology is being published separately in another paper, which will be authored by [Lepore et al.]

By doing a screening of foam production of various thermoplastic polymers, the feasibility of the process could be verified with a series of tests as example the properties and morphologies obtained for various polymers are shown in the following figure.

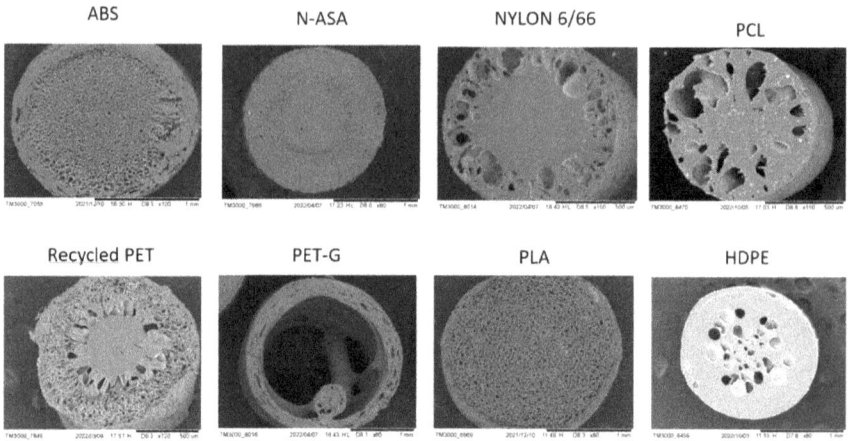

Figure 4 Different thermoplastic polymers foamed with FAM technology

Conclusions

In conclusion, this paper presented FAM as a novel method for 3D printing of foams. FAM has the distinct advantage of producing gradient foams with the highest level of free-form design allowed by additive manufacturing.

- FAM allows for the creation of complex structures with intricate internal geometries that would be difficult or impossible to achieve with traditional manufacturing methods.
- Precise control over printing parameters is necessary to produce high-quality objects with FAM. This includes the extrusion temperature, pressure, speed, and die geometry, which can significantly affect the expansion of the polymer and the resulting foam morphology.
- Future research in FAM should focus on further improving control over the printing parameters and developing new materials and printing techniques to expand the range of possible foam structures and properties.

Overall, FAM is a promising avenue for the production of foams with tailored structures and properties and has the potential to revolutionize the field of foam manufacturing.

References

[1] Lee, S-T., and Chul B. Park, eds. Foam extrusion: principles and practice. CRC press, 2014.

[2] Ambekar, R. S., Kushwaha, B., Sharma, P., Bosia, F., Fraldi, M., Pugno, N. M., & Tiwary, C. S. (2021). Topologically engineered 3D printed architectures with superior mechanical strength. Materials Today, 48, 72-94.

[3] Lakes, R. (1993). Materials with structural hierarchy. Nature, 361(6412), 511-515.

[4] Perricone V et al. 2022 Hexagonal Voronoi pattern detected in the microstructural design of the echinoid skeleton. J. R. Soc. Interface 19: 20220226.

[5] Kwang-Min Park, Kyung-Sung Min1and Young-Sook Roh, 2021, Design Optimization of Lattice Structures under Compression: Study of Unit Cell Types and Cell Arrangements

[6] Tammaro D, Detry ALHS, Landolfi L, et al. Bio-Lightweight Structures by 3D Foam Printing. 2021: 47-51.

[7] Nofar M, Utz J, Geis N, Altstädt V, Ruckdäschel H. Foam 3D Printing of Thermoplastics: A Symbiosis of Additive Manufacturing and Foaming Technology. Advanced Science 2022; 9(11): 2105701

[8] Bellini A, Gu ceri S, Bertoldi M. Liquefier Dynamics in Fused Deposition. Journal of Manufacturing Science and Engineering 2004; 126(2): 237-246.

[9] Behdani, B.; Senter, M.; Mason, L.; Leu, M.; Park, J. Numerical Study on the Temperature-Dependent Viscosity Effect on the Strand Shape in Extrusion-Based Additive Manufacturing. J. Manuf. Mater. Process. 2020, 4, 46. https://doi.org/10.3390/jmmp4020046

[10] Fan C, Wan C, Gao F, et al. Extrusion foaming of poly(ethylene terephthalate) with carbon dioxide based on rheology analysis. Journal of Cellular Plastics. 2016;52(3):277-298. https://doi.org/10.1177/0021955X14566085

[11] De Rosa, S.; Tammaro, D.; D'Avino, G. Experimental and Numerical Investigation of the Die Swell in 3D Printing Processes. Micromachines 2023, 14, 329. https://doi.org/10.3390/mi14020329

[12] Wong, S., Lee, J.W.S., Naguib, H.E. and Park, C.B. (2008), Effect of Processing Parameters on the Mechanical Properties of Injection Molded Thermoplastic Polyolefin (TPO) Cellular Foams. Macromol. Mater. Eng., 293: 605-613. https://doi.org/10.1002/mame.200700362

[13] https://bewi.com/products/biofoam/

Italian Manufacturing Association Conference - XVI AITeM   Materials Research Forum LLC
Materials Research Proceedings 35 (2023) 118-126   https://doi.org/10.21741/9781644902714-15

Exploiting laser-direct energy deposition for customized components: H13 and 316L functionally graded materials

Alessia Teresa Silvestri[1,a]*, Paolo Bosetti[2,b], Matteo Perini[3,c] and
Antonino Squillace[1,d]

[1] Department of Chemical, Materials and Production Engineering, University of Naples "Federico II", P.le Tecchio 80, 80125, Naples, Italy

[2] Department of Industrial Engineering, University of Trento, Via Sommarive 9, 38123 Trento, Italy

[3] PROM Facility, Via Fortunato Zeni 8, 38068 Rovereto, TN, Italy

[a]alessiateresa.silvestri@unina.it, [b]paolo.bosetti@unitn.it, [c]matteo.perini@trentinosviluppo.it, [d]antonino.squillace@unina.it

Keywords: Direct Energy Deposition (DED), Steel, FGM

Abstract. Among the Additive Manufacturing processes, one of the most intriguing technologies to produce customized components is Laser-Direct Energy Deposition (L-DED), which uses a focused laser beam to fuse metallic powders as they are deposited onto a substrate. L-DED can be used to manufacture new parts, apply coatings, repair damaged components, and produce items with tailored compositions and unique properties. The latter are called Functionally Graded Materials (FGMs), a new class of high-performing materials that is gaining great attention in both industry and academia due to their potential, for example, in the field of tooling, offering the possibility to repair damaged molds instead of replacing them. This research aims to enrich the landscape of L-DED production of FGM components with custom and tailored properties. H13 tool steel and 316L stainless steel are suitable for these purposes and are the materials used. Samples were designed, printed and characterized, obtaining satisfactory results.

Introduction

Functionally graded materials (FGMs) have garnered significant interest in recent years due to their unique properties and potential applications in a variety of fields, including aerospace engineering, nuclear power generation, sensors, biomedical implants, optoelectronic devices and energy absorption systems [1]–[3]. FGMs are a class of materials that exhibit graded variation in composition, microstructure, or properties along at least one direction, resulting in tailored properties. The idea arose from the observation of structures composed of graded components that exhibit graded properties that exceed those of the individual constituents and occur in nature, e.g., bone, teeth, wood [4]. One of the most promising technologies for the manufacturing of FGMs is Laser-directed energy deposition (L-DED), which uses a focused laser beam to fuse metallic powders as they are deposited onto a substrate [5], offering the ability to control the chemical composition layer by layer and, so, to create customized components [6].

In this paper, L-DED was used to produce different FGMs made of 316L stainless steel and H13 tool steel. Both materials are widely used in industry, such as in the die and mould industry, making them an attractive combination for FGMs [7].

In this work, functionally graded materials were designed, produced, and analyzed. Four kinds of FGM samples were deposited, with two different scanning strategies and varying percentages of feedstock materials, i.e., 316L and H13 powders. On one side, there are gradual changes in the chemical composition (hereinafter graded structure), on the other side, an alternation of the two materials layer by layer (hereinafter wafer structure). The effects of the variation of composition and deposition strategies were investigated on the microstructural and mechanical properties using optical microscopy, scanning electron microscopy, Vickers microhardness analysis, and three-point bending tests. Our results demonstrate the potential of L-DED to fabricate high-performance FGMs with tailored properties. The findings of this study could have implications for the design and manufacture of FGMs for various applications.

Materials and Methods

Materials. The feedstocks used are gas-atomized powders of hot-work tool steel H13 and 316L stainless steel (supplied by Sandvik Osprey Ltd., U.K), the chemical compositions of which are reported in Table 1. The substrate used is a 316L stainless steel plate with a dimension of 200x80x20 mm³. It was sandblasted and degreased with acetone before the deposition process.

Table 1. Chemical composition of feedstock and substrate (wt%).

Element	C	Cr	Mo	Mn	Si	V	Ni	Fe
Composition (H13)	0.4	5.2	1.6	0.5	1.0	1.2	-	Bal.
Composition (316L)	0.03	17.0	2.5	2.0	1.0	-	12.0	Bal.

Experimental campaign. For all laser deposition experiments, the DMG MORI LASERTEC 65 3D hybrid machine (LT 65 3D hybrid, DMG MORI AG, Pfronten, Germany) was used in collaboration with the ProM Facility Laboratory. The spot diameter was 3 mm and the focal length was 13 mm. High-purity argon gas was used for both carrier and shielding gas with flow rates of 5 l/min and 6 l/min, respectively. Blocks of functionally graded materials measuring 120x50x8 mm³ were deposited to extract samples of 116x10x6 mm³ and specimens for the material characterization (samples were extracted via electrical discharge machining, i.e., EDM). Blocks are composed of 10 layers, with 0.8 mm layer thickness. The parameters recommended by the manufacturer were applied for 316L. For H13, the main process parameters, i.e. the initial laser power, powder feed rate, scanning speed, and overlap, were determined based on previous studies on single and multi-track depositions and are listed in Table 2. These parameters were also used as initial values for the fabrication of functionally graded materials.

Table 2. Process Parameters used for FGMs depositions.

Laser Power [W]	Scanning speed [mm/min]	Powder Feed Rate [g/min]	Overlap [%]
2000	1200	15	60

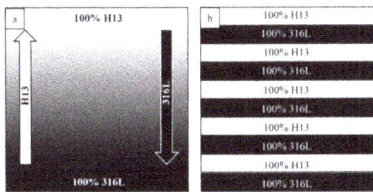

Fig. 1. Schematization of Functionally Graded Materials specimens: a) graded, b) wafer.

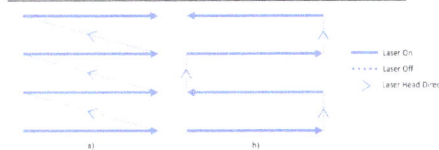

Fig. 2. Deposition strategies: a) Zig (or unidirectional); b) ZigZag (or bidirectional).

Laser power was reduced layer by layer until it reached 1200W. Two types of FGM samples were produced: Graded structure (G), characterized by gradual changes in the material composition, from 100% 316L to 100% H13 (scheme in Fig.1a), and Wafer structure (W), which had an alternating layer-by-layer pattern of 316L and H13 with the first layer being 316L (scheme in Fig. 1b). Two deposition strategies were used: zig (or unidirectional) and zigzag (or bidirectional), shown in Fig. 2a and 2b respectively.

Italian Manufacturing Association Conference - XVI AITeM   Materials Research Forum LLC
Materials Research Proceedings 35 (2023) 118-126   https://doi.org/10.21741/9781644902714-15

The variations in powder feed used throughout the experimental campaign are listed in Table 3, starting from the first layer and ending with the 10th layer.

Table 3. Powders percentages variation in Graded-Zig and Graded-ZigZag, Wafer-Zig and Wafer-ZigZag.

Layer no	Graded (Zig and ZigZag)		Wafer (Zig and ZigZag)	
	Powder Feed 316L	Powder Feed H13	Powder Feed 316L	Powder Feed H13
1	100%	0%	100%	0%
2	100%	0%	0%	100%
3	85%	15%	100%	0%
4	70%	30%	0%	100%
5	55%	45%	100%	0%
6	40%	60%	0%	100%
7	25%	75%	100%	0%
8	10%	90%	0%	100%
9	0%	100%	100%	0%
10	0%	100%	0%	100%

In the following, the samples will be referred to as GZ, GZZ, WZ, and WZZ, indicating Graded samples printed with Zig and ZigZag strategies, and Wafer samples printed with Zig and ZigZag strategies. To provide a comprehensive overview, the following figures illustrate the H13 single-track and multi-track depositions obtained with the parameters listed in Table 2. Fig. 3a shows the cross-section of a single track. It is characterized by a cladding angle greater than 100°, an Aspect Ratio greater than 5, a dilution of about 29%, and a microhardness of about 700 HV. Fig. 3b and Fig. 3c show the surface 3D images acquired by confocal microscopy of multi-track depositions printed, respectively, with a Zig and ZigZag strategy. These depositions exhibit the lowest waviness and defects content and high microhardness (650 HV) among a previous experimental campaign that involved 30, 45, 60, and 75% overlaps.

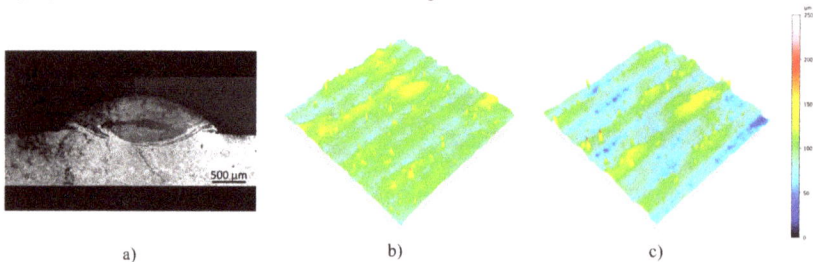

a)   b)   c)

Fig. 3. Singletrack and multi-tracks printed via L-DED: a) Cross-section of the single track, and 3D surface image of multi-tracks printed with the b)Zig and c)ZigZag strategies.

Characterization procedures. Samples were cut using EDM. The specimens were hot mounted and polished according to the metallographic procedure [8]. Microstructure characterization was performed by means of Optical Microscope and Scanning Electron Microscope (SEM) on

Italian Manufacturing Association Conference - XVI AITeM   Materials Research Forum LLC
Materials Research Proceedings 35 (2023) 118-126   https://doi.org/10.21741/9781644902714-15

metallographic cross-sections after chemical etching with Vilella's reagent. Microhardness Vickers measurements were conducted according to the standard [9]. Three-point bending tests were carried out at room temperature using a universal testing machine (Galdabini QUASAR 50, Galdabini SPA, Italy) with a load cell of 50 kN and support holders of 10 mm diameter. The cross-head speed was set at 1 mm/min. Three specimens of size 116x10x6 mm^3 were tested for each experimental condition. The support span was equal to 96 mm, respecting the 16:1 span-thickness ratio [10], with an over-span of 10 mm on each side.

Results and discussion

Metallurgical analysis. For Graded specimens, images showing the microstructure were acquired from the bottom to the upper area. The following are observations made by SEM of two FGM specimens both made by the same process parameters except for the scanning strategy, in the first case, the so-called Zig strategy was used, while in the other, the ZigZag strategy was used (Fig 4 and Fig 5). The grain morphology changes from the bottom to the top, from equiaxed grains of 316L to equiaxed grains of H13, through columnar/dendritic grains in the middle mixed layers. In the 316L layers, austenitic-ferrite microstructures are found, progressively replaced by austenitic-martensitic microstructure of the H13.

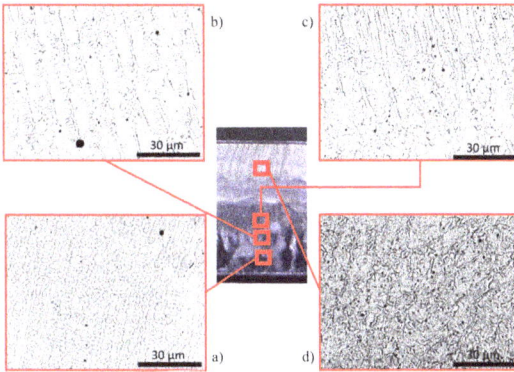

Fig. 4. Micrographs of Graded Zig sample: a)bottom (316L layer), b) and c) middle (316L-H13 mixed layers), and d) top (H13 layer).

Moreover, even though the overall morphology is the same, the grain size is different. The microstructures are finer in the Zig samples compared to the Zig Zag one. These results are consistent with the microhardnesses measured for the different samples, which are discussed in the next section.

The different grain sizes are related to the different cooling rates involved. Specifically, higher values of cooling rates led to a finer microstructure and lower cooling rates led to a coarser one [11]. In the case of the ZigZag strategy, the cooling rate is lower compared to the Zig one. This can be explained considering that a Zig-type scanning strategy is characterized by the deposition of material exclusively along straight trajectories that are all parallel; each segment of the scanning strategy in question is

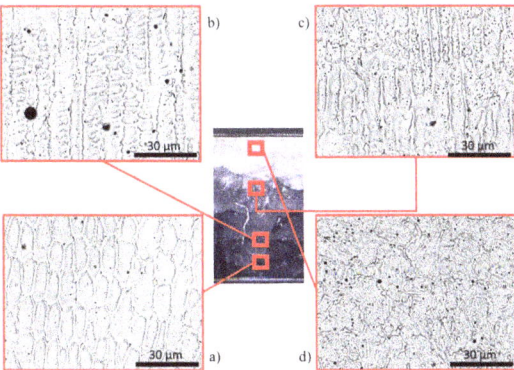

Fig. 5. Micrographs of Graded ZigZag sample: a)bottom (316L layer), b) and c) middle (316L-H13 mixed layers), and d) top (H13 layer).

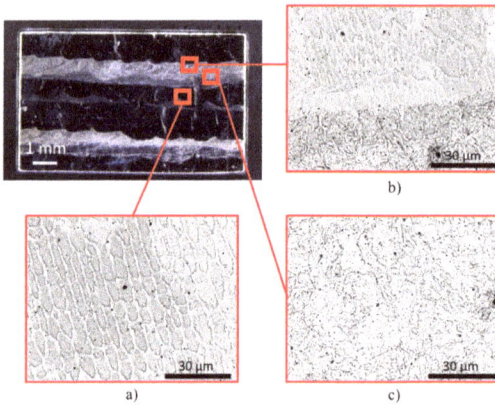

Fig. 6. Micrographs of Wafer sample: a)316L, b) interface between 316L and H13, c) H13.

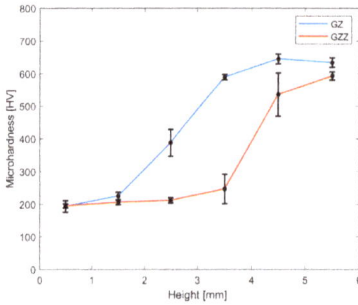

Fig. 7 Microhardness of Graded structures (G) versus the height of the sample, printed with the Zig (blue line) and the ZigZag (red line)

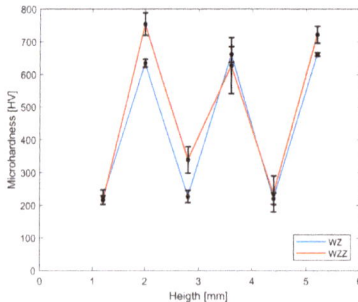

Fig. 8 Microhardness of Wafer structures (W) versus the height of the sample, printed with the Zig (blue line) and the ZigZag (red line) deposition strategies.

traversed twice by the laser head: the first time, the deposition of material and energy along the line takes place, then the head retraces the same section in the opposite direction with the laser off while continuing to insufflate gas that will cool the previously deposited trace, thus preventing the grains from enlarging due to the residual heat that is found in the volume of material even after it has solidified. Similar observations can be made for the Wafer samples (Fig. 6), with the difference that the transition from one material microstructure to another was not gradual. In Fig. 6b), the difference between the microstructures of 316L (up) and H13 (down) is clearly visible.

Microhardness analysis. A microhardness analysis was conducted on cross-sections of samples of the experimental campaign, and the mean values with the relative standard deviation are displayed in Fig. 7 and Fig. 8. The results for the graded samples (Fig. 7), both GZ and GZZ, show an increasing trend from the bottom to the top. Specifically, at the bottom, when the percentage of H13 is zero (in the first layer), the hardness of the 316L is about 200HV, a value higher than that of conventional wrought manufacturing 316L [12]. As the H13 percentage increases, the hardness also increases, reaching a very high value of approximately 700HV. Using the Zig deposition strategy, the outcome achieved is a gradual and continuous rise in hardness, also demonstrated by a central hardness value of approximately 450HV, which sits between the hardness values of 316L and H13. This occurs in the middle part of the sample, where there is 55% of 316L and 45% of H13. Concerning the difference in microhardness in the internal graded layers between the Zig and ZigZag deposition strategies, they are probably due to the

Italian Manufacturing Association Conference - XVI AITeM   Materials Research Forum LLC
Materials Research Proceedings 35 (2023) 118-126   https://doi.org/10.21741/9781644902714-15

repetitive passes of the heat source for the GZZ case, as explained before, that led to a major blending of H13 and 316L in the melting zone, due to the impact forces and to the Marangoni and Buoyancy effects, in successive layers in the ZigZag samples in comparison to the Zig cases, where the earlier solidification allow to respect the chemical composition defined according to Table 3. Fig. 8 shows the microhardness along with the height of WZ and WZZ, and it is possible to notice that the alternation of the layers is respected, resulting in an alternation of microhardness.

Three-point bending analysis. Six types of three-point bending conditions have been carried out, summarized in Table 4.

Table 4. Three-point bending tests conditions

Condition no	Samples	Condition characteristic	Acronyms
1	GZ	100% H13 layer as the outward lower surface	GZ-H13
2	GZ	100% 316L layer as the outward lower surfaces	GZ-316L
3	GZZ	100% H13 layer as the outward lower surface	GZZ-H13
4	GZZ	100% 316L layer as the outward lower surface	GZZ-316L
5	WZ	-	WZ
6	WZZ	-	WZZ

The results of the tests related to the Graded structures are displayed in Fig. 9. In the first condition, concerning GZ-H13 (continuous blue line, Fig. 9), the specimens show deformations without breaking up until displacements of 5 mm, the breaking load exceeds 2500N, reaching the first peak near 3000N. This is followed by brittle failure that propagates for most of the thickness of the specimen starting from the extrados surface and the force decreases below 500N. The crack propagation is blocked by encountering the closest layers to the intrados made of predominantly 316L steel. Therefore, the material shows an increase in fracture toughness which requires to increase the load further to advance the fracture, so a second peak load occurs. At this point, the failure of the specimen progresses in a ductile mood, between the layers mainly composed of 316L steel, which is capable of accumulating large plastic deformations. The same behaviour was also observed in samples printed with the ZigZag strategy, i.e., GZZ-H13 (yellow continuous, Fig. 9). The maximum force reaches about 3000N at a failure displacement of about 7 mm, which is also larger than the corresponding failure displacement for the GZ-H13 specimens, suggesting a higher fracture toughness in comparison with GZ-H13. Also in this case, the failure propagates in two steps. The first part is represented by a brittle failure that starts from the above-mentioned failure displacement and continues from the extrados surface through the layers with a predominant composition of H13, resulting in a reduction of the load from 3000N to 1200N. In the second part, characterized by ductile failure, the crack propagation slow down by encountering the layers near the intrados made of predominantly 316L steel.

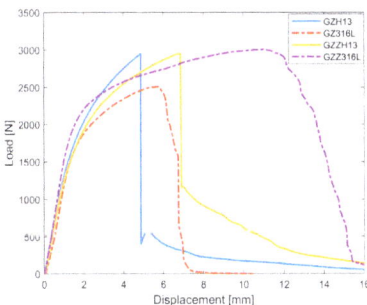

Fig. 9 Three- point bending tests: Load-displacement curves in Graded Structures.

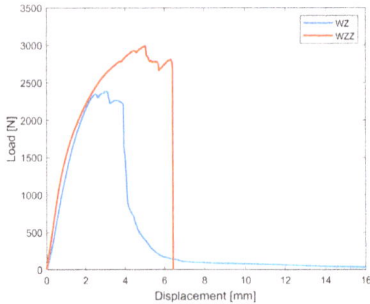

Fig.10 Three- point bending tests: Load-displacement curves in Wafer Structures.

A different mechanical behaviour was observed in specimens printed with the same two deposition strategies but with the 316L at the extrados surface (dotted lines, Fig. 9), where the maximum stresses act. As above mentioned, the 316L is a material able to deform more before breakage than steel H13, indeed, in these cases, a ductile fracture occurs, with large deformations that progress with the applied load. For Wafer samples, different considerations can be drawn with respect to the Graded ones. In the case of WZ (blue line, Fig. 10), the maximum force reaches 2200N (in general, it ranges from 2200 to 2600N, with a mean value of 2500N), with a related displacement of about 2 mm.

From this point, a blended failure mode occurs, it can be considered both brittle and ductile. In the range between 2 and 4 mm, there are failures of the successive layers that make up the volume of the specimen in which the propagation of the fracture is blocked when it encounters the interface between two layers. If the layer is mainly composed of 316L, small increases in the load are necessary to bring the fracture to expand to the following interface, if the layer is mainly composed of H13, there is a collapse of resistance to deformation of the specimen, and the force is sharply reduced. The same considerations can be made for WZZ samples, with the difference of a greater maximum load reached, i.e. 3000N. However, it is important to highlight that in all the conditions performed, despite the effects due to the shear, characteristic of the three-point bending tests and accentuated by the non-homogeneous characteristics of the tested material, the delamination

Fig. 11 Bending failure in FGMs produced via L-DED.

effects do not occur, the fracture propagates between the layers in the direction of the thickness of the specimen and not in the direction of the interface between the various layers of graded and, above all, sandwich items, as shown in Fig. 11, this suggests that there is a robust metallurgical bond at the interface between the layers, that acts as a novel single material.

Conclusions

In the present work, two types of Functionally Graded Materials samples were deposited by using Laser-Directed Energy Deposition, called Graded and Wafer, characterized by two different variations in the chemical composition, and printed with two deposition strategies, i.e., Zig and ZigZag. In the Graded components, there is a gradual decrease of 316L content and an increase of H13 from the bottom to the top. In the Wafer structures, layers made of 316L and H13 are alternated. All specimens were characterized and mechanically tested through microhardness, SEM, and three-point bending tests, and the following conclusions can be drawn:

i) It was demonstrated that the L-DED technology is a promising technology for manufacturing Graded and Wafer parts. The properties of the new 316L-H13 material obtained are compliant with the metallurgical and mechanical requirements. Indeed, after the correct balance of the most influent process parameters, it is possible to change the chemical composition to obtain tailored microstructure in different zones of the same

Italian Manufacturing Association Conference - XVI AITeM   Materials Research Forum LLC
Materials Research Proceedings 35 (2023) 118-126   https://doi.org/10.21741/9781644902714-15

          component, resulting in customized properties, and this is the nature of Functionally Graded Materials.

ii)   The microstructural evolution has involved the transition from austenitic 316L to H13, predominantly made up of lath martensite and austenitic grain boundaries, from the bottom to the top in the Graded structures and at the interface between layers in Wafer components. Microhardness analysis reflects the result previously obtained regarding microstructure: the hardness has gradually changed from about 200HV, a typical hardness of 316L, to about 650HV, the L-DED hardness value of H13, with intermediate values in the Graded structures' transition zone. In the Wafer structures, microhardness alternates the above-reported values. In Graded samples, higher hardness values were measured in cases where a Zig scanning strategy was adopted, in comparison to specimens obtained by adopting a ZigZag scanning strategy, a result consistent with the metallurgical analysis that showed a smaller grain size for specimens produced with a Zig strategy.

iii)   Finally, the above hybrid materials were tested by means of three-point bending tests. Despite the effects due to shear, which are characteristic of the three-point bending test and accentuated by the inhomogeneity characteristics of the tested material, the delamination effects do not occur, but rather the fracture propagates in the direction of the thickness of the specimen and does not in the direction of the interface between the layers, for both Wafer and Graded specimens. The FGMs performed brittle and ductile failures, depending on 316L and H13 in the extrados layers in the Graded components, and a mix of brittle and ductile failures in Wafer components.

References

[1]   F. Hengsbach et al., "Inline additively manufactured functionally graded multi-materials: microstructural and mechanical characterization of 316L parts with H13 layers," *Progress in Additive Manufacturing*, 2018. https://doi.org/10.1007/s40964-018-0044-4

[2]   A. Reichardt et al., "Advances in additive manufacturing of metal-based functionally graded materials," *International Materials Reviews*, 2021. https://doi.org/10.1080/09506608.2019.1709354

[3]   I. M. El-Galy, B. I. Saleh, and M. H. Ahmed, "Functionally graded materials classifications and development trends from industrial point of view," *SN Applied Sciences*. 2019. https://doi.org/10.1007/s42452-019-1413-4

[4]   W. Zhang, M. Soshi, and K. Yamazaki, "Development of an additive and subtractive hybrid manufacturing process planning strategy of planar surface for productivity and geometric accuracy," *International Journal of Advanced Manufacturing Technology*, vol. 109, no. 5–6, pp. 1479–1491, 2020. https://doi.org/10.1007/s00170-020-05733-9

[5]   I. Gibson, D. Rosen, and B. Stucker, "Directed Energy Deposition Processes," 2015, pp. 245–268. https://doi.org/10.1007/978-1-4939-2113-3_10

[6]   A. T. Silvestri, S. Amirabdollahian, M. Perini, P. Bosetti, and A. Squillace, "Direct Laser Deposition for Tailored Structure," *ESAFORM 2021*, 2021. https://doi.org/10.25518/esaform21.4124

[7]   M. Ostolaza, J. I. Arrizubieta, A. Lamikiz, and M. Cortina, "Functionally graded AISI 316L and AISI H13 manufactured by L-DED for die and mould applications," *Applied Sciences (Switzerland)*, 2021. https://doi.org/10.3390/app11020771

[8]   T. Materials, ASM Handbook, Volume 9, Metallography and Microstructures. 2004.

[9] ASTM E384 - 17, "Standard Test Method for Microindentation Hardness of Materials, ASTM International, West Conshohocken, PA," *Book of ASTM Standards*, vol. 03.01, 2017.

[10] ASTM INTERNATIONAL, "Standard Test Methods for Flexural Properties of Unreinforced and Reinforced Plastics and Electrical Insulating Materials. D790," *Annual Book of ASTM Standards*, pp. 1–12, 2002 https://doi.org/ 10.1520/D0790-17.2

[11] T. DebRoy *et al.*, "Additive manufacturing of metallic components – Process, structure and properties," *Progress in Materials Science*, vol. 92. 2018. https://doi.org/10.1016/j.pmatsci.2017.10.001

[12] G. Roberts, G. Krauss, and R. Kennedy, "Tool Steels: 5th Edition," *Book*, 1998.

Italian Manufacturing Association Conference - XVI AITeM
Materials Research Proceedings 35 (2023) 127-134

Materials Research Forum LLC
https://doi.org/10.21741/9781644902714-16

LPBF process of Zn-modified NiTi alloy with enhanced antibacterial response

Carlo Alberto Biffi[1,a*], Jacopo Fiocchi[1,b], Francesca Sisto[2,c], Chiara Bregoli[1,d], Ausonio Tuissi[1,e]

[1] National Research Council of Italy - Institute of Condensed Matter Chemistry and Technologies for Energy, Via Previati 1E, 23900 Lecco, Italy

[2] University of Milan, Department of Biomedical, Surgical and Dental Sciences, Milan, Italy

[a]carloalberto.biffi@cnr.it, [b]jacopo.fiocchi@cnr.it, [c]francesca.sisto@unimi.it, [d]chiara.bregoli@icmate.cnr.it, [e]ausonio.tuissi@cnr.it

Keywords: Additive Manufacturing, Laser Processes, Shape Memory Alloy

Abstract. In this work the use of Laser Powder Bed Fusion (LPBF) process enabled the development of customized implants with advanced functional materials for the biomedical sector. In details, Ni rich NiTi Shape Memory Alloy (SMA) powder, mixed with Zn powder, were used for building samples, able to couple the typical superelasticity of the initial material with the antibacterial response, offered by the dopant element. The main parameters of the LPBF process were revised for achieving full dense parts. Further, the functional performances of the novel NiTiZn SMA were analyzed and compared with the ones of the reference material. Finally, the antibacterial response against different bacteria was tested. It was found a promising antibacterial response against Staphylococcus aureus of the NiTiZn SMA, while the addition of Zn into NiTi did not supress the martensitic transformation of the NiTi alloy and allowed to maintain the superelastic recovery.

Introduction

Shape Memory Alloys (SMAs) are smart and functional materials, which are well known for their unique thermo-mechanical properties, namely shape memory effect (SME) and superelasticity (SE) [1]. Among SMAs, the near-equiatomic intermetallic NiTi compound exhibits optimal SME and SE characteristics and it finds extensive application fields, particularly the biomedical one [2]. Ni-rich NiTi alloys offer some suitable properties for developing of implantable elements, including good biocompatibility, adjustable elastic modulus and the ability of supporting early ingrowth of bone [2,3]. However, a dangerous issue that may trigger the failure of metallic implants, including the use of NiTi alloys, regards the occurrence of post-surgery bacterial infections [4]. This makes the development of antibacterial implants highly auspicious in the view of limiting the occurrence of post-surgery infections troubles. The choice of bioactive surface modifications of NiTi implants have been developed, which often exploit the inherent antibacterial properties of some inorganic elements, such as Ag, Cu, Zn and Ga [5]. Among these elements, Ag is the most investigated as coating in the form of nanoparticles on NiTi surfaces [6-8], whereas Zn seems to be highly promising, since it couples excellent osteogenic ability and favourable antibacterial response [9-10].

All these opportunities in the use of advanced and novel materials need to be well connected with the best choice of manufacturing process routes. Nowadays, bone implants can be manufactured through Additive Manufacturing (AM) technology, which lead to high degree of shape complexity and customization, very important for improving the implant performance and to treat better and faster the patience issue. The use of AM for manufacturing NiTi SMAs have been largely investigated in literature. Several works demonstrated the feasibility of Ti rich and Ni

Italian Manufacturing Association Conference - XVI AITeM   Materials Research Forum LLC
Materials Research Proceedings 35 (2023) 127-134   https://doi.org/10.21741/9781644902714-16

rich NiTi alloys and the investigation regarding the evolution of their functional behaviour with respect to the conventional method of manufacturing. In particular, among AM technologies the most challenging one for processing NiTi powders is Laser Powder Bed Fusion (LPBF), in which a laser beam is adopted for promoting a rapid melting of the powder bed. Probably the most critical aspect to be mentioned for the LPBF process of these smart alloys regards the requirement of precise control of both the chemical composition (i.e. Ni/Ti ratio) and the obtained microstructure: these have a huge impact on the functional behaviour [11]. Therefore, the modification of the composition of 3D built parts for tuning the functional properties is a very challenging topic but just few works have been done in this area.

Under this light, the present work is oriented to investigate the effect of the addition of Zn into Ni rich NiTi SMA powder on the LPBF processability and on the functional and antibacterial response of the built samples.

Experimental

Materials

Two types of powders, obtained from Ni56Ti46 (wt.%) and pure Zn powders (see Figure 1), were used for the LPBF process: (i) Ni rich NiTi powder; and (ii) NiTi mixed with 2% in weight of Zn, from here named as NiTiZn. Such mixture was prepared with a powder mixer working for 2 hours under argon atmosphere; its chemical homogeneity was checked by compositional analysis.

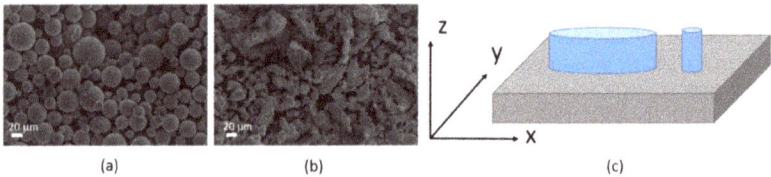

Figure 1: NiTi (a) and Zn (b) powders; schematic depicting the built samples (c).

LPBF process

Samples with NiTi and NiTiZn powders were manufactured by means of a LPBF system (mod. AM400 from Renishaw), equipped with a pulsed wave (PW) laser with a maximum power of 400 W and a reduced build volume (75 mm x 75 mm x 50 mm). The schematic of the PW emission mode is depicted in Figure 2, where the principal energetic and spatial features are reported.

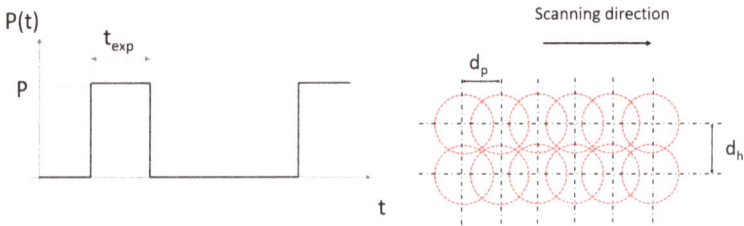

Figure 2: Schematic of the temporal power profile and spatial pulses path (b) in SLM performed with PW emission mode [14].

Italian Manufacturing Association Conference - XVI AITeM   Materials Research Forum LLC
Materials Research Proceedings 35 (2023) 127-134   https://doi.org/10.21741/9781644902714-16

LPBF process conditions, which are reported in Table 1, were investigated for studying the feasibility of the NiTiZn powder, having the aim of maximizing the relative density. The same process conditions were also investigated for printing NiTi powder, as reference. A full factorial design was adopted at varying both laser power and exposure time, as suggested in Table 1; three replicas for each process condition were carried out for estimating the variability of the LPBF process. Cylindrical samples (3 mm in diameter and 5 mm in height) were realized for density measurements, metallographic analyses and compressive testing. Moreover, disc samples (10 mm in diameter and 3 mm thick) for antibacterial testing were also produced on the NiTi platform, as show in Figure 1c.

Table 1: Variable and fixed process parameters used for printing NiTi and NiTiZn samples.

Parameters		Values
Variable parameters	Power	150-160-170 W
	Exposure time	75-100-125 μs
Fixed parameters	Scanning strategy	Meander
	Atmosphere	Argon
	Layer thickness	30 μm
	Hatch distance	50 μm
	Point distance	50 μm
	Laser spot size	65 μm
	Platform temperature	30°C
	Tilting angle	67°
	Oxygen level	< 20 ppm

Density and metallurgical characterization
Part relative density was measured by Archimedes's principle using a Gibertini E50S2 precision digital balance for each cylindrical sample, considering a full density of 6.45 g/cm3.

X-ray micro Computed Tomography (μ-CT) was performed on selected cylindrical samples, having the highest relative density value for each powder batch. A XTH225 –ST system, from Nikon, having an X-ray Gun of 225kV and 16 bits flat panel Varex 4343CT as detector, was adopted to highlight the defects within the entire volume of the sample.

Martensitic transformation temperatures of the two alloys were investigated by differential scanning calorimetry (DSC, TA Instrument Q25) at 10 °C/min heating/cooling scan rate. Compositional analysis was carried out by Scanning Electron Microscopy (SEM, mod. LEO), coupled with energy Dispersive X-ray Spectroscopy (EDS) on the external surface and the polished surface, respectively. Compressive tests were conducted at 25°C by means of an MTS Exceed E45 machine, equipped with extensometer, at strain rate of 0.01 min-1.

Antibacterial characterization
Finally, antibacterial testing was carried out on the built discs, by agar slurry according with the standard protocol [15]. Staphylococcus aureus strain (Gram positive, ATCC 29213) and Pseudomonas aeruginosa (Gram negative, ATCC 19660), were used: these bacteria were selected because they are the main causes of prosthetic infections. Three replicas of the test were performed and studied to ensure the repeatability of the study. The bacteria were cultured on agar media and incubated at 37°C for 24 hours. Then, an inoculum of 0.5 McFarland (about 1.5 x 108 Colony Forming Unit/ml - CFU/ml) was prepared in broth and further diluted to obtain a suspension of about 105 CFU/ml in agar slurry medium [16].

Results

LPBF processability

During the laser scan of the powder bed, the energy provided to the powder bed during the LPBF process, namely also fluence, is described, as follow:

$$F = \frac{P * t_{on}}{d_p * d_h * s}$$

where P is the laser power, ton is the exposure time, dp is the point distance, dh is the hatch spacing, and s is the layer thickness.

The typical trend of the relative density versus the laser fluence can be recognized for both the powders' batches, as shown in Figure 3. In details, the relative density exhibits a rapid increase at low laser fluence values, then it can reach its maximum point, in correspondence of the full densification of the powder; upon another increase of the laser fluence, a decrease of the relative density can be found. In details, for NiTi-Zn powder, as low relative density as 95.5% was achieved from 33 J/mm3; a rapid increase was observed up to 99.6% at 44 J/mm3, then also a rapid decrement of the relative density was appreciated suddenly from 98.3% at 47 J/mm3 down to 95% at 62 J/mm3 (see Figure 3a). For NiTi powder, the relative density increased from 98.3% at 33-35 J/mm3 up to 99.4% at 38 J/mm3, then it decreased slowly down to 99% at 47 J/mm3 and quickly down to 96% at 60 J/mm3 (see Figure 3b). It can be stated that the Zn powder shape was irregular, but it did not influence the sample quality during the LPBF process; in fact, the presence of Zn did not counter the high relative density in the building of NiTiZn parts.

The curves reported in Figure 3 represent the main trend on the measurements, and they can be used as a support for the discussion of the achieved results.

Moreover, the laser fluence for producing full dense NiTi-Zn samples is higher than the one for full densification of NiTi powder; this can be explained because of the higher energy required by the complete melting of NiTi and the vaporization of Zn achieved once the NiTi is melted. In fact, the compositional analysis, carried out with SEM-EDS, revealed that no evident traces of Zn were found in the NiTiZn sample. On the contrary, the external surfaces of the sample showed the presence of Zn, probably just partially melted by the laser beam scan, because of different process parameters adopted for the realization of the sample border.

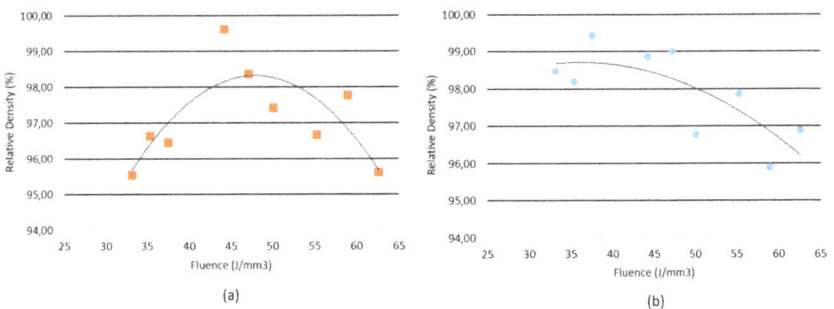

Figure 3: Relative density vs fluence curves for NiTiZn (a) and NiTi (b) powders.

The defects analysis was completed with micro-CT scans of the NiTiZn and NiTi samples. Figure 4 shows the 3D reconstruction of the two samples, in which the detectable defects are highlighted with different colors in function of their size. It can be seen that the NiTiZn sample

Italian Manufacturing Association Conference - XVI AITeM   Materials Research Forum LLC
Materials Research Proceedings 35 (2023) 127-134   https://doi.org/10.21741/9781644902714-16

exhibited limited amount of defects, just few as large as 300 μm around and characterized by pretty irregular shape: this suggests that the Zn evaporation may induce this type of defects. On the contrary, the NiTi sample showed high number of defects having smaller size and spherical.

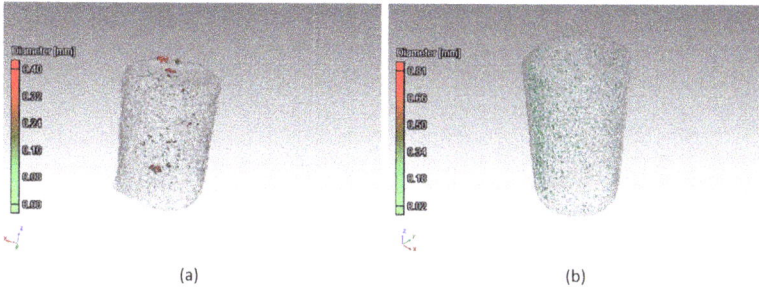

(a)   (b)

Figure 4: Micro-CT analysis of the NiTiZn (a) and NiTi (b) samples, built with the selected process conditions.

Both the built samples show the typical phase transformation of the SMAs [17]. Figure 5 depicted the DSC scans where peaks upon heating and cooling were visible. The transformation temperature of the martensitic transformation suggested that both the samples are austenitic at body temperature (i.e. 37°C), indicating that these can show the SE. This result can also indicate that the Ni/Ti ratio was almost not varied by the scanning of the laser beam, avoiding a sensible change of the chemical composition. Additionally to the DSC of the built samples, the one of the initial powder was also analyzed. Here it can be seen that the shape of the peaks of the phase transformation are pretty different: the powder shows a multistage phase transformation, while single stage for the built samples. This could depend on lack of chemical and microstructural uniformity (change of Ni/Ti ratio, residual stresses, change in grain size). Anyways, the transformation temperatures are not slightly change from the power to the built samples, again.

The SE behaviour was tested on the samples, subjected to a low temperature heat treatment, also namely aging, typically carried out on SMAs for promoting the functional performances. In this case, compression testing was carried out at 30°C; each mechanical test was composed by a loading stage up to 6% as maximum strain, followed by an unloading down to 0%. The results are depicted in Figure 6.

It can be clearly seen that the two samples exhibited largely different mechanical behaviour. In particular, the NiTiZn can reach as low stress as 450 MPa at 6% in strain, while the residual strain upon unloading was 1%. This indicates that a strain recovery of 5% was achieved.

Italian Manufacturing Association Conference - XVI AITeM

Materials Research Forum LLC

Materials Research Proceedings 35 (2023) 127-134

https://doi.org/10.21741/9781644902714-16

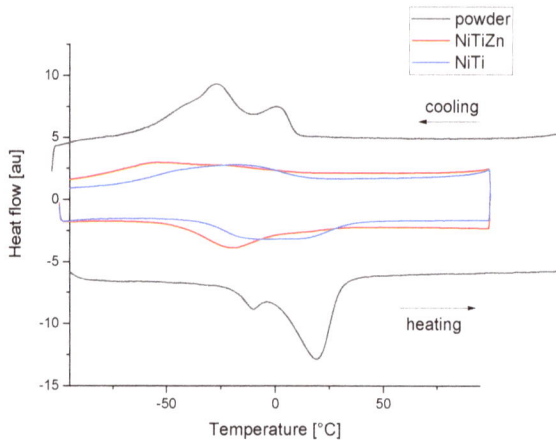

Figure 5: DSC scans of the NiTiZn and NiTi built samples, and the NiTi powder.

On the contrary, the NiTi sample appeared to be less SE, due to a limited strain recovery up to 3%. Moreover, higher maximum stress was achieved up to 1000 MPa for NiTi, while 450 MPa in the case of NiTiZn. These results suggest a higher performance of the NiTiZn sample than the undoped NiTi one.

Finally, the effect of the Zn presence was tested with antibacterial testing. Figure 7 shows the evolution of the bacteria population, Pseudomonas aeruginosa and Staphylococcus aureus, from the beginning of the test (time zero, T0) until 6 hours (T6).

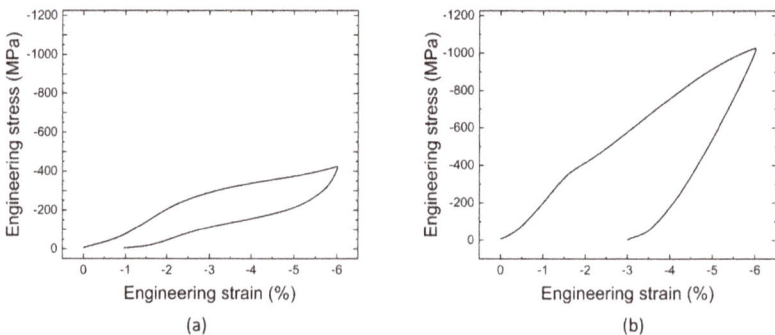

Figure 6: Stress-strain curves of NiTiZn (a) and NiTi (b) built samples, subjected to heat treatment.

It can be seen that the NiTi sample promoted a proliferation of the Pseudomonas aeruginosa bacterium after 6 hours, while the bacterium concentration was constant after the same time (see Figure 7a). On the contrary, the presence of Zn could alter largely the action against the Staphylococcus aureus bacterium after 6 hours, as shown from Figure 7b [17]. Like the previous

Italian Manufacturing Association Conference - XVI AITeM
Materials Research Proceedings 35 (2023) 127-134

Materials Research Forum LLC
https://doi.org/10.21741/9781644902714-16

case, undoped NiTi sample promoted a large proliferation of the Staphylococcus aureus after 6 hours, while its population was decreased on the surface of the NiTiZn sample. This result suggest an evident antibacterial action of the Zn with respect to Staphylococcus aureus; at the second stage, it is also clear that no bacteria growth was allowed n the case of Pseudomonas aeruginosa in presence of Zn.

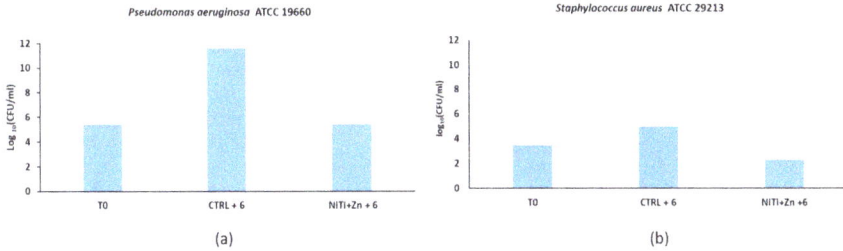

Figure 7: Antibacterial response of the NiTiZn samples, compared with the control (the reference was NiTi) within 6 hours with pseudomonas aeruginosa (a) and Staphylococcus aureus (b).

Conclusions

Laser Powder Bed Fusion process confirmed to be a suitable Additive Manufacturing technology for producing full dense or near full dense NiTi Shape Memory Alloy parts with functional properties. The addition of Zn into NiTi powder allowed to achieve high relative density values, even though its vaporization takes place during the scanning of the powder bed with the laser beam. However, the presence of Zn powder on the border allowed to promote an antibacterial behaviour with higher performances with respect to the undoped NiTi alloy. The operating temperatures were comparable in the samples built with NiTiZn and NiTi powders, while an improvement of the superelasticity was appreciated in the NiTiZn samples then in the NiTi samples. It can be concluded that Zn can be potentially used as antibacterial element to be added to NiTi powder for manufacturing advanced implants. The type of bacterium showed different antibacterial response of the NiTi-Zn alloy; this effect suggests that further investigations are required for better understanding of the biological response of this novel functional alloy for realizing advanced permanent implants.

Acknowledgements

The authors would like to acknowledge Massimo Arosio and Vincenza Ragone for their precious discussions. The authors would like to thank Mayur Nitin Jagdale from Politecnico di Milano, Nicola Bennato and Enrico Bassani from CNR ICMATE for their support in the experimental activity.

References

[1] Funakubo, H.: Shape Memory Alloys, Gordon and Breach Science Publishers, Amsterdam, (1987).

[2] S.A. Shabalovskaya, Biomed. Mater. Eng. 12 (2002) 69-109.

[3] J.X. Xiong,, Y.C. Li, X.J. Wang, P.D. Hodgson, C.E. Wen, J. Mech. Behav. Biomed. Mater. I (2008) 269-273. https://doi.org/10.1016/j.jmbbm.2007.09.003

[4] S. Thomas, Y. Grohens, N. Ninan, in: A.R. Unnithan, R.S. Arathyram, C.S. Kim (Eds.), Scaffolds With Antibacterial Properties, Elsevier Inc., New York (2015), pp. 103-120. https://doi.org/10.1016/B978-0-323-32889-0.00007-8

[5] Erlin Zhang, Xiaotong Zhao, Jiali Hu, Ruoxian Wang, Shan Fu, Gaowu Qin,Antibacterial metals and alloys for potential biomedical implants,Bioactive Materials,Volume 6, Issue 8,2021,2569-2612. https://doi.org/10.1016/j.bioactmat.2021.01.030

[6] M. Saugo, D.O. Flamini, L.I. Brugnoni, S.B. Saidman, Silver deposition on polypyrrole films electrosynthesised onto Nitinol alloy. Corrosion protection and antibacterial activity, Materials Science and Engineering: C, Volume 56, 2015, 95-103, ISSN 0928-4931, https://doi.org/10.1016/j.msec.2015.06.014

[7] Yongkui Yin, Ying Li, Xu Zhao, Wei Cai, Jiehe Sui, One-step fabrication of Ag@Polydopamine film modified NiTi alloy with strong antibacterial property and enhanced anticorrosion performance, Surface and Coatings Technology, Volume 380, 2019, 125013, ISSN 0257-8972. https://doi.org/10.1016/j.surfcoat.2019.125013

[8] Pipattanachat, S., Qin, J., Rokaya, D. et al. Biofilm inhibition and bactericidal activity of NiTi alloy coated with graphene oxide/silver nanoparticles via electrophoretic deposition. Sci Rep 11, 14008 (2021). https://doi.org/10.1038/s41598-021-92340-7

[9] J. Ye, B. Li, M. Li, Y. Zheng, S. Wu, Y. Han, Acta Biomater 107 (2020) 313-324. https://doi.org/10.1016/j.actbio.2020.02.036

[10] K. Yusa, O. Yamamoto, M. Iino, H. Takano, M. Fukuda, Z. Qiao, T. Sugiyama, Arch. Oral Biol. 71 (2016) 162-169. https://doi.org/10.1016/j.archoralbio.2016.07.010

[11] Bassani, P.; Fiocchi, J.; Tuissi, A.; Biffi, C.A. Investigation of the Effect of Laser Fluence on Microstructure and Martensitic Transformation for Realizing Functionally Graded NiTi Shape Memory Alloy via Laser Powder Bed Fusion. Appl. Sci. 2023, 13, 882. https://doi.org/10.3390/app13020882

[12] Farber, E., Zhu, J. N., Popovich, A., & Popovich, V. (2020). A review of NiTi shape memory alloy as a smart material produced by additive manufacturing. Materials Today: Proceedings, 30, 761-767. https://doi.org/10.1016/j.matpr.2020.01.563

[13] Safaei, K., Abedi, H., Nematollahi, M., Kordizadeh, F., Dabbaghi, H., Bayati, P., Reza Javanbakht, Ahmadreza Jahadakbar, Mohammad Elahinia Poorganji, B. (2021). Additive manufacturing of NiTi shape memory alloy for biomedical applications: review of the LPBF process ecosystem. Jom, 73, pages3771-3786 (2021). https://doi.org/10.1007/s11837-021-04937-y

[14] A.G. Demir, P. Colombo, B. Previtali, From pulsed to continuous wave emission in SLM with contemporary fiber laser sources: effect of temporal and spatial pulse overlap in part quality, Int. J. Adv. Manuf. Technol. (2017) 1-14. https://doi.org/10.1007/s00170-016-9948-7

[15] Mahalakshmi S, Hema N, Vijaya P.P., In Vitro Biocompatibility and Antimicrobial activities of Zinc Oxide Nanoparticles (ZnO NPs) Prepared by Chemical and Green Synthetic Route- A Comparative Study. BioNanoScience (2020) 10:112-121 https://doi.org/10.1007/s12668-019-00698-w

[16] ASTM E2180-07 (2012): Test method for evaluation of the activity of antimicrobial agent in polymeric or hydrophobic material.

[17] Biffi, C. A., Fiocchi, J., Sisto, F., Bregoli, C., & Tuissi, A. (2023). Enhanced antibacterial response in Zn-modified additively manufactured NiTi alloy. Materials Letters, 335, 133749. https://doi.org/10.1016/j.matlet.2022.133749

Italian Manufacturing Association Conference - XVI AITeM
Materials Research Proceedings 35 (2023) 135-142

Materials Research Forum LLC
https://doi.org/10.21741/9781644902714-17

Effect of layer and raster orientation on bending properties of 17-4 PH printed via material extrusion additive manufacturing technology

Alessandro Pellegrini[1,a] *, Maria Grazia Guerra[1,b,] Fulvio Lavecchia[1,c] and Luigi Maria Galantucci[1,d]

[1] Dipartimento di Meccanica Matematica e Management, Politecnico di Bari, Bari, Italy

[a]alessandro.pellegrini@poliba.it, [b]mariagrazia.guerra@poliba.it, [c]fulvio.lavecchia@poliba.it, [d]luigimaria.galantucci@poliba.it

Keywords: Material Extrusion, Stainless steel, Debinding and Sintering, Bending property

Abstract. Material Extrusion (MEX) is one of the most popular Additive Manufacturing technologies. Over the years, the material portfolio has expanded and nowadays, it covers metals such as stainless steels, copper and titanium alloys. The mechanical behaviour of metal parts realized by MEX is of great interest to understand both the potentialities and the limits of the technology. In the present work, a commercial filament of 17-4 PH stainless steel was used as feedstock material to realize four groups of bending specimens obtained by varying the printing direction and the infill line strategy. The main goal of the paper was to evaluate the effect of the above-mentioned factors on the flexural properties. With this purpose, a three-points bending test was performed and results were analysed using the one-way ANOVA approach. The density of the parts was also evaluated.

Introduction

The main Additive Manufacturing (AM) methods to realize metal components are referred to the Laser Powder Bed Fusion (L-PBF) and Directed Energy Deposition (DED). These technologies are energy-intensive, time-consuming, and require high investment costs. Industrial-ready binder-based AM technologies as Material Extrusion Additive Manufacturing (MEX) and Binder Jetting (BJ) have been starting to come an economic alternative to the powder-based technologies [1,2]. The extrusion-based processes allow to avoid raw material loss during the process and to avoid risks for human health due to the release of respirable small particles, because the metal powder is embedded in a filament [3]. Moreover, a lower initial investment for the equipment is required [4]. Metal MEX is a hybrid technology based on the combination of the traditional MEX for polymers and Metal Injection Molding (MIM). From MIM, it inherited the feedstock, which is a mixture of a polymeric binder and metal powder, and the two subsequent phases, named Debinding and Sintering (D&S), for polymer removal and powder sintering, fundamental to obtain a full metal part. After the printing step, the obtained part is defined as green part. This part is a mixture of thermoplastic polymer and metal powder. After the debinding, the part is called brown part and after the last phase, the sintered metal part is obtained [5]. The feedstock, in form of a filament, is composed by a high content of metal powder (from 55 to 90 wt.%) and a polymeric matrix. This latter is constituted by three different components: a main binder (i.e., Polyoxymethylene (POM)), a backbone binder as Polypropylene (PP) and also additives like stearic acid [6]. In literature, the main metallic materials investigated for MEX are stainless steels 316L [4] and 17-4 PH [7] and, more recently, copper [8] and titanium alloys [9] are being investigated.

The entire process chain including the printing, debinding and sintering strongly influences the mechanical performance of parts realized by extrusion-based processes. The difficulty in achieving uniform particle distribution and strong adhesion between the metal powders and polymer matrix [2], the elevated anisotropy due to the printing orientation [10], the high porosity due to presence

Italian Manufacturing Association Conference - XVI AITeM   Materials Research Forum LLC
Materials Research Proceedings 35 (2023) 135-142   https://doi.org/10.21741/9781644902714-17

of voids occurred during the printing process [11], and the sintering parameters [7] has been highlighted in literature as the main problems of this technology.

Considering the mechanical performance, tensile properties has been extensively investigated [4,10], while the bending properties and the influence of process parameters have been not fully investigated in literature. More in details, Carminati et al. [2] and Thompson et al. [12] tested specimens with previously optimized parameters in order to obtain the best results of flexural stress (σ_f) and deflection. Gonzalez Gutierrez et al. [8] realized bending specimens made of pure copper with different infill percentages and infill patterns. Suwanpreecha and Manonukul [13] and Henry et al. [14] studied the bending proprieties of 17-4 PH parts realized by using the Atomic Diffusion Additive Manufacturing (ADAM) technology. Due to the limitations of a closed software architecture, such as the Markforged one, the comparison was limited to the consideration of different printing orientations. Moreover, in [13] a comparison between as-printed specimens and as-sintered ones was performed.

Referring to the previous works and to the knowledge of the authors, in this work, a commercial 17-4 PH filament was used to manufacture four groups of bending test specimens with a rectangular cross-section, varying the printing direction and the infill line strategy. Two printing directions, flat (XY) and upright (ZX) were selected and two unidirectional infill line strategy (raster direction) were used, 0° and 90°. Three repetitions were realized for each group of specimens. Thus, three-point bending tests were executed on the as-sintered specimens and the data were collected to obtain the average and standard deviation of the investigated flexural properties. The impact of the printing direction and infill line strategies on the flexural properties and density was then evaluated using the one-way ANOVA statistical approach.

Material and Methods
The material used for the manufacture of the specimens was the BASF Ultrafuse® 17-4 PH (BASF 3D Printing Solutions GmbH, Germany). The feedstock is a mixture of a high content of 17-4 PH powder (≈ 90 wt.%) and a blend of POM+PP. A consumer 3D printer (Henan Creatbot Technology Limited, China) with a ruby nozzle tip of 0.6 mm of diameter was used to manufacture the specimens. The printing parameters, previously optimized, are shown in Table 1 .

Table 1 Printing parameters

Parameters	Value
Infill density (%)	100
Infill line strategy (°)	0 or 90
Wall lines (n°)	2
Printing speed (mm/s)	30
Layer height (mm)	0.15
Flow (%)	120
Nozzle temperature (°C)	260
Bed temperature (°C)	100

To avoid the warpage of the specimens Magigoo® Pro Metal glue stick was adopted. The bending specimens were oversized to compensate the shrinkage that occur after D&S. As reported by the BASF guidelines, the expected shrinkage is 16% for X and Y axes and 20% for Z axis. The dimensions of the sintered parts were reported in Fig. 1a. Fig.1b showed the directions of the infill lines. For a comparison with the literature, two specimens with an alternate infill line of +/-45°, according to the two printing orientation, were printed and sintered

Italian Manufacturing Association Conference - XVI AITeM
Materials Research Proceedings 35 (2023) 135-142

Materials Research Forum LLC
https://doi.org/10.21741/9781644902714-17

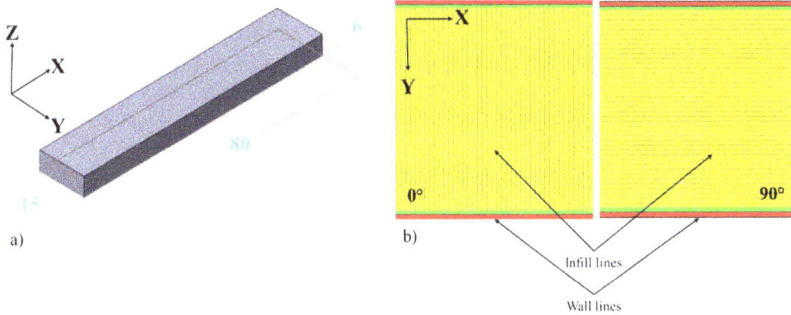

Fig. 1 *a) Dimension of the sintered parts. 2D Scheme of the configuration of the two kinds of infill lines strategy*

Once printed, the specimens' dimensions, such as length (l), width (w) and thickness (t), were measured with using a 3D structured light scanner with a resolution of 0.04 mm and a 100x lens (GOM ATOS Q, Zeiss Corp., Germany). scanner to acquire the main dimensions as length (l), width (w) and thickness (t). The 3D model of the entire specimen (comprising bottom and top faces) was obtained, and it was used for the measurement of volume in the density analysis.

The D&S process was performed according to the manufacturer guidelines (Table 2) and all specimens were debound and sintered on XY plan (Fig. 2).

Table 2 *Debinding and Sintering parameters for BASF Ultrafuse® 17-4 PH*

Phase	Temperature	Holding time	Atmosphere
Debinding	120 °C	Up to 10.5% of weight loss is reached	Nitric acid
Sintering	From room temperature to 600 °C (5°C/min)	1 hours	Hydrogen
	From 600 °C to 1300 °C (5°C/min)	3 hours	

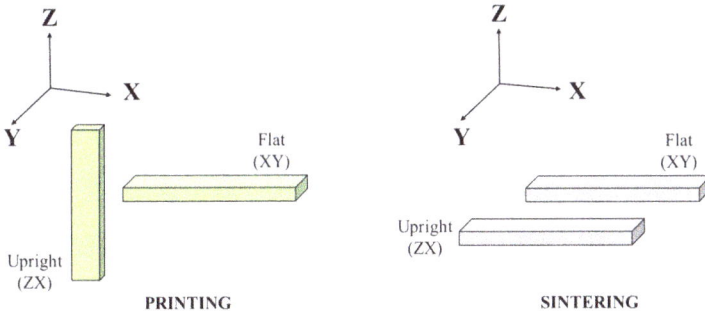

Fig. 2 *Configuration to explain the direction of printing and sintering*

The universal testing machine, Galdabini SUN 10, with a load cell of 100 kN was used for the three-point bending test. A cross head speed of 1 mm/min was selected and a span length of 30 mm was adopted according to the ISO 7438. The contact area between the nose of the punch and the specimens was the last layer (i.e., top layer) for the flat group, instead for the upright specimens was the wide external surface (**Fig. 2**, the plan XY in the "Sintering" configuration).

Italian Manufacturing Association Conference - XVI AITeM
Materials Research Forum LLC
Materials Research Proceedings 35 (2023) 135-142
https://doi.org/10.21741/9781644902714-17

Results and Discussions

Density evaluation

Starting from the obtained 3D model and using the GOM Inspect software, the volume of each specimen was evaluated, while the mass was measured using a precision balance. The ratio between these two parameters (mass/volume) enable to obtain the density both for the green parts and sintered parts. In Table 3 the green and sintered density were reported. For a comparison, other two specimens called XY-45 and ZX-45 were added for comparison.

Table 3 Density values for specimens in green and sintered condition

Group	Green density (g/cm^3)	Sintered density (g/cm^3)
XY-0	4.31±0.02	7.12±0.04
XY-90	4.27±0.02	6.87±0.10
XY-45	4.29	7.08
ZX-0	4.35±0.01	7.13±0.02
ZX-90	4.47±0.01	7.37±0.01
ZX-45	4.40	7.23

From a first analysis, the densities evaluated in each condition did not differ significantly. In the as-printed condition, the average value was between 4.3-4.5 g/cm^3. In the as-sintered condition, instead, except for the XY-90 group, the density was above 7.0 g/cm^3. As a general comment, the upright specimens were denser than flat ones. Using the Pareto chart (Fig. 3a-b), it was possible study deeper this topic. In **Fig. 3a-b** was confirmed the influence of the printing direction both on the green parts and sintered parts, followed by the interactions between printing direction and infill line strategy. Less relevant the effect of the infill line strategy on the sintered density, with respect to the green one, where it was appeared influencing. The difference between the results obtained green and sintered density can be related to the sintering process. After sintering the shrinkage of the parts could form some defects not present in the green condition as delamination between wall lines and infill lines [8], as observed in the following section, or some defect already existent in the green condition as the printing voids induced by the extrusion process [10]. All these defects negatively influence the density and the mechanical performance of the sintered parts.

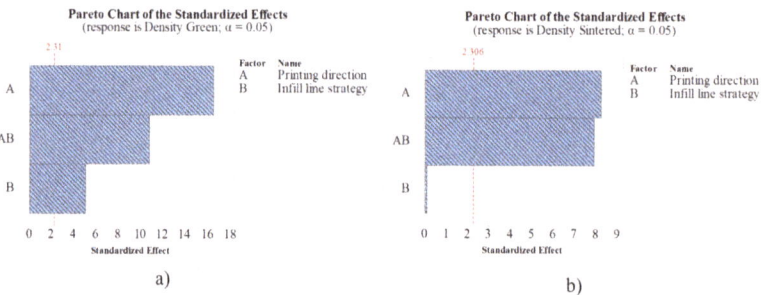

Fig. 3 Pareto chart for a) Green Density b) Sintered Density

The **Table 4** was reported as summary for the parameters investigated (green density and sintered density) with the different p-value obtained from the ANOVA.

Table 4 Summary table of the p-value for each parameter investigated.

Term	p-value Green density	p-value Sintered density
Printing direction	0.000	0.000
Infill line strategy	0.000	0.916
Printing direction*Infill line strategy	0.001	0.000

Italian Manufacturing Association Conference - XVI AITeM   Materials Research Forum LLC
Materials Research Proceedings 35 (2023) 135-142   https://doi.org/10.21741/9781644902714-17

Bending properties

The data obtained from the bending test were the load (F) and the stroke of crosshead. Using Eq.2 it was possible derive the flexural stress (σ_f). The thickness and width used for the calculation of σ_f were derived from the analyses with GOM Inspect, instead the span length (s_l) was defined by the standard ISO 7438 equal to 30 mm.

$$\sigma_f = \frac{3Fs_l}{2wt^2} \tag{2}$$

a)
b)

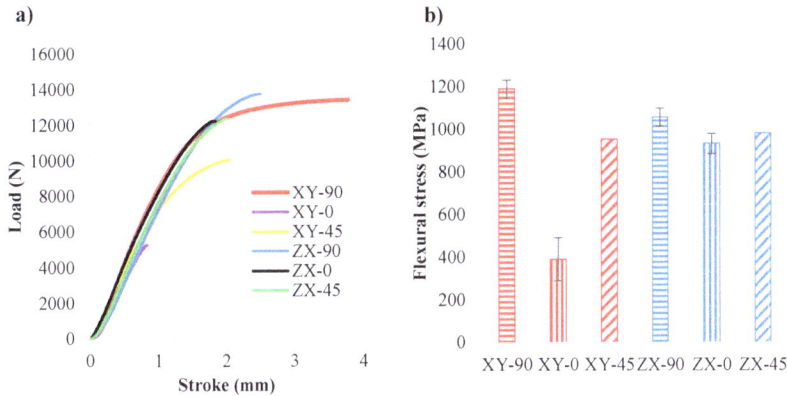

Fig. 4 *a) Curves load-stroke b) Comparison of the average maximum flexural stress and their standard deviation*

In Fig. 4a were reported the comparison of the average maximum σ_f obtained from the tests. In Fig. 4b was selected and reported the curves load-stroke of the specimens that with the highest load for each group. Fig. 4b showed how XY-90 was the better configuration, with a maximum flexural stress of 1188.3±41.7 MPa, followed by ZX-90 with a σ_f of 1056.3±41.6 MPa. The weakest group, as was the XY-0 (390.1±100.4 MPa). This was due to the orientation of the raster that negatively affected the specimens causing a delamination. All the other groups, independently from printing direction, reported a σ_f above the 950 MPa. The flat specimens did not break at the end of the test but reported a crack at the bottom surface where the tensile force acted (Fig. 5a-b). Differently for the upright, where the end of the test occurred when the specimens broke with a delamination of the layers (Fig. 5c-d). More evident the differences between XY-0 (a) and XY-90 (b), with the first one undeformed-like compared to the second one. The presence of lacks adhesion between infill and wall lines for ZX-0 specimen (c) were well showed. This defect could be the main cause to a lower flexural stress compared to the ZX-90 (d), where the cross-section appeared denser, also confirmed by the values of density in Table 3 (7.4 g/cm^3 > 7.1 g/cm^3).

Fig. 5 *Bended specimens: a) XY-0 b) XY-90 and fracture surfaces of c) ZX-0 d) ZX-90*

More in detail the effect of the printing direction and infill line strategy on the bending properties a statistical analysis was investigated. The one-way ANOVA approach was considered using a confident interval of 95% ($\alpha=0.05$).

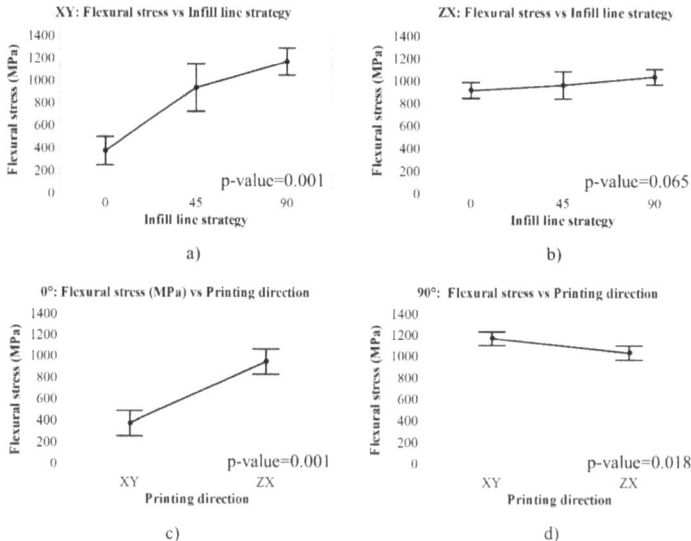

Fig. 6 *Boxplot for a) XY specimens b) ZX specimens of the maximum flexural stress for the different infill line strategy and c) 0° direction and d) 90° direction of the maximum flexural stress for the printing direction*

Italian Manufacturing Association Conference - XVI AITeM   Materials Research Forum LLC
Materials Research Proceedings 35 (2023) 135-142   https://doi.org/10.21741/9781644902714-17

In Fig. 6a and Fig. 6b were showed the trend of flexural stress for the infill line strategies. The specimens +/-45° were insert for comparison, even if they were a single repetition. In Fig. 6a the trend reported for flat specimens highlighted how varying the infill line strategy from unidirectional 0° to +/-45° up to unidirectional 90° has led to increase the bending stress of about 3 times. This was supported by the results of one-way ANOVA, where the p-value was lower than 0.05. On the other hand, the specimens printed in upright (Fig. 6b), were not influenced by the infill line strategy, and confirmed by a p-value bigger than 0.05. In Fig. 6c and Fig. 6d were showed the boxplot of flexural stress for the two printing direction. Fig. 6c, showed how considering 0° line direction, the upright specimens obtained the highest values (σ_f of ZX≃2.5 times the σ_f of XY). The p-value ($0.001 \ll \alpha$) suggested how the printing orientation significantly affected the flexural stress for this type of infill line strategy. The specimens printed with an infill line strategy of 90°, appeared more similar as confirmed by the boxplot of Fig. 6d. However, the one-way ANOVA confirmed also in this case an influence of the printing orientation on the flexural stress.

Conclusions

In the present work, a commercial 17-4 PH filament was used to produce four kinds of bending specimens in order to evaluate the influence of the variation of printing direction and infill line strategy on the density of the printed and sintered part and on flexural stress of the sintered parts. The results obtained were reported below:

- The densities in the green and sintered condition of the parts with a unidirectional infill line strategy were lower than +/-45° specimens. The Pareto chart confirmed the influence of the printing direction on the green and sintered density. The sintered density was influenced by defects induced by the sintering process.
- The bending test revealed a significant influence of the parameters on the flexural stress. The best group was the XY-90 with 1188.3±41.7 MPa of flexural stress, instead the worst one was the XY-0 with 390.1±100.4 MPa. For the upright specimens not relevant differences, as confirmed by the one-way ANOVA for the infill line, emerged with a σ_f of 1056.3±41.6 MPa for ZX-90 and 932.5±47.0 MPa for ZX-0. The printing direction also influenced the bending stress, mainly when the infill line strategy 0° was considered.

References

[1] P. Parenti, D. Puccio, B.M. Colosimo, Q. Semeraro, A new solution for assessing the printability of 17-4 PH gyroids produced via extrusion-based metal AM, J. Manuf. Process. 74 (2022) 557–572. http://doi.org/10.1016/j.jmapro.2021.12.043

[2] M. Carminati, M. Quarto, G. D'urso, C. Giardini, G. Maccarini, Mechanical Characterization of AISI 316L Samples Printed Using Material Extrusion, Appl. Sci. 12 (2022). http://doi.org/10.3390/app12031433

[3] W. Lengauer, I. Duretek, M. Fürst, V. Schwarz, J. Gonzalez-gutierrez, S. Schuschnigg, C. Kukla, M. Kitzmantel, E. Neubauer, C. Lieberwirth, V. Morrison, Fabrication and properties of extrusion-based 3D-printed hardmetal and cermet components, Int. J. Refract. Metals Hard Mater. 82 (2019) 141–149. http://doi.org/10.1016/j.ijrmhm.2019.04.011

[4] M. Sadaf, M. Bragaglia, F. Nanni, A simple route for additive manufacturing of 316L stainless steel via Fused Filament Fabrication, J. Manuf. Process. 67 (2021) 141–150. http://doi.org/10.1016/j.jmapro.2021.04.055

[5] L.M. Galantucci, A. Pellegrini, M.G. Guerra, F. Lavecchia, 3D printing of parts using Metal Extrusion: an overview of Shaping Debinding and Sintering technology, Adv. Technol. Mater. 47 (2022) 25–32. http://doi.org/10.24867/ATM-2022-1-005

[6] J. Gonzalez-Gutierrez, S. Cano, S. Schuschnigg, C. Kukla, J. Sapkota, C. Holzer, Additive manufacturing of metallic and ceramic components by the material extrusion of highly-filled polymers: A review and future perspectives, Materials (Basel). 11 (2018). http://doi.org/10.3390/ma11050840

[7] F. Lavecchia, A. Pellegrini, L.M. Galantucci, Comparative study on the properties of 17-4 PH stainless steel parts made by metal fused filament fabrication process and atomic diffusion additive manufacturing, Rapid Prototyp. J. 29 (2023) 393–407. http://doi.org/10.1108/RPJ-12-2021-0350

[8] J. Gonzalez-gutierrez, S. Cano, J.V. Ecker, M. Kitzmantel, F. Arbeiter, C. Kukla, C. Holzer, Bending Properties of Lightweight Copper Specimens with Different Infill Patterns Produced by Material Extrusion Additive Manufacturing , Solvent Debinding and Sintering, Appl. Sci. (2021). http://doi.org/10.3390/app11167262

[9] P. Singh, V.K. Balla, S. V. Atre, R.M. German, K.H. Kate, Factors affecting properties of Ti-6Al-4V alloy additive manufactured by metal fused filament fabrication, Powder Technol. 386 (2021) 9–19. http://doi.org/10.1016/j.powtec.2021.03.026

[10] M.Á. Caminero, A. Romero, J.M. Chacón, P.J. Núñez, E. García-Plaza, G.P. Rodríguez, Additive manufacturing of 316L stainless-steel structures using fused filament fabrication technology: mechanical and geometric properties, Rapid Prototyp. J. 27 (2021) 583–591. http://doi.org/10.1108/RPJ-06-2020-0120

[11] G. Singh, J.M. Missiaen, D. Bouvard, J.M. Chaix, Copper additive manufacturing using MIM feedstock: adjustment of printing, debinding, and sintering parameters for processing dense and defectless parts, Int. J. Adv. Manuf. Technol. 115 (2021) 449–462. http://doi.org/10.1007/s00170-021-07188-y

[12] Y. Thompson, J. Gonzalez-Gutierrez, C. Kukla, P. Felfer, Fused filament fabrication, debinding and sintering as a low cost additive manufacturing method of 316L stainless steel, Addit. Manuf. 30 (2019) 100861. doi:10.1016/j.addma.2019.100861.

[13] C. Suwanpreecha, A. Manonukul, On the build orientation effect in as-printed and as-sintered bending properties of 17-4PH alloy fabricated by metal fused filament fabrication, Rapid Prototyp. J. 28 (2022) 1076–1085. http://doi.org/10.1108/RPJ-07-2021-0174

[14] T.C. Henry, M.A. Morales, D.P. Cole, C.M. Shumeyko, J.C. Riddick, Mechanical behavior of 17-4 PH stainless steel processed by atomic diffusion additive manufacturing, Int. J. Adv. Manuf. Technol. 114 (2021) 2103–2114. http://doi.org/10.1007/s00170-021-06785-1

Italian Manufacturing Association Conference - XVI AITeM   Materials Research Forum LLC
Materials Research Proceedings 35 (2023) 143-153   https://doi.org/10.21741/9781644902714-18

Non-Newtonian, non-isothermal three-dimensional modeling of strand deposition in screw-based material extrusion

Alessio Pricci[1,2 a *], Gianluca Percoco[2,3 b]

[1] Department of Electrical and Information Engineering (DEI), Polytechnic University of Bari, Via E. Orabona 4, 70125 Bari, Italy

[2] Interdisciplinary Additive Manufacturing (IAM) Lab, Polytechnic University of Bari, Viale del Turismo 8, 74100 Taranto, Italy

[3] Department of Mechanics, Mathematics and Management (DMMM), Polytechnic University of Bari, Via E. Orabona 4, 70125 Bari, Italy

[a]alessio.pricci@poliba.it, [b]gianluca.percoco@poliba.it

Keywords: Additive Manufacturing, Pellet Extrusion, Virtual Modeling

Abstract. Material extrusion (MEX) is one of the most widespread additive manufacturing techniques. Among the MEX processes, pellet additive manufacturing (PAM) is of primary interest in industry 4.0 scenario, mainly because of the lower unit cost, energy consumption and waste production, together with the wider range of printable materials. Mechanical properties are related to the intra and inter layer bonding, which in turn depends on the strand geometry. For the first time, the relationship between PAM processing parameters and layer morphology has been studied by means of non-Newtonian, non-isothermal three-dimensional numerical simulations; the influence on mass flow rate and strand shape has been investigated. A very good correspondence between experiments and numerical computations of layer shape was found. Thermal contact area increases at lower layer heights, but counterpressure limits the extruded mass flow rate. This effect can be mitigated by choosing higher barrel temperatures and screw speed.

Introduction

In internet of things era, the production has become much more flexible, efficient and automated than in the past. In the framework of the upcoming industry 4.0 revolution (4IR), additive manufacturing (AM) represents one of the key pillars [1–3].

The most beneficial aspects of AM for 4IR are the reduction of wastes, energy requested [4] and prototyping times [5], together with an increase in product customization, the change of supply chain and its management [6], and the possibility to digitalize the manufacturing activity, creating a digital twin of the process [7].

Among material extrusion (MEX) techniques, both fused filament fabrication (FFF) and pellet additive manufacturing (PAM) are of primary interest. The first consists in the gradual melting of thermoplastic material, given in the form of a solid filament, and its deposition on a build plate. Instead, in PAM a screw extruder melts and convey a pelletized feedstock [8]. Some of the main applications are in automotive [9], tooling [10], renewable energies [11], and buildings [12].

In recent years, computational fluid dynamics (CFD) has bene established as a very promising way to gain a better insight of MEX processes [13–28].

The first studies were dedicated to FFF; the main works are briefly described, because they are the basis for PAM modeling. In [13] the strand morphology has been analysed at varying layer height and printing speed; this work was experimentally validated in [14]. The numerical investigation was extended in [15,16] to study the corner MEX according to different strategies. Moreover, in [17,18] the interlayer contact between deposited strand has been analysed, yet disregarding the filament coalescence; in addition, the influence of temperature has been neglected. A different approach for single strand deposition was proposed in [19], where an improvement of the results of [13] is found by considering energy equation. The successive layer deposition has been simulated in [20,21], to predict the cooling time and inter-layer contact area. In [22] both single and multi-layer deposition in FFF, together with a study of the solidification process, are presented under the assumption of considering a Newtonian fluid.

More recently, CFD has proved as a powerful tool to study PAM process. Some of the main perspectives are the prediction of final part properties, manufactured both with net and composite materials, and the individuation of the optimal process parameters [23].

In [24] the strand deposition for different rheological models was simulated: the most important aspect which affects strand morphology in PAM is the inclusion of the shear-thinning behaviour in simulations. In [25] the effect of successive layer deposition has been studied with a remeshing technique performed via the Comsol-MATLAB LiveLink [26]. Despite the authors propose a two-dimensional simulation, it gives important insights in layer coalescence, through non-isothermal simulations. Moreover, a good agreement with optical micrographs is shown. Thermodynamic aspects related to reheating of the deposited strand caused by the heated nozzle have been addressed in [27]; here, heat transfer has been coupled with fluid flow equations and studied numerically. Infrared thermography provided a very good correspondence with FEM predictions. Up to now, the three-dimensional investigation of the strand deposition has not been done, when dealing with PAM, even if one considers the deposition of net polymers.

In this work the strand deposition in PAM extrusion has been investigated with both experiments and CFD simulations. A Direct 3D pellet extruder has been used to print consistent layers under different values of the most critical printing parameters, that are the dimensionless printing speed (V^*), layer height (D^*), and nozzle temperature (T^*).

CFD analyses have been performed by means of non-Newtonian, non-isothermal three-dimensional simulations; the main aim was to model the effect of the abovementioned processing parameters on strand morphologies.

At first, the material properties were introduced in section "Material". The experimental setup has been detailed in section "Experimental validation", where the values assigned to the process parameters have been stated. In section "Numerical investigations" full details about the multiphase simulations are given. The effect of process parameters on extruded mass flow rate is investigated in section "Mass flow rate"; then, the effect of dimensionless parameters on strand cross-section has been explored in section "Strand morphology". Finally, conclusions and further works have been outlined.

Materials and methods
In this section, the relevant material's properties have been introduced. Then, full details on experimental and CFD investigations have been given.

Material.
In this study, a particular grade of polylactic acid (PLA) manufactured by NatureWorks (NatureWorks Ingeo 3251D) has been used to investigate the first layer deposition on a heated build plate in PAM extrusion. The material is initially given in the form of almost spherical pellet of around 3 [mm] diameter.
The parameters involved in CFD simulations are:
- Rheological properties
  - Dynamic viscosity, which is a function of both temperature and shear rate, $\eta(T; \dot{\gamma})$, and
  - Glass transition temperature, T_g.

Italian Manufacturing Association Conference - XVI AITeM
Materials Research Proceedings 35 (2023) 143-153

Materials Research Forum LLC
https://doi.org/10.21741/9781644902714-18

- Thermophysical properties
  - Density at solid (ρ_s) and molten (ρ_m) states,
  - thermal conductivity, $\lambda(T)$, and
  - heat capacity $c_p(T)$,

Temperature-dependent behavior of abovementioned properties was modeled with Moldflow software (Moldflow Plastics Labs. Ithaca, NY 14850, USA).

Experimental investigation.
A Direct 3D PAM extruder (Direct 3D s.r.l.) has been employed to study the first layer deposition under a wide set of process parameters.

The extruder consists of a constant pitch screw placed inside a heated barrel. At first, the pelletized material is conveyed in a hopper. Then, it is gradually heated above T_g. Finally, the molten polymer is extruded through a nozzle, to be deposited layer-by-layer, as in FFF.

The mechanical torque is provided by a Nema17 HS4401 stepper motor, to be delivered to the vertical screw by means of a timing belt.

The pressure starts to rise in the solid-conveying zone, up to the last screw vanes. Then, it drops in the nozzle and deposited layer, up to the atmospheric value [29].

The most important extrusion parameters in MEX layer deposition are nozzle outlet diameter (D), layer height (h), printing speed (V_p), flow rate (fr%) and both nozzle (T_n) and build plate (T_b) temperatures.

To be said, for a given flow rate, the screw peripheral speed (V_s) is set; for that reason, it will be referred directly to this parameter.

On this basis, the minimal set of dimensionless parameters are:
- $H^* = h/D$,
- $T^* = T_n/T_b$, and
- $V^* = V_s/V_p$

In all investigations, V_p and T_b were fixed to 20 [mm/s] and 60 [°C], respectively. Moreover, $D = 1.2$ [mm].

Two nozzle temperatures (190 and 210 [°C]), two flow rate values (500 and 1000 %, that result in 30 and 60 [rpm], respectively) and three layer heights (0.3, 0.6 and 0.9 [mm]) have been examined.

A total of 12 investigations were carried out (Table 1):

Table 1. Full set of experimental dimensionless printing conditions

Index	T^*	D^*	V^*	Index	T^*	D^*	V^*
1	3.15	0.25	0.95	7	3.5	0.25	0.95
2	3.15	0.25	1.89	8	3.5	0.25	1.89
3	3.15	0.5	0.95	9	3.5	0.5	0.95
4	3.15	0.5	1.89	10	3.5	0.5	1.89
5	3.15	0.75	0.95	11	3.5	0.75	0.95
6	3.15	0.75	1.89	12	3.5	0.75	1.89

Italian Manufacturing Association Conference - XVI AITeM   Materials Research Forum LLC
Materials Research Proceedings 35 (2023) 143-153   https://doi.org/10.21741/9781644902714-18

The experimental workflow for a generic printing condition has been reported in Fig.1:

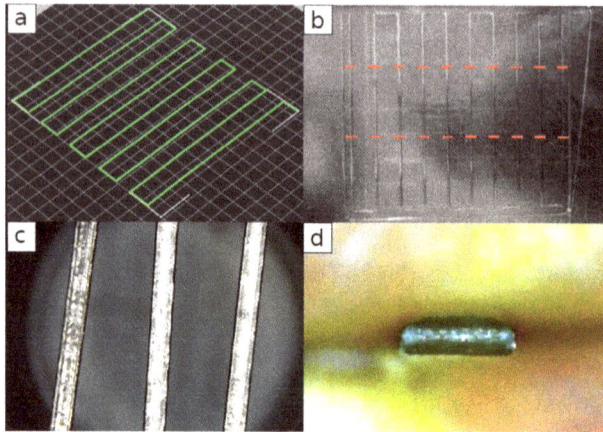

Fig. 1. a.) Serpentine geometry; b.) Printed samples with indication of trim start and stop; c) trimmed samples at microscope; d.) Best sample cross-section

At first, a serpentine geometry was printed (Fig.1a): a total of 10 straight layers were cut (Fig.1b) and their widths were analyzed to provide replications (Fig.1c); a Tomlov DM11 7'' optical microscope was used for investigating strands' geometries.

For a given printing condition, the mean and standard deviation of strand width were estimated, together with the coefficient of variation (COV): a low COV (<10) was found in all printing conditions, which confirmed the consistency of first strand deposition.

Then, the relative deviation of each strand width from mean value was calculated, to find the best one; its cross-section (Fig.1d) was observed through the microscope.

The cross-sectional area (A_s) was calculated with a MATLAB subroutine and the mean velocity in the nozzle calibration zone was evaluated by mass flow rate conservation from nozzle outlet to the deposited strand:

$$u_m = 4 \frac{A_s}{\pi D^2} V_p \frac{\rho_s}{\rho_m} \tag{1}$$

In previous equation, ρ_s and ρ_m are solid and molten polymer densities, respectively. The mean velocity values were the input boundary conditions for CFD studies.

Numerical investigations.

The CFD analyses were performed with the commercial finite volume method (FVM) software Ansys Fluent.

The computational domain consists of the nozzle calibration zone, the air gap between the nozzle outlet and build plate. The nozzle tip geometry was taken out from the original rectangular domain (Fig.2).

The screw tip was not included in the computational domain because swirling motion does not influence the flow field, at least for net polymers [30].

A structured mesh with hexahedral elements was adopted; their number was gradually inflated near the nozzle outlet and build plate, to fully capture the local boundary layer regions.

For instance, the CFD modeling of the $D^* = 0.25$ case has been reported in Fig.2:

Italian Manufacturing Association Conference - XVI AITeM   Materials Research Forum LLC

Materials Research Proceedings 35 (2023) 143-153   https://doi.org/10.21741/9781644902714-18

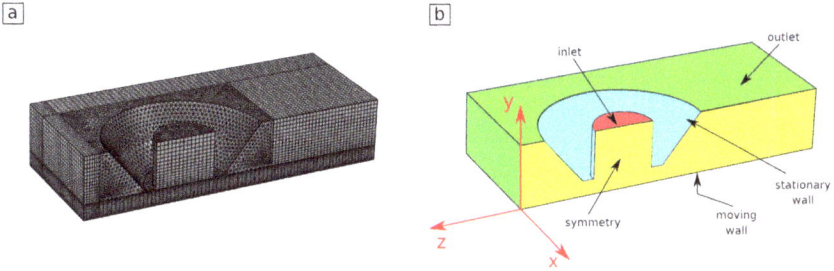

Fig. 2. a.) Meshed domain and b.) boundary conditions and reference system

For CFD simulations, only half of the geometry was considered due to the domain symmetry. The modeling of flow deposition is three-dimensional, non-isothermal and aimed at capturing the non-Newtonian flow behavior. In fact, it has been shown [24] that the most important factor which affects the strand shape is the inclusion of the non-Newtonian behavior, instead of the actual rheological model.

The boundary conditions have been highlighted in Fig.2b. The printing speed is assigned to the build plate instead of the nozzle, to avoid domain remeshing.

An implicit numerical scheme with interfacial anti-diffusion and implicit body force was adopted; the interface between the molten polymer and surrounding air was captured by means of the volume of fluid (VOF) method, because of its robustness in free-surface tracking.

A second order upwind scheme was adopted to discretize momentum and energy equations, to lower numerical diffusion, while the compressive method was used for the continuity equation of the volume fraction of the VOF method.

Only the volume fraction threshold residual was lowered to 1e-6 (default value was 1e-5), to achieve a better approximation of the interface.

The time-stepping was automatically set to guarantee a Courant number of 0.25, so to enhance numerical stability. The overall study stops after finding the steady state extrudate profile.

Results and discussion

In this section, the comparison between experiments and CFD in all extrusion conditions has been made.

Fluent has been widely used for this purpose, when dealing with sub millimetric nozzle diameters and conventional FFF [13,14].

In other works, a 2D approach has been detailed [30–33], to deal with screw-based thermoplastic MEX. In [24] a 3D model has been proposed, but the study does not account for the effect of temperature, and it is limited to a single printing condition.

Instead, the interaction of the minimum set of dimensionless parameters is here considered (Table 1): a full-factorial experimental and numerical design of experiment has been adopted for this purpose.

Italian Manufacturing Association Conference - XVI AITeM   Materials Research Forum LLC
Materials Research Proceedings 35 (2023) 143-153   https://doi.org/10.21741/9781644902714-18

Mass flow rate.

Before proceeding with the investigation of CFD results, the effect of nozzle temperature, layer height and screw speed on mass flow rate has been investigated (Fig.3).

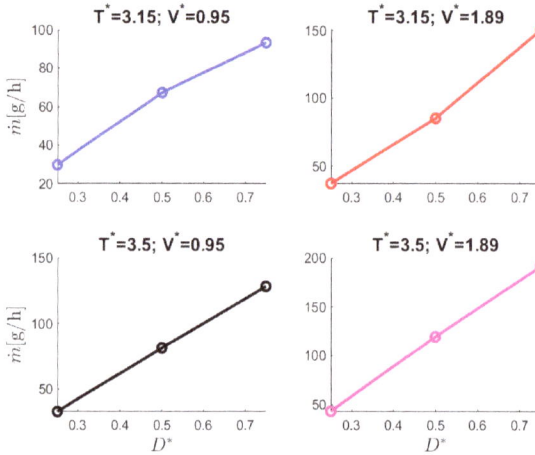

Fig. 3. Mean extruded mass flow rate at different values of the operating parameters; $\dot{m}$: mass flow rate

At increasing D^*, mass flow rate rises almost linearly; this is due to the lower counterpressure which develops in PAM extrusion [29].

The same trend applies for T^*, but it is due to the lower viscosity in the nozzle calibration zone. In addition, mass flow rate increases with V^*, no matter of the nozzle temperature and layer height.

Strand morphology.

The analysis starts with the mean inlet speed evaluation (see: Eq. 1), calculated through the experimental measurement of cross-sectional areas. Then, the mean flow velocity in the nozzle calibration zone was used as inlet condition (Fig.2b) in CFD computations.

A difference with conventional FFF is that the layer width set in the slicing software is different from that found experimentally, as will be seen in next figure; this is caused by the operating mechanism of the Direct 3D PAM extruder, where the flow rate percentage directly controls the screw rotation, and so the mass flow rate.

Instead, in FFF the mass flow rate calculation is straightforward; for prescribed layer height and width, mass flow rate can be evaluated by one-dimensional continuity principle.

In Fig.4 the comparison between numerical and experimental cross-sectional profiles has been done; X^* and Y^* are the absolute reference system coordinates (Fig.2), made dimensionless with respect to the nozzle outer diameter.

Italian Manufacturing Association Conference - XVI AITeM   Materials Research Forum LLC
Materials Research Proceedings 35 (2023) 143-153   https://doi.org/10.21741/9781644902714-18

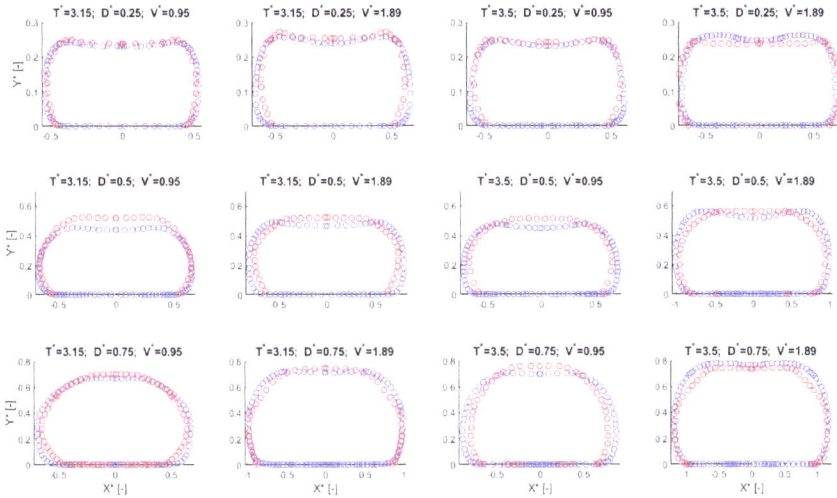

Fig. 4. Experimental (red) and numerical (blue) cross-section profiles of the deposited strands at different processing conditions (see Table 1)

Overall strand shape is fully captured by the chosen numerical setup. The maximum strand width in CFD is slightly lower than in experiments, especially at $D^* = 0.5$. A possible reason can be related to having disregarded the actual viscoelastic behavior.

By lowering D^*, a more oblong cross-sectional shape is found [13], which determines a broader interface for thermal diffusion, which is necessary to create strong inter-layer bonds.

On the other hand, the circular-like shape which arises at high D^* is well captured by FVM.

Moreover, T^* impact deeply layer shape [19]; an increase of this parameter leads to broader layer widths, together with a central collapse of the upper part of the layer, probably caused by the Newtonian-like behavior of PLA at high temperatures.

The effect of V^* on cross-section shape is the most noticeable; an increase in V^* means a higher mass flow rate and strand width increases consequently; a quantitative analysis based on CFD computations has been reported in Table 2:

Italian Manufacturing Association Conference - XVI AITeM   Materials Research Forum LLC
Materials Research Proceedings 35 (2023) 143-153   https://doi.org/10.21741/9781644902714-18

Table 2. Increase in maximum layer width (W_{max}) when switching from low to high V^*

D^*	T^*	Low V^*	W_{max} at low V^* [um]	High V^*	W_{max} at high V^* [um]	Increase in W_{max} switching from low to high V^* [%]
0.25	3.15	0.95	1290	1.89	1552	20.31
0.25	3.5	0.95	1460	1.89	1702	16.57
0.5	3.15	0.95	1572	1.89	1933	22.96
0.5	3.5	0.95	1754	1.89	2400	36.83
0.75	3.15	0.95	1612	1.89	2237	38.77
0.75	3.5	0.95	2100	1.89	2820	34.28

The maximum layer width increases with layer height and temperature. The first trend is related to the lower counterpressure (and consequently, higher mass flow rate, see Fig.3) which develops at high layer heights [29]. The second one is motivated by the more Newtonian-like behavior of the deposited material; it undergoes a larger radial squeezing at higher nozzle temperatures because of the lower viscosity.

Conclusions and further works

A numerical and experimental workflow for PAM extrusion has been presented and validated with respect to different operating conditions.

A systematic method for the evaluation of the cross-section of the first deposited strand has been proposed. The method differs from the ones conventionally proposed for FFF, where the mass flow rate can be calculated by the flow parameter set in the slicing software. In PAM extrusion, the flow parameter is related to the screw peripheral speed, which is the key factor affecting mass flow rate and layer width: the exact layer width imposed in slicing software can't be generally met, because a variation in screw speed results in a different extruded mass flow rate.

The effect of dimensionless layer height (D^*), temperature (T^*) and speed (V^*) on mass flow rate and strand shape was investigated.

The D^* parameter has a deep impact on the mechanical properties of the printed part; in fact, by lowering D^*, higher surface finish can be generally reached; this is caused by the higher thermal contact area provided by the resulting oblong strand shape. However, the mass flow rate conveyed by the screw-barrel system at small D^* is low because of the higher counterpressure. Nevertheless, the reduction in mass flow rate at small D^* can be mitigated by choosing higher barrel temperatures.

On the other hand, higher D^* produce more circular-like shaped strands, with very low thermal contact areas. In addition, higher productivity can be achieved because of a very low counterpressure.

Dimensionless temperature T^* was varied on two levels in the suggested interval for PLA; increasing T^* lead to higher layer width and lower height (top collapse), together with higher mass flow rates. This is probably caused by the Newtonian-like flow behavior, which shows up at higher T^*.

Finally, higher V^* lead to wider radial squeezing, when layer is deposited on the build plate because of the higher mass flow rate which is delivered by the screw-barrel system.

Italian Manufacturing Association Conference - XVI AITeM   Materials Research Forum LLC
Materials Research Proceedings 35 (2023) 143-153   https://doi.org/10.21741/9781644902714-18

In general, it has been shown a good agreement between numerical and experimental results, proving CFD as a milestone in establishing a digital twin for PAM.

In further work, the investigation will be carried out for fiber-reinforced polymers, which are of remarkable interest for PAM because of the improved parts' mechanical properties, such as the overall strength and surface accuracy, driven by the flow fiber alignment.

References

[1] U.M. Dilberoglu, B. Gharehpapagh, U. Yaman, M. Dolen, The Role of Additive Manufacturing in the Era of Industry 4.0, Procedia Manuf. 11 (2017) 545–554. https://doi.org/10.1016/j.promfg.2017.07.148

[2] A. Haleem, M. Javaid, Additive Manufacturing Applications in Industry 4.0: A Review, Journal of Industrial Integration and Management. 04 (2019) 1930001. https://doi.org/10.1142/S2424862219300011

[3] J. Butt, Exploring the Interrelationship between Additive Manufacturing and Industry 4.0, Designs (Basel). 4 (2020) 13. https://doi.org/10.3390/designs4020013

[4] Post B. K., Lind R. F., Lloyd P. D., Kunc V., Linhal J. M., Love L. J., The Economics of Big Area Additive Manufacturing, in: Solid Freeform Fabrication 2016: Proceedings of the 26th Annual International Solid Freeform Fabrication Symposium – An Additive Manufacturing Conference, 2016.

[5] S. Chong, G.-T. Pan, J. Chin, P. Show, T. Yang, C.-M. Huang, Integration of 3D Printing and Industry 4.0 into Engineering Teaching, Sustainability. 10 (2018) 3960. https://doi.org/10.3390/su10113960

[6] M. Attaran, Additive Manufacturing: The Most Promising Technology to Alter the Supply Chain and Logistics, Journal of Service Science and Management. 10 (2017) 189–206. https://doi.org/10.4236/jssm.2017.103017

[7] L. Zhang, X. Chen, W. Zhou, T. Cheng, L. Chen, Z. Guo, B. Han, L. Lu, Digital twins for additive manufacturing: A state-of-the-art review, Applied Sciences (Switzerland). 10 (2020) 1–10. https://doi.org/10.3390/app10238350

[8] A. Pricci, M.D. de Tullio, G. Percoco, Analytical and Numerical Models of Thermoplastics: A Review Aimed to Pellet Extrusion-Based Additive Manufacturing, Polymers, Vol. 13, Page 3160. 13 (2021) 3160. https://doi.org/10.3390/POLYM13183160

[9] F. Talagani, C. Godines, Numerical Simulation of Big Area Additive Manufacturing (3D Printing) of a Full Size Car, 2015.

[10] B.K. Post, P.C. Chesser, R.F. Lind, A. Roschli, L.J. Love, K.T. Gaul, M. Sallas, F. Blue, S. Wu, Using Big Area Additive Manufacturing to directly manufacture a boat hull mould, Virtual Phys Prototyp. 14 (2019) 123–129. https://doi.org/10.1080/17452759.2018.1532798

[11] B.K. Post, B. Richardson, R. Lind, L.J. Love, P. Lloyd, V. Kunc, B.J. Rhyne, A. Roschli, J. Hannan, S. Nolet, K. Veloso, P. Kurup, T. Remo, D. Jenne, Big Area Additive Manufacturing Application in Wind Turbine Molds, 2017.

[12] J. Pasco, Z. Lei, C. Aranas, Additive Manufacturing in Off-Site Construction: Review and Future Directions, Buildings. 12 (2022) 53. https://doi.org/10.3390/buildings12010053

[13] R. Comminal, M.P. Serdeczny, D.B. Pedersen, J. Spangenberg, Numerical modeling of the strand deposition flow in extrusion-based additive manufacturing, Addit Manuf. 20 (2018) 68–76. https://doi.org/10.1016/j.addma.2017.12.013

[14] M.P. Serdeczny, R. Comminal, D.B. Pedersen, J. Spangenberg, Experimental validation of a numerical model for the strand shape in material extrusion additive manufacturing, Addit Manuf. 24 (2018) 145–153. https://doi.org/10.1016/j.addma.2018.09.022

[15] R. Comminal, M.P. Serdeczny, D.B. Pedersen, J. Spangenberg, Motion planning and numerical simulation of material deposition at corners in extrusion additive manufacturing, Addit Manuf. 29 (2019) 100753. https://doi.org/10.1016/j.addma.2019.06.005

[16] R. Comminal, M.P. Serdeczny, D.B. Pedersen, J. Spangenberg, Numerical modling of the material deposition and contouring in fused deposition modeling, in: Solid Freeform Fabrication 2018: Proceedings of the 29th Annual International (Ed.), Solid Freeform Fabrication 2018: Proceedings of the 29th Annual International, 2018.

[17] M.P. Serdeczny, R. Comminal, D.B. Pedersen, J. Spangenberg, Numerical simulations of the mesostructure formation in material extrusion additive manufacturing, Addit Manuf. 28 (2019) 419–429. https://doi.org/10.1016/j.addma.2019.05.024

[18] R. Comminal, S. Jafarzadeh, M. Serdeczny, J. Spangenberg, Estimations of Interlayer Contacts in Extrusion Additive Manufacturing Using a CFD Model, in: Industrializing Additive Manufacturing, Springer International Publishing, Cham, 2021: pp. 241–250. https://doi.org/10.1007/978-3-030-54334-1_17

[19] B. Behdani, M. Senter, L. Mason, M. Leu, J. Park, Numerical Study on the Temperature-Dependent Viscosity Effect on the Strand Shape in Extrusion-Based Additive Manufacturing, Journal of Manufacturing and Materials Processing. 4 (2020) 46. https://doi.org/10.3390/jmmp4020046

[20] J. Du, Z. Wei, X. Wang, J. Wang, Z. Chen, An improved fused deposition modeling process for forming large-size thin-walled parts, J Mater Process Technol. 234 (2016) 332–341. https://doi.org/10.1016/j.jmatprotec.2016.04.005

[21] H. Xia, J. Lu, G. Tryggvason, Fully resolved numerical simulations of fused deposition modeling. Part II – solidification, residual stresses and modeling of the nozzle, Rapid Prototyp J. 24 (2018) 973–987. https://doi.org/10.1108/RPJ-11-2017-0233

[22] E.P. Furlani, V. Sukhotskiy, A. Verma, V. Vishnoi, P. Amiri Roodan, Numerical Simulation of Extrusion Additive Manufacturing: Fused Deposition Modeling, TechConnect Briefs. 4 (2018) 118–121.

[23] T.G. Crisp, Weaver Jason M., Review of Current Problems and Developments in Large Area Additive Manufacturing (LAAM), in: Solid Freeform Fabrication 2021: Proceedings of the 32nd Annual International Solid Freeform Fabrication Symposium – An Additive Manufacturing Conference, 2021.

[24] R. Comminal, M.P. Serdeczny, N. Ranjbar, M. Mehrali, D.B. Pedersen, H. Stang, J. Spangenberg, Modelling of material deposition in big area additive manufacturing and 3D concrete printing, in: Joint Special Interest Group Meeting between Euspen and ASPE, 2019. www.euspen.eu

[25] A.-D. Le, B. Cosson, A.C.A. Asséko, Simulation of large-scale additive manufacturing process with a single-phase level set method: a process parameters study, The International Journal of Advanced Manufacturing Technology. 113 (2021) 3343–3360. https://doi.org/10.1007/s00170-021-06703-5

[26] Comsol Multiphysics, LiveLink TM for MATLAB ® User's Guide, 2009. www.comsol.com/blogs

[27] B. Cosson, A.C. Akué Asséko, L. Pelzer, C. Hopmann, Radiative Thermal Effects in Large Scale Additive Manufacturing of Polymers: Numerical and Experimental Investigations, Materials. 15 (2022) 1052. https://doi.org/10.3390/ma15031052

[28] A. Pricci, M.D. de Tullio, G. Percoco, Semi-analytical models for non-Newtonian fluids in tapered and cylindrical ducts, applied to the extrusion-based additive manufacturing, Mater Des. 223 (2022) 111168. https://doi.org/10.1016/j.matdes.2022.111168

[29] A. Pricci, M.D. de Tullio, G. Percoco, Modeling of extrusion-based additive manufacturing for pelletized thermoplastics: Analytical relationships between process parameters and extrusion outcomes, CIRP J Manuf Sci Technol. 41 (2023) 239–258. https://doi.org/10.1016/j.cirpj.2022.11.020

[30] Z. Wang, D.E. Smith, Numerical analysis of screw swirling effects on fiber orientation in large area additive manufacturing polymer composite deposition, Compos B Eng. 177 (2019) 107284. https://doi.org/10.1016/J.COMPOSITESB.2019.107284

[31] Z. Wang, D.E. Smith, Rheology Effects on Predicted Fiber Orientation and Elastic Properties in Large Scale Polymer Composite Additive Manufacturing, Journal of Composites Science. 2 (2018) 10. https://doi.org/10.3390/jcs2010010

[32] Z. Wang, D.E. Smith, A Fully Coupled Simulation of Planar Deposition Flow and Fiber Orientation in Polymer Composites Additive Manufacturing, Materials. 14 (2021) 2596. https://doi.org/10.3390/ma14102596

[33] Z. Wang, D.E. Smith, Finite element modelling of fully-coupled flow/fiber-orientation effects in polymer composite deposition additive manufacturing nozzle-extrudate flow, Compos B Eng. 219 (2021) 108811. https://doi.org/10.1016/j.compositesb.2021.108811

Italian Manufacturing Association Conference - XVI AITeM   Materials Research Forum LLC
Materials Research Proceedings 35 (2023) 154-162   https://doi.org/10.21741/9781644902714-19

Influence of the energy density on the Young modulus and fatigue strength of Inconel 718 produced by L-PBF

Michele Abruzzo[1,a] *, Giuseppe Macoretta[1,b] , Luca Romoli[1,c] and Gino Dini[1,d]

[1]Università di Pisa, Dipartimento di ingegneria Civile e Industriale, Largo Lucio Lazzarino 1, Pisa, PI, Italy, 56122

[a]michele.abruzzo@phd.unipi.it, [b]giuseppe.macoretta@unipi.it, [c]luca.romoli@unipi.it, [d]gino.dini@unipi.it

Keywords: Powder Bed Fusion, Nickel Alloys, Mechanical Testing Equipment

Abstract. The present work analyses the influence of the energy density on the Young modulus and the fatigue strength of specimens obtained by Laser-Powder Bed Fusion (L-PBF) in Inconel 718. The specimen production and the process parameters taken as variables are described. The bulk and surface properties of the material are studied through static mechanical tests, surface roughness measurements, and fatigue tests. In addition, an approach based on ping tests and laser interferometry is proposed as a more efficient way to calculate the Young modulus of the specimens. The proposed method does not require any preparation of the specimens and allows for a quick and accurate evaluation of the material's Young modulus. The results obtained highlight the influence of the process parameters on the Young modulus and the fatigue strength, suggesting a different usage of the material based on the productivity parameters adopted.

Introduction

Laser-Powder Bed Fusion (L-PBF) is a widely used additive manufacturing technique that allows the production of high-precision and complex three-dimensional objects. The applications of L-PBF are vast, ranging from aerospace to biomedical and automotive industries. However, the mechanical properties of the produced components are highly dependent on the process parameters used. Studies have shown that the process parameters significantly affect the mechanical properties of the produced components [1-7]. For instance, high laser power and low scanning speed result in the formation of larger grains, which increase the ductility and toughness of the material. On the contrary, higher scanning speeds and lower powder bed temperature result in the formation of smaller grains, leading to higher mechanical strength.

In this context, the present work investigates the effects of the energy density on the mechanical properties of Inconel 718. The feasible region of the process, which describes the occurrence of the principal defects characterizing the L-PBF technology (namely lack of fusion and meltpool instability [8, 9]), was defined based on the analytical model proposed by Moda [10]. Several mechanical tests were used to study the Young modulus and High Cycle Fatigue (HCF) behavior of the material. In addition, the influence of the process parameters combination on the surface roughness of the flat specimens was investigated and discussed. Compared to similar studies and standard measurement methods, the study proposes the ping test as an effective and efficient test to quantify the effects of the energy density on the Young modulus of the material, providing a deeper insight into the correlation between the mechanical properties studied and the process parameters adopted. Under the hypothesis that the bulk and surface properties of the material depend on different combinations of physical phenomena involved in the technological process, the aim of the study is to quantify the influence of the process parameters on the mechanical properties of the material, identifying the combinations required for obtaining the desired features.

Italian Manufacturing Association Conference - XVI AITeM   Materials Research Forum LLC
Materials Research Proceedings 35 (2023) 154-162   https://doi.org/10.21741/9781644902714-19

Specimen production

The specimens were produced using a Renishaw RenAM 500S Flex SLM machine. The machine can deliver a maximum laser beam power equal to 500 W, and several laser beam modulation parameters can be adopted. The material used was a standard Inconel 718 powder, compliant with the ASTM F3055 standard. The powder features a Particle Size Distribution (PSD) ranging from 13 μm to 53 μm (D10 ÷ D90).

All the specimens were printed within a unique batch using a vertical building direction. The hatching region was realized using a stripe scanning strategy and a layer rotation of 67°, while the outer region was realized by employing a standard contour scanning path. The built plate was pre-heated at 170°C to minimize the residual stresses, and the process chamber was filled with argon. The oxygen concentration obtained was lower than 7 ppmw, guaranteeing the absence of oxidation phenomena.

Three different geometries of the specimens were used to realize the mechanical tests. In particular, the round specimens employed for the tensile tests were designed in accordance with the ASTM E8-16 standard, with a gauge diameter of 6 mm. The round specimens used for the High Cycle Fatigue (HCF) tests were designed according to the ASTM E466 standard. Finally, flat specimens with rectangular section (8x3x80 mm) were used for the ping tests to simplify the printing process, reduce the powder consumption, and simplify the test execution.

Process parameters selection.

The process parameters used for the specimen production were obtained by exploiting an extension of the Rosenthal solution [12] for modeling the thermal field produced by the L-PBF process [10]. The present formulation has been successfully validated by Macoretta et al. [11], who demonstrated that process parameters belonging to the so defined L-PBF feasible region produce a material presenting a material density greater than 99.5% in the worst case.

Taking advantage of the mathematical model adopted, the feasible region of the L-PBF process parameters (i.e., the parameters combination leading to a full dense material) can be defined as a function of two dimensionless parameters, namely the normalized speed V^* and the normalized power P^* defined in [10]. In particular, the L-PBF feasible region for the Inconel 718 alloy is reported in Fig. 1. The lower bound of the feasible region is given by the Lack of Fusion (LoF) curve, while the upper bound is given by the keyhole region (extrapolated from literature data). The dashed lines represent meltpools having the same aspect ratio (Iso-Ar).

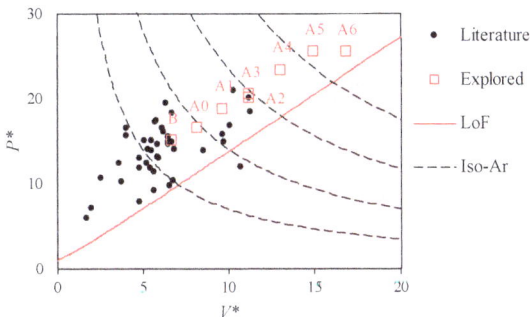

Figure 1: L-PBF process feasible region (white hatched region). Literature data and A_2, A_3, A_5 are obtained from [10] and [11], respectively.

For each parameter combination inside the feasible region, the productivity of the L-PBF process can be calculated as the nominal melted volume per unit of time, neglecting the fixed dwell time due to the recoater motion. Moreover, the process build rate B_r can be defined as $B_r = t \cdot h \cdot v$ [11], in which t is the layer thickness, h the hatch distance, and v the laser scan speed.

The productivity parameters adopted can also be condensed into an additional parameter and represented using the energy density $E_d = \frac{P}{t \cdot h \cdot v}$, which is defined as the ratio between the laser power P and the nominal melted volume per unit of time.

The final process parameters adopted for the specimen production were obtained starting from the baseline values suggested by Renishaw and already investigated in other studies [13, 14] (hereafter defined as baseline) and moving along a straight line having a similar margin with respect to the lack of fusion region, maintaining a constant layer thickness (hereafter defined as A_l). A total of 8 different configurations were considered (Table 1), with the only exceptions in the proposed logic given by the A_3 and the A_6 process parameters, which were obtained with a higher hatch distance and increasing the scan speed at the maximum deliverable power, respectively. Parameters A_2, A_3, and A_5 are the same adopted in [11]. It is interesting to notice that the process productivity can reach a 150% increase compared to the baseline case.

Table 1. *L-PBF process parameters, and process build rate.*

	P [W]	v [m/s]	h [μm]	t [μm]	E_d [J/mm³]	B_r [mm³/s]
Baseline	280	0.90	90	60	57.6	4.9
A_0	310	1.10	90	60	51.9	5.9
A_1	350	1.30	90	60	49.9	7.0
A_2	375	1.50	90	60	46.3	7.8
A_3	460	1.50	135	60	45.4	10.1
A_4	435	1.75	90	60	46.0	9.5
A_5	475	2.00	90	60	44.0	10.8
A_6	475	2.25	90	60	39.1	12.2

Surface roughness measurements

The surface roughness of each specimen was measured by a profilometer (Jenoptic Waveline W812R) using a stylus having a tip radius of 2 μm. The measurements were carried out on the top side of the flat specimens (neglecting the effect of the section shape interpolation on surface roughness) at five different positions along their length considering a total measuring window of 15 mm. According to ISO standard 4288:1988, the primary profile obtained was filtered using a Gaussian filter with a cut-off wavelength of 2.5 mm. After filtering, the measured profiles were compared using the arithmetical mean deviation of the profile heights Ra, the mean roughness depth Rz, the skewness Rsk, and the kurtosis Rku, which are reported in Table 2.

As expected, the baseline parameters correspond to the lower value of surface roughness obtainable. On the contrary, increasing the productivity of the SLM process leads to a noticeable increase in surface roughness. Considering the values reported in Table 2, Ra and Rz reach a maximum value for the A_4 process parameters and then undergo a further reduction (plausibly due to the high laser power and scan speed). The skewness is almost null for all the considered specimens, indicating an almost perfectly symmetric distribution of the measured roughness profiles. Depending on the measuring window considered, a higher skewness value (about 1) can be obtained for $E_d = 57.6$ J/mm³ (baseline) and is manifested by a few high-intensity peaks, which are possibly due to residual powder inclusions. The same conclusions can be drawn from the kurtosis analysis of the measured profiles. As the productivity of the process increases, the profile shape shifts from leptokurtic ($R_{ku} > 3$) to mesokurtic ($R_{ku} \approx 3$), indicating that, even if the overall surface roughness is increased, peaks and valleys are evenly distributed.

Italian Manufacturing Association Conference - XVI AITeM | Materials Research Forum LLC
Materials Research Proceedings 35 (2023) 154-162 | https://doi.org/10.21741/9781644902714-19

Table 2. *Surface roughness parameters, mean value and associated standard deviation.*

	Baseline	A_0	A_1	A_2	A_3	A_4	A_5	A_6
Ra [μm]	7.2 ± 0.8	10.3 ± 1.4	11.4 ± 0.9	14.1 ± 1.1	16.2 ± 1.2	17.0 ± 1.3	13.3 ± 0.9	12.8 ± 1.8
Rz [μm]	50.9 ± 2.8	71.2 ± 4.3	71.7 ± 7.3	84.1 ± 7.6	91.6 ± 3.2	98.6 ± 8.5	92.1 ± 10.1	80.2 ± 12.0
Rsk [-]	0.41 ± 0.36	0.21 ± 0.03	0.03 ± 0.05	0.31 ± 0.15	0.23 ± 0.15	0.30 ± 0.10	0.25 ± 0.08	0.39 ± 0.13
Rku [-]	3.56 ± 0.71	3.27 ± 0.39	2.75 ± 0.18	2.88 ± 0.29	2.67 ± 0.21	2.72 ± 0.17	3.08 ± 0.40	3.04 ± 0.21

Mechanical tests

A series of mechanical tests were used to characterize the mechanical properties of the specimens obtained by L-PBF. In particular, the Young modulus of the material E obtained with several process parameters was measured using tensile and ping tests, using the energy density E_d as the main process parameter for the results comparison. The fatigue strength was calculated through a series of High Cycle Fatigue (HCF) tests.

The results of the tensile and HCF tests refer to [11]. Compared to the aforementioned study, the proposed work proposes a new method for identifying the Young modulus, featuring higher accuracy and sensitivity. Furthermore, four additional process parameter combinations are considered, and the relationship between all the mechanical properties of the printed material is investigated, highlighting the coherence of the results obtained.

Tensile tests.

Tensile tests were performed in displacement control on an MTS servo-hydraulic machine (load capacity of 50 kN) at room temperature. At least 3 specimens were tested for each combination of the technological parameters considered. The deformation of the specimens was measured using an extensometer featuring a gauge length of 10 mm (MTS 634.21-F25). In the tensile tests, the Young modulus of the material was calculated as the angular coefficient of the line tangent to the elastic region of the stress-strain curve (up to a maximum stress value of 1/5 of the yield strength).

Ping tests.

The ping test consists in the application of an impulsive excitation to the specimen through an impact tool and in the measurement of the subsequent dynamic response. The analysis of the frequency-response-function (FRF) allows the quantification of the frequency of the stimulated natural modes and, consequently, the identification of the dynamic Young modulus of the material without the need of complex support systems, elaborate setups, or alignments (ASTM E1876-21).

Figure 2: Ping test experimental setup.

Italian Manufacturing Association Conference - XVI AITeM   Materials Research Forum LLC
Materials Research Proceedings 35 (2023) 154-162   https://doi.org/10.21741/9781644902714-19

In the present study, an impulsive excitation was applied through a lab hammer, and the dynamic response of the specimen was measured using a laser optic-fiber vibrometer (Polytec OFV 551, maximum sampling rate 50 kHz). The specimen was instrumented using a reflective stamp and was supported using a foam bed, allowing it to vibrate freely according to any combination of its natural modes (Fig. 2). The test was repeated 5 times for each specimen considered. The dynamic response was processed using a Fast Fourier Transform (FFT) to identify the natural frequency of the first bending mode f_r, which can be directly correlated to the Young modulus.

The Young modulus was identified using an accurate 3-D Finite Element Model (FEM) reproducing the first bending mode of the specimen. The material was modeled as linear elastic, homogeneous, and isotropic, taking the measured density ($\rho = 7968$ kg/m^3) and the nominal Poisson's ratio ($\nu = 0.3$) as fixed inputs. The proposed approach neglects the effect of the material porosity, usually limited when evaluating a bulk property such as the Young modulus, and of the well-known anisotropic behavior due to the building direction of the L-PBF process, which should not affect the evaluation of E. Once the FEM was set up, the Young modulus was obtained through a parametric analysis to match the natural frequency calculated experimentally for each specimen.

Fatigue tests.
The HCF tests consisted of axial fatigue tests performed at room temperature on a RUMUL Mikrotron resonant testing machine. The load ratio considered was equal to 0.05 for all the tests performed, regardless of the productivity parameters adopted. The tests were conducted on round specimens in the surface as-built conditions, designed according to the ASTM E466 standard. The main result of HCF tests was given by the fatigue resistance at 10^6 cycles $\Delta\sigma$.

Results and discussion
Tensile tests results.
The comparison between the tensile tests of the specimens obtained with different process parameters in as-built conditions is reported in Fig. 3. As E_d is reduced (i.e., the productivity is increased), the material exhibits a less ductile behavior (Fig. 3 (b)). In addition, all the tensile curves are characterized by the absence of a well-defined elastic part (evident for small strains), plausibly due to the high residual stresses present in the specimens (of the order of GPa), making the estimation of the Young modulus particularly complex (difference between the secant and tangent modulus in the elastic region), as well as dispersed. Additionally, it is difficult to distinguish between the dispersion of the Young modulus due to the measurement method and due to the production process using the tensile tests.

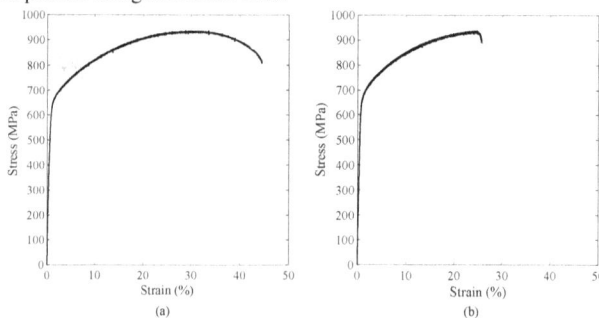

Figure 3: Tensile test results with different process parameters [11]. (a) Baseline. (b) A₅.

Table 3. *Young modulus obtained using tensile tests [11].*

	Baseline	A_2	A_3	A_5
E [GPa]	139 ± 7	142 ± 8	138 ± 7	135 ± 5

Table 3 shows the average Young modulus, together with the associated standard deviation, obtained using tensile tests as a function of the process parameters. It can be noticed that the Young modulus slightly decreases as the E_d is reduced, despite the difference between the baseline and the A_5 parameters being 2.8% of the average value.

Ping tests results.
The Young modulus estimated using the ping tests is represented as a function of the energy density in Fig. 4 and shown in Tab. 4.

In the as-built conditions, the Young modulus shows a high dependency on the energy density and increases as the energy density decreases. Moreover, the Young modulus reaches a plateau for $E_d < 46$ J/mm^3, corresponding to a 33% increase compared to the baseline value. The specimens obtained at lower energy density show a reduced Young modulus dispersion due to the manufacturing process, while the Young modulus obtained with the baseline parameters is affected by a high variability linked to the manufacturing process, with a maximum variation equal to approximately 7% of the mean value.

The peculiar static behavior obtained can be explained by considering the microstructure variations related to the production process: adopting lower energy density values realizes a more severe thermal cycle (i.e., higher thermal transfer and faster cooling) which produces a finer, and consequently stiffer, microstructure [6]. Two examples of metallographies are reported in Fig. 10, showing that a full dense material can be obtained even at higher productivity and (qualitatively) the different influence of columnar structures.

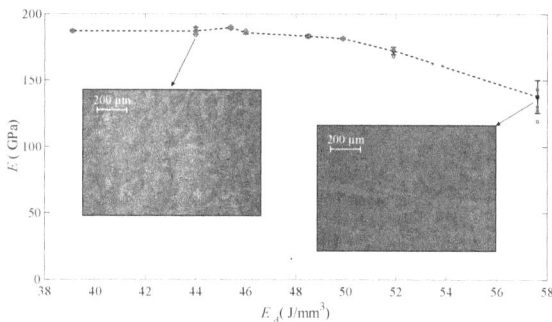

Figure 4: Young modulus obtained using ping tests and metallographies for two process parameters sets.

Table 4. *Young modulus obtained using ping tests.*

	Baseline	A_0	A_1	A_2	A_3	A_4	A_5	A_6
E [GPa]	137.9 ± 12.3	172.4 ± 2.9	181.6 ± 0.6	183.4 ± 1.2	189.7 ± 0.9	185.9 ± 0.9	187.2 ± 2.5	187.2 ± 0.6

Static results comparison.
The static results obtained show a different trend produced by the Young modulus estimation methods. In the case of tensile tests, the estimate is intrinsically affected by a strong dispersion due to the measurement uncertainties and the linear part stress limit introduced in the fitting

procedure, producing a lower and almost constant estimate of E characterized by high measurement dispersion. On the contrary, the ping tests produce a more reliable estimation of the Young modulus by exploiting the tangent at the origin to the tensile curve (small oscillations), and the measurement dispersion is null. The tensile tests allow to measure several material properties, but are inadequate to study the Young modulus variations and dispersion due to the technological parameters adopted (i.e., the elastic modulus of the specimens is measured instead).

Fatigue results comparison.
In Fig. 5, the fatigue strength at 10^6 cycles (Table 5) is compared with Young's modulus obtained using the ping tests and the arithmetical mean deviation of the profile heights as a function of the energy density. Apart from the baseline parameters, E and $\Delta\sigma$ show a reduced dispersion at each combination of the productivity parameters adopted. Notwithstanding the increased surface roughness at higher productivity parameters, $\Delta\sigma$ is almost constant with E_d (maximum variation with respect to the mean value 6%) and has an opposite trend compared to E. $\Delta\sigma$ reaches a minimum point (instead of a maximum) at $E_d = 45.4$ J/mm^3. On the other hand, Young's modulus obtained in as-built conditions is highly dependent on the energy density (maximum variation 30%).

The results obtained through static and HCF tests suggest the possibility of using the material in as-built conditions, achieving excellent mechanical performances depending on the productivity parameters adopted. The fatigue strength of the material shows a limited dependence on the energy density, while Young's modulus has higher values at lower energy densities. In other words, in the case of stiffness-designed components, it is possible to use higher productivity parameters to obtain a more rigid material with almost no influence on its fatigue strength.

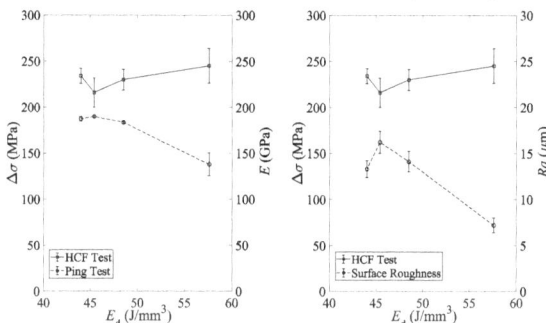

Figure 5: *Fatigue results comparison as a function of E_d. (a) $\Delta\sigma$ and E. (b) $\Delta\sigma$ and Ra.*

Table 5. *Fatigue strength at 10^6 cycles for several productivity parameters [11].*

$\Delta\sigma$ [MPa]	Baseline	A_2	A_3	A_5
As-built	245 ± 18.8	230 ± 11.3	216 ± 15.8	234 ± 8.2

Conclusions
In the present study, the influence of the energy density on the mechanical properties of specimens obtained by Laser-Powder Bed Fusion (L-PBF) is characterized. The process parameters adopted were estimated using first-order analytical models to avoid defects such as lack of fusion, keyhole, or melt pool instability. Furthermore, two dimensionless parameters were derived to describe the feasible region of the process and choose the parameter combinations to be studied.

The surface roughness parameters were measured using a profilometer, and three mechanical tests were used to identify the Young modulus and fatigue strength of the printed specimens as a function of the energy density. The results obtained show that, as the energy density decreases, the material presents different features compared to those obtained using the baseline process parameters. In particular, adopting lower values of energy density (corresponding to more severe thermal cycles) leads to a worse surface finish and a finer material microstructure. From a mechanical point of view, this leads to a significant Young modulus increase, quantified using the ping tests in the as-built conditions, accompanied by a negligible fatigue strength reduction. Several combinations of process parameters can be adopted to increase the process productivity while achieving different mechanical properties.

In conclusion, the mechanical properties of the material obtained through L-PBF depend on several physical phenomena. The Young modulus mainly depends on the material microstructure, while the fatigue strength shows a limited dependence on the surface roughness and the contour parameters adopted. Therefore, it becomes fundamental to choose the most suitable combination of process parameters not only to avoid classic defects or increase the process productivity but also to emphasize the effects of certain consolidation processes. Without constraints on the surface roughness or fatigue strength, the energy density adopted can be reduced, obtaining stiffer components at higher process productivity. The higher and lower energy densities produce the same fatigue strength, and the eventual worsening of the surface finish can be compensated with the refinement of the material's microstructure. Finally, the surface roughness can be minimized by adopting the highest energy density values.

References

[1] Liu S., Qin C., Zong R., Fang X.: The effect of energy density on texture and mechanical anisotropy in selective laser melted Inconel 718. Materials and Design 191 (2020). https://doi.org/10.1016/j.matdes.2020.108642

[2] Balbaa M., Mekhiel S., Elbestawi M., McIsaac J.: On selective laser melting of Inconel 718: Densification, surface roughness, and residual stresses. Materials and Design 193 (2020). https://doi.org/10.1016/j.matdes.2020.108818

[3] Tucho W., Cuvillier P., Sjolyst-Kverneland A., Hansen V.: Microstructure and hardness studies of Inconel 718 manufactured by selective laser melting before and after solution heat treatment. Materials Science and Engineering A 689 (2017). https://doi.org/10.1016/j.msea.2017.02.062

[4] Calandri M., Yin S., Aldwell B., Calignano F., Lupoi R., Ugues D.: Texture and microstructural features at different length scales in Inconel 718 produced by selective laser melting. Material 12 (2019). https://doi.org/10.3390/ma12081293

[5] Facchini M., Magalini L., Robotti E., Molinari A., Höges S., Wissenbach K.: Ductility of a Ti-6Al-4V alloy produced by selective laser melting of prealloyed powders. Rapid Prototyping Journal 16 (2010). https://doi.org/10.1108/13552541011083371

[6] Guo S., Li Y., Gu J., Liu J., Peng Y., Wang P., Zhou Q., Wang K.: Microstructure and mechanical properties of Ti6Al4V/ B4C titanium matrix composite fabricated by selective laser melting (SLM). Journal of Materials Research and Technology 23 (2023). https://doi.org/10.1016/j.jmrt.2023.01.126

[7] Liverani E., Toschi S., Ceschini L., Fortunato A.: Effect of selective laser melting (SLM) process parameters on microstructure and mechanical properties of 316L austenitic stainless steel. Journal of Materials Processing Technology 249 (2017). https://doi.org/10.1016/j.jmatprotec.2017.05.042

[8] Wu H., Ren J., Huang Q., Zai X., Liu L., Chen C., Liu S., Yang X., Li R.: Effect of laser parameters on microstructure, metallurgical defects and property of AlSi10Mg printed by selective laser melting. Journal of Micromechanics and Molecular Physics 2 (2017). https://doi.org/10.1142/S2424913017500175

[9] Scime L., Beuth J.: Melt pool geometry and morphology variability for the Inconel 718 specimens produced by additive manufacturing including notch effects. Fatigue and Fracture of Engineering Materials and Structures 43 (2020). https://doi.org/10.1016/j.addma.2019.100830

[10] Moda M., Chiocca A., Macoretta G., Monelli BD., Bertini L.: Technological implications of the Rosenthal solution for a moving point heat source in steady state on a semi-infinite solid. Materials & Design 223 (2022). https://doi.org/10.1016/j.matdes.2022.110991

[11] Macoretta, G., Bertini, L., Monelli, B.D., Berto, F.: Productivity-oriented SLM process parameters effect on the fatigue strength of Inconel 718. International Journal of Fatigue 168 (2022). https://doi.org/10.1016/j.ijfatigue.2022.107384

[12] Rosenthal D.: The Theory of Moving Sources of Heat and Its Application to Metal Treatments. Transactions ASME 43 (1946). https://doi.org/10.1115/1.4018624

[13] Ravichander BB., Amerinatanzi A., Moghaddam NS.: Study on the effect of powder-bed fusion process parameters on the quality of as-built In718 parts using response surface methodology. Metals 10 (2020). https://doi.org/10.3390/met10091180

[14] Ma XFF., Zhai HLL., Zuo L., Zhang WJJ., Rui SSS., Han QNN.: Fatigue short crack propagation behavior of selective laser melted Inconel 718 alloy by in-situ SEM study: Influence of orientation and temperature. International Journal of Fatigue 139 (2020). https://doi.org/10.1016/j.ijfatigue.2020.105739

Italian Manufacturing Association Conference - XVI AITeM   Materials Research Forum LLC
Materials Research Proceedings 35 (2023) 163-172   https://doi.org/10.21741/9781644902714-20

Development of a 3D printer optimized for rapid prototyping with continuous fiber fabrication technology

Matteo Benvenuto[1,a] *, Enrico Lertora[1,b], Chiara Mandolfino[1,c],
Luigi Benvenuto[1,d], Alberto Parmiggiani[2,e], Mirko Prato[3,f], Marco Pizzorni[1,g]

[1] Department of Mechanical Engineering, Polytechnic School of University of Genoa, Via opera pia 15, 16145, Genoa, Italy

[2] Centre for Robotics and Intelligent Systems, MSW Facility, Istituto Italiano di Tecnologia, Via San Quirico 19D, 16163, Genoa, Italy

[3] Materials Characterization Facility, Istituto Italiano di Tecnologia, Via Morego 30, 16163, Genoa, Italy

[a]matteo.benvenuto@edu.unige.it, [b]enrico.lertora@unige.it, [c]chiara.mandolfino@unige.it, [d]luigi.benvenuto@edu.unige.it, [e]alberto.parmiggiani@iit.it, [f]mirko.prato@iit.it, [g]marco.pizzorni@unige.it

Keywords: Additive Manufacturing, Material Extrusion, Composites

Abstract. In the industrial field, Additive Manufacturing is a production concept that is increasingly gaining ground. The secret of its success lies in its definition: being able to produce an object by the progressive deposition of material instead of its removal as for the traditional machining. In this way, problems such as waste quantities, complex geometries and machining changes are greatly reduced, making these processes particularly useful and effective for rapid prototyping and the production of small series of objects. This experimental work examines the changes made to a 3D printer initially set up for the use of polymeric materials with Fused Deposition Modelling technology. Going into more detail, this machine was modified in a manner that would make it compatible with the introduction of polymer filaments reinforced with continuous carbon fiber according to the main principles of Continuous Fiber Fabrication technology. The changes made also involved electronics and informatics so that the printer could be easily operated through the platform.

Introduction

Additive Manufacturing (AM) is currently the most ground-breaking manufacturing technology capable, at least potentially, of changing traditional manufacturing paradigms. Revolutionary is the idea of considering objects as a superposition of a number of sections of extremely limited thickness. Exploiting this concept, it is possible to obtain the desired piece by depositing a series of layers of certain materials (typically thermoplastic polymer-based) using appropriate devices [1,2]. Specifically for this reason, AM contrasts with classical production methods based on plastic deformation of the material, such as molding, or centred on the removal of material, as in the case of turning and milling [2].

The development of AM has already reached such a level that it can be used in several industrial sectors (including aerospace, automotive, robotics, biomedical industry) [3,4,5], and its progress is expected to continue in the future, allowing it to become even more widespread [6].

AM is compatible with various types of materials, although the most widely used are undoubtedly thermoplastic polymers, e.g. polyamide (PA), polylactic acid (PLA), acrylonitrile butadiene styrene (ABS) and polyether-ether-ketone (PEEK). In fact, this class defines a good compromise between ease of production, low cost, and mechanical performance of the manufactured product [7]. Frequently, these materials are used in the form of filaments, a

Italian Manufacturing Association Conference - XVI AITeM   Materials Research Forum LLC
Materials Research Proceedings 35 (2023) 163-172   https://doi.org/10.21741/9781644902714-20

characteristic that makes them suitable for certain additive manufacturing processes such as, for example, Fused Deposition Modelling (FDM), also known as Fused Filament Fabrication (FFF) [8].

An evolution of this manufacturing process has involved the introduction of reinforcements within the filament to increase the overall performance of polymers [9]. In this way, the structures obtained combine stiffness, toughness with light weight and corrosion resistance properties [10], aspects that make them highly coveted in sectors like prototyping or production of small series objects (e.g., in the automotive and biomedical fields).

The most widespread AM technology for these composite materials is certainly Continuous Fibre Fabrication (CFF), in which thermoplastic filaments containing carbon (but also glass or Kevlar) reinforcement in the form of short (SF) or continuous fibres (CF) are used [11,12]. This technology has experienced significant growth in the last few years since it is only from 2015 that several companies started to develop systems to process thermoplastic polymers reinforced with carbon fibres. Some of these devices are based on fibre deposition using 6-DOF robotic arms (e.g., Continuous Composites, Moi Composites), but most of the proposed systems involve the use of desktop 3D printers (e.g., Markforged, Anisoprint, Desktop Metal) [13].

Certainly, the degree of complexity of CFF technology is higher compared to FFF for neat-thermoplastic polymers since the composite has several elements, each of which has its own physical properties. Moreover, unlike traditional manufacturing technologies for composites in which strength and orthotropy characteristics are imparted to the part based on the weave of the reinforcing fabric, in CFF these are strictly dependent on the process set-up, particularly the deposition mode adopted and the orientation imposed on the fiber on each subsequent layer [11,12].

The high potential of this new composite manufacturing technology has also attracted the research community in recent years. To date, there are several studies on 3D-printed composites in the literature, most of which report mechanical characterizations of composites made with commercial 3D printers (generally, as a function of printing parameters) [11,14,15]. However, it is evident that most of the criticalities highlighted are related to hardware and software limitations of commercial printers. In this context, as there have only recently been significant developments in this field, the literature on 3D printing of carbon fibre composites still lacks information and does not offer solutions on aspects that are crucial to definitely improve the quality of printed parts [16].

Based on the above, this paper summarises the modifications made to a commercial CubePro Duo 3D printer by 3D System, originally designed for rapid prototyping of pure polymer parts, to make it suitable to process polymer filaments reinforced with both short and continuous carbon fibres. The scope of the modification is indeed to allow the use of commercial reinforced filaments (in particular, Onyx and CFR filaments traded by Markforged) by means of a device that is improved in terms of process conditions (e.g., heating of chamber) and software flexibility.

3D printer set up for neat polymer and short-fibre reinforced filaments

The transformation process described in this paper considered a CubePro Duo™ (3D Systems, Rock Hill, SC, USA), an Additive Manufacturing 3D printer based on the principles of Fused Filament Fabrication. It is a closed-chamber machine with an overall build envelope of 578mm x 578mm x 591mm, capable of producing parts with two different polymer materials. From a kinematic point of view, this printer falls into the Cartesian category as it operates along the three Cartesian axes X, Y and Z. Specifically, the horizontal movements in the X-Y plane are carried out by stepper motors which move the extrusion head by means of a toothed belt drive, while the vertical movements (along the Z axis) are performed by the glass printing plate through a threaded rod. Consequently, the print volume of the machine is limited both by the size of the printing plate (275 mm x 265 mm) and by the maximum vertical excursion allowed (230 mm). One of the

Italian Manufacturing Association Conference - XVI AITeM   Materials Research Forum LLC
Materials Research Proceedings 35 (2023) 163-172   https://doi.org/10.21741/9781644902714-20

strengths of this printer is undoubtedly the heating of the internal chamber using a heater that can raise the internal temperature to ~70°C. In fact, this feature leads to a considerable improvement in performance, as working in a heated environment reduces the effects of warping, an anaesthetic printing defect that causes strains due to uncontrolled cooling and excessive temperature changes to which the newly melted material is subjected. In addition, internal heating improves the quality of adhesion between subsequent layers of the workpiece [18,19].

The CubePro Duo is a machine with a certain flexibility, especially in terms the materials that can be used. Depending on the requirements and the final characteristics of the part to be produced, it is indeed possible to choose the most suitable filament from polyamide (PA), polylactic acid (PLA) or acrylonitrile butadiene styrene (ABS). However, the range of materials that can be used does not include composites. For this reason, given the scope of this work, it was necessary to replace the printhead with one able to process fibre-reinforced polymer filaments regardless of the fiber length (short or continuous). Specifically, the filaments to be introduced are Onyx and CFR, both traded by Markforged, whose main properties are listed in Table 1.

Table 1 Material properties of filaments by Markforged Inc [17].

Property	CFR	Onyx
Density [g/cm^3]	1,4	1,2
Tensile Strength [MPa]	800	40
Tensile Modulus [GPa]	60	2,4
Flexural Strength [MPa]	540	71
Flexural Modulus [GPa]	51	3,0
Compressive Strength [MPa]	420	N/A
Compressive Modulus [MPa]	62	N/A
Izod Impact-notched [J/m]	960	330
Heat Deflection Temp [°C]	105	145

The choice therefore fell on a Mark Two model (Markforged® Inc., Watertown, MA, USA), a head technology that represents the latest updated version of the most diffused commercial desktop printer for CF-reinforced composites (Fig. 1a).

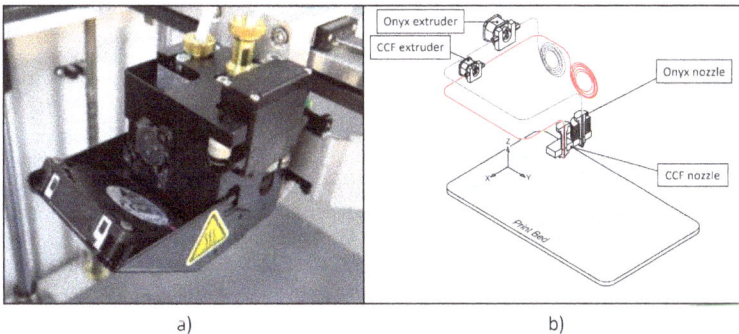

a)   b)

Fig. 1 Markforged® Mark Two printhead: (a) specific view and (b) scheme of the printing structure.

Italian Manufacturing Association Conference - XVI AITeM   Materials Research Forum LLC
Materials Research Proceedings 35 (2023) 163-172   https://doi.org/10.21741/9781644902714-20

Obviously, although it has managed to retain the Cartesian structure of the machine (Fig. 1b), the replacement of this latter device has required inevitable changes to the structure of the movable crossbar on which the extrusion head is fixed. In fact, due to different geometry of the original and new extrusion heads, it was necessary to design the support entirely, the final rendering of which is illustrated in Fig. 2. The solution adopted makes possible the exploitation of the space between the two sleeves of the moving crossbar in such a way as to limit as much as possible the overall dimension in the vertical direction, with limited repercussions on the size of the print volume.

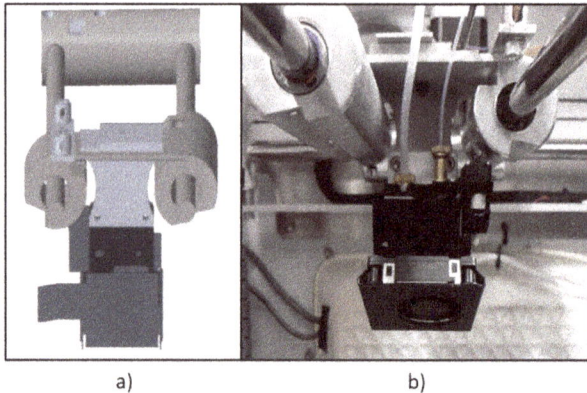

a)   b)

Fig. 2 New printhead fixing system: (a) CAD view and (b) specific view

Another aspect to consider is the positioning of the stepper motors that manage the filament feed since, contrary to the original layout of CubePro Duo, the extruders cannot be fixed directly on the printhead. As a result, the consequent reduction in volume and weight of the printhead makes it possible to reduce the inertial mass terms with obvious advantages, particularly, in terms of maximum values assumed by accelerations and jerks. This results in a reduction in printing times, especially those related to stages where there are no limits imposed by the filament printing parameters.

Obviously, all these advantages were only gained after having defined a new position for the extruders inside the printer. A convenient location must be selected to guarantee for periodic maintenance of the motor, and, at the same time, not to create any impediments to the movements of the printhead. Among the various solutions considered, it was decided to install these electrical devices on a support to be designed and inserted in place of a portion of the upper case of the CubePro Duo. In addition, this layout ensures a short path between the printhead and the actual extruders, significantly reducing problems associated with filament jamming inside the polytetrafluoroethylene (PTFE) tubes that connect the two components.

The proposed support roof cover (represented in red in Fig. 3a) envisages a profile obtained by thermoforming and able to lead to an increase in the internal chamber due to the genesis of a new internal compartment in which the extruders are placed and fixed through a bolted joint (Fig. 3b).

Italian Manufacturing Association Conference - XVI AITeM
Materials Research Proceedings 35 (2023) 163-172

Materials Research Forum LLC
https://doi.org/10.21741/9781644902714-20

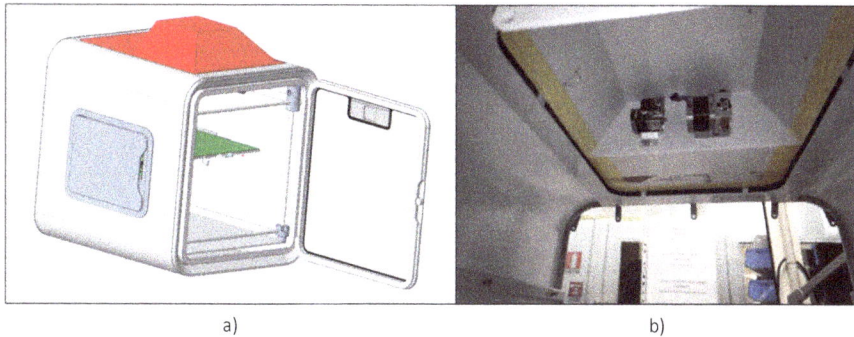

a)   b)

Fig. 3 New roof cover design: (a) CAD external view and (b) extruder fixing seat

As visible in Fig. 3b, the increase in volume of the inner chamber affects only the front portion of the machine. This decision was made to limit the increase in chamber volume as much as possible. Indeed, being located above the printhead, this portion of space is not part of the print volume. Moreover, an excessive volumetric increase could have been detrimental from a thermal standpoint, leading to both a lengthening of the chamber heating time and a slight decrease in the maximum temperature reached inside it.

It is worth noting that a support cover with such a profile was also advantageous because it provided sufficient space behind the extruder compartment to be able to attach the spool holders. In addition, this location makes replacement of the spools very easy and fast, and limits the distance travelled by filaments to reach the extruders.

Electronic and IT operations carried out on the 3D printer
The modified printer is able to properly move the head within the printing volume. Therefore, attention then turned to the electronics of the system, particularly the motherboard. In fact, the original motherboard of the CubePro Duo has some criticalities, among which it is worth noting the impossibility of implementation with auxiliary boards and its incompatibility with the newly introduced printhead.

A new motherboard was therefore needed to solve all the problems encountered. The choice fell on an MB6HC model from the Duet 3 series, a next-generation control board that can be used with a wide range of machines, including 3D printers, CNCs, laser cutters and other devices. The main goal of the Duet 3 series is to achieve maximum flexibility in machine design through high-capacity motherboards, expansion boards, and custom expansion modules. Configuration flexibility and advanced functionality are provided by the RepRapFirmware running on the motherboard and the DuetSoftwareFramework running instead on a Single Board Computer (SBC; in this case, a Raspberry Pi 4). The overall hardware requirements and the operating limits of the MB6HC are compatible with the equipment on the modified CubePro [20]. In addition, not all connectors on the board are used so that there are free terminals useful for future insertion of other devices (e.g., insertion of LED strips or platter heaters).

Once the motherboard was properly introduced, attention turned to the inclusion of the Raspberry Pi 4, i.e., an SBC with which to implement the performance of the Duet 3 MB6HC. In fact, this addition offers several advantages, including higher network transfer speeds, support for plug-ins that require more than the RepRapFirmware present on the Duet Web Control (DWC) platform, and the ability to connect devices such as screens with HDMI ports, keyboards, mouse, and USB flash drives. As one can imagine, these two new electronic components cannot be placed

Italian Manufacturing Association Conference - XVI AITeM   Materials Research Forum LLC
Materials Research Proceedings 35 (2023) 163-172   https://doi.org/10.21741/9781644902714-20

in the place where the old motherboard originally was. Hence, it was decided to place both devices inside a compartment on the right side of the printer, already equipped with brackets for attachment (Fig. 4).

Both case and relative cover were made of polylactic acid (PLA) and were fabricated additively

Fig. 4 Side case containing Duet 3 motherboard and
Raspberry Pi 4.

by Selective Laser Sintering (SLS) technology. A series of side holes were made for the passage of cables. To protect the electronic devices placed inside this compartment, the support is fitted with a cover with four magnets to ensure the compartment is closed properly. This prevents any short circuits due to unwanted contact of the back of the board with the backing plate.

Once the new board was installed, the focus was on the various electrical connections to be made with all the electronic components of the modified CubePro Duo. It was then sufficient to connect the components (i.e., nozzles, extruders, stepper motors, and limit switches) with the MB6HC of the Duet 3D following the wiring diagram provided by the motherboard producer [21].

It is worth noting that during the wiring phase of the machine, a series of technical problems arose, then completely solved. Among these, it is worth to mentioning those relating to the connections of the SBC with all the interface devices required to guarantee the correct operation of the CubePro Duo. In fact, the Raspberry Pi 4 has four USB ports that under normal printer operating conditions cannot be used as they are located inside the chamber described before. Therefore, disregarding the inconvenient hypothesis of redesigning the electronic compartment, it was decided to reuse the original slot working in two ways depending on the device to be connected. As regards the interface devices with the Raspberry and the interface platform (mouse and keyboard), it was decided to use Bluetooth wireless devices, as this leads to the advantage of having no connecting cables but only small receivers that can be easily plugged in before the machine is switched on.

Obviously, the solution just described was not feasible for connection with a USB mobile unit (necessary for the introduction of the CAD files of the object to be printed) due to obvious difficulties in inserting the unit under normal operating conditions. This led to the need to use USB ports originally found on the external surface of the CubePro Duo, a requirement that was solved by designing an extension cable compatible with the Raspberry's USB ports, of sufficient length to reach the CubePro's side sockets. For this application, four-core Belden cables with twisted wires were used: together with the external shielding layer, this solution makes it possible to reduce electrical noise, thus providing a stable signal that is little affected by the presence of other electrical equipment placed nearby.

Given the positive result achieved, this approach was also used for the other connections to the external environment, namely the network cable and the HDMI video cable for the screen. However, unlike the USB sockets, the CubePro does not have a designated location for such connections to the outside environment. For this reason, it was opted to install a plate with the appropriate modules (e.g., RJ45 and HDMI sockets) near the compartment for the SBC.

Once the phase of wiring all the electronics was finished, the firmware for the motherboard (called RepRapFirmware) was configured. Specifically, this stage involves writing (also with the help of appropriate programs) the configuration files used to define the operating and functional conditions of the printer (such as defining print volume limits) and all the electronic devices of the machine (e.g., stepper motors, heaters, limit switches, temperature sensors).

3D printer set up for continuous-fibre reinforced filaments
Firmware installed on the Duet 3 motherboard ensures that the printer will operate properly with filaments made of pure polymer or, at the limit, reinforced with short fibres (e.g., Onyx filament), but, in this base condition, it still does not allow the use of Continuous Fibre Reinforced (CFR) filaments. This limitation is imposed by the conditions under which the newly deposited material detaches from the filament still contained in the hotend nozzle when the filament feed is interrupted. Under these conditions, detachment occurs independently only if the fibres contained in the filament are short. In contrast, when CFR filaments are used, the separation process is prevented by the fibre itself, which, being unaffected by heating, remains continuous. Therefore, in this case, the filament separation must be achieved by a cutting operation performed by a servomotor fixed on the printhead. This additional operation therefore required writing a special macro in the firmware that will manage the activation of the servomotor whenever it is necessary to cut the CFR filament (typically, between one deposited layer and the next).

At this stage the printer can execute all operations and can be switched on to carry out the preliminary printing tests, based on which to adjust the process parameters and optimize printing. Therefore, the first printing tests were focused on the generation of increasingly complex flat figures using only Onyx filament (Fig. 5), to obtain feedback on the good coordination of the movement systems along the X and Y axes. The execution of this initial work has made it possible to solve some typical problems at the start of printing, such as, for example, the height of the first layer, which is critical due to aspects relating to the correct adhesion of the extruded material to the printing plate, and the correction of the extrusion factor to obtain a correct filling of the printed figures without excess material.

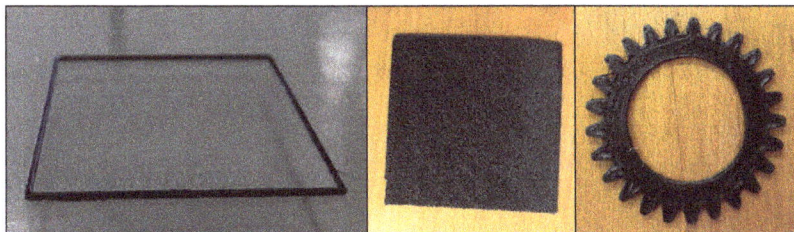

Fig. 5 First 2D objects obtained with printer converted to Continuous Fibre Fabrication (CFF) technology

Italian Manufacturing Association Conference - XVI AITeM   Materials Research Forum LLC
Materials Research Proceedings 35 (2023) 163-172   https://doi.org/10.21741/9781644902714-20

Then, having completed the optimisation of 2D figures, the first attempts at the genesis of 3D objects were carried out, also employing continuous carbon fibre in the filling stages. As an example, Fig. 6 shows a series of specimens prepared for tensile tests in full compliance with the ISO standard valid for composite materials [22].

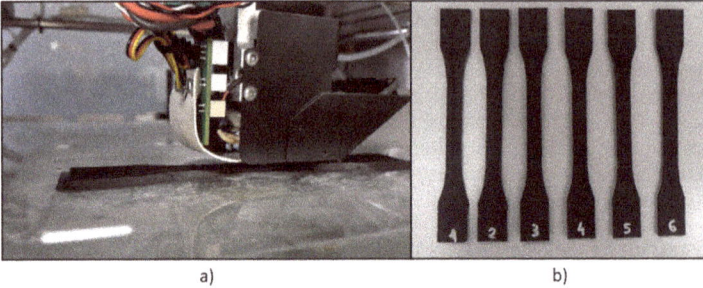

a)   b)

Fig. 6 3D printing process of a tensile test samples (a) and a comparison of the different parts produced (b)

The final parts thus obtained met all the dimensional tolerances of the standard with maximum deviations from the nominal value within 0.2 mm. Furthermore, as can be seen in Fig. 6, the surface finish of the parts thus obtained was satisfactory and comparable to those found in the literature [23].

Conclusions

The experience described in this paper was aimed at converting a 3D printer based on FDM technology to the use of commercial composite filaments with short or continuous carbon fibre reinforcements (Onyx and CFR, respectively). The conversion process needed both various modifications to the original printer's components and implementation with dedicated components (e.g., the replacement of the printhead with a Mark Two by Markforged). In parallel with this stage of re-building, the printer's informatics was also updated by introducing a new motherboard with enhanced performance than the original one. This device also made it possible to connect a Single Board Computer, a Raspberry Pi 4, that improved and simplified the machine's work management.

Attention was then turned to the electronics (e.g., complete re-wiring of the machine), and to the machine programming, through which all the codes necessary to ensure proper operations with reinforced filaments were written. The final phase of the conversion process was the setting up of the device, which involved calibration steps (heaters and stepper motors), movement tests, and extrusion tests. Tables 2-3 summarise the main technical operating specifications of the converted 3D printer.

Table 2 Main technical specifications of CubePro Duo modified with Mark Two printhead.

Parameter	3D printer
Technology	Continuous Fiber Fabrication (CFF)
Printer dimensions [mm]	578 (w) x 578 (l) x 591 (h)
Maximum build size [mm]	242,9 (w) x 270,4 (l) x 230 (h)
Z axis resolution [mm]	0,10
Chamber heating	Yes, up to 70°C
Bed heating	No

Table 3 Main printing features of CubePro Duo modified with Mark Two printhead.

Parameter	CFR	Onyx
Filament diameter [mm]	0,35	1,75
Extrusion temperature [°C]	270	265
Extrusion feed rate [mm/min]	600	1200
Filament drying	Recommended	Recommended
Nozzle diameter [mm]	0,35	0,40
Layer thickness [mm]	0,15	0,15

At this stage, the modified 3D printer can work with both short and continuous carbon fibre-reinforced filaments, under process conditions comparable with those of a Mark Two commercial printer. It is noteworthy that, unlike the latter, the implementation of the heating system in the chamber, based on some preliminary tests, is proving its effectiveness in limiting deformations of the produced part (due to lower thermal gradients between cooled deposit and fused filament), and should also improve adhesion between successive layers.

Certainly, this result leaves wide space to further optimization, and is to be considered as a starting point for future developments inherent to technological, process, and final product aspects, most of which, at the time of writing, are already under investigation (for example, the implantation with a heated printing bed to further improve the printed product's quality).

References

[1]   S. D. Nath and S. Nilufar, "An overview of additive manufacturing of polymers and associated composites," Polymers, vol. 12, no. 11. 2020. http://doi.org/10.3390/polym12112719

[2]   D. L. Bourell, "Perspectives on Additive Manufacturing," Annual Review of Materials Research, vol. 46. 2016. http://doi.org/10.1146/annurev-matsci-070115-031606

[3]   O. Abdulhameed, A. Al-Ahmari, W. Ameen, and S. H. Mian, "Additive manufacturing: Challenges, trends, and applications," Advances in Mechanical Engineering, vol. 11, no. 2, 2019. http://doi.org/10.1177/1687814018822880

[4]   A. Paolini, S. Kollmannsberger, and E. Rank, "Additive manufacturing in construction: A review on processes, applications, and digital planning methods," Additive Manufacturing, vol. 30. 2019. http://doi.org/10.1016/j.addma.2019.100894

[5]   M. Salmi, "Additive manufacturing processes in medical applications," Materials, vol. 14, no. 1. 2021. http://doi.org/10.3390/ma14010191

[6]   G. Liu et al., "Additive manufacturing of structural materials," Materials Science and Engineering R: Reports, vol. 145. 2021. http://doi.org/10.1016/j.mser.2020.100596

[7]   Y. Zheng, W. Zhang, D. M. B. Lopez, and R. Ahmad, "Scientometric analysis and systematic review of multi-material additive manufacturing of polymers," Polymers, vol. 13, no. 12. 2021. http://doi.org/10.3390/polym13121957

[8]   S. Hasanov et al., "Review on additive manufacturing of multi-material parts: Progress and challenges," Journal of Manufacturing and Materials Processing, vol. 6, no. 1. 2022. http://doi.org/10.3390/jmmp6010004

[9]   A. El Moumen, M. Tarfaoui, and K. Lafdi, "Additive manufacturing of polymer composites: Processing and modeling approaches," Compos B Eng, vol. 171, pp. 166–182, Aug. 2019. http://doi.org/10.1016/j.compositesb.2019.04.029

[10] D. K. Rajak, D. D. Pagar, R. Kumar, and C. I. Pruncu, "Recent progress of reinforcement materials: A comprehensive overview of composite materials," Journal of Materials Research and Technology, vol. 8, no. 6, 2019. http://doi.org/10.1016/j.jmrt.2019.09.068

[11] M. Galati, M. Viccica, and P. Minetola, "A finite element approach for the prediction of the mechanical behaviour of layered composites produced by Continuous Filament Fabrication (CFF)," Polym Test, vol. 98, 2021. http://doi.org/10.1016/j.polymertesting.2021.107181

[12] D. Jiang and D. E. Smith, "Anisotropic mechanical properties of oriented carbon fiber filled polymer composites produced with fused filament fabrication," Addit Manuf, vol. 18, 2017. http://doi.org/10.1016/j.addma.2017.08.006

[13] M. Pizzorni, E. Lertora, and A. Parmiggiani, "Adhesive bonding of 3D-printed short- and continuous-carbon-fiber composites: An experimental analysis of design methods to improve joint strength," Compos B Eng, vol. 230, 2022. http://doi.org/10.1016/j.compositesb.2021.109539

[14] M. Araya-Calvo et al., "Evaluation of compressive and flexural properties of continuous fiber fabrication additive manufacturing technology," Addit Manuf, vol. 22, 2018. http://doi.org/10.1016/j.addma.2018.05.007

[15] H. Oberlercher et al., "Additive manufacturing of continuous carbon fiber reinforced polyamide 6: The effect of process parameters on the microstructure and mechanical properties," in Procedia Structural Integrity, Elsevier B.V., 2021, pp. 111–120. http://doi.org/10.1016/j.prostr.2021.12.017

[16] S. M. F. Kabir, K. Mathur, and A. F. M. Seyam, "A critical review on 3D printed continuous fiber-reinforced composites: History, mechanism, materials and properties," Composite Structures, vol. 232. 2020. http://doi.org/10.1016/j.compstruct.2019.111476

[17] Information on https://www-objects.markforged.com/craft/materials/CompositesV5.2.pdf

[18] D. Song, A. M. C. Baek, J. Koo, M. Busogi, and N. Kim, "Forecasting warping deformation using multivariate thermal time series and k-nearest neighbors in fused deposition modeling," Applied Sciences (Switzerland), vol. 10, no. 24, 2020. http://doi.org/10.3390/app10248951

[19] C. Casavola, A. Cazzato, D. Karalekas, V. Moramarco, and G. Pappalettera, "The effect of chamber temperature on residual stresses of FDM parts," in Conference Proceedings of the Society for Experimental Mechanics Series, Springer Science and Business Media, LLC, 2019, pp. 87–92. http://doi.org/10.1007/978-3-319-95074-7_16

[20] Information on https://docs.duet3d.com/Duet3D_hardware/Duet_3_family/Duet_3_Mainboard_6HC_Hardware_Overview

[21] Information on https://docs.duet3d.com/en/How_to_guides/Wiring_your_Duet_3

[22] ISO 527-4, "International Standard International Standard - ISO 527-4," Iso, vol. 2012, 2012.

[23] R. Maier, S. G. Bucaciuc, and A. C. Mandoc, "Reducing Surface Roughness of 3D Printed Short-Carbon Fiber Reinforced Composites," Materials, vol. 15, no. 20, 2022. http://doi.org/10.3390/ma15207398.

Italian Manufacturing Association Conference - XVI AITeM   Materials Research Forum LLC
Materials Research Proceedings 35 (2023) 173-181   https://doi.org/10.21741/9781644902714-21

Reverse bending fatigue of 316L stainless steel components produced by laser powder bed fusion

Stefano Guarino[1,2,a], Emanuele Mingione[1,2,b], Gennaro Salvatore Ponticelli[1,2,c,*], and Simone Venettacci[1,2,d]

[1]University Niccolò Cusano, Department of Engineering, Via Don Carlo Gnocchi 3, 00166 Rome, Italy

[2]ATHENA European University

[a]stefano.guarino@unicusano.it, [b]emanuele.mingione@unicusano.it, [c]gennaro.ponticelli@unicusano.it, [d]simone.venettacci@unicusano.it

Keywords: Laser Powder Bed Fusion, 316L Stainless Steel, Fatigue Life

Abstract. The freedom to manufacture metal components with very complex geometries using additive manufacturing techniques, such as laser powder bed fusion (LPBF), has opened new possibilities to produce innovative solutions with a high technological impact. It is therefore pivotal to have a detailed knowledge of the performance characteristics, both in the short and in the long term. Within this framework, this study firstly highlights the monotonic tensile properties of the LPBF samples by changing the laser scanning speed, the layer thickness, and the building orientation. Then, within the same process conditions, the fatigue life is investigated through reverse bending loading tests. The results verify an improved resistance, a reduced rigidity, and a strong anisotropy for the LPBF specimens if compared to the bulk material. The dependence on the orientation, together with the porosity of the LPBF samples, are the primarily responsible for the reduction of the fatigue limit.

Introduction

Stainless steels are metallic materials of high mechanical properties, good machinability, excellent corrosion resistance and low production costs. These properties promote their wide application in numerous engineering sectors, from biomedical, to automotive, aerospace, and so forth [1]. However, the traditional processes for producing this alloy, based on the subtraction of material, are characterised by low production flexibility, severe limitations in terms of complexity of the final part, as well as considerable investment costs and high resource consumption. The recent development and industrialisation of innovative non-conventional technologies have made it possible to overcome these barriers, providing more sustainable solutions than traditional processes [2].

Additive Manufacturing (AM) technologies have allowed to eliminate limitations on the complexity of geometry, materials, and level of customization, greatly increasing the process flexibility and prototyping capabilities [3], and, at the same, time reducing manufacturing waste and increasing process automation and sustainability [4]. Laser Powder Bed Fusion (LPBF) represents the most widely used and studied AM technology in industry and research. LPBF processes apply selective fusion of metal powders through the action of a high-power laser beam in an inert chamber, thus generating layer by layer a finished object with excellent mechanical performance and a high level of precision [5]. The current major limitations of LPBF processes are low productivity and high uncertainty regarding the quality and mechanical performance of the produced components [6], mainly due to the presence of defects such as trapped gas, unmelted material, oxides, etc. [7]. Therefore, a detailed knowledge of both the physics and the effect of process parameters on microstructure, internal defects, and mechanical performance, is necessary to guarantee long-lasting and good reliability in service of the products [8].

In this context, the characterisation of the fatigue life of LPBF-ed 316 L components is a subject worthy of much more in-depth study. This work therefore proposes to first analyse the mechanical and surface properties of the LPBF-ed samples, then to investigate their fatigue performance, by comparing the results with conventional 316L specimens. To this end, tests were first carried out under static tensile loading, then by applying a cyclic loading-unloading force, and finally, by means of reverse bending for fatigue life evaluation.

Experimental

The research study aims at evaluating the mechanical properties of 316L stainless steel components produced by using the laser powder bed fusion technique and compared with traditionally hot rolled laminates cut by laser (named "bulk" in the following). The experimental approach consisted in three main steps: (i) static monotonic tensile tests, to evaluate the fundamental mechanical properties and the initial data set to define the procedure of the following steps; (ii) load-unload tensile cyclic tests; (iii) reverse bending fatigue tests.

Materials and sample preparation

The material adopted to fabricate the samples by LPBF is a commercial metal powder with an average diameter of 32.4 μm supplied by Sandvik Osprey Ltd. and processed with the SLM 280HL machine by SLM Solutions Group. While the bulk samples have been cut starting from a 3 mm thick sheet laminate supplied by Hans-Erich Gemmel & Co. by using the CO_2 laser cutting machine TruLaser Cell 7020 by Trumpf. Table 1 and Table 2 show the main characteristics of the starting materials and the main features of the processing machines, as declared by the suppliers. Table 3 summarizes the design of the experiments based on the main process parameters, i.e. laser power, laser scanning speed, hatch distance, layer thickness, and building orientation. The choice was made according to preliminary studies aimed at obtaining full-dense samples for LPBF and minimizing burr formation for laser cutting. The geometry of the samples was the same for all the tests, according to the standards ASTM E8/E8M and ISO 3928 (Fig. 1).

Table 1 – Main characteristics and chemical composition of the starting materials.

Characteristic	LPBF				Bulk			
Melting point [°C]	1371-1399				1385-1400			
Density [g/cm³]	7.87				7.91			
Chemical composition [wt%]	Cr	16.9	P	0.03	Cr	17.5-19.5	P	0.045
	Ni	10.5	C	0.014	Ni	8.5-10.5	C	0.07
	Mo	2.3	S	0.005	Mo	-	S	0.03
	Mn	0.99	N	-	Mn	2	N	0.11
	Si	0.66	Fe	Bal.	Si	1	Fe	Bal.

Table 2 – Machine configurations for LPBF and laser cutting.

Feature	SLM 280HM	TruLaser Cell 7020
Working volume XYZ [mm³]	280×280×365	2000x1500x750
Max. laser power [W]	400	4000
Layer thickness [μm]	20-90	-
Laser beam focus diameter [μm]	80-115	50
Max. laser scanning speed [m/s]	10	2.5
Average gas consumption in process [L/min]	2.5 (Ar)	348 (N_2)

Italian Manufacturing Association Conference - XVI AITeM   Materials Research Forum LLC
Materials Research Proceedings 35 (2023) 173-181   https://doi.org/10.21741/9781644902714-21

Table 3 – Process parameters adopted to produce the samples.

Parameter	LPBF			Bulk
Laser power [W]	175			2800
Hatch distance [µm]	100			-
Laser scanning speed [mm/s]	750			50
Layer thickness [µm]	30			-
Building orientation [°]	0	45	90	-

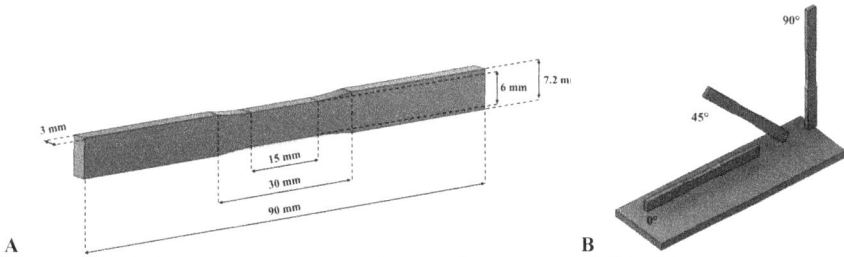

Figure 1 – Schematic representation of A) sample geometry and B) building orientation.

Characterization tests

The study of the mechanical properties was carried out by performing three different tests. At first, the fundamental mechanical properties, as elastic modulus (E), yield strength (Y_s), and ultimate tensile strength (UTS) were evaluated through monotonic quasi-static tensile tests by using the 50 kN MTS Insight Electromechanical Testing System with a crosshead speed set at 1.2 mm/min according to the ASTM E8/E8M standard. Then, the load-unload tests were completed on the same machine by applying an increasing load of 500 N every cycle, while the pure reverse bending tests were performed by using the 3 kN MTS Acumen Electrodynamic Testing System with a sinusoidal load and a frequency of 5 Hz. The latter tests were conducted on ad hoc designed bending system, shown in Fig. 2, in which is also schematized the bending moment distribution along the sample surface, being constant and resulting from the vertical load transferred by the top grip. The tests were considered valid over $8 \cdot 10^3$ cycles and marked as run-out after $2 \cdot 10^6$ cycles [9]. It is worth noting that all the tests were carried out at ambient temperature after preparing the samples by wire-cutting and grinding them to remove any unwanted protrusions.

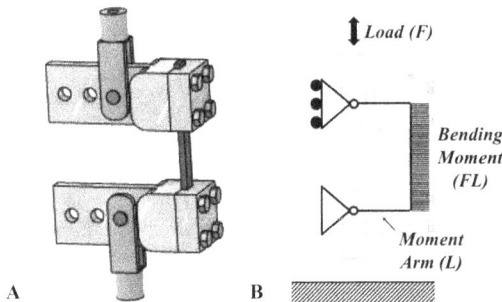

Figure 2 – Schematic representation of A) bending system and B) moment distribution.

After the tests, the failure surfaces were analysed with the scanning electron microscope SEM Leo SUPRA 35 by ZEISS. Moreover, the surface roughness was quantified by means of the

Italian Manufacturing Association Conference - XVI AITeM   Materials Research Forum LLC
Materials Research Proceedings 35 (2023) 173-181   https://doi.org/10.21741/9781644902714-21

arithmetical mean height of the surface parameter (Sa) according to the standard ISO 25178. The measurements, three for each sample, were performed before testing by using the 3D surface profiling system Talisurf CLI 2000 and further elaborated with the software MountainsMap® 7 by Digital Surf.

Results and Discussion

Quasi-static tensile characterization

The first step of the characterization dealt with the evaluation of the mechanical properties under quasi-static load and the comparison with bulk samples produced through conventional casting process. In this way, it was possible to define the range of loads which will be used during the following fatigue life tests. Moreover, the inspection of the surface quality in terms of *Sa* was performed. The main results for each building orientation are summarised in Table 4.

Table 4 – Mechanical and surface properties of the LPBF fabricated samples.

Property	LPBF			Bulk
	0°	45°	90°	
UTS [MPa]	696 ± 25.3	631 ± 12.9	584 ± 12.4	538 ± 12.0
Ys [MPa]	502 ± 14.2	486 ± 15.9	484 ± 10.1	274 ± 6.5
E [GPa]	168 ± 6.2	165 ± 8.8	167 ± 8.0	190 ± 4.4
Sa [μm]	9.86 ± 0.59	9.49 ± 2.97	8.04 ± 0.44	0.26 ± 0.05

Results from the quasi-static characterization highlight an anisotropic effect on both mechanical and surface properties due to the different building orientations along which the samples were fabricated. From the tensile strength values obtained it is notable that the 0° orientation allows the highest values of *UTS* and *Ys*, while for the modulus *E* the difference is negligible. It is worth to note the variability of the results which is quantified through standard deviation (shown in Table 4 after the plus/minus sign). Those results can be explained because larger and more frequent pores, as well as unmelted particles and inclusions, are more frequent at the border and in between two consecutive layers (Fig. 3), thus acting as stress concentrations and inhibiting the mechanical performances [10]. In fact, the 90° samples have layers orthogonally oriented with respect to the applied tension. Therefore, since they have more layer interfaces, a not favourable load orientation, and being the layer interfaces less cohesive areas, this orientation is characterized by the lowest values of *UTS* and *Ys*.

Figure 3 – SEM images of the fracture surface after tensile test of a 45° oriented sample.

It is worth noting that the *Ys* values of the bulk material (274 ±6.5 MPa) are almost the half of those obtained from the 0° oriented samples (502 ±14.2 MPa). Moreover, also *UTS* is improved, as it increases from an average of 538 ± 12.0 MPa of the bulk samples up to 696 ± 25.3 MPa of the horizontal LPBF samples. These results are ascribed to the Hall-Petch phenomena [11], for which the finer the grain size, the higher the number of grain boundaries, the greater the energy needed by a dislocation to move to another grain, and the higher the mechanical strength. In addition, the pile-up of the dislocations near the grain boundaries increases their density and

Italian Manufacturing Association Conference - XVI AITeM   Materials Research Forum LLC
Materials Research Proceedings 35 (2023) 173-181   https://doi.org/10.21741/9781644902714-21

decreases the free path for their movement [12]. For these reasons, the LPBF samples are more resistant to an external force application.

Despite such improvements, the elastic response of the material changes. The elastic modulus lowers from 190 ±4.4 GPa of the bulk samples down to an average modulus of 169 ±7.5 GPa for the LPBF. According to the literature [13], the reduction of stiffness depends on different factors such as the preferential orientation of the grains along one direction, the presence of porosities into the samples and the higher dislocation density and segregation effect.

The results on *Sa* highlights the most critical aspect of the LPBF process, with values up to 50 times higher. There is not a marked difference between the building orientations since the samples were produced with the same process parameters, however, it can be noted that the standard deviation of the tilted samples is 6 times higher than the others. This can be ascribed to the staircase effect which increase the overall variability of the external surface.

Load-unload tensile characterization
Load-unload tests were performed to evaluate a correlation with the fatigue behaviour of the samples in terms of accumulated damage. The main results of load-unload tests are summarized in Fig. 4, in which are showed the plots for the stress and the displacement (calculated through the extensometer) against time. The tests were carried out in the elastic regime. In fact, despite the increasing load steps at each cycle, the total stress is always lower than the calculated yield strength for each sample. Moreover, the maximum strain is always lower than 0.2% regardless of the experimental condition.

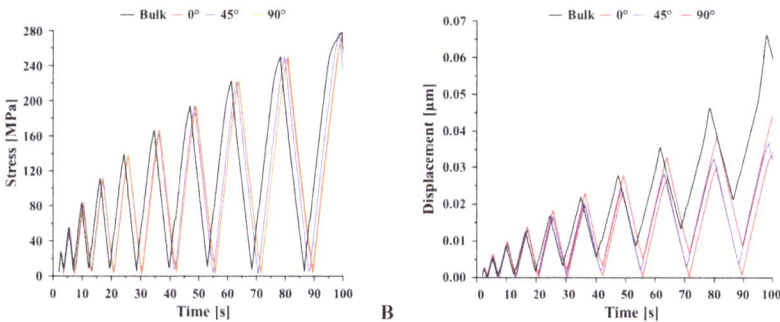

Figure 4 – A) Stress and B) strain over time calculated in load-unload tests.

To evaluate the difference in deformation of the bulk compared to LPBF samples, the maximum and minimum values at each load-unload step were plotted in Fig. 5A. As shown in the latter, the sample with the highest deformation during the loading and unloading phase is the bulk one, followed respectively by the orientations 90°, 45° and 0°. Fig. 5B shows the difference between the maximum and minimum displacement at each cycle, indicated as Δstrain. In particular, a higher value of Δstrain indicates a higher springback of the material. It is worth noting that the bulk is the one with the lowest springback, followed by the LPBF oriented at 90°, 45° and 0°.

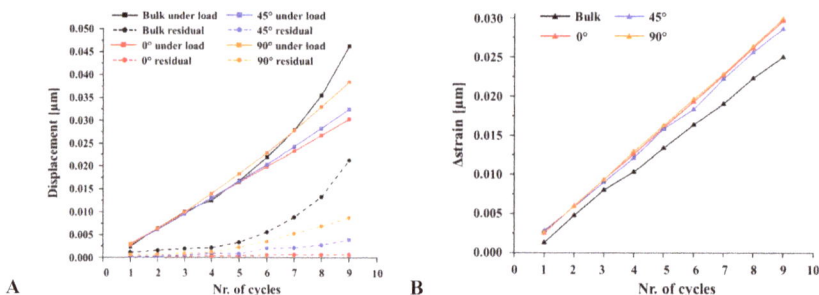

Figure 5 – A) Max. and min. displacement and B) Δstrain induced by load-unload cycles.

However, these results are not indicative to predict the fatigue behaviour since does not consider the damage induced for each cycle. With this aim, the variation of the slopes of the load curves at each cycle were calculated according to Eq. 1, thus, determining the damage induced by the test [14], where E_1 is the elastic modulus at the first cycle, and E_n is the apparent elastic modulus at the n-th cycle. Fig. 6 shows the results.

$$\text{Damage} = (1 - E_n/E_1). \tag{1}$$

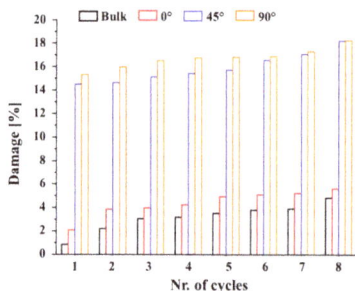

Figure 6 – Damage percentage at each load-unload step.

The LPBF samples with the 90° and 45° orientations present increasingly higher damage values compared to the 0° and the bulk. This result is attributable to the presence of more porosities in the LPBF material, since the most porous zones act as a stress amplifier and a preferential zone in which the crack propagation can start. The difference between the 0° samples compared to the others can be explained since the major porosities and inclusions are present within the printing layers which are oriented in the same direction as the load, as it is for the tilted and vertical samples.

Reverse bending characterization

To confirm the hypothesis from the analysis of the percentage damage, reverse bending fatigue tests were performed. The tests were carried out at varying stress magnitudes spanning the entire finite life region, i.e. from $8 \cdot 10^3$ to $2 \cdot 10^6$ cycles. Based on the previous investigation, the fatigue limit of the specimens is expected to be within the range 0.35 to 0.60 of the tensile strength [15]. To build up the fatigue curves, as a first guess, the reverse bending stress was set equal to around half the ultimate strength from the static tests. The main results are shown in Fig. 7.

Italian Manufacturing Association Conference - XVI AITeM   Materials Research Forum LLC
Materials Research Proceedings 35 (2023) 173-181   https://doi.org/10.21741/9781644902714-21

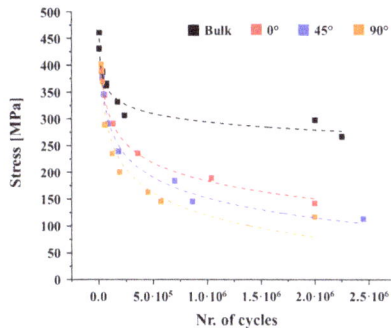

Figure 7 – Fatigue curves comparison.

All the LPBF fabricated specimens have similar low-cycle fatigue responses than the bulk material. This is usually ascribed to high surface roughness, tensile residual stresses, and the presence of pores and other defects that promote crack initiation [16]. The fatigue response of LPBF specimens is further found to be affected by the building orientation. The estimated fatigue limits range from 98.0 MPa for the 90° samples up to 168.3 MPa for 0°, i.e., around 17% and 24% of *UTS*, respectively. The obtained fatigue limits are therefore much lower than the expected fatigue strengths. Moreover, regardless of the parameters' combination, the fatigue strength is less than half of the conventionally processed samples.

The fatigue strength of the horizontal specimens is the highest among the building orientations considered within this study, while the vertical one is characterized by the lowest fatigue limit. This finding is in agreement within the results of the previous damage analysis and can be addressed to the orientation of the deposited layers relative to the applied load, which for the vertical samples is normal to the stress induced by the bending moment. In fact, consecutive layers and their boundaries are characterized by the presence of larger and more frequent pores, as well as unmelted particles and inclusions, therefore, they represent stress concentrators and crack initiators that reduce the fatigue life of the samples [17].

Conclusions

This study investigated the possibility to predict the fatigue behaviour of laser powder bed fused 316L stainless steel samples and traditionally bulk laminates cut by laser, with incremental load-unload tensile cyclic tests. The results obtained were compared with reverse bending fatigue tests on the same samples. The main conclusions which can be drawn are the following:

- During quasi-static tensile tests, the horizontal samples show a doubled yield strength and a 1.3 times ultimate tensile strength than the bulk ones, due to the refining of the grain size through the Hall-Petch phenomena.
- During the incremental load-unload tests, LPBF samples with the 90° and 45° orientations present increasingly higher damage values compared to the 0° and the bulk. This result is attributable to the presence of more porosities within the printing layers in the LPBF material.
- Load-unload tests also confirm the anisotropy highlighted during monotonic tensile tests, showing a worsening trend moving from horizontal to vertical in terms of damage percentage, which increases from approximately 6% up to around 18%.
- The reverse bending fatigue tests confirmed the damage prediction from the incremental load-unload tests since the fatigue behaviour of the horizontal specimens is the best among the building orientations while the vertical one is characterized by the lowest fatigue limit.

References

[1] Shin, W.S.; Son, B.; Song, W.; Sohn, H.; Jang, H.; Kim, Y.J.; Park, C. Heat Treatment Effect on the Microstructure, Mechanical Properties, and Wear Behaviors of Stainless Steel 316L Prepared via Selective Laser Melting. Materials Science and Engineering: A 806 (2021). https://doi.org/10.1016/j.msea.2021.140805

[2] Mehrpouya, M.; Vosooghnia, A.; Dehghanghadikolaei, A.; Fotovvati, B. The Benefits of Additive Manufacturing for Sustainable Design and Production. Sustainable Manufacturing (2021) 29-59. https://doi.org/10.1016/B978-0-12-818115-7.00009-2

[3] Ponticelli, G.S.; Tagliaferri, F.; Venettacci, S.; Horn, M.; Giannini, O.; Guarino, S. Re-Engineering of an Impeller for Submersible Electric Pump to Be Produced by Selective Laser Melting. Applied Sciences (Switzerland) 11 (2021). https://doi.org/10.3390/app11167375

[4] Gardner, L. Metal Additive Manufacturing in Structural Engineering - Review, Advances, Opportunities and Outlook. Structures 47 (2023) 2178-2193. https://doi.org/10.1016/j.istruc.2022.12.039

[5] Yasa, E. Selective Laser Melting: Principles and Surface Quality. Addit Manuf (2021). https://doi.org/10.1016/B978-0-12-818411-0.00017-3

[6] Ponticelli, G.S.; Venettacci, S.; Giannini, O.; Guarino, S.; Horn, M. Fuzzy Process Optimization of Laser Powder Bed Fusion of 316L Stainless Steel. Prog. Addit. Manuf. (2022). https://doi.org/10.1007/s40964-022-00337-z

[7] Liu, Y.; Zhang, M.; Shi, W.; Ma, Y.; Yang, J. Study on Performance Optimization of 316L Stainless Steel Parts by High-Efficiency Selective Laser Melting. Opt Laser Technol 138 (2021). https://doi.org/10.1016/j.optlastec.2020.106872

[8] Blinn, B.; Ley, M.; Buschhorn, N.; Teutsch, R.; Beck, T. Investigation of the Anisotropic Fatigue Behavior of Additively Manufactured Structures Made of AISI 316L with Short-Time Procedures PhyBaL LIT and PhyBaL CHT. Int J Fatigue 124 (2019) 389-399. https://doi.org/10.1016/j.ijfatigue.2019.03.022

[9] Riemer, A.; Leuders, S.; Thöne, M.; Richard, H.A.; Tröster, T.; Niendorf, T. On the Fatigue Crack Growth Behavior in 316L Stainless Steel Manufactured by Selective Laser Melting. Eng Fract Mech 120 (2014) 15-25. https://doi.org/10.1016/j.engfracmech.2014.03.008

[10] Casati, R.; Lemke, J.; Vedani, M. Microstructure and Fracture Behavior of 316L Austenitic Stainless Steel Produced by Selective Laser Melting. J Mater Sci Technol 32 (2016) 738-744. https://doi.org/10.1016/j.jmst.2016.06.016

[11] Tucho, W.M.; Lysne, V.H.; Austbø, H.; Sjolyst-Kverneland, A.; Hansen, V. Investigation of Effects of Process Parameters on Microstructure and Hardness of SLM Manufactured SS316L. J Alloys Compd 740 (2018), 740, 910-925. https://doi.org/10.1016/j.jallcom.2018.01.098

[12] Kocks, U.F.; Mecking, H. Physics and Phenomenology of Strain Hardening: The FCC Case. Prog Mater Sci 48 (2003) 171-273. https://doi.org/10.1016/S0079-6425(02)00003-8

[13] Saeidi, K.; Akhtar, F. Subgrain-Controlled Grain Growth in the Laser-Melted 316 L Promoting Strength at High Temperatures. R Soc Open Sci 5 (2018). https://doi.org/10.1098/rsos.172394

[14] Lemaitre, J.; Desmorat, R. Engineering Damage Mechanics: Ductile, Creep, Fatigue and Brittle Failures; 1st ed.; Springer: Berlin (2005).

[15] ASM International Fatigue. In Elements of Metallurgy and Engineering Alloys; ASM International, Ed.; ASM International: Ohio (2008) 243-264. https://doi.org/10.31399/asm.tb.emea.t52240243

[16] Pellizzari, M.; AlMangour, B.; Benedetti, M.; Furlani, S.; Grzesiak, D.; Deirmina, F. Effects of Building Direction and Defect Sensitivity on the Fatigue Behavior of Additively Manufactured H13 Tool Steel. Theoretical and Applied Fracture Mechanics 108 (2020). https://doi.org/10.1016/j.tafmec.2020.102634

[17] Fotovvati, B.; Namdari, N.; Dehghanghadikolaei, A. Fatigue Performance of Selective Laser Melted Ti6Al4V Components: State of the Art. Mater Res Express 6 (2018). https://doi.org/10.1088/2053-1591/aae10e

Italian Manufacturing Association Conference - XVI AITeM
Materials Research Proceedings 35 (2023) 182-190

Materials Research Forum LLC
https://doi.org/10.21741/9781644902714-22

Process parameters optimization in fused deposition modeling of polyether ether ketone

Emanuele Vaglio[1,2,a] *, Erica Billè[2,b], Marina Franulović[3,c],
Alessandro Gambitta[4,d], David Liović[3,e], Alfredo Rondinella[1,f],
Marco Sortino[1,g], Giovanni Totis[1,h]

[1]Polytechnic Department of Engineering and Architecture, University of Udine, Via delle Scienze 206, 33100 Udine, Italy

[2]Department of Engineering and Architecture, University of Trieste, Via Alfonso Valerio 6/1, 34127 Trieste, Italy

[3]Faculty of Engineering, University of Rijeka, Vukovarska 58, 51000 Rijeka, Croatia

[4]Elettra-Sincrotrone Trieste, Area Science Park, 34149 Basovizza, Trieste, Italy

[a] emanuele.vaglio@uniud.it, [b] erica.bille@studenti.units.it, [c] marina.franulovic@riteh.hr,
[d] alessandro.gambitta@elettra.eu, [e] dliovic@riteh.hr, [f] alfredo.rondinella@uniud.it,
[g] marco.sortino@uniud.it, [h] giovanni.totis@uniud.it

Keywords: Additive Manufacturing, Material Extrusion, Process Parameters Optimization

Abstract. Fused Deposition Modeling is increasingly used for producing high-performing, creep-resistant, biocompatible, fireproof, highly-stable parts from polyether ether ketone. However, the knowledge on this process is still poor and fragmented, and the lack of relevant data inhibits many applications. In this paper, the effects of the nozzle temperature, nozzle speed and layer thickness on the properties of PEEK processed by Fused Deposition Modeling were investigated by performing indentation, tensile, Scanning Electron Microscope, Computer Tomography and Energy Dispersive X-ray Spectroscopy tests on as-built samples. The outgassing behavior was also analyzed, while the synchrotron radiation was used to characterize the structure of selected samples on a hitherto unexplored scale. The samples morphology was finally used to identify the optimal process window. The results provided new insights on the process and novel data enabling new applications.

Introduction

Polyether ether ketone (PEEK) is a semi-crystalline thermoplastic polymer belonging to the polyaryl ether ketone (PAEK) family. This fully recyclable material has excellent mechanical properties, chemical and radiation stability, biocompatibility and can withstand exceptionally high temperatures [1]. These unique properties make PEEK a highly appropriate material for multiple high-performance applications in several fields, including medical, aerospace, electrical, and chemical [2]. However, the processability of materials through additive techniques is today a fundamental requirement in these fields.

For a long time, Selective Laser Sintering (SLS) was the prevailing method used for additive processing of PEEK due to its high melting temperature. However, this approach has important drawbacks [3] that led scientists and industry to explore alternative solutions, such as Fused Deposition Modeling (FDM). Recently, Ding et al. [3] showed that increasing nozzle temperature (T_n) improves strength of PEEK parts produced by FDM due to decreased porosity and enhanced layer bonding. Sikder et al. [4] showed that heating the build chamber also improve the mechanical properties of the material, while heating the build platform enhances the adhesion of the material to the build surface but does not positively affect its mechanical properties. In the same study, it

Italian Manufacturing Association Conference - XVI AITeM   Materials Research Forum LLC
Materials Research Proceedings 35 (2023) 182-190   https://doi.org/10.21741/9781644902714-22

was found that higher nozzle speed (v_n) and lower layer thickness (t_l) results in improved mechanical properties. Wu et al. [5] surprisingly observed that the best mechanical properties are obtained at intermediate values of t_l, and that also the raster angle has a significant effect on the tensile, compression, and three-point bending behavior. Vaezi et al. [6] proved that using a heated build platform and heated build chamber together can effectively mitigate the effects of the exceptional thermal stresses caused by the high melting temperature of PEEK and prevent the material's deformation and delamination. The authors noted that too high T_n results in low accuracy due to the extruded bead deformation and in material degradation, while too low T_n causes delamination due to insufficient bonding with the previous layers and nozzle clogging. Lee et al. [7] found that many factors influence the cooling rate and, in turn, crystallinity, including the build platform and build chamber temperature, v_n, the raster angle, and the number and geometry of parts produced simultaneously, which impact the time interval between successive layers. Yang et al. [8] showed that T_n influences multiple phenomena, including crystal melting, crystallization, bonding between extruded beads, and polymer degradation. The authors pointed out that performing thermal treatments may be the most effective way to improve the crystallinity and the mechanical properties of PEEK processed by FDM. El Magri et al. [9] found that the annealing temperature significantly influences the crystallization processes and that the treated material is stronger but less ductile. Zhao et al. [10] showed that none of the FDM parameters influence the chemical composition of the material.

Although noteworthy advancements in optimizing the FDM technology for processing PEEK have been made in recent years, the knowledge in this field remains scattered and incomplete. Therefore, further progress is needed to apply this technology in both conventional and advanced sectors that would benefit the most from its potential. In this paper, the effects of T_n, v_n, and t_l on the properties of parts produced from PEEK by FDM were analyzed to identify the optimal process condition and demonstrate the suitability of the process for critical applications.

Materials and methods

The samples tested in this work were prepared using a commercial PEEK filament of 1.75 mm diameter. Before use, the filament was analyzed by X-ray Computed Microtomography (X-ray μCT) and by Scanning Electron Microscope (SEM). No defects, impurities or pores were observed.

Subsequently, 18 cubic samples 20x20x20 mm in size were produced on an Intamsys Funmat HT machine. The cubes consisted of a core and a contour volume that overlapped with each other by 0.25 mm, and were produced one at a time by varying the process parameters according to a full factorial design of experiments. In more details, the nozzle temperature (T_n) was varied on 3 levels from 380 °C to 420 °C by discrete increments of 20 °C, the core nozzle speed (v_{nb}) was varied on 3 levels from 15 mm/s to 65 mm/s by increments of 25 mm/s, and the layer thickness (t_l) was varied on 2 levels corresponding to 0.1 mm and 0.2 mm. The core infill ratio was 100 % and it was obtained by depositing the material according to a bi-directional and rotated pattern resulting in a ±45 ° raster angle. The v_{nb} was reduced by 20 % during the deposition of the 6 top and bottom layers, according to the recommendations of the machine manufacturer. The contour was obtained instead by depositing 3 perimeter outlines at a reduced nozzle speed $v_{nc}=0.5v_{nb}$. In this way, real operating conditions were reproduced, which is crucial for optimizing the FDM process parameters for real applications. The nozzle diameter was kept fixed and equal to $d_n=0.4$ mm, while the building plate temperature and the building chamber temperature were $T_{bp}=145$ °C and $T_{bc}=90$ °C, respectively. The fan speed was set at 50 % of the maximum value admitted by the machine.

The so obtained samples were measured with a Mitutoyo ABS AOS digital caliper to assess the dimensional conformity, and they were analyzed with a Sensofar S neox confocal microscope to study the average surface roughness (R_a). The roughness profiles were extracted perpendicular to the grooves formed between layers or beads, and a cut-off wavelength of 0.8 mm was used for computing R_a. The top horizontal surfaces of the samples were also examined with a Zeiss Evo 40

Italian Manufacturing Association Conference - XVI AITeM   Materials Research Forum LLC
Materials Research Proceedings 35 (2023) 182-190   https://doi.org/10.21741/9781644902714-22

Scanning Electron Microscope to characterize the defects caused by different process conditions and to analyze the chemical composition of the material by Energy Dispersive X-ray Spectroscopy (EDXS).

The samples produced using the extreme process configurations of the experimental design, representing the most and least favorable condition for obtaining fully dense parts, underwent X-ray μCT at the TomoLab station of Elettra Sincrotrone Trieste research center to investigate internal porosity. The voltage of the X-ray tube was set to 70 kV, and a 0.25 mm thick aluminum filter was positioned between the source and the samples. The system resolution was 18 μm. The porosities of the samples produced at high T_n were further analyzed on a resolution scale of 0.88 μm by using the SYRMEP Synchrotron X-ray μCT of Elettra Sincrotrone Trieste research center. The samples were subsequently sectioned and polished to obtain flat surfaces for hardness testing, which were performed according to the EN ISO 2039-1 standard with an applied load of 358 N. In details, 10 measurements were taken for each sample using a DuraJet G5 tester equipped with a 5 mm diameter spherical indenter. Eventually, 5 tensile test samples complying to the ISO 527-5A specimen type were produced using each set of the selected process parameters and tested using the strain rate of 1% min^{-1}, as recommended in ISO 521-1. The load cell of 25 kN was used for load measurements, while the Epsilontech 3442-010M-050M-ST extensometer with the gauge length of 20 mm was used for strain measurements during the tensile tests.

The Ultra-High Vacuum (UHV) behavior of the most promising sample was also evaluated by Residual Gas Analysis (RGA) performed using a mass spectrometer HAL 201 RC after bake-out at 120 °C and 200 °C.

Results and discussion

Dimensional conformity. Fig. 1 shows that the dimensional error parallel to the building plane increased with increasing T_n and t_l while decreasing v_n. Perpendicular to the building plane, the dimensional error was instead primarily influenced by T_n which again led to larger dimensional error by causing the extrusion of expanded and overly fluidized material that was spread out around the nozzle sides forming surface defects. Nevertheless, a weak non-monotonic effect of v_n was also observed. Interestingly, t_l did not affect the dimensional error in the vertical direction. The analysis of variance confirmed the significance of all the factors analyzed, except for t_l in relation to the dimensional error in the vertical direction.

Surface roughness. Major adverse effects on the quality of both horizontal and vertical surfaces were observed with increasing t_l, while minor favorable effects resulted from increasing v_n, presumably because the cooling efficiency was higher when the hot nozzle rapidly left the solidification area. No clear trends were observed for T_n. Instead, a sharp increase in R_a of horizontal surfaces due to the formation of visible defects was noticed above a specific threshold temperature. The analysis of variance confirmed the statistical validity of these findings.

Surface defects. The SEM inspections revealed that the horizontal surfaces of the samples produced at high T_n and low v_n (Fig. 2 (a)) exhibit burrs caused by over-extrusion (low speed) of expanded and overly fluidized (high temperature) material that is subsequently spread out by the nozzle. The severity of this flaw increased with increasing t_l due to a higher extrusion rate, and decreased with increasing v_n. However, it remained partially visible even in the samples produced at intermediate v_n. The horizontal surfaces of the samples produced at low T_n and high v_n (Fig. 2 (b)) exhibit instead lack of fusion voids between the deposited beads, and small elongated craters aligned in the direction of nozzle motion on their surface. These craters were likely a consequence of the high viscosity of the material, which reduces wettability [11] and promotes the formation of tears in the extruded and stretched material. It was found that craters and voids decreased both in number and size with increasing t_l and decreasing v_n. However, significant reductions in v_n are required to avoid this flaw since it was observed even in samples produced at intermediate v_n. The horizontal surfaces of the samples produced at low T_n and low v_n (Fig. 2 (c)) finally appeared to be uniform and flawless, especially when thin layers were processed.

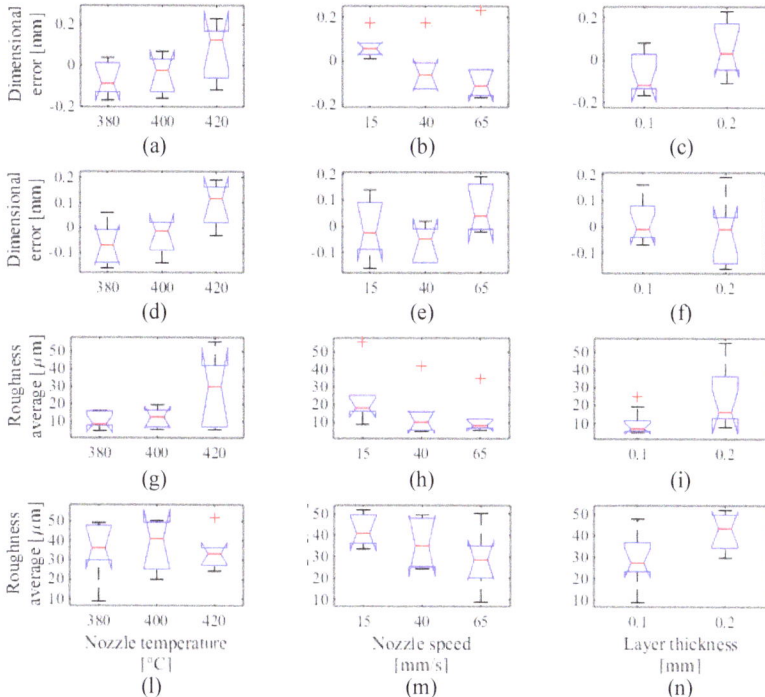

Fig. 1: Effect of the process parameters on (a), (b), (c) the dimensional error parallel to the building plane, (d), (e), (f) the dimensional error perpendicular to the building plane, (g), (h), (i) the roughness average of the surface parallel to the building plane and (l), (m), (n) the surface roughness of the surfaces perpendicular to the building plane.

Density. The samples produced at high v_n and high t_l were found to be exceedingly porous. Under low T_n condition (Fig. 3 (e)) the pores were formed due to under-extrusion, which resulted in air gaps between the deposited beads. Nevertheless, some spherical pores within the beads were also observed. Under high T_n condition (Fig. 3 (b)) the pores were instead uniformly distributed and had a rounded shape suggesting the entrapment of gas. This type of porosity is shown in more detail in Fig. 3 (f) where a complex structure of morphologically regular pores belonging to a wide dimensional spectrum reaching the nanoscale can be observed. The origin of these pores is still not clear. They are usually attributed to melt flow and solidification phenomena, or to gas voids primarily generated during the filament fabrication [12]. However, the preliminary tests performed on the filament used in this study proved the absence of pores.

The samples produced at low v_n and low t_l exhibited considerably higher density ρ. Under low T_n condition (Fig. 3 (d)), the initial layers of the sample were free of pores, while the upper layers contained a non-negligible amount of under-extrusion pores located between the deposited beads. This effect was attributed to the heated platform, which contributes to the energy input in the initial stages of the process and is crucial for interpreting the results of mechanical tests carried out on thin tensile samples. Under high T_n condition (Fig. 3 (a)), the sample was instead practically free of pores, although the Synchrotron X-ray μCT revealed the presence of some isolated defects (Fig. 3 (c)).

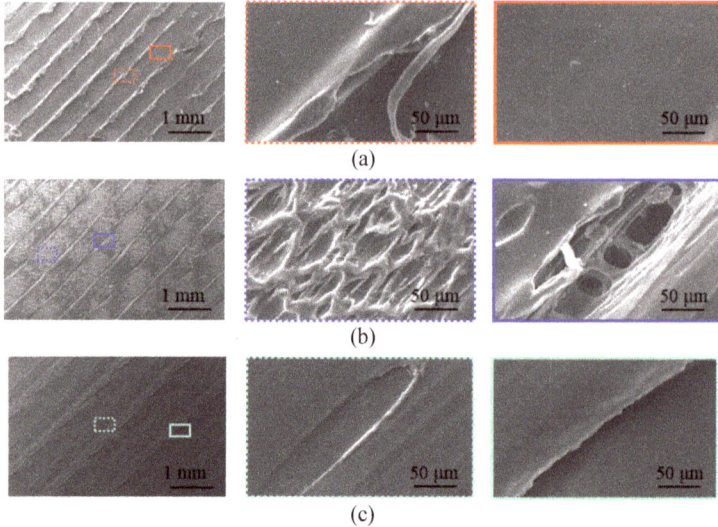

Fig. 2: SEM view of defects on the horizontal surfaces of the samples produced at (a) $T_n=420$ °C, $v_{nb}=15$ mm/s and $t_l=0.1$ mm, (b) $T_n=380$ °C, $v_{nb}=65$ mm/s and $t_l=0.1$ mm, (c) $T_n=380$ °C, $v_{nb}=15$ mm/s and $t_l=0.1$ mm.

Finally, no pores were found in the samples contour, except when the sample produced at high T_n, high v_n and high t_l was examined. Under these conditions, several under-extrusion defects formed in limited areas. However, these defects were attributed to casual factors since pores were localized and not detected in the sample produced at a lower T_n. Some isolated pores caused by insufficient interpenetration between core and contour were detected in all the samples.

Hardness. The mean and standard deviation of the hardness measured on the selected samples are reported in Table 1. Regardless of T_n, the combination of high v_n and high t_l resulted in very low hardness, while the combination of low v_n and low t_l resulted in better outcomes. Overall, a robust correlation with the internal porosity of the material was observed. It is worth noting that the hardness measured on the fully dense sample was slightly lower than the hardness of the conventionally processed PEEK.

Mechanical properties. The mechanical properties on the selected samples are reported in the Table 1. The lowest mean value of ultimate tensile strength (UTS) were obtained from the samples produced at low T_n, high v_n, and high t_l, which were also characterized by the lowest maximum elongation. Conversely, the highest mean value of the maximum elongation was found on samples produced at high T_n, low v_n, and low t_l. The mean values of the Young's modulus (E) were similar and within the statistical uncertainty. These results are in accordance with the density results shown in Fig. 3. However, it was not possible to establish a direct correlation between them since a large portion of the tensile samples consisted of contours, and the entire volume was deposited in close proximity to the building platform.

Table 1: Hardness and tensile test results given as mean value (standard deviation).

T_n [°C]	v_n [mm/s]	t_l [mm]	HB [MPa]	UTS [MPa]	E [MPa]	Max. elong. [%]
420	15	0.1	166 (20)	84 (3)	3161 (359)	126 (54)
380	15	0.1	147 (13)	83 (3)	3284 (277)	27 (8)
420	65	0.2	53 (4)	83 (1)	3267 (105)	17 (2)
380	65	0.2	53 (15)	77 (1)	3221 (272)	5 (1)

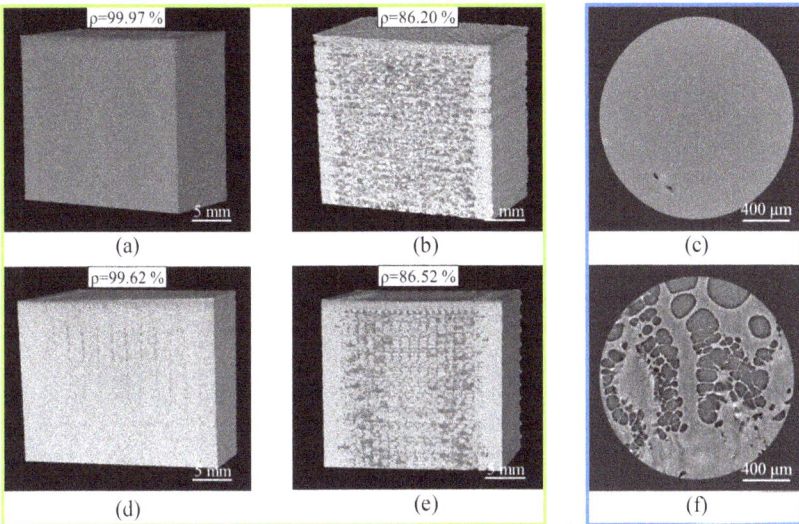

Fig. 3: X-ray μCT of the samples produced (a) T_n=420 °C, v_{nb}=15 mm/s and t_l=0.1 mm, (b) T_n=420 °C, v_{nb}=65 mm/s and t_l=0.2 mm, (d) T_n=380 °C, v_{nb}=15 mm/s and t_l=0.1 mm, (e) T_n=380 °C, v_{nb}=65 mm/s and t_l=0.2 mm, and Synchrotron X-ray μCT of the samples produced at (c) T_n=420 °C, v_{nb}=15 mm/s and t_l=0.1 mm, (f) T_n=420 °C, v_{nb}=65 mm/s and t_l=0.2 mm.

Chemical properties. No contaminants were found by EDSX inspections, while statistical analysis of the data showed that none of the investigated process parameters significantly influenced the carbon-oxygen weight ratio, which on average was C/O=3.57.

Ultra-High Vacuum behavior. The UHV behavior of the sample produced at T_n=380°C, v_n=15 mm/s and t_l=0.1 mm is shown in Fig. 4. No significant deviations from the background mass spectrum after bake-out at 120 °C were found, except for a small increase in O_2 and H_2, and a further reduction in partial pressure throughout the spectrum was obtained after bake-out at 200 °C. Therefore, it can be inferred that good levels of UHV can be achieved within a short time frame already at low bake-out temperature. This confirmed the suitability of PEEK for UHV applications as no significant outgassing level was recorded.

Italian Manufacturing Association Conference - XVI AITeM

Materials Research Forum LLC

Materials Research Proceedings 35 (2023) 182-190

https://doi.org/10.21741/9781644902714-22

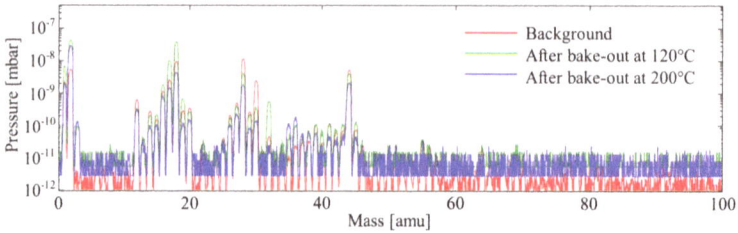

Fig. 4: Residual Gas Analysis under Ultra-High Vacuum condition of the samples produced at T_n =380 °C, v_{nb}=15 mm/s and t_l=0.1 mm.

(a)

(b)

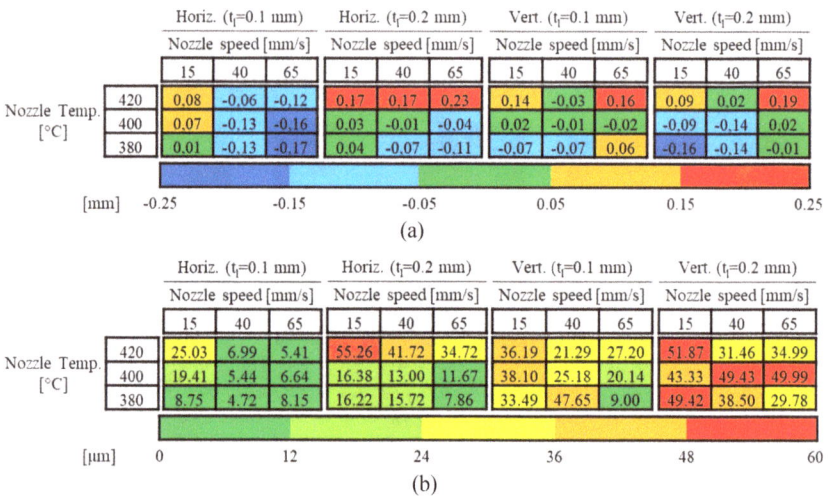

Fig. 5: Process maps obtained by analyzing the variability range of (a) the dimensional error and (b) the average surface roughness.

Optimal process parameters. Fig. 5 shows the process map obtained by classifying the dimensional errors and the average surface roughness according to Sturges' formula [13]. The optimal process conditions for each corresponding property are highlighted in dark green. The samples produced at T_n=380 °C, v_{nb}=15 mm/s and t_l=0.1 mm achieved the highest relative evaluation (two first-class ratings, one second-class rating, and one third-class rating). This set of parameters proved the capacity of providing the best surface quality, high hardness, and suitable parts for UHV applications, but they caused the formation of some under-extrusion pores. Other five sets of parameters achieved the same rating. However, three of them involved a high v_n, which proved to be unsuitable for producing dense and resistant parts. The remaining two sets involve intermediate v_n. These parameter sets would increase productivity, but additional tests are required to punctually examine the process output in this v_n range.

Overall, advanced optimization algorithms can be used to process this data from time to time to identify the optimal process parameters for specific multi-objective requirements. When doing this, productivity should also be considered.

Summary
In this paper, the influence of the nozzle temperature, nozzle speed and layer thickness on the properties of polyether ether ketone processed by Fused Deposition Modeling were analyzed with the aim of identifying the optimal process parameters. Results showed that geometrical conformity and surface quality can be generally improved by decreasing T_n and t_l, and by increasing v_n. However, highly porous material is obtained at high v_n and t_l either due to under-extrusion (low T_n) or due to flow and solidification phenomena (high T_n) resulting in complex structure of regularly-shaped pores. It was observed that the mechanical properties of the samples were primarily influenced by the material density, but even in fully dense conditions they were slightly lower compared to conventional material. Fully dense PEEK was instead proved to be suitable for UHV applications. However, further investigation in the intermediate v_n range is necessary to determine the absolute optimal process conditions.

Acknowledgements
The authors are grateful to Dott. D. Dreossi for performing μCT analysis, to Dott. L. Novinec for performing UVH tests and to Ing. I. Cudin for the helpful support. Part of the experimental work has been performed using the equipment acquired through the Croatian Science Foundation project IP-2019-04-3607 and by University of Rijeka under project uniri-tehnic-18-34. Elettra-Sincrotrone Trieste and the Laboratory for Advanced Mechatronics - LAMA FVG - of the University of Udine are also gratefully acknowledged for providing the equipment used for the experimental work and for the technical support. Emanuele Vaglio is grateful for funding under the REACT EU Italian PON 2014–2020 Program – Action IV.4 – Innovation (DM 1062, 10/08/2021).

References

[1] S.M. Kurtz, PEEK biomaterials handbook, second ed., William Andrew Publishing, 2019.

[2] R. Dua, Z. Rashad, J. Spears, G. Dunn and M. Maxwell, Applications of 3d-printed peek via fused filament fabrication: A systematic review, Polymers 13 (2021) 4046. https://doi.org/10.3390/polym13224046

[3] S. Ding, B. Zou, P. Wang and H. Ding, Effects of nozzle temperature and building orientation on mechanical properties and microstructure of PEEK and PEI printed by 3D-FDM, Polym. Test. 78 (2019) 105948. https://doi.org/10.1016/j.polymertesting.2019.105948

[4] P. Sikder, B.T. Challa and S.K. Gummadi, A comprehensive analysis on the processing-structure-property relationships of FDM-based 3-D printed polyetheretherketone (PEEK) structures, Materialia 22 (2022) 101427. https://doi.org/10.1016/j.mtla.2022.101427

[5] W. Wu, P. Geng, G. Li, D. Zhao, H. Zhang, and j. Zhao, Influence of layer thickness and raster angle on the mechanical properties of 3D-printed PEEK and a comparative mechanical study between PEEK and ABS, Materials 8 (2015) 5834-5846. https://doi.org/10.3390/ma8095271

[6] M. Vaezi and S. Yang, Extrusion-based additive manufacturing of PEEK for biomedical applications, Virtual Phys. Prototyp. 10 (2015) 123-135. https://doi.org/10.1080/17452759.2015.1097053

[7] A. Lee, M. Wynn, L. Quigley, M. Salviato and N. Zobeiry, Effect of temperature history during additive manufacturing on crystalline morphology of PEEK, Adv. Ind. Manuf. Eng. 4 (2022) 100085. https://doi.org/10.1016/j.aime.2022.100085

[8] C. Yang, X. Tian, D. Li, Y. Cao, F. Zhao and C. Shi, Influence of thermal processing conditions in 3D printing on the crystallinity and mechanical properties of PEEK material, J. Mater. Process. Technol. 248 (2017) 1-7. https://doi.org/10.1016/j.jmatprotec.2017.04.027

Italian Manufacturing Association Conference - XVI AITeM   Materials Research Forum LLC
Materials Research Proceedings 35 (2023) 182-190   https://doi.org/10.21741/9781644902714-22

[9] A. El Magri, K. El Mabrouk, S. Vaudreuil, H. Chibane and M.E. Touhami, Optimization of printing parameters for improvement of mechanical and thermal performances of 3D printed poly (ether ether ketone) parts, J. Appl. Polym. Sci. 137 (2020) 49087. https://doi.org/10.1002/app.49087

[10] F. Zhao, D. Li and Z. Jin, Preliminary investigation of poly-ether-ether-ketone based on fused deposition modeling for medical applications, Materials 11 (2018) 288. https://doi.org/10.3390/ma11020288

[11] K. Okumura, Y. Tanaka and K. Iwai, Effect of Viscosity and Surface Roughness on Improvement of Solid-liquid Wettability by Ultrasonic Vibration, ISIJ Int. 62 (2022) 2217-2224. https://doi.org/10.2355/isijinternational.ISIJINT-2022-268

[12] E.A. Papon, A. Haque and S.B. Mulani, Process optimization and stochastic modeling of void contents and mechanical properties in additively manufactured composites, Compos. B: Eng. 177 (2019) 107325. https://doi.org/10.1016/j.compositesb.2019.107325

[13] H.A. Sturges, The choice of a class interval, J. Am. Stat. Assoc. 21 (1926) 65-66. https://doi.org/10.1080/01621459.1926.10502161

Italian Manufacturing Association Conference - XVI AITeM
Materials Research Proceedings 35 (2023) 191-197

Materials Research Forum LLC
https://doi.org/10.21741/9781644902714-23

Experimental analysis of FDM structures in shape memory polylactic acid

Maria Pia Desole[1,a], Annamaria Gisario[1,b*], Franco Maria Di Russo[1,c], Massimiliano Barletta[2,d]

[1] Sapienza Università di Roma, Dipartimento di Ingegneria Meccanica e Aerospaziale, Via Eudossiana 18, 001884 Roma (Italy)

[2] Università degli Studi Roma Tre, Dipartimento di Ingegneria Industriale, Elettronica e Meccanica, Via della Vasca Navale 79, 00146 Roma (Italy)

[a]mariapia.desole@uniroma1.it, [b]annamaria.gisario@uniroma1.it, [c]francomaria.dirusso@uniroma1.it, [d]massimiliano.barletta@uniroma3.it

Keywords: Additive Manufacturing, Energy Absorption, Shape Recovery

Abstract. The behavior of solid cellular structures in polylactic acid (PLA) manufactured by Fused Deposition Modeling (FDM) is herein investigated. In particular, the manuscript investigates the capability of permanently deformed PLA structures to restore their starting shapes, once a thermal stimulus is applied on them. In this study, a structure called Rototetrachiral was produced, which originates from Rotochiral and Tetrachiral. The latter was tested to verify its mechanical response and its ability to absorb energy when subjected to a compression stress, repeated over several cycles. The experimental results showed a close connection between the structure's ability to absorb energy and its extent of damage, which gradually increases with the number of cycles. Microscopic analysis shows that the central cells are the most deformed. However, the applied thermal stimulus allows to recover the deformation, ensuring good performance of the structure for a certain number of cycles.

1. Introduction

Shape memory materials are materials that once programmed into a temporary form, following an external stimulus, for example thermal, can return to their original configuration [1]. In the last decade, the applications in which they are involved are many: from the robotics sector [2], to the biomedical sector [3,4] and their presence is also evident in the civil engineering sector [5] and textiles [6,7]. Shape memory materials can be either polymeric [8], or metal alloys [9]. These materials combine well with the growing interest in additive technologies, in particular 4D printing [10,11]. 4D printing starts from the fundamentals of 3D printing, taking into account a fourth dimension, namely time [12,13]. By using additive technology and exploiting the properties of shape memory materials, we have seen how it is possible to study the behavior of certain types of reticular structures, for example chiral, anti-chiral, bio-inspired or auxetic [14]–[16]. Often such structures are studied from the mechanical point of view, but in reality they act as good energy absorbers, also thanks to the shape return properties of the material chosen for the manufacture. It has been seen in some studies as the design of the cell and the process parameters, influence in an important way the maximum compression load and energy absorption [17], [18]. PLA is more powerful than other polymers commonly used in FDM technology [19]. Changing printing parameters affects the compression behavior and the shape recovery of the cell structure. By varying the thickness of the layer, the printing speed, the temperature of the nozzle and the thermal stimulus of the product, it is clear that the latter is decisive for the trigger of the recovery of shape [20]. Until now, the reticular structures have been analysed, carrying out a single cycle of deformation-recovery of form, without further investigating what happens as the number of cycles

Italian Manufacturing Association Conference - XVI AITeM   Materials Research Forum LLC
Materials Research Proceedings 35 (2023) 191-197   https://doi.org/10.21741/9781644902714-23

increases. In addition, there is no analysis of the state of damage to structures and how this affects the capacity of the structure to absorb energy. The following scientific paper analyses a chiral structure not present in the literature: the Rototetrachiral, inspired by the Tetrachiral and Rotochiral structures. These geometries were studied in [21]. Rototetrachiral has been designed as it has an intermediate absorption behaviour between Rotochiral and Tetrachiral. Wanting to carry out a series of cycles on the structure it was preferred to design a new structure with intermediate characteristics and with medium-high absorption capacity. Manufacturing was carried out using FDM additive technology and PLA was chosen as the material. Once it has been subjected to compression, its shape has been recovered by external thermal stimulus. The procedure was repeated over several cycles in order to assess the energy absorption capacity of the geometry. By means of a microscope visual analysis, it was analysed how the state of damage affects the mechanical properties of the structure, as well as the energy absorption of the structure.

2. Materials and Methods
2.1 Definition of geometry and material

The Rototetrachiral is a structure that has 4 adjacencies, that is 4 elements called arms, that originate from the circular element, arranged at 90°. The relative density chosen at the design stage is 0.153, lower than the geometries referred to. The pattern was then recreated in the CAD environment through the software "Autodesk Inventor 2021". Figure 1 shows the CAD model of the geometry (a), a front section of the geometry (b) and a detail of the cell (c).

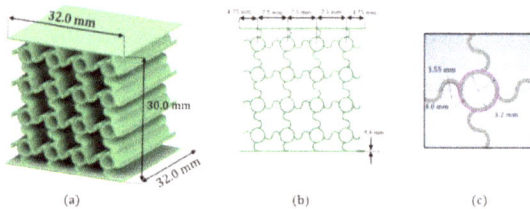

Figure 1. CAD model of the geometry (a), front section (b) and a detail of the cell (c).

ùThe 30x30x30 mm^3 cube is delimited by two layers (top and bottom respectively) of 4 mm thickness and 32x32 mm^2 dimension. The geometry was produced in PLA with green color. The 1 mm increase in layer size facilitates the increase of the adhesion surface with the plates and a more gradual stress distribution.

2.2 3D printing process

The production of the samples was done using FDM technology, with an Ultimaker 3D printer, model S5. The Software "Ultimaker Care" allowed you to set the print parameters.

In particular, attention was paid to the thickness of the layer and the printing speed of 0.2 mm and 50 mm/s, respectively chosen to obtain aesthetically appreciable products and not to induce states of residual stress [22]. The printing temperature was set to 200 °C as per material datasheet, while the filling density of 100% was chosen to exhibit maximum mechanical strength. These parameters made sure to have three specimens dimensionally identical in height equal to 32 mm. The step of 1 mm in both the top and bottom layer is used to correctly realize the geometry, without problems of thermal expansion of the layers in contact with the plane. In this regard, a hydrophilic support in polyvinyl alcohol is inserted, removed as a result of the printing process.

Italian Manufacturing Association Conference - XVI AITeM
Materials Research Proceedings 35 (2023) 191-197

Materials Research Forum LLC
https://doi.org/10.21741/9781644902714-23

2.3 Compression tests

Downstream of the production of the specimens they are subjected to the compression test, with the load applied perpendicular to the two layers. The machine used for the test is Shimadzu, model "Autograph AGS-X series" with 5 kN load cell and guaranteed maximum error of 1%. The standard considered for the test is the ASTM C365, specific to test compression sandwich structures made of polymer. The tests were carried out with a maximum displacement of 8 mm and the prescribed lowering speed of the plate of 3 mm/min. To ensure a certain reliability of the results and in the mechanical behaviour of the geometry, three replicates are made.

2.4 Springback and shape recovery

Springback is measured by means of a slide gauge, with measurements being taken for one hour every 5 minutes. For the recovery of form, on the other hand, the specimens are immersed in a bath at a controlled temperature of 75 °C, higher than the glass transition of polylactic acid, which is about 62 °C. A tracking software has been used to monitor the recovery of shape. Two markers are placed at the ends of the top layer and the movement has been traced, through the acquisition of recordings made with full HD JVC camera, model GZ-E205.

2.5 Cycle analysis and subsequent shape recovery

The above steps from the compression test to the shape return have been repeated for a total of 6 cycles. At the sixth cycle the structure suffered a permanent failure, to be no longer usable, so it was decided to finish the trial at the sixth cycle, leaving out the springback and form recovery phases for the sixth cycle.

3. Results and discussion
3.1 Compression tests 8 mm – 1st cycle

Figure 2a shows the Force-Displacement diagram of the 8 mm displacement compression test. From the graph you can see how the maximum load is reached near the displacement of 2 mm and is repeated, in a similar way for movements close to 4 and 6 mm. There are three peaks whose ascending phases correspond to the longitudinal deformation of the rows of cells that make up the sample. The first peak refers to the deformation of the central files, while the other two refer to the external files in contact with the layers. The descending phases instead correspond to the collapse of the row of cells.

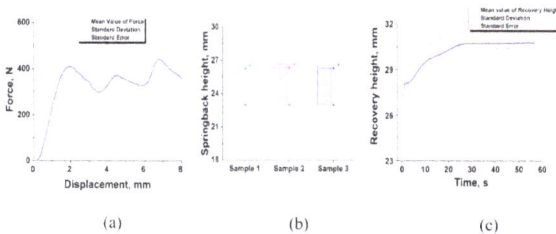

(a)   (b)   (c)

Figure 2. Force-Displacement diagram of the compression test at 8 mm (a), springback for the three samples (b) and the shape recovery (c)

The maximum load borne by the structure is slightly higher than 400 N, lower than other chiral structures, due to its low relative density [23].

Italian Manufacturing Association Conference - XVI AITeM   Materials Research Forum LLC
Materials Research Proceedings 35 (2023) 191-197   https://doi.org/10.21741/9781644902714-23

3.2 Springback and shape recovery

For springback at zero time, the height value at the end of compression is considered. While at time t = 1 min, this refers to the height value at the instant after the plate separation. The springback height undergoes minimal variation in the three test pieces, with an average value of approximately 26 mm, indicating good replicability of the tests, as shown in Figure 2b. Figure 2c shows the average Rototetrachiral shape recovery model, measured over a time interval of 60 s. Recovery ends at 25 s and is not instantaneous due to geometric damage.

3.3 Compression tests 8 mm - up to the 6th cycles

Once a complete test cycle has been carried out, the following cycles have been carried out, until the geometry has been completely broken. Figure 3a shows the average trend of the Force-Displacement curves of the compression test for the 6 cycles envisaged. The graph shows that there is a strong load drop between the first and the second cycle, with the maximum load being reduced by about 38%. Between the second and third cycles, however, there is an almost overlap of the Force-Displacement curves and a limited load drop. A justification for this phenomenon can be found in the fact that the second and third cycles are carried out 48 hours apart, resulting in aging of the material. Similarly, the situation is repeated between the fifth and sixth cycles.

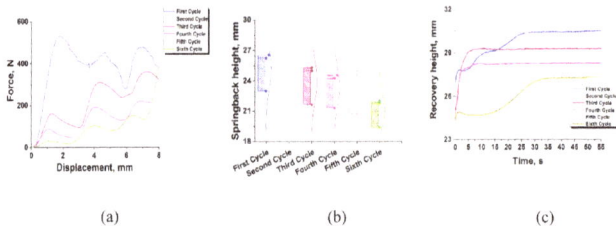

(a)   (b)   (c)

Figure 3. Force-Displacement diagram of the compression test up to the 6th cycles (a), springback for 6 cycles (b) and shape recovery for 6 cycles (c)

3.4 Springback and shape recovery for 6 cycles

Figure 3b shows the trend of springback that in the last cycle, unlike the first, is contained. The reduction of the average springback height between first and last cycles is greater than 5 mm and is due to increased mobility in the links between arms and circumferences, caused by repeated deformations. In the sixth cycle the structure is highly compromised, which justifies the limited springback. Figure 3c shows the recovery height average trend, highlighting how in the last cycles the recovery is slower, always because of the high damage of the structure.

3.5 Energy Absorption

The absorption energy is calculated using the Specific Energy Absorption (SEA) parameter, which normalizes the absorption energy measured from the subtended area of the Force-Displacement curve to the densification condition, relative to the mass of the structure [24].

Figure 4a shows the histogram describing the trend of normalized energy as cycles change. It is noted that in the first cycles the structure has a good level of energy absorption, averaging between 1180 mJ/g and 729 mJ/g. Absorption in the last three cycles amounts to an average value of 441 mJ/g, 340 mJ/g and 298 mJ/g respectively, with a significant decrease due to increased damage. There is no net change between the second and third cycles for the reasons described above. In the last cycle, despite the obvious fractures and injuries present in the structure, the level of energy absorption does not decrease significantly. Therefore, it can be shown that the structure is a good energy absorber, even in the presence of permanent damage. Figure 4b shows the trend of the maximum peak as cycles change. In this case, the maximum load decreases, except between the second and third cycles and between the fifth and sixth, due to the aging of the material. As for the absorption in the last two cycles, there is a low decrease in the maximum load which is equal to an average value of about 154 N.

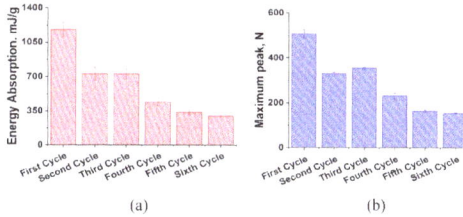

Figure 4. Energy absorption for 6 cycles (a) and maximum peak (b)

3.6 Microscopic analysis

Figure 5 shows the microscopic analysis of the structure after the first and sixth cycles.

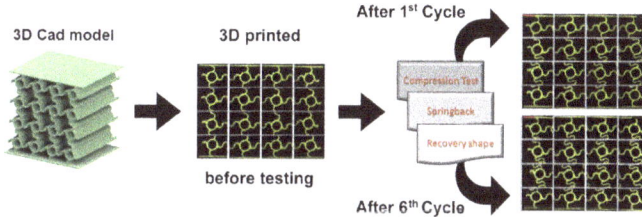

Figure 5. Microscopic analysis of the structure after the first and sixth cycles

Compared to the non-deformed configuration, the geometry is more damaged in the central region, particularly at the connection level between arms and circumferences.

4. Conclusions

In this paper, the behaviour of the Rototetrachiral cell structure has been evaluated, both mechanically and in terms of absorbed energy. The geometry was subjected to a static compression load and subsequent recovery of the shape repeated for six cycles. After the compression tests, the springback height was analysed and then the recovery of the specimen shape. The deformations induced by the subsequent compressive stresses can lead to a reduction in mechanical strength, albeit not substantially. The analysis shows that the geometry gives the structure a good ability to act as an energy absorber, particularly in the first cycles, when the sustainable load from the specimen is high. As number of cycles increase, absorption energy is reduced, with a decrease that can be estimated at 70% after the sixth cycle. For springback seems to be a correlation between damage and recovered height, with a decrease in height recovered elastically between first and sixth upper cycle of 16%. The same considerations can be extended to the recovery of shape.

Microscopic analysis shows that the central cells are the most damaged, while in the peripheral cells the arms connected respectively to the top and to the bottom layer do not show lesions. The application of a thermal stimulus shows how the deformation can be recovered, partly restoring the bonds within the structure and also affecting the mechanical performance of the same. Finally, the adoption of three replicates was sufficient as the variability on the tests was found to be low and estimated to be less than 7%.

References

[1] A. Melocchi *et al.*, «Shape memory materials and 4D printing in pharmaceutics», *Adv. Drug Deliv. Rev.*, vol. 173, pp. 216–237, giu. 2021. https://doi.org/10.1016/j.addr.2021.03.013

[2] Xiaonan Huang, M. Ford, Z. J. Patterson, M. Zarepoor, C. Pan, e C. Majidi, «Shape memory materials for electrically-powered soft machines», *J. Mater. Chem. B*, vol. 8, fasc. 21, pp. 4539–4551, giu. 2020. https://doi.org/10.1039/D0TB00392A

[3] N. Sabahi, W. Chen, C.-H. Wang, J. J. Kruzic, e X. Li, «A Review on Additive Manufacturing of Shape-Memory Materials for Biomedical Applications», *JOM*, vol. 72, fasc. 3, pp. 1229–1253, mar. 2020. https://doi.org/10.1007/s11837-020-04013-x

[4] R. Sarvari *et al.*, «Shape-memory materials and their clinical applications», *Int. J. Polym. Mater. Polym. Biomater.*, vol. 71, fasc. 5, pp. 315–335, mar. 2022. https://doi.org/10.1080/00914037.2020.1833010

[5] I. Abavisani, O. Rezaifar, e A. Kheyroddin, «Multifunctional properties of shape memory materials in civil engineering applications: A state-of-the-art review», *JOBE*, vol. 44, p. 102657, dic. 2021. https://doi.org/10.1016/j.jobe.2021.102657

[6] M. C. Biswas, S. Chakraborty, A. Bhattacharjee, e Z. Mohammed, «4D Printing of Shape Memory Materials for Textiles: Mechanism, Mathematical Modeling, and Challenges», *Adv. Funct. Mater.*, vol. 31, fasc. 19, p. 2100257, 2021. https://doi.org/10.1002/adfm.202100257

[7] M. O. Gök, M. Z. Bilir, e B. H. Gürcüm, «Shape-Memory Applications in Textile Design», *Procedia Soc.*, vol. 195, pp. 2160–2169, lug. 2015. https://doi.org/10.1016/j.sbspro.2015.06.283

[8] M. Mehrpouya, A. Azizi, S. Janbaz, e A. Gisario, «Investigation on the Functionality of Thermoresponsive Origami Structures», *Adv. Funct. Mater.*, vol. 22, fasc. 8, p. 2000296, 2020. https://doi.org/10.1002/adem.202000296

[9] H. E. Karaca, E. Acar, H. Tobe, e S. M. Saghaian, «NiTiHf-based shape memory alloys», *Mater. Sci. Technol.*, vol. 30, fasc. 13, pp. 1530–1544, nov. 2014. https://doi.org/10.1179/1743284714Y.0000000598

[10] I. Akbar, M. El Hadrouz, M. El Mansori, e D. Lagoudas, «Toward enabling manufacturing paradigm of 4D printing of shape memory materials: Open literature review», *Eur. Polym. J.*, vol. 168, p. 111106, apr. 2022. https://doi.org/10.1016/j.eurpolymj.2022.111106

[11] A. Subash e B. Kandasubramanian, «4D printing of shape memory polymers», *Eur. Polym. J.*, vol. 134, p. 109771, lug. 2020. https://doi.org/10.1016/j.eurpolymj.2020.109771

[12] S. Joshi *et al.*, «4D printing of materials for the future: Opportunities and challenges», *Applied Materials Today*, vol. 18, p. 100490, mar. 2020. https://doi.org/10.1016/j.apmt.2019.100490

[13] E. Pei e G. H. Loh, «Technological considerations for 4D printing: an overview», *Prog Addit Manuf*, vol. 3, fasc. 1, pp. 95–107, giu. 2018. https://doi.org/10.1007/s40964-018-0047-1

[14] A. Alderson *et al.*, «Elastic constants of 3-, 4- and 6-connected chiral and anti-chiral honeycombs subject to uniaxial in-plane loading», *Compos Sci Technol*, vol. 70, fasc. 7, Art. fasc. 7, lug. 2010. https://doi.org/10.1016/j.compscitech.2009.07.009

[15] A. Sorrentino, D. Castagnetti, L. Mizzi, e A. Spaggiari, «Bio-inspired auxetic mechanical metamaterials evolved from rotating squares unit», *Mech. Mater.*, vol. 173, p. 104421, ott. 2022. https://doi.org/10.1016/j.mechmat.2022.104421

[16] A. Papadopoulou, J. Laucks, e S. Tibbits, «Auxetic materials in design and architecture», *Nat Rev Mater*, vol. 2, fasc. 12, Art. fasc. 12, dic. 2017. https://doi.org/10.1038/natrevmats.2017.78

[17] M. Mehrpouya, T. Edelijn, M. Ibrahim, A. Mohebshahedin, A. Gisario, e M. Barletta, «Functional Behavior and Energy Absorption Characteristics of Additively Manufactured Smart Sandwich Structures», *Adv. Eng. Mater.*, vol. 24, fasc. 9, Art. fasc. 9, 2022. https://doi.org/10.1002/adem.202200677

[18] T. Li, J. Sun, J. Leng, e Y. Liu, «Quasi-static compressive behavior and energy absorption of novel cellular structures with varying cross-section dimension», *Compos. Struct.*, vol. 306, p. 116582, feb. 2023. https://doi.org/10.1016/j.compstruct.2022.116582

[19] A. P. Valerga, M. Batista, J. Salguero, e F. Girot, «Influence of PLA Filament Conditions on Characteristics of FDM Parts», *Mater.*, vol. 11, fasc. 8, Art. fasc. 8, ago. 2018. https://doi.org/10.3390/ma11081322

[20] M. Barletta, A. Gisario, e M. Mehrpouya, «4D printing of shape memory polylactic acid (PLA) components: Investigating the role of the operational parameters in fused deposition modelling (FDM)», *JMP*, vol. 61, pp. 473–480, gen. 2021. https://doi.org/10.1016/j.jmapro.2020.11.036

[21] A. Forés-Garriga, G. Gómez-Gras, e M. A. Pérez, «Mechanical performance of additively manufactured lightweight cellular solids: Influence of cell pattern and relative density on the printing time and compression behavior», *Mater. Des.*, vol. 215, p. 110474, mar. 2022. https://doi.org/10.1016/j.matdes.2022.110474

[22] W. Zhang *et al.*, «Characterization of residual stress and deformation in additively manufactured ABS polymer and composite specimens», *Compos Sci Technol*, vol. 150, pp. 102–110, set. 2017. https://doi.org/10.1016/j.compscitech.2017.07.017

[23] L. J. Gibson, «Cellular Solids», *MRS Bulletin*, vol. 28, fasc. 4, Art. fasc. 4, apr. 2003. https://doi.org/10.1557/mrs2003.79

[24] A. Yousefi, S. Jolaiy, M. Lalegani Dezaki, A. Zolfagharian, A. Serjouei, e M. Bodaghi, «3D-Printed Soft and Hard Meta-Structures with Supreme Energy Absorption and Dissipation Capacities in Cyclic Loading Conditions», *Adv. Eng. Mater.*, p. 2201189, nov. 2022. https://doi.org/10.1002/adem.202201189

Italian Manufacturing Association Conference - XVI AITeM
Materials Research Forum LLC

Materials Research Proceedings 35 (2023) 198-205
https://doi.org/10.21741/9781644902714-24

Dimensional and geometric deviations of parts in PA12 manufactured by selective laser sintering: numerical and experimental analyses

Valentina Vendittoli[1,a] *, Achille Gazzerro[1,b], Wilma Polini[1,c] and Luca Sorrentino[1,d]

[1]Department of Civil and Mechanical Engineering, University of Cassino and Southern Lazio, via G. di Biasio 43, 03043 Cassino, Italy

[a]valentina.vendittoli1@unicas.it, [b]achille.gazzerro@unicas.it, [c]polini@unicas.it, [d]sorrentino@unicas.it

Keywords: Additive Manufacturing, Inspection, Polymer

Abstract. Selective Laser Sintering (SLS) uses a laser to sinter powdered polymeric materials, such as Polyamide 12 (PA12). Industrially, it is commonly used as a mixture of virgin and aged powder. The aged powder has undergone various thermal cycles without being sintered. This work aims to evaluate the differences in the dimensional and geometrical deviations of parts in PA12 obtained through SLS by virgin and aged powder. A numerical approach was used to simulate the SLS software to foresee these dimensional deviations as a function of the powder's physical-chemical properties and the process parameters. The obtained results were validated through an experimental approach. Parallelepiped-shaped specimens were manufactured using an SLS printer and measured with a Coordinate Measuring Machine (CMM). The numerical results agree with the experimental ones. It seems that the differences between the dimensional deviations of the parts manufactured through virgin and aged powders are very small.

Introduction

In recent years, with the advent of Additive Manufacturing (AM), many technologies and materials have been implemented. These technologies favor the realization of layer-by-layer objects. Selective Laser Sintering (SLS) is an AM technique that uses the heat of a laser to sinter the portion of the material in a building chamber to obtain the required object layer-by-layer [1].

Most of the used materials are polymers [2]. Those plastic materials need low processing temperatures and laser power. The most commonly used material is Polyamide 12 (PA12). However, due to the printing process itself, since not all the powder is commonly sintered, 80% of it remains unused [3]; hence the material undergoes a thermal cycle, that changes its chemical and physical properties, leading to degradation effects affecting dimensional accuracy and mechanical performances [4, 5].

The printing process parameters, such as laser power, laser speed, scan spacing, and layer thickness, strongly influence the outcome, considering dimensional accuracy and mechanical performance. The energy density (ED) is defined as the energy concentrated in the area or in the volume of the part [6]; it takes into account the main printing parameters, and it is fundamental to achieve the required quality. A low ED provides weak sintering which means higher porosity and higher roughness. Moreover, the mechanical properties are lower. On the opposite side, using a high ED provides printed parts where the powder particles are bonded together, thus involving a low roughness and better mechanical properties. However, the dimensional accuracy is influenced by the higher shrinkage [7, 8]. Along the three-building direction, the x-direction maintains a higher dimensional accuracy regardless of the energy density. The accuracy along the y-direction deteriorates dramatically at high energy density levels. Finally, Z-direction accuracy was the lowest, with the highest values obtained at low energy densities. When energy density increased, accuracy steadily decreased [9].

Italian Manufacturing Association Conference - XVI AITeM   Materials Research Forum LLC
Materials Research Proceedings 35 (2023) 198-205   https://doi.org/10.21741/9781644902714-24

When the effect of the laser power is analyzed, a drastic increase leads to a wide increase in shrinkage [10]. When comparing different layer thicknesses, the structural characteristics revealed obvious changes in the crystal structure of the polymer. Progressive crystal development with increasing layer thickness occurs in all construction orientations. In fact, dimensional accuracy and mechanical properties increase with smaller thicknesses [11].

However, it is also fundamental to recycle unsintered powder in the process in order to decrease manufacturing costs. This powder can be mixed in ratio with virgin one, to create a mixed powder, that can be reprocessed. Although being impacted by a variety of process factors, the dimensional accuracy of components printed with aged PA12 powders diminishes when compared to the initial item produced with virgin powder and the same settings [12]. Controlling the SLS process parameters, such as energy density and laser scanning approach, can help to minimize surface roughness and remove the "orange peel" effect [13].

According to different works, the change in mechanical performance is conflicting. This is most likely due to the numerous times the powder was utilized in these studies, the setting, and the used printer. Tensile strength improves somewhat throughout the first five buildings but drops by around 25% after the sixth build [14, 15].

This study intends to describe the dimensional deviations from the nominal of specimens printed in PA12 through the SLS process using PA12 virgin powder using both a numerical and an experimental approach. To accomplish this goal, ten parallelepiped-shaped specimens were designed, and once defined the values of the process parameters used to print, were manufactured numerically through simulation software. The obtained models of the manufactured parts were evaluated through inspection software to analyze the dimensional accuracy.

An experimental set-up was carried out to validate the numerical result. Further experimental tests were carried out using PA12 aged powder; the dimensional and geometrical deviations from nominal were measured through a coordinate measuring machine and the obtained results were compared with those due to the virgin powder.

The following is how the paper is organized: section 1 discusses the material properties, the printing machine, the used process parameters, and the numerical approach used to simulate the printing process. Section 2 presents the experimental setup and the measuring methods used to quantify specimen measurements. Section 3 presents the findings.

Material and methods

Following the ASTM D790-17 standard for flexural mechanical properties [16], parallelepiped-shaped specimens were designed, visible in Fig.1a, and all surfaces were nominated as shown in Fig.1b. Nylon 12 produced by Sintratec was used to manufacture the benchmarks and the main properties are shown in Table 1. Particularly, the set of specimens was manufactured using virgin powder and more than 5 times reused PA12 powder through the same values of the process parameters. The powder was subjected to consecutive printing processes without undergoing powder refreshing.

The numerical approach involves two steps: the simulation of the selective laser sintering process and that of the inspection process (see Fig.2). The SLS process was simulated through the Digimat® software package, Release 2022.1 [17] on which the reference input parameters on the type of the used printer and powder were imported.

In order to carry out a study able to significantly analyze the repeatability and potential of the simulation process, ten specimens were manufactured, five for two levels of the building volume (see Fig.3).

Inside the software, four steps were performed: definition, manufacturing, simulation, and outcomes analysis. The initial stage involves entering printer parameters, importing a benchmark, and specifying material.

Italian Manufacturing Association Conference - XVI AITeM   Materials Research Forum LLC
Materials Research Proceedings 35 (2023) 198-205   https://doi.org/10.21741/9781644902714-24

For this purpose, the Sintratec-Kit printer was used [18], and it was equipped with a laser power of 2000 mW. The maximum buildable volume is 100 mm x 100 mm x 110 mm, while the maximum suggested volume is 90 mm x 90 mm x 90 mm. The machine is equipped with software that allows controlling the printing process. The parameters' set is constituted by a scan speed of 550 mm/s, a scan spacing of 0.1mm, and a layer height of 0.1mm. Furthermore, the chamber heating temperature was set at 140°C, the powder surface heating temperature at 150°C, and the melting temperature at 170 °C.

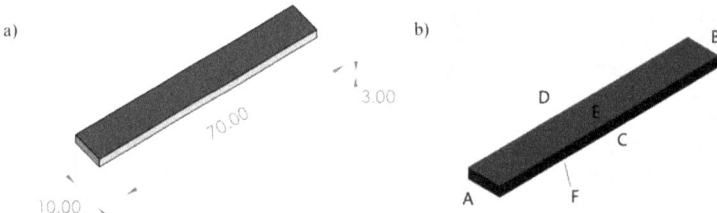

Fig. 1. (a) Benchmark dimensions in mm; (b) Surfaces.

Table 1. Nylon 12 thermo-physical properties.

Parameter	Value
Colour	Grey
Melting Point	176 °C, 348.8 F
Stable Temp (Max)	130 °C, 266 F
Bending Stress (Max)	43.1 MPa
Tensile Stress (Max)	47.8 MPa
Tensile Modulus (Max)	1750 MPa
Particle Size	0.06 mm (60 micron)

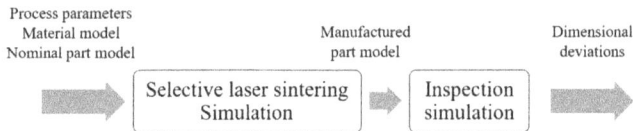

Fig. 2. Numerical approach.

For carrying out the simulation, the benchmark was discretized into a mesh voxel. The software recommended a voxel size between two and ten times the layer's height. For this work, a mesh size of 1 mm was used corresponding to the upper limit of the voxelization.

The output is saved as an STL file and loaded into the GOM Inspect software suite [19].

By analyzing the mesh, it was possible to compare the nominal CAD with the actual part to check the dimensional deviations. The step of the analysis consisted of creating reference elements through the Gaussian Best Fit method.

Italian Manufacturing Association Conference - XVI AITeM   Materials Research Forum LLC
Materials Research Proceedings 35 (2023) 198-205   https://doi.org/10.21741/9781644902714-24

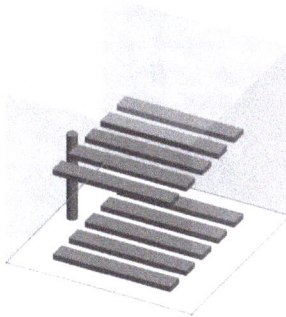

Fig. 3. Distribution of the specimens in the building chamber.

Experimental tests

Ten specimens placed on two levels were printed through the Sintratec-Kit printer as Fig.3, using the same process parameters of the numerical simulation.

The flexural specimens were measured using the coordinate measuring machine (CMM) of ZEISS. In order to perform this dimensional analysis, a type of fixture, visible in Fig.4a, b was chosen that does not geometrically deform the specimens.

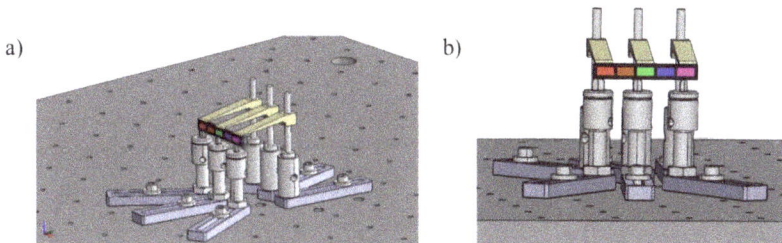

Fig. 4. (a, b) Benchmark's position on the CMM.

The Calypso software was used to analyse all the deviations presented in the specimens [20].

On the specimen, 5 areas were detected on the widest face, 2 on the side face, and 2 on the ends of the specimen. The coordinate measuring machine measured a set of points for each area; specifically on the broad face, 40 points were acquired for each area, while on the side and end areas, 8 points. From the measured points, the thickness, width, and length of the specimens were calculated. Finally, the averages were calculated since no significant difference was detected between the measured areas of each plane.

Result and discussion

The simulation result was superimposed on the reference geometry in the centerline to estimate any deviations in mm, that are shown in Fig.5. The deviations, presented in different colours, ranging from blue (smallest deviations) to red (largest deviations), are the dimensional changes of the part from the middle of the nominal geometry. By summing up the deviations on the endpoints, it will be possible to estimate the total deviation for each measurement.

Italian Manufacturing Association Conference - XVI AITeM   Materials Research Forum LLC
Materials Research Proceedings 35 (2023) 198-205   https://doi.org/10.21741/9781644902714-24

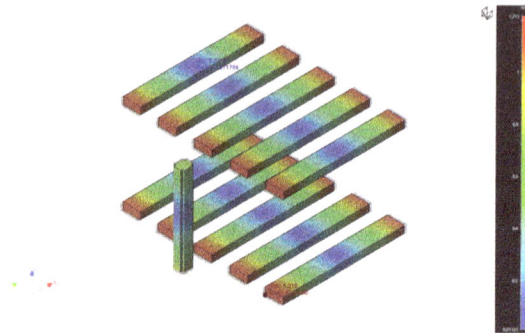

Fig. 5. Output from the numerical simulation.

For the ten specimens, the mean and standard deviation were calculated by the areas. From the results, it was observed that there was no significant difference among the different areas of the same plane for thickness, width, and length values. Therefore, all the data related to the different areas were put together. All the dimensions in the experimental, numerical, and nominal values were plotted in Fig.6.

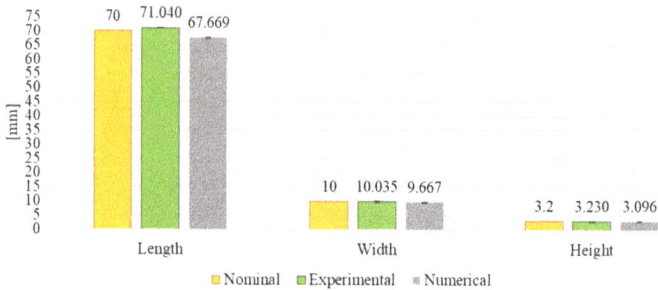

Fig. 6. Dimensional values using virgin PA12: comparison of experimental and numerical results.

From Fig.6 it can be seen that the experimental results are constantly bigger than the nominal ones. However, according to the specification of the printer, the obtained differences are inside the 5% accuracy declared by the printer builder.

Furthermore, the numerical results are smaller compared to the other results. This could be caused by the non-considered stress relaxation phenomena in the numerical simulation, so the shrinkage is higher [21, 22]. The percentage difference between the experimental and nominal results ranges from 0.35% to 1.49%, these values are inside the accuracy of the printer.

From 3% to 4% is the range of the percentage difference between experimental and numerical results, while the numerical results are smaller than the nominal ones of 3%, which is also confirmed in Fig.5.

The printing set-up (see Materials and methods) for the flexural specimens was repeated using aged PA12 powder, so further consideration can be given based on the used powder. A

Italian Manufacturing Association Conference - XVI AITeM
Materials Research Proceedings 35 (2023) 198-205

Materials Research Forum LLC
https://doi.org/10.21741/9781644902714-24

comparison of the dimensional and geometrical deviations between virgin and recycled powder is listed in Fig.7 and Fig.8.

Fig.7 shows the average value of the dimensions of the ten specimens since there is no significant variation between specimens constructed on the bottom plane as well as those built on the top plane, as previously detected by manufacturing using virgin powder.

From the experimental results, there is no significant difference in the dimensional accuracy between the two manufacturing processes. Thus, the variance between virgin and n-times recycled powder is irrelevant, considering a variation in the difference of a maximum of 2%.

Further analysis where made, considering the geometric deviations of the specimens. In particular, the flatness of the planes of every specimen and the perpendicularity between the couples of planes were evaluated. At first, it was observed that there is no significant difference among the ten considered positions of the specimen on the printer volume. Therefore, the average for all ten positions was evaluated and reported in Fig.8.

It is possible to notice that the flatness values obtained experimentally range from 0.02 mm to 0.06 mm, while the perpendicularity values range from 0.04 mm to 0.07 mm.
It seems that in most measures there is no significant difference between virgin and recycled powder. It appears a large dispersion of the data, that increase of about 10% when the aged powder is used. However, there is some exception that requires further analysis.

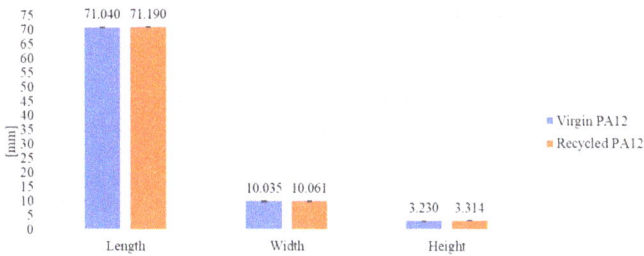

Fig. 7. Dimensional comparison between powders: experimental results.

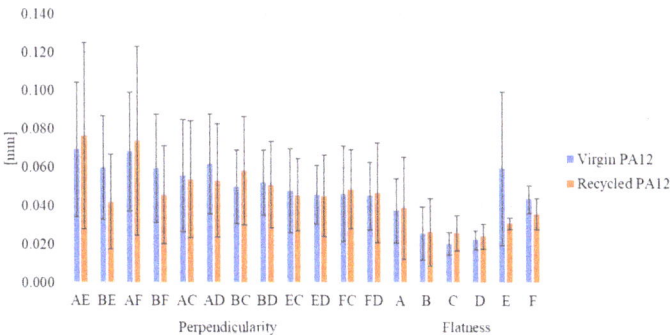

Fig. 8. Geometrical comparison between powders: experimental results.

Conclusion
The following work aimed to evaluate the dimensional deviations of components obtained through SLS with a 3D printer, making use of the Digimat® software.

Experimentally, ten flexural specimens made with the Sintratec Kit and the same process parameters as the simulated ones were manufactured. For such experimental data, a coordinate measuring machine was used to acquire multiple measurement points and extract the measurements of interest, and finally compare them with those obtained numerically.

The percentage variation between experimental and nominal results ranges from 0.35% to 1.49%, which is within the printer's declared accuracy of 5%.

The percentage difference between experimental and numerical findings ranges from 3% to 4%, whereas numerical results are 3% smaller than the nominal results. Additionally, the numerical findings are less than the other outcomes. This might be due to the numerical simulation failing to account for stress relaxation events, resulting in greater shrinkage.

Finally, the experimental procedure was repeated by changing the powder, and adopting a recycled PA12 of the same brand as the previous studies, and the obtained results were compared with those obtained previously.

The testing findings show that there is no statistically significant variation in dimensional accuracy between the two manufacturing procedures. Hence, with a maximum difference of 2%, the difference between virgin and n-times recycled powder is meaningless.

Further analyses were made on the geometric deviation connected with the parts manufactured. The experimentally achieved flatness values vary from 0.02 mm to 0.06 mm, whereas the perpendicularity values range from 0.04 mm to 0.07 mm.

Because the range in geometric outcomes is minor, geometric deviations are unaffected by the position in the construction chamber or the powder employed. In most cases, it appears that there is no discernible difference between virgin and recycled powder. However, there are several exceptions that demand an additional investigation.

Future studies should focus on the software characterization of aged powders, and residual stresses present within components in order to improve the reproducibility of parts made in SLS and investigate the mechanical performance to define the complete degradation. In addition, geometric deviations should be the subject of future studies to go to investigate future assemblies in objects of interest, experimentally and numerically.

Acknowledgments

This research did not receive any specific grant from funding agencies in the public, commercial, or not-for-profit sectors.

References

[1] S. Impey, P. Saxena, K. Salonitis, Selective Laser Sintering induced residual stresses: Precision measurement and prediction, Journal of Manufacturing and Materials Processing. 5 (2021) 101. https://doi.org/10.3390/jmmp5030101

[2] A. Awad, F. Fina, A. Goyanes, S. Gaisford, A.W. Basit, 3D printing: Principles and pharmaceutical applications of Selective Laser Sintering, International Journal of Pharmaceutics. 586 (2020) 119594. https://doi.org/10.1016/j.ijpharm.2020.119594

[3] L. Wang, A. Kiziltas, D.F. Mielewski, E.C. Lee, D.J. Gardner, Closed-loop recycling of polyamide12 powder from Selective Laser Sintering into sustainable composites, Journal of Cleaner Production. 195 (2018) 765–772. https://doi.org/10.1016/j.jclepro.2018.05.235

[4] K. Dotchev, W. Yusoff, Recycling of polyamide 12 based powders in the Laser Sintering Process, Rapid Prototyping Journal. 15 (2009) 192–203. https://doi.org/10.1108/13552540910960299

[5] D.T. Pham, K.D. Dotchev, W.A. Yusoff, Deterioration of polyamide powder properties in the Laser Sintering Process, Proceedings of the Institution of Mechanical Engineers, Part C: Journal of Mechanical Engineering Science. 222 (2008) 2163–2176. https://doi.org/10.1243/09544062jmes839

Italian Manufacturing Association Conference - XVI AITeM   Materials Research Forum LLC
Materials Research Proceedings 35 (2023) 198-205   https://doi.org/10.21741/9781644902714-24

[6] A. Wegner, G. Witt, Correlation of process parameters and part properties in laser sintering using response surface modeling, Physics Procedia. 39 (2012) 480–490. https://doi.org/10.1016/j.phpro.2012.10.064

[7] E.C. Hofland, I. Baran, D.A. Wismeijer, Correlation of process parameters with mechanical properties of laser sintered PA12 parts, Advances in Materials Science and Engineering. 2017 (2017) 1–11. https://doi.org/10.1155/2017/4953173

[8] A. Wegner, C. Mielicki, T. Grimm, B. Gronhoff, G. Witt, J. Wortberg, Determination of robust material qualities and processing conditions for laser sintering of Polyamide 12, Polymer Engineering & Science. 54 (2013) 1540–1554. https://doi.org/10.1002/pen.23696

[9] T. Czelusniak, F.L. Amorim, Influence of energy density on polyamide 12 processed by SLS: From physical and mechanical properties to microstructural and crystallization evolution, Rapid Prototyping Journal. 27 (2021) 1189–1205. https://doi.org/10.1108/rpj-02-2020-0027

[10] J. Choren, V. Gervasi, T. Herman, S. Kamara, J. Mitchell, SLS powder life study, International Solid Freeform Fabrication Symposium. (2001).

[11] A. Liebrich, H.-C. Langowski, R. Schreiber, B.R. Pinzer, Effect of thickness and build orientation on the water vapor and oxygen permeation properties of laser-sintered polyamide 12 sheets, Rapid Prototyping Journal. 27 (2021) 1030–1040. https://doi.org/10.1108/rpj-05-2020-0101

[12] P. Chen, M. Tang, W. Zhu, L. Yang, S. Wen, C. Yan, et al., Systematical mechanism of polyamide-12 aging and its micro-structural evolution during Laser Sintering, Polymer Testing. 67 (2018) 370–379. https://doi.org/10.1016/j.polymertesting.2018.03.035

[13] W.A. Yusoff, The application of scanning electron microscope and melt flow index for Orange Peel in Laser Sintering process, Indonesian Journal of Electrical Engineering and Computer Science. 6 (2017) 615. https://doi.org/10.11591/ijeecs.v6.i3.pp615-622

[14] K. Kozlovsky, J. Schiltz, T. Kreider, M. Kumar, S. Schmid, Mechanical properties of reused nylon feedstock for powder-bed additive manufacturing in Orthopedics, Procedia Manufacturing. 26 (2018) 826–833. https://doi.org/10.1016/j.promfg.2018.07.103

[15] K. Wudy, D. Drummer, F. Kühnlein, M. Drexler, Influence of degradation behavior of polyamide 12 powders in laser sintering process on produced parts, AIP Conference Proceedings. (2014). https://doi.org/10.1063/1.4873873

[16] ASTM D 790, Plastics (I). Standard test methods for flexural properties of unreinforced and reinforced plastics and electrical insulating materials, Annual book of ASTM standards. American Society for Testing and Materials; 2017

[17] Digimat, Hexagon. https://www.mscsoftware.com/it/product/digimat (accessed February 22, 2023)

[18] Sintratec Kit. https://sintratec.com/product/sintratec-kit/ (accessed February 22, 2023)

[19] Gom Inspect software. https://www.gom.com/en/products/gom-suite/gom-inspect-pro (accessed February 22, 2023)

[20] Zeiss Calypso. https://www.zeiss.com/metrology/products/software/calypso-overview/calypso.html (accessed February 22, 2023)

[21] A. Al Rashid, M. Koç, Experimental validation of numerical model for thermomechanical performance of material extrusion additive manufacturing process: Effect of process parameters, Polymers. 14 (2022) 3482. https://doi.org/10.3390/polym14173482

[22] M.Q. Shaikh, P. Singh, K.H. Kate, M. Freese, S.V. Atre, Finite element-based simulation of metal fused filament fabrication process: Distortion Prediction and experimental verification, Journal of Materials Engineering and Performance. 30 (2021) 5135–5149. https://doi.org/10.1007/s11665-021-05733-0

Italian Manufacturing Association Conference - XVI AITeM
Materials Research Proceedings 35 (2023) 206-215

Materials Research Forum LLC
https://doi.org/10.21741/9781644902714-25

Mechanical properties and dynamic response of 3D printed parts in PLA/P(3HB)(4HB) blends

Daniele Almonti[1,a] *, Clizia Aversa[1,b], Alessandra Piselli[2,c],
Massimiliano Barletta[1,d]

[1] Dipartimento di Ingegneria, Università degli Studi Roma Tre, Via Vito Volterra 62, 00146, Roma, Italy

[2] Dipartimento di Ingegneria Meccanica ed Aerospaziale, Sapienza Università degli Studi di Roma, Via Eudossiana 18, 00184 Rome, Italy

[a]daniele.almonti@uniroma3.it, [b]clizia.aversa@uniroma3.it, [c]alessandra.piselli@uniroma1.it, [d]massimiliano.barletta@uniroma3.it

Keywords: Polymers, Material Extrusion, 3D Printing

Abstract. Oil-based plastics can meet several technical requirements in different industrial applications at a reasonable cost, but they can cause high environmental impact. Bioderived polyesters like PLA and PHAs, are, instead, eco-friendly alternatives to reduce the environmental burden caused by conventional plastics. In this respect, dynamic response and mechanical properties of 3D printed parts made in (PLA)/Poly-3-hydroxybuutyrate-4-hydroxybutyrate(P(3HB)(4HB)) were investigated. The blends were achieved by extrusion compounding of different amount of P(3HB)(4HB) (0%, 10%, 20% and 30%) in PLA. The resulting compounds were extruded to achieve customized self-made filaments, which were reprocessed by Fused Filament Fabrication (FFF) to get the final parts. Hence, the 3D printed parts were tested to evaluate their performance, all of them showing good compromise between mechanical strength and flexibility as well as valuable dynamic response, with high potential in many fields. In particular, it has been observed that the addition of 10% of P(3HB)(4HB) is the most performing solution because it allows to obtain a 50% increase relative to the Young's Modulus.

Introduction

Plastic waste is a major environmental concern, with millions of tons of plastic ending up in landfills, oceans, and other natural habitats every year. This waste not only takes up valuable space but also poses a serious threat to wildlife, ecosystems, and human health [1, 2]. To address these challenges, there is a growing need for sustainable alternatives to conventional plastics. One promising solution is the use of biodegradable polymers, which are designed to break down into natural components that can be safely assimilated by the environment [3-6]. These materials have the potential to significantly reduce the negative impact of plastic waste, as they can degrade much more quickly than traditional plastics and do not accumulate in the environment [7-9]. Furthermore, biodegradable polymers offer several other advantages over traditional plastics such as strength, flexibility, and water resistance, making them suitable for a wide range of applications. Despite these potential benefits, there are also challenges associated with the use of biodegradable polymers, including issues related to their production, disposal, and performance [10-13].

PHAs (Polyhydroxyalkanoates) are a family of biodegradable polymers that can be produced by certain microorganisms through fermentation of organic matter. PHAs have gained attention as a potential alternative to traditional petroleum-based thermoplastics due to their biodegradability and renewability. One type of PHA, called PHB (Polyhydroxybutyrate), has similar properties to PLA and can be blended with PLA to form a biodegradable polymer blend called PHA/PLA.

Italian Manufacturing Association Conference - XVI AITeM   Materials Research Forum LLC
Materials Research Proceedings 35 (2023) 206-215   https://doi.org/10.21741/9781644902714-25

PHACT (PHB-4HB-co-4HBr) is a specific type of PHA that is often used in blends with PLA. PHACT has a similar structure and properties to PHB but has improved thermal and mechanical properties due to the incorporation of 4-hydroxybutyrate (4HB) and 4-hydroxybutyrate-co-4-hydroxyhexanoate (4HBr) monomers. The addition of PHACT to PLA can improve its strength, toughness, and thermal stability while maintaining biodegradability. Overall, adding PHACT to PLA can result in a biodegradable polymer blend with improved properties compared to pure PLA, making it an attractive option for a range of applications where sustainability and biodegradability are important factors.

Additive manufacturing, also known as 3D printing, is a rapidly growing technology that enables the creation of complex structures and products through layer-by-layer deposition of material. One of the most popular 3D printing technologies is fused filament fabrication (FFF), which uses a thermoplastic filament as the printing material [14-16]. This has led to a wide range of applications, including prototyping, customized manufacturing, and biomedical engineering. However, there are also limitations to FFF technology with PLA. For example, PLA is not suitable for high-stress or high-temperature applications, as it can deform or melt under these conditions. Despite these limitations, FFF technology with PLA has significant potential for a wide range of applications, particularly in industries where sustain- ability and biocompatibility are important factors [17-21]. PLA can be relatively brittle, particularly at low temperatures, which can lead to cracking and failure of printed parts under certain conditions. It is important to note that ongoing research and development in additive manufacturing with PLA is helping to address some of these limitations [22]. For example, there are efforts to improve the heat resistance and toughness of PLA through modifications of the material or the printing process [23, 24]. Additionally, the development of new composite materials, such as PLA reinforced with carbon fibers or nanoparticles, can improve the mechanical and thermal properties of the material [25].

The purpose of this work is to investigate the manufacture and use of polymer blendes based on PLA with P(3HB)(4HB) as a secondary phase in order to reduce the mechanical fragility of PLA alone to obtain a more flexible material. It was chosen to obtain the polymer blend directly with a reactive extrusion process as this is a process of wide use in the industrial field and then it was passed to the phase of production of the wire. The specimens were printed according to two main orientations in order to observe extreme conditions in the print layers.

Materials and methods
Reactive extrusion process
To prepare the PLA/P(3HB)(4HB) blends, PLA Luminy LX175 was chosen. It is a high-viscosity, fully biobased PLA homopolymer suitable for injection molding, supplied by Total Corbion PLA DV, Gorinchem, Netherlands. It features 96% L-isomer polylactic acid (PLA), slow to crystalize. It has a density of $1.24 g/cm^3$, a melting peak temperature of about 155°C, a glass transition temperature of about 60°C, an elastic modulus of 3500 MPa and a tensile strength of 50 MPa. Table 1 shows formulation developed in this work.

Table 1 Composition of the P(3HB)(4HB)/PLA blends realized.

Sample ID	P(3HB)(4HB) [wt. %]	PLA [wt. %]
CL00	0	100
CL10	10	90
CL20	20	80
CL30	30	70

After drying for 6 h at 55 °C (Drymax E60, Wittmann Bottenfeld, Wien, Austria), the formulations were compounded by a corotating twin screw extruder with a screw diameter of 27

Italian Manufacturing Association Conference - XVI AITeM | Materials Research Forum LLC
Materials Research Proceedings 35 (2023) 206-215 | https://doi.org/10.21741/9781644902714-25

mm (ZSE 27 MAXX, Leistritz Extrusionstechnik GmbH, Nuremberg, Germany) equipped with two gravimetric feeders (Flexwall Plus Feeder FW 40/5, Brabender Technologie GMBH & Co, Duisburg, Germany) and one volumetric feeder (EC30 M, BHT Srl, Camposanto (MO), Italy) to dose pellets. The barrel of the extruder is divided into 10 controlled temperature sections. The volumetric metering feeder with paddle-massaged flexible hopper for pellets (Flexwall Plus Feeder, Brabender Technologie GMBH & Co, Duisburg, Germany) is located in the first zone. The twin-screw corotating extruder is equipped with the following auxiliary apparatus: a temperature-controlled strand die head flange mounted to the last barrel of the extruder with 3 bores 2 mm in diameter for the manufacturing of the plastic strands; a cooling bath fitted with several strand guide rolls to guide the strands in the water for cooling purposes; a strand blowing unit for strand pre-drying; and a speed-controlled strand pelletizer (Haake Fisions PP1 pelletizer POSTEX, Thermo Fisher Scientific, Waltham, MA, USA) with frequency converter and range adjustment of pellets length for pellet cutting. Table 2 summarizes the setting of the processing parameters of the twin-screw extruder.

Table 2 Setting of the twin-screw extruder processing parameters.

Parameters	Value
Temperature, zone 1 [°C]	170
Temperature, zone 2 [°C]	180
Temperature, zone 3 [°C]	180
Temperature, zone 4 [°C]	175
Temperature, zone 5 [°C]	170
Temperature, zone 6 [°C]	168
Temperature, zone 7 [°C]	165
Temperature, zone 8 [°C]	162
Temperature, zone 9 [°C]	160
Temperature, die head [°C]	185
Screw speed [rpm]	550
Volumetric feeding [%]	31
Gravimetric feeding [kg/h]	7.95

Filament maker process
The pellets obtained in the previous phase of reactive extrusion were then used inside a machine that allows to obtain filaments (Precision series 450, 3devo, Netherlands). Given the small size of the control stations of this machine and the difference in size with the previous one, it was necessary to scale the process in an adequate way, in order to obtain filaments that had a dimensional tolerance lower than 0.05 mm, as higher variations give rise to phenomena of obstruction of the power supply system of the 3D printer or different conditions respect the previsions of slicer software. Table 3 shows setting parameters adopted during filament extruder process.

Additive process
For the realization of the specimens, given the preliminary nature of the experimentation, it was decided to realize for each type of test piece two modes of orientation with respect to the printing plan. the first with the test piece oriented longitudinally with respect to the printing surface (called V∥), the second with the test piece oriented perpendicular to the printing surface (called V⊥).

Figure 1 shows the two orientations during printing for the respective types of specimens. For each pair of material/test, 5 replicas were made, each specimen was printed individually and by placing the model in the middle of the printing platform. This choice was made to avoid any edge effects present on the printing surface and to obtain samples with similar characteristics. This condition allows to avoid different timing of deposition and cooling between successive layers that would occur in the case of printing with more models. Moreover, the realization of the samples was carried out randomizing the order of realization in order to reduce any defects obtained on the filament. An Ultimaker s5 (Ultimaker, Netherlands) printer was used to make the samples. The samples were made by adopting an extrusion temperature equal to 175 °C, the layer thickness was set to 0.15 mm with a filling of 100%.

Table 3 Setting of the filament maker extruder processing parameters

Parameters	Value
Temperature, zone 1 [°C]	195
Temperature, zone 2 [°C]	210
Temperature, zone 3 [°C]	200
Temperature, die head [°C]	180
Screw speed [rpm]	4.0
Cooling fan [%]	50
Filament diameter [mm]	2.85

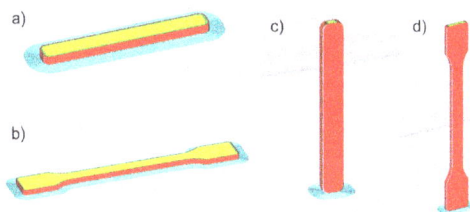

Figure 1 Specimen orientation: a) bending and Izod test V‖; b) tensile test V‖; c) bending and Izod test V⊥; d) tensile

Mechanical tests

Tensile test was conducted in accordance with ISO 527-2. For each formulation, 10 specimens were prepared and tested (5 for each printed orientation) (Figure 2). The stress–strain measurements of samples were performed using a universal testing machine (Shimadzu, China) at room temperature. Cross- head speed was set at 5 mm/min for samples. Bending test was conducted on in accordance with ISO 178:2019 (Figure 2). The test conditions are the same of tensile test. Izod test was performed on samples prepared with the previously mentioned printing conditions with PLA/P(3HB)(4HB) blends listed above. Izod impact tests were performed by using AMSE XJUD equipment with a pendulum impact energy of 5 J and impact velocity of 3.5 m/s, in the unnotched configuration, according to ISO 180:2000 standard. The measurements were performed at room temperature and the results were reported as average values of the five specimens.

Italian Manufacturing Association Conference - XVI AITeM   Materials Research Forum LLC
Materials Research Proceedings 35 (2023) 206-215   https://doi.org/10.21741/9781644902714-25

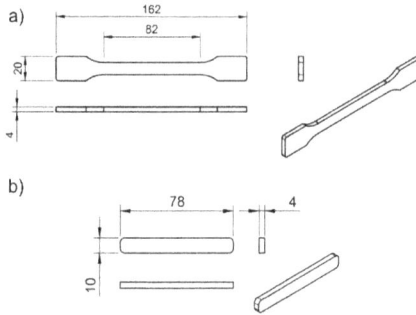

Figure 2 Specimen dimensions: a) tensile test specimen; b) bending and izod test specimen

Results and Discussion

Both extrusion processes were stable with the parameters used. In the compounding process, no degradation of the material has occurred, this has highlighted the goodness of the parameters used. In the process of making the filament for the 3D printer, there were no particular difficulties in managing the process parameters. What was observed was the higher rigidity of the pure material (without PHACT). This phenomenon has led to greater criticality in the realization of the coils (spooling process) as subsequently, the wire has presented more breaking phenomena related to the power system of the 3D printer. As regards the dimensional tolerances of the wire obtained, variations in diameter not exceeding 0.05 mm were recorded. To evaluate this, sample measurements were made on a wire sample of 3 m made before the production of the print filaments. During the production of the samples by FFF, there were no particular problems such as failure to adhere to the printing plan or jams in the extruder feed. The realization of the samples was carried out by means of a production plan that randomized the sequence of production of the samples in order to avoid any systematic errors or to distribute any defects related to the material or filament on the samples realized. Tensile tests have excellent repeatability. It was noted that the V‖ specimens presented a dynamic of the rupture that first affected the outer layers of the geometry (as made as a continuous contour during printing) and then the inner part (Figure 3). The V⊥ samples, on the other hand, showed a resistance linked to the conditions of adhesion between the layers. In fact, in the first case, the break happened in an exhausting way while in the second case, a break was observed between two consecutive layers. In both configurations, the trend of the elastic modulus found presents a maximum, but this material related trend is also affected by the printing orientation as in the V⊥ specimens it is noted that the perpendicular stress to the print layers tends to present the peak at lower P(3HB)(4HB) concentrations. This is also observable by the maximum force achieved in individual tests, as in V⊥ specimens the values achieved are very similar to each other regardless of the percentage of P(3HB)(4HB). As for the bending tests, in this case, the specimens made with V‖ orientation have presented a rupture due to the detachment between successive layers due to the different deformations of the same (Figure 3). While in the case of V⊥ auditions, the break happened in a clear way between two consecutive layers. This rupture behavior of the specimen is reconnected in the measured quantities. In fact, the V⊥ test pieces show a break that does not suffer significantly from the percentage of PHACT in the formulation. What is observed is that in the iteration between the different layers, the test pieces without P(3HB)(4HB) show a better adhesion against the greater fragility of the material. The addition of P(3HB)(4HB) tends to increase the overall elasticity of the specimen but reduces the adhesion resistance between consecutive layers. In test piece V‖, in which the layer arrangement is perpendicular to the type of stress, it is observed that the deformation increases up to a maximum of 50% compared to specimens without PHACT.

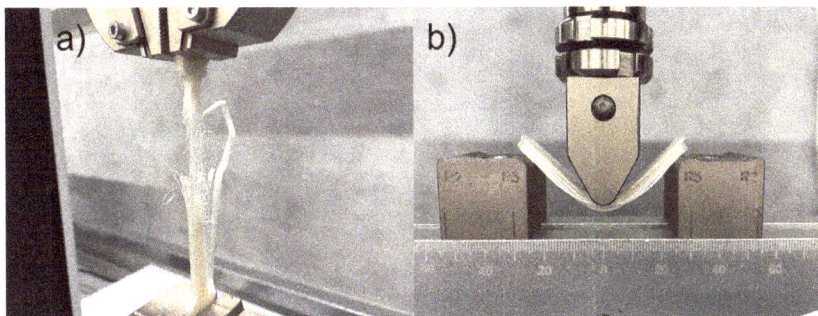

Figure 3 Broken Specimen: a) tensile test specimen Vǀ; b) bending test specimen Vǀ.

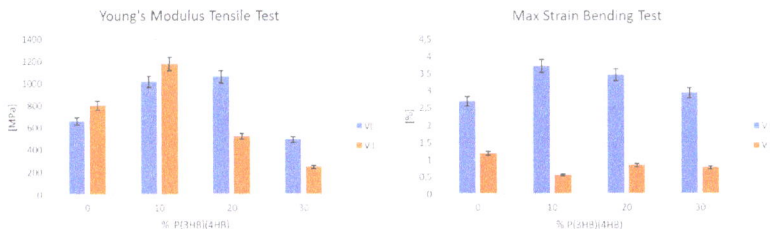

Figure 4 Results mechanical tests: a) Young's Modulus from tensile test; b) Maximum strain from bending test.

Table 4 Mechanical tests results

PHACT %	Young's Modulus [MPa]		Maximum Strain Bending [%]	
	Vǁ	V⊥	Vǁ	V⊥
0	661	803	2,69	1,18
10	1019	1180	3,72	0,55
20	1064	524	3,46	0,84
30	494	249	2,94	0,77

Italian Manufacturing Association Conference - XVI AITeM   Materials Research Forum LLC
Materials Research Proceedings 35 (2023) 206-215   https://doi.org/10.21741/9781644902714-25

Izod tests showed that the orientation of the test pieces in the additive process has an important influence on the material's impact response. Particularly in V∥ specimens, it was observed that the increase in P(3HB)(4HB) gives the specimen an increase in the energy required to break an order of magnitude. While compared to the increase in P(3HB)(4HB) we note a linear trend, this behavior could be related to the iteration between the two phases that even if you do not have a total mixing (more evident in the samples made in CL30) tend to exhibit greater dissipative capacity. As for the samples made with V⊥ orientation, the impact resistance is completely dependent on the adhesion between the layers and therefore not affected in this case by the presence of the secondary phase.

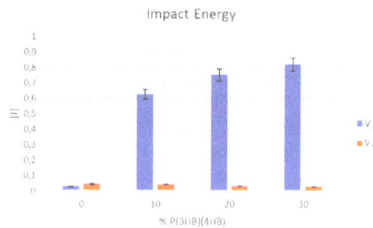

Figure 5 Results Izod tests.

SEM observations showed that the sections of the specimens made without P(3HB)(4HB) clearly present fragile rupture zones in line with the macroscopic observations found in the mechanical tests. Figure 6 shows the sections of a test piece characterized by traction, Figure 6 a) shows the section of a test piece without PHACT made with orientation V∥, while Figure 6 b) shows the section of a test piece made with the same material but printed with orientation V⊥. In both sections are evident the fragile rupture zones characterize the mechanical response of the PLA. Increasing the composition of PHACT shows that the rupture sections have less and less rupture propagation due to the reduced fragility of the formulation with P(3HB)(4HB).

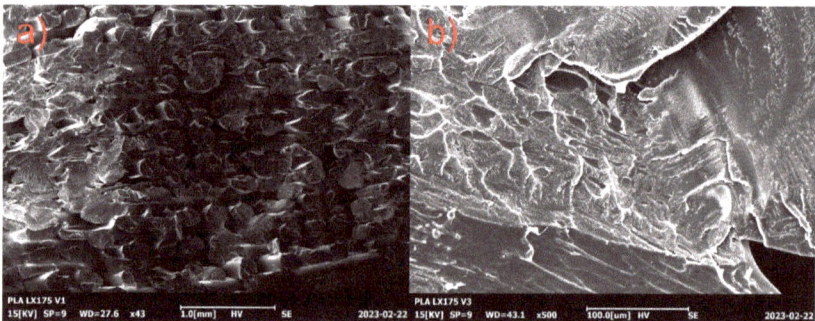

Figure 6 SEM observation of pure PLA samples: a) V∥; b) V⊥.

Figure 7 shows the section of a sample made with 30% of P(3HB)(4HB), you can notice the secondary phase of the blend. In this case, the presence of P(3HB)(4HB) particles tends not to confer particular advantages to the base material as the parameters used during the compounding phase should probably be optimized in a performant way compared to the percentage of phase.

Italian Manufacturing Association Conference - XVI AITeM   Materials Research Forum LLC
Materials Research Proceedings 35 (2023) 206-215   https://doi.org/10.21741/9781644902714-25

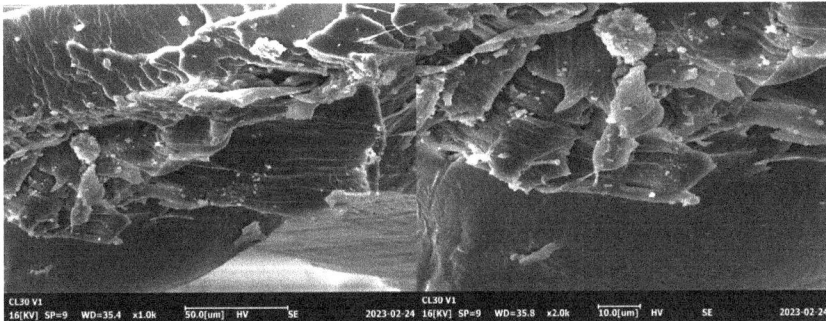

Figure 7 SEM observation of CL30 VI sample.

Conclusions

In this work, the influence on the mechanical properties of blends having the primary PLA phase and the secondary P(3HB)(4HB) phase was observed. The secondary phase has been increased to 30%. Pellets of different polymer blends were made and subsequently filaments were made for additive. Several samples were made varying their orientation on the platform of printing. Subsequently, a characterization phase of the printed samples with polymer blends was carried out and the increase in mechanical properties was observed due to the addition of P(3HB)(4HB). Mainly, a reduction in PLA fragility and a considerable increase in shock absorption were observed. It has also been observed that orientation during printing has influenced the observed results. In particular, the following results were obtained:

- Increase of Young's Module by about 50% with the addition of 10% by weight to PLA;
- Increase of the maximum bending deformation of about 38% with the addition of 10% by weight to the PLA;
- Increase of an order of magnitude of impact energy by adding PHACT to PLA.

It has also been observed that the orientation of the specimens with respect to the printing plan substantially influences the mechanical characteristics obtained.

References

[1] N. Burgos, I. Armentano, E. Fortunati, F. Dominici, F. Luzi, S. Fiori, F. Cristofaro, L. Visai, A. Jiménez, J. M. Kenny, Functional properties of plasticized bio-based poly (lactic acid)_poly (hydroxybutyrate)(pla_phb) films for active food packaging, Food and Bioprocess Technology 10 (2017) 770-780. https://doi.org/10.1007/s11947-016-1846-3

[2] J. Jiang, Q. Dong, H. Gao, Y. Han, L. Li, Enhanced mechanical and antioxidant properties of biodegradable poly (lactic) acid-poly (3-hydroxybutyrate-co-4-hydroxybutyrate) film utilizing α-tocopherol for peach storage, Packaging Technology and Science 34 (3) (2021) 187-199. https://doi.org/10.1002/pts.2553

[3] H. Zhao, Y. Bian, Y. Li, Q. Dong, C. Han, L. Dong, Bioresource-based blends of poly (3-hydroxybutyrate-co-4-hydroxybutyrate) and stereocomplex polylactide with improved rheological and mechanical properties and enzymatic hydrolysis, Journal of Materials Chemistry A 2 (23) 215 (2014) 8881-8892. https://doi.org/10.1039/c4ta01194e

[4] N. Loureiro, J. Esteves, J. Viana, S. Ghosh, Mechanical characterization of polyhydroxyalkanoate and poly (lactic acid) blends, Journal of Thermoplastic Composite Materials 28 (2) (2015) 195-213. https://doi.org/10.1177/0892705712475020

[5] J. Ostrowska, W. Sadurski, M. Paluch, P. Tyński, J. Bogusz, The effect of poly (butylene succinate) content on the structure and thermal and mechanical properties of its blends with polylactide, Polymer International 68 (7) (2019) 1271-1279 https://doi.org/10.1002/pi.5814

[6] I. Armentano, E. Fortunati, N. Burgos, F. Dominici, F. Luzi, S. Fiori, A. Jiménez, K. Yoon, J. Ahn, S. Kang, et al., Processing and characterization of plasticized pla/phb blends for biodegradable multiphase systems.

[7] Y. Li, C. Han, Y. Yu, D. Huang, Uniaxial stretching and properties of fully biodegradable poly (lactic acid)/poly (3-hydroxybutyrate-co-4-hydroxybutyrate) blends, International journal of biological macromolecules 129 (2019) 1-12. https://doi.org/10.1016/j.ijbiomac.2019.02.006

[8] H. Li, X. Lu, H. Yang, J. Hu, Non-isothermal crystallization of p (3hb-co-4hb)/pla blends: Crystallization kinetic, melting behavior and crystal morphology, Journal of Thermal Analysis and 230 Calorimetry 122 (2015) 817-829. https://doi.org/10.1007/s10973-015-4824-5

[9] C. M. Chan, L.-J. Vandi, S. Pratt, P. Halley, Y. Ma, G.-Q. Chen, D. Richardson, A. Werker, B. Laycock, Understanding the effect of copolymer content on the processability and mechanical properties of polyhydroxyalkanoate (pha)/wood composites, Composites Part A: Applied Science and Manufacturing 124 (2019) 105437. https://doi.org/10.1016/j.compositesa.2019.05.005

[10] Z. Guo, Z. Wang, Y. Qin, J. Zhang, Y. Qi, B. Liu, W. Pan, Fabrication of biodegradable nanofibers 250 via melt extrusion of immiscible blends, e-Polymers 22 (1) (2022) 733-741. https://doi.org/10.1515/epoly-2022-0059

[11] Z. Chen, Z. Zhao, J. Hong, Z. Pan, Novel bioresource-based poly (3-hydroxybutyrate-co-4-hydroxybutyrate)/poly (lacticacid) blend fibers with high strength and toughness via meltspinning, Journal of Applied Polymer Science 137 (32) (20 https://doi.org/10.1002/app.48956

[12] Y.-X. Weng, L. Wang, M. Zhang, X.-L. Wang, Y.-Z. Wang, Biodegradation behavior of p (3hb, 255 4hb)/pla blends in real soil environments, Polymer testing 32 (1) (2013) 60-70. https://doi.org/10.1016/j.polymertesting.2012.09.014

[13] B. Imre, B. Pukánszky, Compatibilization in bio-based and biodegradable polymer blends, European polymer journal 49 (6) (2013) 1215 https://doi.org/10.1016/j.eurpolymj.2013.01.019

[14] D. Chaidas, J. D. Kechagias, An investigation of pla/w parts quality fabricated by FFF, Materials and Manufacturing Processes 37 (5) (2022) 582-590. https://doi.org/10.1080/10426914.2021.1944193

[15] J. Kechagias, N. Vidakis, M. Petousis, N. Mountakis, A multi-parametric process evaluation of the mechanical response of pla in fff 3d printing, Materials and Manufacturing Processes (2022) 1-13 https://doi.org/10.1080/10426914.2022.2089895

[16] E. García Plaza, P. J. N. López, M. Á. C. Torija, J. M. C. Muñoz, Analysis of pla geometric properties processed by fff additive manufacturing: Effects of process parameters and plate extruder precision motion, Polymers 11 (10) (2019) https://doi.org/10.3390/polym11101581

[17] S. Bardiya, J. Jerald, V. Satheeshkumar, Effect of process parameters on the impact strength of fused filament fabricated (fff) polylactic acid (pla) parts, Materials Today: Proceedings 41 (2021) 1103-1106. https://doi.org/10.1016/j.matpr.2020.08.066

[18] J. Butt, R. Bhaskar, V. Mohaghegh, Non-destructive and destructive testing to analyse the effects 280 of processing parameters on the tensile and flexural properties of fff-printed

graphene-enhanced pla, Journal of Composites Science 6 (5) (2022) 148.
https://doi.org/10.3390/jcs6050148

[19] V. Cojocaru, D. Frunzaverde, C.-O. Miclosina, G. Marginean, The influence of the process parameters on the mechanical properties of pla specimens produced by fused filament fabrication-a review, Polymers 14 (5) (2022) 886. https://doi.org/10.3390/polym14050886

[20] B. Arifvianto, Y. B. Wirawan, U. A. Salim, S. Suyitno, M. Mahardika, Effects of extruder temperatures and raster orientations on mechanical properties of the fff-processed polylactic-acid (pla) material, Rapid Prototyping Journal 27 (10) (2021) 1761-1775.
https://doi.org/10.1108/RPJ-10-2019-0270

[21] A. Kallel, I. Koutiri, E. Babaeitorkamani, A. Khavandi, M. Tamizifar, M. Shirinbayan, A. Tcharkhtchi, Study of bonding formation between the filaments of pla in fff process, International Polymer Processing 34 (4) (2019) 434-444. https://doi.org/10.3139/217.3718

[22] N.-A. Masarra, M. Batistella, J.-C. Quantin, A. Regazzi, M. F. Pucci, R. El Hage, J.-M. Lopez Cuesta, Fabrication of pla/pcl/graphene nanoplatelet (gnp) electrically conductive circuit using the fused filament fabrication (fff) 3d printing technique, Materials 15 (3) (2022) 762.
https://doi.org/10.3390/ma15030762

[23] Z. C. Kennedy, J. F. Christ, Printing polymer blends through in situ active mixing during fused filament fabrication, Additive Manufacturing 36 (2020) 101233.
https://doi.org/10.1016/j.addma.2020.101233

[24] N. Bhardwaj, H. Henein, T. Wolfe, Mechanical properties of thermoplastic polymers in fused filament fabrication (fff), The Canadian Journal of Chemical Engineering 100 (11) (20
https://doi.org/10.1002/cjce.24562

[25] A. Frengkou, E. Gkartzou, S. Anagnou, D. Brasinika, C. Charitidis, 3d printed pla-based blends with improved interlayer bonding.

Italian Manufacturing Association Conference - XVI AITeM   Materials Research Forum LLC
Materials Research Proceedings 35 (2023) 216-224   https://doi.org/10.21741/9781644902714-26

In-process inspection of lattice geometry with laser line scanning and optical tomography in fused filament fabrication

Michele Moretti[1,a*], Arianna Rossi[1,b] and Nicola Senin[1,c]

[1]Department of Engineering, University of Perugia, Via G. Duranti 67, 06125 Perugia, Italy

[a]michele.moretti@unipg.it, [b]arianna.rossi2@studenti.unipg.it, [c]nicola.senin@unipg.it

Keywords: Additive Manufacturing, Material Extrusion, Lattice Manufacturing, In-Process Measurement

Abstract. One of the challenges of lattice manufacturing by fused filament fabrication is to achieve geometric accuracy of the internal reticular structures. In this work a solution for in-process inspection is presented, based on combining a custom laser scanner system, mounted into the fabrication machine, and a method for optical tomography. The scanner allows for 2.5D layer measurement, with superior detection of layer edges with respect to 2D optical imaging. Optical tomography is then achieved by vertical stacking of reconstructed layer boundaries, leading to a full volumetric reconstruction of the lattice as a voxel model. Inspection can be performed layer-wise, by comparing the current slice measured by the laser scanner with a reference virtual layer obtained by simulation of the deposition process, or on entire portions of reconstructed 3D geometry, by performing voxel-wise comparisons in 3D, to identify local missing or excess deposited material. The proposed solution proves capable of monitoring an evolving 3D part geometry, allowing also the observation of internal regions, invisible when using conventional optical, post-process imaging methods.

Introduction

Amongst additive manufacturing technologies, *fused filament fabrication* (FFF) is gaining traction beyond rapid prototyping, and is increasingly adopted to fabricate high-value added parts of significant geometric complexity [1]. FFF is a material extrusion process based on thermal reaction bonding (MEX-TRB [2]) where the material (polymer of composite with polymer matrix) is provided in form of a filament. FFF allows the realization of hollow structures as well as trabecular, lattice and otherwise reticular patterns, through layer-based fabrication. However, as geometric requirements become increasingly more stringent, geometric inspection becomes more challenging, in particular as it is impossible to access the internal structures for measurement with contact or non-contact optical means. The only current approach to measure the internal regions of a hollow or reticular structure is via *X-ray CT scanning* [3-5]. Apart from the issues of metrological uncertainty and the costs/complexity of the approach, measurement can only be executed post-process, thus any imperfection originated in the early stages of fabrication can be detected only after the part has been completed, leading to significant waste of time, energy and material in case of part rejection. A viable alternative is *optical tomography*, a method of in-process measurement where 2D images of each layer are taken using optical cameras, typically observing from above [6-10]. Image processing can be used to reconstruct the arrangement of material and void regions within each image. The advantage is that anomalies can be detected as they occur within each layer. Moreover, if required when the part is complete, images can be vertically stacked on top of each other as if they were slices of the part, thus allowing to obtain a full 3D dataset depicting external and internal structures [6]. Optical tomography is however, itself characterized by significant, currently unsolved challenges. The reconstruction of the spatial arrangement of material and voids in each layer is based on image processing, which in turn is influenced by illumination, optical properties of the material and optical property of the camera

Italian Manufacturing Association Conference - XVI AITeM   Materials Research Forum LLC
Materials Research Proceedings 35 (2023) 216-224   https://doi.org/10.21741/9781644902714-26

[11]. Self and projected shadows, focus issues, and inherent complexity of observing structures on a layer which are barely discernible from those of the layers underneath, make the identification of the internal and external layer contours very challenging. Recent work [6] has shown that contour identification based on edge detection can be improved by using a digital twin whose purpose is to narrow down the regions within the image that should be searched for edges. The method consists of using a simulation of the FFF extrusion process to generate a virtual copy of the image that a camera is expected to see if observing from above. As the virtual image is semantically labelled, the camera can use it to extract the regions where the real edge is supposedly located, and search for edges only within those regions [6].

Objective of this work is to provide an alternative to the approach illustrated in [6] by exploring:

- the use of laser line scanning to replace optical 2D imaging. The transition to 2.5D topography acquisition is designed to provide a more robust means to identify internal and external layer contours within each layer;
- the use of digital geometric models replicating expected layer geometry, to allow for a direct comparison between expected and measured results for each layer, so that warnings can be immediately generated upon identification of discrepancies.

The use of laser line scanning technologies to observe the part in FFF has been attempted before [12-14]. The proposed method presents specific elements of distinction with respect to the state of the art:

- there is no need to perform a registration process between the geometry measured by laser scanning and the reference geometry used for comparison. Our laser scanning solution is integrated within the FFF machine, uses the same axes, and relies on the same axis encoders. The two datasets exist within the same reference frame, and alignment features the same positioning accuracy achieved in fabrication;
- our laser line scanning solution is not configured to perform a 2.5D measurement of the layer, but rather works as a robust, 2D detector of material boundaries, by recognising the points where material falls below a reference band representing the expected layer height. Whilst this approach does not let us investigate phenomena that require the reconstruction of the entire 2.5D map of layer heights (like in [14] for warpage), it provides an effective boundary detection functionality, superior to edge detection in 2D layer images, which can instead be influenced by shadows and focusing issues [6].

Materials and Methods

Hardware. A laser line scanner was built in-house, and mounted on a prototype FFF machine being developed at our labs (Fig. 1). The laser source (80mW@520nm) [15] is pointed vertically downwards onto the current layer and is observed by a tilted optical microscope (5 megapixels, 10× - 220× magnification - Dinolite AM7915MZT [16]). The scanner is rigidly connected to the machine z-axis and placed so the when the laser line is projected on the layer surface, it falls exactly in the middle of the field of view and at focal range of the tilted optical microscope. The scanner is operated by a dedicated Arduino controller [17]. The laser line is aligned to the x-axis. When the measurement is triggered, the y-axis of the worktable is automatically operated to achieve the scanning motion. The y-axis positioning error, measured through a rotary optical encoder, results to be lower than <5E-3 μm [18].

Italian Manufacturing Association Conference - XVI AITeM

Materials Research Forum LLC

Materials Research Proceedings 35 (2023) 216-224

https://doi.org/10.21741/9781644902714-26

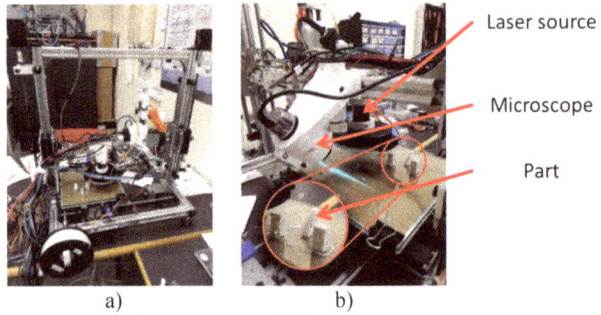

Figure 1. a) FFF machine prototype; b) laser line scanner.

Part program pre-processing. The part program g-code is automatically analysed to retrieve the extrusion path for each layer. Additional code is inserted for the realisation of two sacrificial structures at the sides of the layer: these are needed to reinstate steady-state extrusion conditions after the extrusion process is interrupted to measure the completed layer. A geometric analysis of the extrusion path for each layer is also carried out to determine the scan path for the laser line scanner in order to achieve full coverage of the layer. The scan path is automatically converted into new blocks of g-code and inserted in the part program, for execution between consecutive layers.

Part program execution. When the FFF machine executes the part program to fabricate the part, the following operations are repeated for each layer. The extrusion path is retrieved from g-code and used by a simulator to estimate the layout of deposited material in the x,y plane of the current layer. Width and length of each deposited strand are computed using previously developed extrusion models [6], taking into account extrusion trajectory, traversal velocity of the extrusion nozzle, and filament feed-rate. The layout of deposited material on the layer is encoded as a raster, binary grid (material vs. voids) of user-defined resolution. The resulting grid is used as the nominal reference of where material is supposed to be after layer fabrication, for use by the in-process inspection system. Once layer fabrication is complete, the FFF machine executes the scanning path encoded in the same part program. The primary purpose of laser line scanning is the identification of transitions between material and void in the current layer. Therefore, instead of delivering a complete 2.5D reconstruction of the layer by triangulation, a simpler and quicker approach is implemented as summarised in Fig. 2. For each position of the laser line, the optical microscope acquires an image from a tilted viewpoint. Parts of the laser line that hit layer material will appear located within a predefined horizontal reference band within the image, whilst parts of the laser line hitting void regions (possibly falling through on material at lower layers) will appear outside and below the reference band. A simple binarization process within the image band can be adopted to generate a binary strip encoding material and void regions within the band covered by the laser line. As the imaging system is calibrated, image processing results (in pixels) can be converted to mm.

Italian Manufacturing Association Conference - XVI AITeM   Materials Research Forum LLC

Materials Research Proceedings 35 (2023) 216-224   https://doi.org/10.21741/9781644902714-26

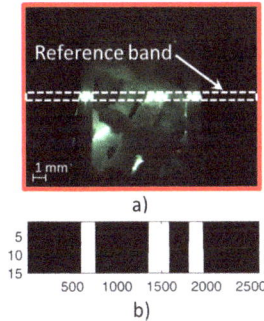

Figure 2. Operating principle of the laser line scanner; a) digital image acquired by the microscope. Laser line regions hitting layer material are located within the reference band. Laser line regions hitting void or lower layers are located outside and below; b) resulting binary strip for the laser line shown (shown in pixel units).

The binarization approach implemented in the laser line scanning method proves to be more robust at finding material boundaries compared to edge detection in 2D images, as it is not as easily confused by shadows or focusing issues, although the analysis of the laser line carries its own series of issues and challenges. Regardless, an entire binary raster map of the layer material/void regions can be obtained by combining strips generated during scanning. In-process inspection then consists of comparing the binary, raster map obtained via simulation (layout of deposited material estimated from g-code) with the binary, raster map obtained via laser line scanning (measure of actual deposited material). Note that, whilst the resolution of the measured raster map is constrained by the properties of the scanning device, the resolution of the simulated raster map is only limited by computational resources, therefore the latter is currently downsampled to 0.1 mm $\times$ 0.1 mm to match the resolving power of the measurement solution, so that a one-to-one "pixel-wise" comparison is possible. Since the laser scanning device is rigidly coupled with the extrusion system and positioned using the same axes and encoders, simulated and measured layer data are intrinsically registered through an offset calculated during calibration of the laser scanning device. Pixel-wise discrepancies calculated in the inspection procedure are then highlighted as excess or missing material pixels (in the measurement map, with respect to the simulated reference), and can be used as the basis to develop in-process monitoring solutions.

Finally, if needed, both the simulated and measured raster maps can be vertically stacked into voxel volumes (using layer thickness as voxel z-width), to respectively generate the volumetric nominal expectation for the part and the actual volumetric measurement of the fabricated part. These models can be compared voxel-wise to infer further quality issues, although post-process.

Results

The test case chosen for this work is illustrated in Fig. 3. It consists of a test geometry featuring an internal lattice-like reticular structure.

Italian Manufacturing Association Conference - XVI AITeM
Materials Research Proceedings 35 (2023) 216-224

Materials Research Forum LLC
https://doi.org/10.21741/9781644902714-26

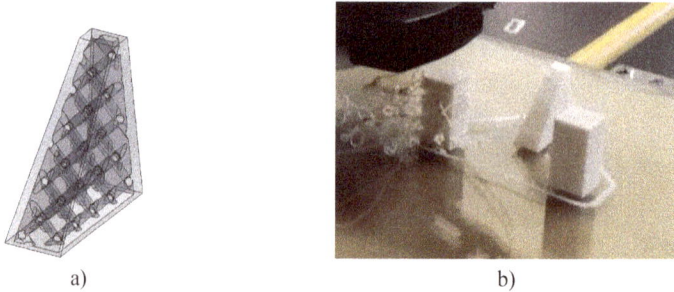

a)   b)

Figure 3. Test case. Material: PLA, approx. size: 20 mm x 8 mm x 20 mm. Lattice structure consisting of a network of struts arranged within cubic cells (5 mm size) and strut thickness equal to 1.2 mm; a) CAD model; b) one of the fabricated specimens, along with the sacrificial structures at the sides.

The process to generate the binary, raster map of material layout by simulation is illustrated in Fig. 4 for one of the layers. Within the in-process inspection solution, this map is effectively a digital twin of the current layer. The result of laser line scanning for the same layer and the comparison map (measured vs simulated) are illustrated in Fig. 5.

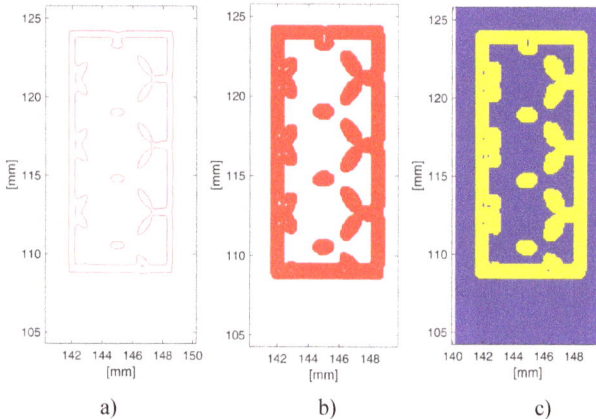

a)   b)   c)

Figure 4. Process to generate (by simulation) the expected binary, raster map of material layout for one layer of the test part; a) extrusion path as extracted from the g-code; b) material layout in the x,y plane using a simulation model to estimate length and width of the deposited strands; c) final material/void raster map, resampled at the resolution of the measured map.

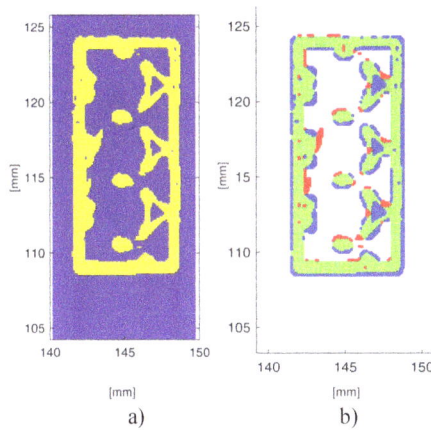

a)   b)

Figure 5. a) raster map resulting from laser line scanning measurement; b) pixel-wise comparison of the measurement map with the simulated map previously shown in Fig. 4. Red pixels: excess material (material present in measurement, but not in the simulated map), blue pixels: missing material (material absent in measurement, but present in the simulated map), green pixels: correct material (present in both measurement and simulation), transparent pixels: correct voids (absent in both measurement and simulation).

From the analysis of excess and missing material pixels, warning signals can be triggered right after layer measurement, for example based on discrepancy count or spatial distribution. The actual triggering criteria are application-dependent.

In Fig. 6 the two three-dimensional voxel models obtained by vertical stacking of simulation and measurement raster maps are shown. Three-dimensional maps of missing and excess material voxels can be generated as well, in case post-process assessment of part quality is needed. The three-dimensional distribution of detected discrepancies may provide further avenues for quality assessment.

Italian Manufacturing Association Conference - XVI AITeM
Materials Research Proceedings 35 (2023) 216-224

Materials Research Forum LLC
https://doi.org/10.21741/9781644902714-26

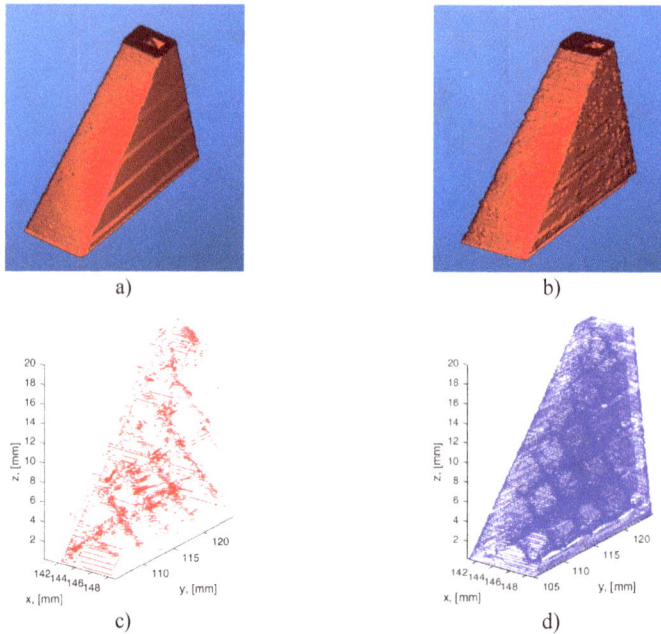

Figure 6. Voxel models of simulated and measured part and voxel-wise analysis of discrepancies. a) voxel model of the simulated part; b) voxel model of the measured part; c) three-dimensional distribution of voxels representing excess material; d) three-dimensional distribution of voxels representing missing material.

Conclusions

An original solution for in-process inspection of fused filament fabrication has been proposed, capable of addressing the realisation of complex geometries featuring hollow and reticular internal features, such as lattice patterns. The in-process inspection solution combines optical tomography, implemented via laser line scanning, with a pixel and voxel-based method for real-time comparison between the fabricated geometry and a nominal reference generated by simulation (digital twin). The main points of strength of the proposed solution are in the original use of laser line scanning to retrieve external and internal layer contours in optical tomography, which overcomes the limitations of edge detection applied to 2D digital images; and the pixel- / voxel-based modelling and comparison of expected and measured geometries, which allows for robust detection of missing and excess material. Further challenges to be addressed are related to determining the performance of the inspection method. The main issue is that, at the moment, we haven't fully characterised the metrological performance of the laser line scanning-based measurement solution, therefore we are unable to quantify how much of the observed discrepancies may be due to measurement error. Metrological characterisation is challenging due to the absence of a reference measurement result to compare our solution against. The use of X-ray CT-scanning is being considered, but CT-scanning itself is hardly a reference, given the currently unsolved issues related to boundary determination and effects of measurement artefacts (e.g., beam-hardening). Moreover, CT-scanning would have to be performed post-process, on a final part which could have changed shape because of warpage, shrinkage or hydrophilic effects.

The second main unsolved issue in reference to the pixel-wise comparison lays on the simulation side, as part of the observed discrepancies may be indeed due to the simulation's inability to accurately reproduce the expected results, for example in relation to the expected thickness of the deposited strand. Work is in progress to further refine and validate the simulation model. Other issues which may influence the comparison are related to the data processing steps and include for example the process of binarising the laser line position in measurement (affected by multiple factors, including the determination of the laser centreline and the determination of the vertical width of the reference band), as well as the steps of rasterization/voxelization, in particular the resolutions adopted for comparing both layer maps and entire 3D volumes. The assessment of inspection performance in relation to the sizes of the most frequent geometric anomalies in FFF-manufactured goods (application dependent) will also need to be addressed.

References

[1] S. Singh, G. Singh, C. Prakash and S. Ramakrishna, "Current status and future directions of fused filament fabrication," J. Manuf. Process., (2020), 288-306 https://doi.org/10.1016/j.jmapro.2020.04.049

[2] International Organization for Standardization, "ISO/ASTM 52900:2021-Additive manufacturing - General principles - Fundamentals and vocabulary" BSI Standards Ltd

[3] X. Wang, L. Zhao, J. Y. H. Fuh and H. P. Lee, "Effect of porosity on mechanical properties of 3D printed polymers: Experiments and micromechanical modeling based on X-ray computed tomography analysis," Polymers (Basel)., (2019), 11(7). https://doi.org/10.3390/polym11071154

[4] A. Thompson, I. Maskery and R. K. Leach, "X-ray computed tomography for additive manufacturing: A review," Meas. Sci. Technol., vol. 27, no. 7, (2016), 072001. https://doi.org/10.1088/0957-0233/27/7/072001

[5] B. M. Colosimo, M. Grasso, F. Garghetti and B. Rossi, "Complex geometries in additive manufacturing: A new solution for lattice structure modeling and monitoring," J. Qual. Technol., 2022, VolL. 54, No. 4, 392-414 https://doi.org/10.1080/00224065.2021.1926377

[6] M. Moretti, A. Rossi and N. Senin, "In-process monitoring of part geometry in fused filament fabrication using computer vision and digital twins," Addit. Manuf., (2020), 101609. https://doi.org/10.1016/j.addma.2020.101609

[7] Y. Wu, K. He, X. Zhou and W. Ding, Machine vision based statistical process control in fused deposition modeling, in Proceedings of the 2017 12th IEEE Conference on Industrial Electronics and Applications, ICIEA. (2017), 936-941.

[8] F. Imani, A. Gaikwad, M. Montazeri, P. Rao, H. Yang and E. Reutzel, Layerwise in-process quality monitoring in laser powder bed fusion, ASME 2018 13th International Manufacturing Science and Engineering Conference, MSEC. 1 (2018). https://doi.org/10.1115/MSEC2018-6477

[9] B. M. Colosimo, F. Garghetti, L. Pagani and M. Grasso, "A novel method for in-process inspection of lattice structures via in-situ layerwise imaging," Manuf. Lett., (2022),Vol 32, 67-72. https://doi.org/10.1016/j.mfglet.2022.03.004

[10] M. Grazia Guerra, M. Lafirenza, V. Errico and A. Angelastro, "In-process dimensional and geometrical characterization of laser-powder bed fusion lattice structures through high-resolution optical tomography," Opt. Laser Technol., (2023), Vol 162, 109252. https://doi.org/10.1016/j.optlastec.2023.109252

[11] S. Nuchitprasitchai, M. Roggemann and J. M. Pearce, Factors effecting real-time optical monitoring of fused filament 3D printing, Prog. Addit. Manuf. 2 (2017) 133-149. https://doi.org/10.1007/s40964-017-0027-x

[12] M. Faes, W. Abbeloos, F. Vogeler, H. Valkenaers, K. Coppens, T. Goedemé and E. Ferraris, Process Monitoring of Extrusion Based 3D Printing via Laser Scanning. International Conference on Polymers and Moulds Innovations (PMI). (2014).

[13] W. Lin, H. Shen, J. Fu and S. Wu, "Online quality monitoring in material extrusion additive manufacturing processes based on laser scanning technology," Precis. Eng., (2019), Vol 60, 76-84. https://doi.org/10.1016/j.precisioneng.2019.06.004

[14] K. Xu, J. Lyu and S. Manoochehri, "In situ process monitoring using acoustic emission and laser scanning techniques based on machine learning models," J. Manuf. Process., (2022), Vol 84, 357-374. https://doi.org/10.1016/j.jmapro.2022.10.002

[15] Information on https://www.berlinlasers.com/it/oem-lab-lasers/laser-line-generator

[16] Information on https://www.dino-lite.eu/it/component/eshop/am7915mzt-edge?Itemid=0

[17] Information on https://docs.arduino.cc/hardware/leonardo

[18] M. Moretti, F. Bianchi and N. Senin, "Towards the development of a smart fused filament fabrication system using multi-sensor data fusion for in-process monitoring," Rapid Prototyp. J., (2020), Vol 36, 7. https://doi.org/10.1108/RPJ-06-2019-0167

Italian Manufacturing Association Conference - XVI AITeM   Materials Research Forum LLC
Materials Research Proceedings 35 (2023) 225-231   https://doi.org/10.21741/9781644902714-27

Effect of process parameters on the thermal properties of material extruded AM parts

Luigi Morfini[1,2,a] *, Nicola Gurrado[1,b] and Roberto Spina[1,2,3,c]

[1]Dipartimento di Meccanica, Matematica e Management (DMMM), Politecnico di Bari, Bari, Italy

[2]Istituto Nazionale di Fisica Nucleare (INFN) - Sezione di Bari, Bari, Italy

[3]Consiglio Nazionale delle Ricerche - Istituto di Fotonica e Nanotecnologie (CNR-IFN), Bari, Italy

[a] luigi.morfini@poliba.it, [b] nicola.gurrado@poliba.it, [c] roberto.spina@poliba.it

Keywords: Material Extrusion, Quality Control, Thermal Properties

Abstract. This research focuses on the thermal characterization of 3D-printed parts obtained via Material Extrusion Additive Manufacturing using various process parameters. Differential Scanning Calorimetry and Thermal Conductivity measurements were used to evaluate the samples' thermal characteristics and heat transport behavior. The experimental results showed a significant influence of some parameters, such as wall layer count and layer thickness, on the thermal behavior of the printed part.

Introduction

Additive manufacturing (AM) is a widespread research technique for engineering and biomedical applications. Unlike traditional machining, AM utilizes either an energy source or an extruder to build parts by selectively melting and solidifying materials. An energy source such as a laser or an electron beam can fuse metal powders or wires, resulting in a solid part. Alternatively, an extruder can dispense feedstock materials like plastic, ceramic, or metal, which are then fused using heat or chemical reactions to form a solid part. In either approach, the material is built up layer-by-layer until the final shape is achieved [1]. This technique allows for close control of the fabrication process and greater design freedom for traditional machining, introducing challenges related to functional properties. While AM has long been used for rapid prototyping, the recent focus has been on building functional parts capable of withstanding thermal/mechanical loads. Filament-based Material Extrusion (MEX), according to ISO/ASTM 52900 classification, is a commonly used AM process relying on an extruder dispensing polymer above its glass transition temperature. However, given the AM complexity, it is essential to understand the fundamental transport processes and the dependence of functional properties on process parameters to optimize the process [2]. MEX-built parts have been reported to have anisotropic mechanical properties with the highest strength in the direction of the material [3][4]. The merging of adjacent polymer lines, known as roads, plays a critical role in determining the microstructure and functional properties of the built part [5]. This merging process depends on various process parameters and material properties, which have been studied previously [2][6]. It is vital to comprehend and enhance this merging process to attain high-quality parts.

While there is significant research on the mechanical properties of MEX parts [2]-[4],[6]-[8], a relatively small amount of work has been done on the thermal properties and their correlation with microstructure and process parameters. Thermal conductivity is a crucial property governing heat flow through the part. It is essential in engineering applications where heat generation and flow are critical. The thermal conductivity of the MEX part differs from that of the original material, expecting to vary between the raster and build directions. For this reason, a direct measurement of the thermal properties of the final part is essential. The in-plane and out-of-plane thermal

Italian Manufacturing Association Conference - XVI AITeM
Materials Research Proceedings 35 (2023) 225-231

Materials Research Forum LLC
https://doi.org/10.21741/9781644902714-27

conductivities of MEX-built parts were measured by Shemelya et al. using the transient plane source method [9]. Defects in the printing process can cause variations in thermal properties. Improper adhesion caused by poorly designed process parameters can determine inadequate heat flow, leading to low thermal conductivity. Chung et al. [10] have conducted measurements to investigate the impact of random voids on thermal properties. Ravoori et al. [11] investigated the effects of process parameters on the thermal properties of AM parts by using a one-dimensional heat flux method to measure the thermal conductivity in the build direction. Their analysis revealed a significant correlation between thermal conductivity, raster speed, and layer thickness, further supported by high-speed imaging of the printing process conducted at varying levels of these parameters. The anisotropy of thermal conductivity in AM components is challenging, as highlighted by Prajapati et al. [12]. Their study reveals the fundamental reason for this anisotropy is the strong interfacial thermal contact resistance in the build direction. They provide valuable data on the dependence of this parameter on process conditions. The measurements show significant anisotropy in thermal conduction, with thermal conductivity in the Z-direction being significantly lower than in the X-direction. This anisotropy is influenced by the air gap between roads during deposition. Thermal conductivity is an essential property of many materials, and various methods have been developed to measure it. Thermal conductivity describes the mode of heat transfer via conduction and can be affected by multiple factors such as temperature, pressure, and material composition. The measured property may be referred to as effective thermal conductivity if other modes of heat transfer, such as convection and radiation, are significant. Thermal conductivity can be measured through steady-state or transient methods using a specimen of simple geometry in contact with a heat source and one or more temperature sensors. Transient methods may be contact or non-contact and involve generating a dynamic temperature field within the specimen and measuring the temperature response over time. Depending on the setup, one or more thermo-physical properties can be obtained. The response is analyzed using a model and a set of solutions designed for the specific geometry and boundary conditions. The Transient Hot Wire (THW) method and the Laser-Flash Analysis (LFA) are commonly used to measure thermal conductivity. In recent research, an excellent agreement was found between THW and LFA values measured on the same components in the same conditions [13][14]. The THW method offers a low-cost alternative to LFA with reduced accuracy. The Temporary Hot Bridge (THB) is an evolution of the THW. It provides fast, precise measurements of thermal conductivity, thermal diffusivity, and volumetric specific heat from a single experiment on solids.

This work aimed to analyze the thermal conductivity of samples made with MEX in Polylactic acid (PLA) using the THB method. The key parameters were the infill percentage, which indirectly affects the air gaps between the lines and the number of top/bottom layers. During the study, the variation of the two factors above was analyzed to investigate their effect on the thermal properties while keeping the other process parameters constant. Additionally, an analysis was conducted to evaluate the surface roughness.

Materials and methods

A commercial black PLA 1.75 mm filament (AzureFilm d.o.o., Slovenia) was used as the material for producing the samples. The filament was dried for 4 hours at 40°C in an oven to prevent issues related to potential moisture presence and then stored in a vacuum-sealed bag before starting the printing process. This study did not consider the influence of additives on base PLA material [15]. The samples were produced on a commercial 3D printer, Flying Bear Ghost 4S (Zhejiang Flying Bear Intelligent Technology Co., Ltd.), with a build volume of $255 \times 210 \times 210$ mm^3. The printer has a 0.4 mm brass nozzle and a Polyetherimide (PEI) print bed to ensure high adhesion while limiting warping phenomena.

The process parameters to investigate were selected after some preliminary testing. A line width equal to 0.4 mm followed the nozzle diameter of 0.4 mm, and the layer height was set to half the line width to maintain dimensional accuracy and acceptable quality. The crosshatch raster strategy with angles ±45° helped slightly contain anisotropy among the top and bottom layers. The wall layer was set to three in the slicer (Ultimaker Cura) to ensure the shell could clamp with a vise. In MEX, the wall layer refers to the outer layer of the printed object. Increasing the number of wall layers can make a part more robust and resistant to external forces. Lower values were considered as not strong enough. During the printing of the samples, the number of top and bottom layers and the infill percentage were adjustable, while the other process parameters remained constant. The top and bottom layers refer to the layers of material that make up the top and bottom surfaces of the printed object, respectively. Infill percentage refers to the material used to fill the object's interior. Table 1 reports the values of the process parameters.

Table 1. **Process parameters of MEX**

Process Parameter	Value	Unit size
Nozzle temperature	210	[°C]
Bed temperature	60	[°C]
Printing speed	60	[mm/s]
Layer height	0.2	[mm]
Line width	0.4	[mm]
Wall layer count	3	-
Infill strategy	Grid	-
Layer raster strategy	-45°/+45°	-

To explore how the number of layers and infill percentage impacted the thermal conductivity of $70 \times 40 \times 10$ mm^3 prismatic parts, a complete factorial Design of Experiments (DoE) was employed, as shown in Table 2.

Table 2. Complete factorial Design of Experiment

Factors	Name	Level 1	Level 2	Level 3	Level 4
Layer number	L_N	2	4	6	8
Infill percentage [%]	I_P	20	40	60	80

As a result of the material's properties, the sample surface was measured with the scanControl 2900-50 BL (Micro-Epsilon Messtechnik GmbH & Co. KG), a 2D/3D blue laser scanner and profilometer designed for measuring semi-transparent, red-hot glowing, and organic materials with a 405 nm wavelength laser. The scanner features a resolution of 4 μm on the Z-axis, considering the object's height growth direction, and 1,280 points/profile resolution on the X-axis. The specimens' top surface roughness Ra was also measured using the Surtronic 3P surface roughness profilometer (Taylor-Hobson, UK) with a cut-off value of 0.8mm to confirm the previous measurements.

The THB instrumentation (Linseis Messgeräte GmbH, Germany) detected the thermal conductivity without sample preparation. The THB method measured the time for a thermal signal to propagate through a material sample. The measurement was performed dynamically by applying a short-duration heat pulse to the thermal bridge, which exchanged heat with the sample. The propagation time of the thermal signal through the sample was then used to calculate the material's thermal properties. The instrumentation consisted of a flexible sensor (THB-B type) pressed manually between two flat surfaces of the specimen (Figure 1-left) to ensure good thermal contact without air inclusions. The specifications of the sensor are as follows: dimensions of 42×22 mm^2, a thermal conductivity range of 0.1÷2 W/m×K, as well as a temperature range of -150÷200 °C. The tests were conducted at room temperature according to UNI-EN ISO 22007 standard, with an execution time of 160 s and a pulse supply current of 50 mA. A non-uniform temperature profile is generated, driving a Wheatstone bridge off-balance to produce an offset-free output. A heater R_H created a time-dependent, distant-dependent temperature field along the sensor and the material under test. Additional resistors were connected to a Wheatstone bridge with a bridge voltage UB dependent on the temperature difference ΔT. Measuring the temperature difference ΔT enabled the calculation of thermal conductivity λ using the following formula:

$$\lambda = \frac{\Phi}{2 \times \pi \times \Delta T} \times k \qquad (1)$$

The heater R_H generated the applied thermal power Φ supplied with current, while k was the sensor calibration factor. The THB evaluation is refined to enable anisotropic materials' direction-dependent thermal diffusivity measurement. This new method requires only plate-shaped specimens and allows the determination of thermal diffusivity in all three directions [16][17]. It is important to note that the tests were conducted on the specimen's top surface, corresponding to the last printed layer, due to warping phenomena observed during the printing process. Running the tests on the top surface ensured more accurate results, using four different infill percentages (Figure 1-right).

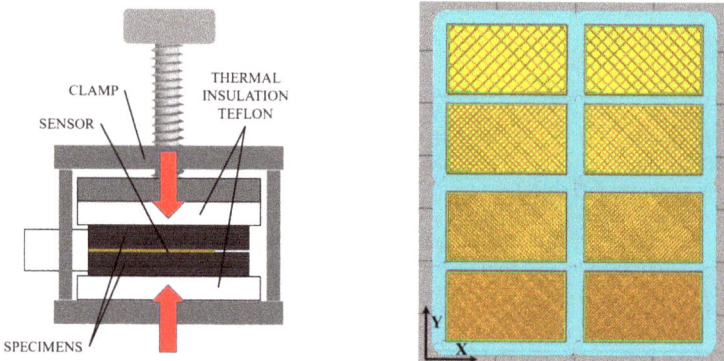

Figure 1. The functioning scheme of the THB system (left) and printing strategies (right).

Results and discussion

Before conducting tests to determine the thermal properties of the specimens, it was necessary to perform some preliminary studies on the roughness of the surfaces. This analysis was required to ensure that the specimen roughness values were within acceptable limits to avoid any potential influence on the measurement of thermal properties, as highlighted in the relevant literature. The samples were inspected as printed. Laser scanning was used to obtain a roughness profile, measured by sampling three sections (20×20 mm²) of the specimen. The sections were taken in areas far from the edges to avoid measurement errors due to edge effects in the 3D-printed specimens. The average roughness values were $R_z = 15.6$ μm and $R_a = 3.61$ μm. To compare these values, measurements were taken with the Surtronic 3P profilometer, which yielded similar R_a values to those recorded with the scanControl 2900-50 BL. All measurements followed the UNI EN ISO 21920 standard to ensure accuracy and consistency. Preliminary tests were carried out on 100% infill specimens to evaluate the impact of roughness on the thermal conductivity values (Table 3). It was found that roughness was an influential factor, as specimens with higher roughness resulting from contact with the rough-textured PEI plate reported worse results in

Italian Manufacturing Association Conference - XVI AITeM   Materials Research Forum LLC
Materials Research Proceedings 35 (2023) 225-231   https://doi.org/10.21741/9781644902714-27

thermal conductivity. Instead, specimens printed on the smooth-textured PEI plate reported the highest values of the three measurements, like those obtained from printed top layers. Another purpose of this surface roughness analysis was to verify the samples' overall quality and test if the roughness values were evenly distributed. Moreover, it should be noted that the roughness values obtained through MEX processes are generally higher than those obtained through traditional manufacturing processes such as injection molding (1.06 μm) [18].

Table 3. Roughness and thermal conductivity values of 100% filled samples.

	Top layer	Smooth PEI	Rough PEI	Unit
Roughness R_a	3.62	2.19	7.96	[μm]
Thermal conductivity λ	0.186	0.189	0.141	[W/m×K]

Table 4. Statistical analysis results

	Estimate	Lower 95%	Upper 95%
Intercept	0.12985	0.12501	0.13468
layer_number(2,8)	0.02726	0.02501	0.0295
layer_number×layer_number	0.00318	-0.0015	0.00782
infill_percentage(20,80)	0.02642	0.02417	0.02867
infill_percentage×infill_percentage	-0.0044	-0.0091	0.00023
layer_number×infill_percentage	-0.0073	-0.0099	-0.0048
R^2	0.9928		
Mean squared error	0.0034		

Another critical aspect considered was the choice of the part infill. Test samples were not printed 100% filled because the main reason for MEX is the ability to obtain complex or lightweight structures. Therefore, for the reduction of material waste and better optimization, it was essential to study the thermal conductivity on lightweight samples with different infill percentages and top/bottom layer count to observe the correlation between the tested values.

The statistical analysis results are graphically in Table 4 and Figure 2-(right). The regression model's coefficient R^2 is 99.28%. The figure shows an analysis of the expected values compared to the observed ones. It can be noticed that these factors were linear, indicating that the model used to generate the predicted values was a good fit for the observed data. This linearity could be seen in the proximity of the data points to the regression line. Additionally, the slope of the regression line suggests that the observed values tend to be slightly higher than the expected values, but the difference was not significant. Figure 2-(left) proves that the model accurately predicted the observed values.

It is observed that there was a linear correlation between the infill percentage and thermal conductivity. Furthermore, the trend line for the data shows a convex increasing structure, indicating that as the infill percentage increased, the rate of change in thermal conductivity also increased at an increasing rate. This result suggested optimizing the infill percentage could significantly improve the material's thermal conductivity. This observation was consistent with the findings reported in the literature and could be explained by air having a very low thermal conductivity. As the infill percentage increased, the volumetric fraction of air increased, which reduced the overall thermal conductivity of the specimen (at the expense of the PLA fraction).

Similarly, as the number of surface layers increased, the thermal conductivity increased because heat encountered the insulating layer of air later. It was worth noting that the trend line of the data exhibits a concave increasing structure, suggesting that the rate of increase in thermal conductivity became progressively smaller as the number of layers increased. Therefore, it could be concluded that the more the component was filled, the better it could transmit heat through its thickness.

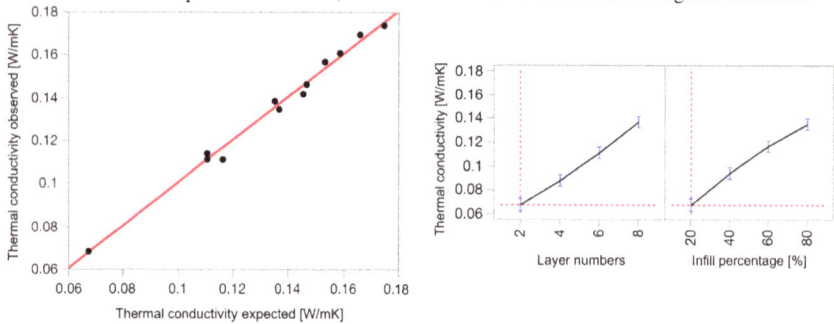

Figure 2. Thermal conductivity expected vs. observed analysis (left); Statistical analysis results (right)

Conclusion

This study investigated the thermal conductivity properties of samples made with filament-based MEX in PLA. The effect of infill percentage and the number of layers on thermal properties was examined while keeping the other process parameters constant. The results indicated a linear correlation between infill percentage and thermal conductivity, as well as between the number of layers and thermal conductivity. Moreover, further investigations are necessary to fully comprehend the impact of anisotropy on thermal properties and its correlation with the number of surface layers. The transient hot-bridge sensor provided fast and accurate measurements of thermal conductivity, thermal diffusivity, and volumetric specific heat from a single experiment. These findings can help optimize the additive manufacturing process to achieve the desired thermal properties in-built parts. This aspect is crucial for engineering applications with significant heat generation and flow.

Acknowledgments

Funder: Project funded under the program Department of Excellence - Law number 232/2016 (Grant No. CUP - D93C23000100001) and the National Recovery and Resilience Plan (NRRP), Mission 4 Component 2 Investment 1.3 - Call for tender No. 341 of 15/03/2022 of the Italian Ministry of University and Research (MUR), funded by the European Union – NextGenerationEU. Award Numbers: Law number 232/2016 (Grant No. CUP - D93C23000100001) and PE00000004, Concession Decree No. 1551 of 11/10/2022 of the Italian Ministry of University and Research, CUP D93C22000920001, MICS (Made in Italy - Circular and Sustainable).

The authors thank Prof Luigi Galantucci and Prof Fulvio Lavecchia of Politecnico di Bari for their suggestions and support.

References

[1] L.M. Galantucci, M.G. Guerra, M. Dassisti, F. Lavecchia, Additive Manufacturing: New Trends in the 4th Industrial Revolution, in Lecture Notes in Mechanical Engineering, Springer International Publishing, 2019: pp. 153-169. https://doi.org/10.1007/978-3-030-18180-2_12
[2] J.M. Chacón, M.A. Caminero, E. García-Plaza, P.J. Núñez, Additive manufacturing of PLA structures using fused deposition modelling: Effect of process parameters on mechanical properties and their optimal selection, Mater Des. 124 (2017) 143-157. https://doi.org/10.1016/j.matdes.2017.03.065

[3] L. Auffray, P.A. Gouge, L. Hattali, Design of experiment analysis on tensile properties of PLA samples produced by fused filament fabrication, International Journal of Advanced Manufacturing Technology. 118 (2022) 4123-4137. https://doi.org/10.1007/s00170-021-08216-7

[4] S.H. Ahn, M. Montero, D. Odell, S. Roundy, P.K. Wright, Anisotropic material properties of fused deposition modeling ABS, Rapid Prototyp J. 8 (2002) 248-257. https://doi.org/10.1108/13552540210441166

[5] S.F. Costa, F.M. Duarte, J.A. Covas, Estimation of filament temperature and adhesion development in fused deposition techniques, J Mater Process Technol. 245 (2017) 167-179. https://doi.org/10.1016/j.jmatprotec.2017.02.026

[6] F. Bähr, E. Westkämper, Correlations between Influencing Parameters and Quality Properties of Components Produced by Fused Deposition Modeling, in: Procedia CIRP, Elsevier B.V., 2018: pp. 1214-1219. https://doi.org/10.1016/j.procir.2018.03.048

[7] S. Garzon-Hernandez, D. Garcia-Gonzalez, A. Jérusalem, A. Arias, Design of FDM 3D printed polymers: An experimental-modelling methodology for the prediction of mechanical properties, Mater Des. 188 (2020). https://doi.org/10.1016/j.matdes.2019.108414

[8] N. Zohdi, R.C. Yang, Material anisotropy in additively manufactured polymers and polymer composites: A review, Polymers (Basel). 13 (2021). https://doi.org/10.3390/polym13193368

[9] C. Shemelya, A. De La Rosa, A.R. Torrado, K. Yu, J. Domanowski, P.J. Bonacuse, R.E. Martin, M. Juhasz, F. Hurwitz, R.B. Wicker, B. Conner, E. MacDonald, D.A. Roberson, Anisotropy of thermal conductivity in 3D printed polymer matrix composites for space based cube satellites, Addit Manuf. 16 (2017) 186-196. https://doi.org/10.1016/j.addma.2017.05.012

[10] S.Y. Chung, D. Stephan, M.A. Elrahman, T.S. Han, Effects of anisotropic voids on thermal properties of insulating media investigated using 3D printed samples, Constr Build Mater. 111 (2016) 529-542. https://doi.org/10.1016/j.conbuildmat.2016.02.165

[11] D. Ravoori, L. Alba, H. Prajapati, A. Jain, Investigation of process-structure-property relationships in polymer extrusion based additive manufacturing through in situ high speed imaging and thermal conductivity measurements, Addit Manuf. 23 (2018) 132-139. https://doi.org/10.1016/j.addma.2018.07.011

[12] H. Prajapati, D. Ravoori, R.L. Woods, A. Jain, Measurement of anisotropic thermal conductivity and inter-layer thermal contact resistance in polymer fused deposition modeling (FDM), Addit Manuf. 21 (2018) 84-90. https://doi.org/10.1016/j.addma.2018.02.019

[13] S. Wang, D. Zhang, G. Liu, W. Wang, M. Hu, Application of hot-wire method for measuring thermal conductivity of fine ceramics, Medziagotyra. 22 (2016) 560-564. https://doi.org/10.5755/j01.ms.22.4.12543

[14] M. Ruoho, K. Valset, T. Finstad, I. Tittonen, Measurement of thin film thermal conductivity using the laser flash method, Nanotechnology. 26 (2015). https://doi.org/10.1088/0957-4484/26/19/195706

[15] R. Spina, Performance analysis of colored PLA products with a fused filament fabrication process, Polymers. 11 (2019). https://doi.org/10.3390/polym11121984

[16] J. Gaiser, M. Stripf, F. Henning, Enhanced Transient Hot Bridge Method Using a Finite Element Analysis, Int J Thermophys. 40 (2019). https://doi.org/10.1007/s10765-018-2476-y

[17] U. Hammerschmidt, V. Meier, New transient hot-bridge sensor to measure thermal conductivity, thermal diffusivity, and volumetric specific heat, Int J Thermophys. 27 (2006) 840-865. https://doi.org/10.1007/s10765-006-0061-2

[18] O. Ozdilli, Comparison of the Surface Quality of the Products Manufactured by the Plastic Injection Molding and SLA and FDM Method, Uluslararası Muhendislik Arastirma ve Gelistirme Dergisi. (2021) 428-437. https://doi.org/10.29137/umagd.762942

Italian Manufacturing Association Conference - XVI AITeM   Materials Research Forum LLC
Materials Research Proceedings 35 (2023) 232-240   https://doi.org/10.21741/9781644902714-28

Robust in-line qualification of lattice structures manufactured via laser powder bed fusion

Bianca Maria Colosimo[1,a], Marco Grasso[1,b *], Federica Garghetti[1,c] and Luca Pagani[2,d]

[1]Politecnico di Milano, Department of Mechanical Engineering, Via La Masa 1, 20156 Milano (Italy)

[2]GOM GmbH, Schmitztrasse 2, 38112 Braunschweig (Germany)

[a]biancamaria.colosimo@polimi.it, [b]marcoluigi.grasso@polimi.it, [c]federica.garghetti@polimi.it, [d]luca.paggy@gmail.com

Keywords: Additive Manufacturing, Quality Modelling, Profile Monitoring, Lattice, In-Situ Sensing

Abstract. The shape complexity enabled by AM would impose new part inspection systems (e.g., x-ray computed tomography), which translate into qualification time and costs that may be not affordable. However, the layerwise nature of the process potentially allows anticipating qualification tasks in-line and in-process, leading to a quick detection of defects since their onset stage. This opportunity is particularly attractive in the presence of lattice structures, whose industrial adoption has considerably increased thanks to AM. This paper presents a novel methodology to model the quality of lattice structures at unit cell level while the part is being built, using high resolutions images of the powder bed for in-line geometry reconstruction and identification of deviations from the nominal shape. The methodology is designed to translate complex 3D shapes into 1D deviation profiles that capture the "geometrical signature" of the cell together with the reconstruction uncertainty.

Introduction

Additive manufacturing (AM) technologies enable a variety of innovative product shapes, performances and functionalities that have been profitably exploited in a continuous increasing number of industrial applications. Among novel and most interesting geometries, lattice structures opened several opportunities in rethinking advanced and high-value-added components in sectors like healthcare, aerospace, space, and racing. Lattices are periodic structures made of unit cells of a predefined geometry that repeats in space, leading to enhanced performances thanks to a more efficient use of the material they are composed of [1]. Local inaccuracies may have a detrimental effect on the functional performance of the whole structure [2 – 4]. As an example, a local dimensional and/or geometrical mismatch between the manufactured part and the nominal geometry may influence the mechanical properties together with the type of failure mechanism [2]. However, there is still a lack of statistical methods to model and monitor the quality of these structures in industry.

A seminal approach proposed in [5] was aimed to transform the 3D deviation between the ex-situ reconstructed geometry of the lattice (via X-ray computed tomography (CT)) and its nominal geometry into 1D deviation profiles that captured a layerwise deviation metric on a cell-by-cell basis. Colosimo et al. [5] showed that the proposed approach enabled the detection of local geometrical distortions in one cell or in one sub-portion of the lattice. An even higher potential for industrial adoption of such lattice structure modelling and monitoring methodology would involve the possibility to move from ex-situ (post-process) to in-situ (in-process) measurements, taking advantage of layerwise image data collection in AM. Exploiting in-situ and in-line data would

Italian Manufacturing Association Conference - XVI AITeM   Materials Research Forum LLC
Materials Research Proceedings 35 (2023) 232-240   https://doi.org/10.21741/9781644902714-28

allow end-users to anticipate the detection of anomalies and defects while the part is being produced, enabling possible remedy actions (e.g., part suppression), reducing both post-process inspection costs and material wastes. In-line geometry reconstruction and monitoring could be also combined with advanced process control and first-time-right strategies, to heal and mitigate deviations from the nominal shape (see for example [6 – 8]). Colosimo et al. [9] introduced an extension of the aforementioned method that replaces ex-situ X-ray CT measurements with in-situ powder bed images. The authors devoted special attention to the challenges imposed by various sources of variability that may affect the in-situ geometry reconstruction. The authors proposed a methodology that combines a robust active contour algorithm for solidified layer image segmentation with a robust fitting technique to enhance the reconstruction accuracy. Another approach was proposed by Guerra et al. [10], where authors used powder bed images to generate a 3D optical tomography reconstruction of the printed structure for dimensional and geometrical characterization. The authors showed the potential use of the method to detect process inaccuracies and geometric distortions.

The present study inherits the robust reconstruction and modelling approach presented in [9], and it extends that work with two additional and novel analyses. The first regards the quantitative evaluation of the reconstructions accuracy that can be achieved via in-situ active contour segmentation of a lattice structure. The second involves the performance analysis of a robust fitting approach applied to 1D deviation profiles, in terms of deviation from the nominal shape. The aim is to demonstrate to what extent in-situ deviation models could be effective in replacing more complex and expensive ex-situ gathered ones.

Differently from Guerra et al. [10], the lattice geometry was reconstructed by means of high-resolution powder bed images rather than using a long-exposure optical tomography image. The proposed in-situ robust active contour segmentation was compared against the CT-based reconstruction used as a reference (or ground truth) to highlight the agreement of dimensional measurements obtained with the two different data sources. As far as the cell-wise 1D deviation profile reconstruction is concerned, a quantitative analysis of the benefits provided by the proposed robust model compared with a traditional least square fitting is presented too. Robust modelling represents a key issue to reduce the uncertainty of deviation estimates for the design of statistical process monitoring tools that may leverage on the proposed in-situ modelling and qualification framework.

A lattice structure production using an industrial laser powder bed fusion (L-PBF) system is presented as real case study. The original camera and lighting setup available as embedded off-the-shelf equipment was used, which allows evaluating the suitability of the proposed approach for implementation on an industrial L-PBF platform with no need for machine modification.

The paper is organized as follows. Section 2 presents the case study. Section 3 briefly describes the methodology, and Section 4 presents the major results. Section 5 concludes the paper.

Case Study

As a real case study, a maraging steel lattice structure produced via L-PBF on an industrial Trumpf TruPrint 3000 system was considered. The structure was composed by 64 equal rhombic cells, with a nominal struct thickness equal to 1.5 mm. The overall specimen dimension was $40 \times 40 \times 40$ mm, with a nominal cubic cell envelope with side equal to 10 mm. The specimen was produced with two lateral walls of thickness equal to 0.6 mm, removed from the analysis of the current work. A gas atomized powder was used and nominal process parameters provided by the machine vendor were applied. Fig. 1 shows the manufactured part, the nominal geometry of the rhombic cell and the process parameters used for the experiment.

Italian Manufacturing Association Conference - XVI AITeM   Materials Research Forum LLC
Materials Research Proceedings 35 (2023) 232-240   https://doi.org/10.21741/9781644902714-28

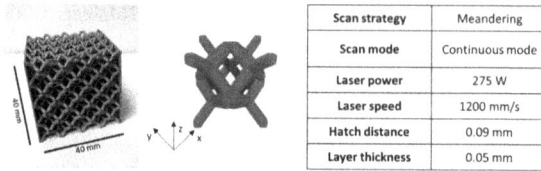

Scan strategy	Meandering
Scan mode	Continuous mode
Laser power	275 W
Laser speed	1200 mm/s
Hatch distance	0.09 mm
Layer thickness	0.05 mm

Fig. 1 Printed part, nominal unit cell and process parameters (the meandering scan direction was rotated by 67° every layer)

The Trumpf TruPrint 3000 machine has an integrated powder bed camera (Basler acA3800-14uc USB 3.0 camera) that allows capturing post-deposition and post-melting images with an instant field of view (spatial resolution) of 100 μm/pixel. The light source was inclined of about 60° with respect to the building plate. It consists of a LED stripe on the ceiling of the chamber.

The specimen was inspected using a North Star Imaging X25 x-ray CT scan system with a voxel size of 33 μm. The 3D shape reconstructed ex-situ was first aligned with respect to the nominal geometry (via Iterative Closest Point (ICP) registration) and then sliced with a slice thickness equal to the layer thickness used during the L-PBF process. This allowed the superimposition of both the in-situ and ex-situ geometry reconstructions in every layer for comparison purpose. The surface determination was performed via the proprietary algorithm "standard classic" implemented in VGStudio Max. The uncertainty of the geometry reconstructed via x-ray CT is mainly affected by the i) surface determination error and ii) the ICP registration error. Minimizing these errors allows enhancing the suitability of the x-ray CT as ground truth. In this study, algorithms implemented in VGStudio Max were used as representative of the current industrial practice. Nevertheless, other state-of-the-art methods can be considered to further enhance the ground truth accuracy [11 - 12]. Generally speaking, x-ray CT represents the current state of the art for the reconstruction of the actual shape, as it is commonly used for product acceptance, and its uncertainty is one order of magnitude lower than the one of in-situ reconstructions with standard powder bed cameras, which makes x-ray CT suitable as a ground truth.

Methodology

Before applying the proposed approach, a pre-processing of powder bed images is need to perform a perspective correction and to align the resulting images against the nominal geometry. Standard algorithms may be used to this aim, see for example [13 – 14]. The methodology then consists of three steps: i) segmentation of each solidified layer, ii) estimation of a deviation metric to quantify the deviation between the in-situ reconstructed shape and the nominal one in each layer, iii) robust modelling of the deviation as a function of the build direction for each single unit cell.

Regarding image segmentation, we advocate the active contour methodology [15 – 16]. In entails an iterative segmentation procedure that starts from a first boundary definition in the form of a closed curve and adapts it by applying shrink/expansion operations until the boundary converges to the final reconstructed contour. In L-PBF, the nominal geometry itself can be used as starting boundary [17 – 19]. A robust variant of the active contour method was shown in [19] to be particularly effective in powder bed image segmentation. It combines edge-based and region-based active contours segmentation operations:

$$\frac{\delta\varphi(x,t)}{\delta t} = w(t)\delta_{region} + (1 - w(t))\delta_{edge} \tag{1}$$

where $\varphi(x)$ is the signed distance function in pixel location x, which is iteratively varied until convergence to the final reconstructed contour; δ_{region} is the region-based term, δ_{edge} is the edge-based term, and $w(t)$ such that $0 \le w(t) \le 1$ is a weight to balance the influence of each term.

As proposed in [19], the weight $w(t)$ may vary as a function of the iteration counter, t, associating higher weight to the region-based term in initial iterations, for rough contour adjustments, and higher weight to the edge-based term in last iterations, for a final fine-tuning of the contour. The reader is referred to [19] for full details. The active contour algorithm can be calibrated and tuned using a reference geometry and its x-ray CT reconstruction as a ground truth. Calibration is envisaged to take into account the specific conditions imposed by the L-PBF machine and its machine vision and illumination setup.

The underlying idea is that by combining region- and edge-based segmentations, a more robust reconstruction of the actual contour is achieved, in a way that is robust to pixel intensity variations within foreground and background areas. The output of the segmentation consists of a binary image, where background pixels have intensity $I = 0$, and foreground pixels (solidified layer) have intensity $I = 1$. Since both the in-situ reconstructed shape and its nominal geometry are represented in binary format, the deviation between them can be computed as:

$$\delta_{i,j,k}(z) = \frac{1}{N}\sum_l \mathcal{J}(I_{insitu}(l) - I_{nominal}(l) \neq 0)_{i,j,k}, \text{ for the } (i,j,k)^{th} \text{ cell} \qquad (2)$$

where z is the build direction, N is the number of pixels within the analyzed region of the powder bed image, $\mathcal{J}(\cdot)$ is the indicator function, $I_{insitu}(l)$ is the value the lth pixel of the binarized powder bed image generated by the active contours segmentation, and $I_{nominal}(l)$ is the value of the lth pixel of the binary image representing the nominal shape. Each unit cell is identified by the indexes (i,j,k), which refers to the location of the cell along the X, Y and Z directions, respectively, inside the lattice structure.

The aim of the proposed approach consists of modelling the $\delta_{i,j,k}(z)$ at unit cell-level. In [5], a B-spline basis defined as:

$$C_{i,j,k}(z) = \sum_{q=1}^{Q+L-1} B_q(z,\tau)P_{i,j,k,q} \qquad (3)$$

where B_q are the B-Spline basis functions of order $Q = 3$, $\boldsymbol{\tau} = \{\tau_l, l = 1,2,\dots,L-1\}$ is the B-spline knot sequence, L is the number of subintervals, and $P_{i,j,k,1}, P_{i,j,k,2}, \dots, P_{i,j,k,Q+L-1}$ are the control points for the $(i,j,k)^{th}$ unit cell. The standard B-spline model is fitted via the least squares (LS) method, and it implies that all levels z, namely all layers, are equally trustworthy. However, when the deviation profile must be estimated from powder bed images, a layerwise variation of pixel intensity patterns may lead to a layerwise variation of the reconstructed contour accuracy. Some layers may exhibit a so-called *bright field* condition, because the solidified layer produces intense light reflections towards the camera: a high intensity of foreground pixels is known to yield poor edge detection results, as they tend to force the segmentation algorithms to isolate the brightest area rather than the whole foreground region. Other layers may exhibit a so-called *dark field* condition, where most of light is reflected in other directions. This condition is known to be the most appropriate for accurate edge detection. Switching from one condition to the other, being fixed the chamber lighting setup, occurs as a consequence of the layerwise varying scan direction. We thus propose a robust approach to give a higher weight to layer where a dark-field pattern is present and lower weight to the ones where a bright-field is present. The final goal is to gather a more accurate prediction $\hat{\delta}_{i,j,k}(z)$ of the 1D deviation from the nominal shape, relying more on layers where more accurate geometry reconstruction is expected.

Since a bright-field condition yields a higher pixel intensity within the foreground, we propose a weighting scheme where the weight $\omega_{i,j,k}(z)$ is proportional to the inverse of the variance $s_{i,j,k}(z)^2$ of pixel intensities within a region of interest Ω such that:

$$\omega_{i,j,k}(z) = \frac{1}{S} \frac{1}{s_{i,j,k}(z)^2} \tag{4}$$

where $S = \sum_k \frac{1}{s_{i,j,k}(k)^2}$ is just a corrective factor. The region of interest Ω can be either the bounding box of the part within the build area, including both the solidified layer and the surrounding powder, or a reduced region that adapts, layer by layer, to the geometry of the part. We advocate this latter approach. Colosimo et al. [9] proposed a method to define such region as an envelope (or a band) that is centred on the contour of the nominal geometry in the layer, and extend by one half within the foreground, and by the other half within the background (loose powder). The same approach was used here to define such region (full details are referred to [9] for sake of space). Since background pixel intensities in one location are assumed to be quite stable from one layer to another, the underlying rationale is the following: a high value of $s_{i,j,k}(z)^2$ within the region of interest adaptively identified in each layer implies a higher intensity of foreground pixels compared to background ones, which is common in bright field conditions. The opposite is expected in the presence of a dark field, where foreground and background pixels are expected to have more similar intensities. Thus, low values of $\omega_{i,j,k}(z)$ penalize layers where a poor reconstruction accuracy is expected because of the bright field. B-spline control points can be computed using the weighted least squares (WLS) method:

$$P_{i,j,k} = \left(B^T W_{i,j,k}(z)B\right)^{-1} B^T W_{i,j,k}(z)\delta_{i,j,k}(z) \tag{5}$$

where B is the model matrix and $W_{i,j,k}(z)$ is a diagonal matrix whose diagonal elements are the weights $\omega_{i,j,k}(z)$. In this study, a simple knot sequence τ composed by 21 equi-spaced knots was used, but the knot sequence may be tailored in principle depending on the specific geometry of the lattice cell. It is worth noticing that: i) due to illumination inhomogeneity, different parts in the same build area may exhibit different bright or dark field conditions, and hence wights can be computed on a part-by-part basis rather than on a layer-by-layer basis, ii) bright and dark field conditions are highly dependent on the scan direction, and hence the a-priori knowledge about the scan direction in each layer may be used to enhance the weighting procedure.

Results
Fig. 2 shows two examples of powder bed images in two different layers, where the raw image was superimposed to three contours, namely the ex-situ reconstruction obtained via X-ray CT, the nominal contour, and the in-situ reconstruction generated by the proposed robust active contour methodology. Fig. 2 shows that the in-situ reconstruction is in good agreement with the other two reference contours.

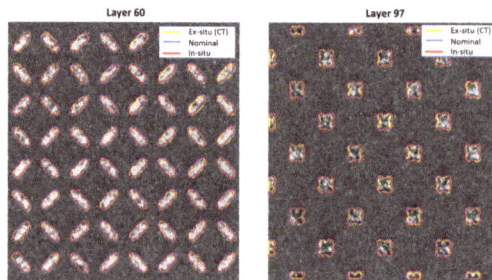

Fig. 2 – Example of two layers with in-situ and ex-situ reconstructions compared to the nominal one

Italian Manufacturing Association Conference - XVI AITeM   Materials Research Forum LLC
Materials Research Proceedings 35 (2023) 232-240   https://doi.org/10.21741/9781644902714-28

To quantitatively determine the accuracy of the in-situ reconstruction with respect to the one achieved via ex-situ CT, Eq. 2 was used by replacing the nominal shape with the CT reference. The powder bed images were clustered into two sets by applying a K-means clustering with $K = 2$ to the values of weights $\omega_{i,j,k}(z)$. The result was a separation of layers with lower weights (corresponding to bright field conditions) from layers with higher weights (corresponding to dark field conditions). For each set, a 95% confidence interval (CI) of the mean deviation between the in-situ and ex-situ geometry reconstructions were computed. Fig. 3 shows two examples of powder bed images belonging to the two clusters, and the 95% CIs of the mean deviations.

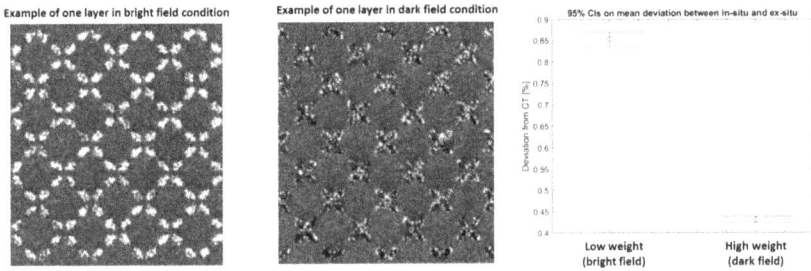

Fig. 3 – Two examples of layers characterized by a bright field and a dark field pattern (right) and 95% CIs on the mean deviation between in-situ and ex-situ reconstructed shapes in these two different conditions

Fig. 3 (right panel) shows that when a dark field pattern is present, the in-situ reconstruction is significantly closer to the CT-reconstructed shape used as a ground truth than in the presence of a bright field condition. Fig. 3 (right panel) also shows that the proposed weighting scheme is effective in separating layers characterized by a poor reconstruction from those characterize by a good one. Finally, the mean deviation between in-situ and ex-situ reconstructed shapes under dark field conditions is quite small (average lower than 0.45%), much less than in bright field conditions (average of about 0.85%). In terms of strut thickness it corresponds to an average deviation between 0.05 and 0.1 mm.

The proposed approach is not simply aimed to yield a good layerwise reconstruction of the part, but also to generate a synthetic representation of the deviation from the nominal geometry that is suitable to capture the salient "signature" of each unit cell while allowing a data format that is easier and more efficient to handle than the full 3D shape. To this aim, the WLS fitting approach was tested on the deviation profiles of all 64 unit cells. Fig. 4 (left panels) shows the WLS fitting and the model residuals for all the unit cells. The proposed WLS model filters out the effect of bright field patterns on the resulting 1D fitted deviation profile, enhancing the whole deviation pattern estimate. In order to quantify the benefits of the proposed WLS approach, a comparison against a traditional LS model based on the same B-spline basis (but without any weighting scheme applied to individual layers) is shown in Fig. 4, right panel. The comparison involves two performance metrics, namely the root mean square (RMS) of the predicted deviation from the nominal, and the variance of the prediction (95% bootstrap confidence intervals are shown).

Italian Manufacturing Association Conference - XVI AITeM   Materials Research Forum LLC
Materials Research Proceedings 35 (2023) 232-240   https://doi.org/10.21741/9781644902714-28

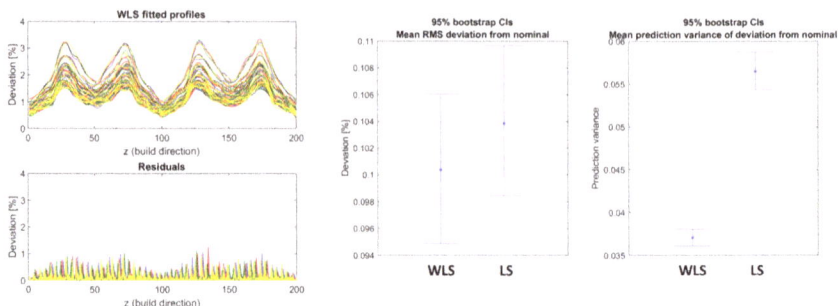

Fig. 4 – Left: WLS fitted deviation profiles and residuals; right: 95% bootstrap CIs for the mean RMS deviation from the nominal and the mean prediction variance of deviation from the nominal for WLS and LS models

Fig. 4 (right panels) shows that, in terms of RMS, the two models are not statistically different, but the WLS model yields a much lower prediction variance, i.e., a more precise estimate of the geometry deviation signature. This latter result is highly relevant for the design of statistical process monitoring methods that may take advantage of in-situ deviation profiles, as a lower associated variance implies a higher power in detecting out-of-control deviations caused by anomalous distortions. It is worth noticing that the pattern of the fitted profiles is highly influenced by the actual shape of the cell, with peaks in correspondence of lattice nodes and a symmetry along Z that inherits the unit cell symmetry. Thus, the deviation profile can be regarded as a signature of the specific lattice geometry.

Conclusions

The increased shape complexity of high-value-added products enabled by AM technologies has opened a variety of new challenges concerning efficient product qualification methodologies and statistical quality monitoring instruments. Post-process ex-situ CT inspections, despite their actual effectiveness, are often too expensive and time-consuming to meet high productivity constraints. In this study, we presented a method to support the in-line qualification of complex shapes like lattice structures, which can be used to anticipate the detection of geometrical distortions and to possibly reduce the need for post-process inspections. A key issue is the accuracy of in-situ geometry reconstructions and estimates, as powder bed images are affected by variability sources that are not present in X-ray CT. The analysis presented in this study highlighted that in presence of favourable pixel intensity patterns, the accuracy of in-situ lattice reconstruction is very close to the one obtained via ex-situ measurements. Moreover, by reconstructing the cell-wise 1D deviation profile with the proposed weighted fitting method, it is possible to filter out nuisance effects caused by layers exhibiting lower reconstruction accuracy, finally leading to a more accurate and precise estimation of the lattice geometrical signature.

An on-going work will extend this analysis to multiple lattice components, exploring different deviation metrics as well as one novel way to take into explicit account the spatial dependence of deviation profiles in unit cells of the same structure and/or of different structures in the same build. The performance analysis of the proposed approach in the presence of actual geometrical distortions is under development too. The authors are working on other promising research directions, including novel solutions to augment the sensing setup to further improve the accuracy of in-situ quality estimates and possible extensions of the proposed framework to more complex geometries and different families of lattice structure. Another development regards the study of spatial correlation among profiles associated to different unit cells, which may further enhance the characterization of the natural variability as well as the defect detection performance.

References

[1] Helou, M., & Kara, S. (2018). Design, analysis and manufacturing of lattice structures: an overview. International Journal of Computer Integrated Manufacturing, 31(3), 243-261. https://doi.org/10.1080/0951192X.2017.1407456

[2] Liu, L., Kamm, P., García-Moreno, F., Banhart, J., & Pasini, D. (2017). Elastic and failure response of imperfect three-dimensional metallic lattices: the role of geometric defects induced by Selective Laser Melting. Journal of the Mechanics and Physics of Solids, 107, 160-184. https://doi.org/10.1016/j.jmps.2017.07.003

[3] Melancon, D., Bagheri, Z. S., Johnston, R. B., Liu, L., Tanzer, M., & Pasini, D. (2017). Mechanical characterization of structurally porous biomaterials built via additive manufacturing: experiments, predictive models, and design maps for load-bearing bone replacement implants. Acta biomaterialia, 63, 350-368. https://doi.org/10.1016/j.actbio.2017.09.013

[4] Dallago, M., Raghavendra, S., Luchin, V., Zappini, G., Pasini, D., & Benedetti, M. (2019). Geometric assessment of lattice materials built via Selective Laser Melting. Materials Today: Proceedings, 7, 353-361. https://doi.org/10.1016/j.matpr.2018.11.096

[5] Colosimo B.M., Grasso, M., Garghetti, F., Rossi, B. (2021), Complex geometries in additive manufacturing: A new solution for lattice structure modeling and monitoring, Journal of Quality Technology. https://doi.org/10.1080/00224065.2021.1926377

[6] Liu, C., Le Roux, L., Ji, Z., Kerfriden, P., Lacan, F., & Bigot, S. (2020). Machine Learning-enabled feedback loops for metal powder bed fusion additive manufacturing. Procedia Computer Science, 176, 2586-2595. https://doi.org/10.1016/j.procs.2020.09.314

[7] Vasileska, E., Demir, A. G., Colosimo, B. M., & Previtali, B. (2020). Layer-wise control of selective laser melting by means of inline melt pool area measurements. Journal of Laser Applications, 32(2), 022057. https://doi.org/10.2351/7.0000108

[8] Colosimo, B. M., Grossi, E., Caltanissetta, F., & Grasso, M. (2020). Penelope: a novel prototype for in situ defect removal in LPBF. Jom, 72, 1332-1339. https://doi.org/10.1007/s11837-019-03964-0

[9] Colosimo, B. M., Garghetti, F., Pagani, L., & Grasso, M. (2022). A novel method for in-process inspection of lattice structures via in-situ layerwise imaging. Manufacturing Letters, 32, 67-72. https://doi.org/10.1016/j.mfglet.2022.03.004

[10] Guerra, M. G., Lafirenza, M., Errico, V., & Angelastro, A. (2023). In-process dimensional and geometrical characterization of laser-powder bed fusion lattice structures through high-resolution optical tomography. Optics & Laser Technology, 162, 109252. https://doi.org/10.1016/j.optlastec.2023.109252

[11] Dewulf, W., Bosse, H., Carmignato, S., & Leach, R. (2022). Advances in the metrological traceability and performance of X-ray computed tomography. CIRP Annals, 71(2), 693-716. https://doi.org/10.1016/j.cirp.2022.05.001

[12] Withers, P. J., Bouman, C., Carmignato, S., Cnudde, V., Grimaldi, D., Hagen, C. K., ... & Stock, S. R. (2021). X-ray computed tomography. Nature Reviews Methods Primers, 1(1), 18. https://doi.org/10.1038/s43586-021-00015-4

[13] Szeliski, R. (2022). Image alignment and stitching. In Computer Vision (pp. 401-441). Springer, Cham. https://doi.org/10.1007/978-3-030-34372-9_8

[14] Avants, B.B., Tustison, N.J., Stauffer, M., Song, G., Wu, B., Gee, J.C. (2014) The insight toolkit image registration framework, Front. Neuroinform. 8(44). https://doi.org/10.3389/fninf.2014.00044

[15] Liu, S., & Peng, Y. (2012). A local region-based Chan-Vese model for image segmentation. Pattern Recognition, 45(7), 2769-2779. https://doi.org/10.1016/j.patcog.2011.11.019

[16] Soomro, S., Munir, A., & Choi, K. N. (2018). Hybrid two-stage active contour method with region and edge information for intensity inhomogeneous image segmentation. PloS one, 13(1), e0191827. https://doi.org/10.1371/journal.pone.0191827

[17] Caltanissetta, F., Grasso, M., Petro, S., & Colosimo, B. M. (2018). Characterization of in-situ measurements based on layerwise imaging in laser powder bed fusion. Additive Manufacturing, 24, 183-199. https://doi.org/10.1016/j.addma.2018.09.017

[18] Aminzadeh, M., & Kurfess, T. (2016, June). Vision-based inspection system for dimensional accuracy in powder-bed additive manufacturing. In International manufacturing science and engineering conference (Vol. 49903, p. V002T04A042). American Society of Mechanical Engineers. https://doi.org/10.1115/MSEC2016-8674

[19] Pagani, L., Grasso, M., Scott, P. J., & Colosimo, B. M. (2020). Automated layerwise detection of geometrical distortions in laser powder bed fusion. Additive Manufacturing, 36, 101435. https://doi.org/10.1016/j.addma.2020.101435

Italian Manufacturing Association Conference - XVI AITeM
Materials Research Proceedings 35 (2023) 241-248

Materials Research Forum LLC
https://doi.org/10.21741/9781644902714-29

Ductility and linear energy density of Ti6Al4V parts produced with additive powder bed fusion technology

Gianluca Buffa[1,a] *, Dina Palmeri[1,b], Gaetano Pollara[1,c], Livan Fratini[1,d],
Alessandro Benigno[2,e],

[1]Dipartimento di Ingegneria, Università degli Studi di Palermo, Viale delle Scienze 90128,
Palermo, Italy

[2]AB Ingegneria, Via Segesta 26, 90141, Palermo, Italy

[a] gianluca.buffa@unipa.it, [b] dina.palmeri@unipa.it, [c] gaetano.pollara@unipa.it,
[d] livan.fratini@unipa.it, [e] alessandrobenigno@libero.it

Keywords: Powder Bed Fusion, Linear Energy Density, Ductility

Abstract. Hybrid metal forming processes involve the integration of commonly used sheet metal forming processes, as bending, deep drawing and incremental forming, with additive manufacturing processes as Powder Bed Fusion. In recent ybears, these integrations have been more developed for manufacturing sectors characterized by components with complex geometries in low numbers, as the aerospace sector. Hybrid additive manufacturing overcomes the typical limitations of additive manufacturing related to low productivity, metallurgical defects and low dimensional accuracy. In this perspective, a key aspect of hybrid processes is the production of parts characterized by high strength and ductility. In the present work, a study was carried out on the influence of process parameters, such as laser power and scanning speed, on material ductility for Ti6Al4V alloy samples produced by Selective Laser Melting. In particular, the material strength and ductility were related to the process linear energy density (LED).

Introduction

Recently, there has been a rise in interest in additive manufacturing (AM) technologies due to their capacity to produce complex shapes. Changes to the part designs can be rapidly and readily incorporated using this technique. Moreover, because no tooling is required, production time is decreased in general [1]. With AM techniques, less process scrap is generated as a three-dimensional component is built layer by layer from the material in the form of wire or powder [2]. It is advisable to use forming procedures that offer quicker production periods when producing large batch sizes due to AM's slow rate of production. Moreover, the use of additive manufacturing (AM) in tight-tolerance and critical applications is constrained by its low resolution, the occurrence of porosity defects, partially melted powder, and residual tensions [3]. Unfortunately, forming procedures are not versatile enough to produce a variety of [c] product variants since they require specific instruments [4].

Due to the possibility to use the unique advantages of each technique to improve part qualities above conventional production, hybrid techniques have become more and more common as a means of overcoming these restrictions. Incorporating more materials, structures, or functionalities into the component is made possible by combining additive manufacturing with other traditional methods, leading to improved new attributes. Economic aspects may be enhanced together with technical benefits. In terms of L-PBF and forming, AM makes it possible to do away with the requirement for a single forming machine configuration for the creation of pre-forms and a single forming tool that needs to be designed for each part geometry [5].

Several process chains have recently been researched with a focus on titanium alloys. In comparison to ordinary wrought material with a lamellar microstructure generally used in the traditional forging of Ti-6Al-4V, it has been demonstrated that AM materials display good formability, lower flow stresses, and activation energy for hot forming [6]. It's crucial to ensure that the printed pieces can withstand the generated distortion before carrying out the subsequent forming procedure. Hence, the paper's goal is to provide a production window to create components with increased ductility and strength for use in the hybrid forming process. The effects of building orientation on the mechanical characteristics of titanium alloy generated by casting, selective laser melting (SLM), and electron beam melting were investigated by Pasang et al (EBM). With the exception of 45° orientation, when SLM samples displayed greater strength, wrought samples produced the greatest results.

The ductility of the wrought alloy consistently outperformed that of SLM and EBM [7]. Surface morphology was chosen by Wang et al. as the quality indicator for SLM parts. They established a relationship between the line energy density and the manufactured samples' surface quality, which influences their ductility [8]. One of the causes of the brittle fractures of selectively melted Ti6Al4V, according to Moridi et al., is the existence of printing flaws during the process [9]. Liu et al. investigated the impact of the process parameters to enhance the printed parts' tensile strength and ductility. It was found that the process parameters have a significant impact on the tensile behaviour, in particular the ductility, of the Ti alloy created by SLM. The significant improvement in mechanical behaviour is principally attributable to the decrease of pores and the limiting of martensite production [10]. In order to reduce these flaws and increase ductility, process parameter optimization is crucial. In the present work, a study was carried out on the influence of process parameters, such as laser power and scanning speed, on material ductility for Ti6Al4V alloy samples produced by Selective Laser Melting. In particular, the material strength and ductility were related to the process linear energy density (LED).

Materials and Method
Ti6Al4V powder with a particle size of 20-60 μm was used in this study. All of the samples were created using the L-PBF process by the SLM 280 HL machine. Argon inert gas was used to completely fill the printing chamber and lower the oxygen content to less than 0.1%. To minimize pore defects and increase the strength of the material, the building orientation was maintained constant and set at 0° [11, 12]. In order to examine the impact of line energy density (LED) on the ductility of the printed parts, hatch distance, layer thickness, and scan strategy were also kept constant at 100 μm, 60 μm, and 0°, respectively. In particular, 39 samples (N=3 duplicates for set of parameters) with varying LED levels between 0,175 and 0,258 J/mm were produced.

The selection of the LED values has been conducted without a uniform increase in order to have more data near the LED value suggested by SLM solution equal to 0,25 J/mm. The other LED values were distributed inside the best-density process window.

Starting with the reference of the parameter identified by the SLM Solution, the range of LED values examined for the identification of the best conditions of ductility and resistance of the material was chosen. In particular, three different ranges of variation of the LED were selected with increasing amplitudes as the values considered move from the reference value of 0.25 J/mm. According to Table 1, four samples with different process parameters were chosen for the LED range values Δ_{LED2} (ID 6-9) and Δ_{LED3} (ID 10-13), whereas five samples were taken into account for the LED range values Δ_{LED1} (ID 1-5).

The investigated range was selected in accordance with earlier experimental campaigns where significant porosity defects were noted for samples printed outside the range used for this investigation. The process parameters employed in this work, including Laser Power (P), Scanning Speed (v), Hatch Distance (h), Scanning Strategy (s), Building Orientation (b), Layer Thickness (t), and LED, may be seen in Table 1.

Figure 1. Dog bone samples used for the tensile tests.

Table 1. Process parameters.

ID	P [W]	v [mm/s]	LED [J/mm]	
1	305	1739	0,175	
2	312	1660	0,188	
3	328	1576	0,208	$\Delta_{LED1}= 0.044$
4	334	1552	0,215	
5	338	1541	0,219	
6	348	1506	0,231	
7	351	1516	0,232	
8	365	1504	0,243	$\Delta_{LED2}= 0.016$
9	366	1483	0,247	
10	380	1520	0,250	
11	373	1491	0,250	
12	377	1502	0,251	$\Delta_{LED3}= 0.008$
13	378	1463	0,258	

On a Galdabini Sun 5, quasi-static tensile tests at a rate of 1 mm/min were performed to assess the mechanical behavior of the samples in terms of ultimate tensile strength (UTS) and elongation to failure (ETF). Printed dog bone specimens (Figure 1) with a 5mm × 3mm section were used for the tensile tests as a reduction of the ASTM E8 standard. Archimedes technique was used to estimate relative density, and KERN balance with 0.001g accuracy was used to weigh the samples in air and water.

The reference density value for the Ti6Al4V alloy was considered equal to 4.43 g/cm^3. Furthermore, optical microscope (OM) inspections were carried out to examine fracture surfaces.

For the above analysis, the extent of the fracture propagation surface (A), in the specimens subjected to tensile stress, was measured by means of proper image analysis software.

The materials capacity to absorb energy before failing was also studied in order to identify the optimal material properties suitable for hybrid metal forming processes. The product of the UTS and ETF values can be used to measure the aforementioned factor.

Results and discussion

In order to provide a process window for the hybrid manufacturing process, a preliminary research on the impact of LED on the mechanical behavior of Ti6Al4V parts produced by L-PBF was reviewed in this work. To do this, tensile tests were conducted on each specimen, which were also weighted to determine the relative density. The results were then compared to the mechanical properties, such as UTS and ETF, that had been determined.

Tensile Strength and Elongation to Failure

Regarding the material's tensile strength and elongation, the results of the tensile tests measured in terms of UTS and ETF, show three distinct patterns, unique for each ΔLED range identified, namely Δ_{LED1}, Δ_{LED2}, and Δ_{LED3}. As can be seen from Figure 2, the UTS values exhibit a decreasing trend in the Δ_{LED1} range, a curved trend with a relative maximum in the Δ_{LED2} range, and an almost constant trend in the Δ_{LED3} range.

The ETF trends, on the other hand, display a growing trend in the Δ_{LED1} range, a curved trend with a relative minimum point in the interval Δ_{LED2}, and a nearly constant trend in the interval Δ_{LED3}. Also, it should be observed that from the specimens ID1 to ID13, both the LED and the laser power values increase.

According to the observed trends, the effect of the scanning speed prevails on that one of the laser power for LED values that deviate more from the optimal value recommended for printing (0.25 J/mm). It follows that for high scanning speeds, even when laser power and LED value rise, causes a refinement of the material structure, increasing its mechanical strength at the expense of ductility. A distinct mechanism which results in an increase in mechanical strength and a loss in ductility characteristics is shown when the laser power is increased. This transition zone is seen in the Δ_{LED2} range, where the effect of the scanning speed still predominates. With the activation of more intense heat fluxes that affect the martensite formation mechanism, the material also starts to strengthen in the previously mentioned LED range.

High laser power levels in the range Δ_{LED3} result in the stabilization of the material's strengthening process linked to heat fluxes, deactivating the refining phenomena carried on by the changes in scanning speed. In fact, it is found that in this range, both the material's ductility and mechanical strength follow a nearly constant trend.

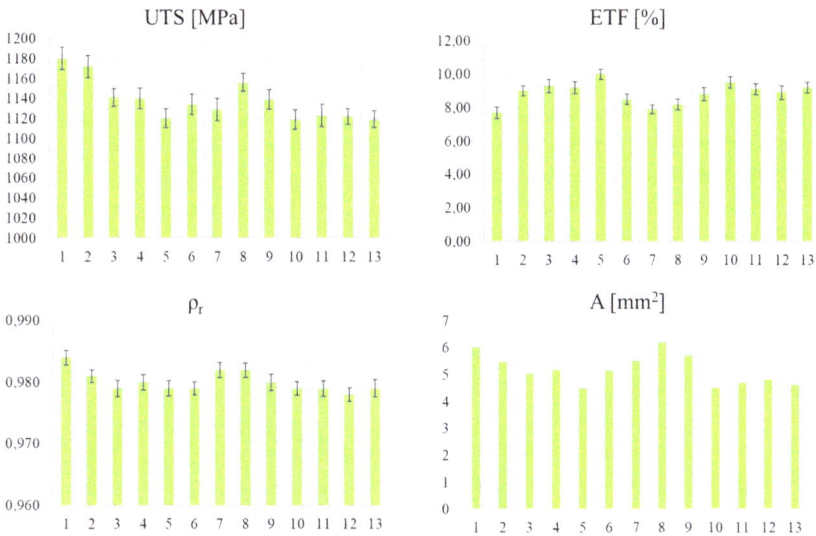

Figure 2. In this figure are shown the results obtained from the tensile tests (UTS, ETF), density measurement (ρ_r), and the fracture surface analysis (A).

The ID 1 sample, which is characterized by the lowest LED and laser power values, showed the best mechanical strength value and the lowest ductility values.

Table 2. Values of UTS×ETF and relative density.

ID	UTS×ETF	ρ_r [%]
1	9086	98,4%
2	10548	98,1%
3	10611	97,9%
4	10488	98,0%
5	11200	97,9%
6	9639	97,9%
7	8919	98,2%
8	9479	98,2%
9	10023	98,0%
10	10631	97,9%
11	10219	97,9%
12	9986	97,8%
13	10295	97,9%

The evaluations conducted in the three ΔLEDs revealed that, in terms of the material capacity to absorb energy before breaking, the best compromise between material strength and ductility

Italian Manufacturing Association Conference - XVI AITeM   Materials Research Forum LLC
Materials Research Proceedings 35 (2023) 241-248   https://doi.org/10.21741/9781644902714-29

was found not so much at the LED values that are optimal for the printing process but rather at the upper end of the ΔLED1 range, i.e. for the process parameters that characterize sample ID5. The value of the index of the material's ability to absorb energy before failure is displayed for all samples in Table 2 above.

By measuring the size of the fracture propagation area in the necking section, image analysis (Figure 3) was used to analyze the fracture surfaces of the tensile tested specimens. The pattern discovered supports previous tensile strength observations. The image analysis for all the samples (ID1-ID13) is displayed in Figure 3.

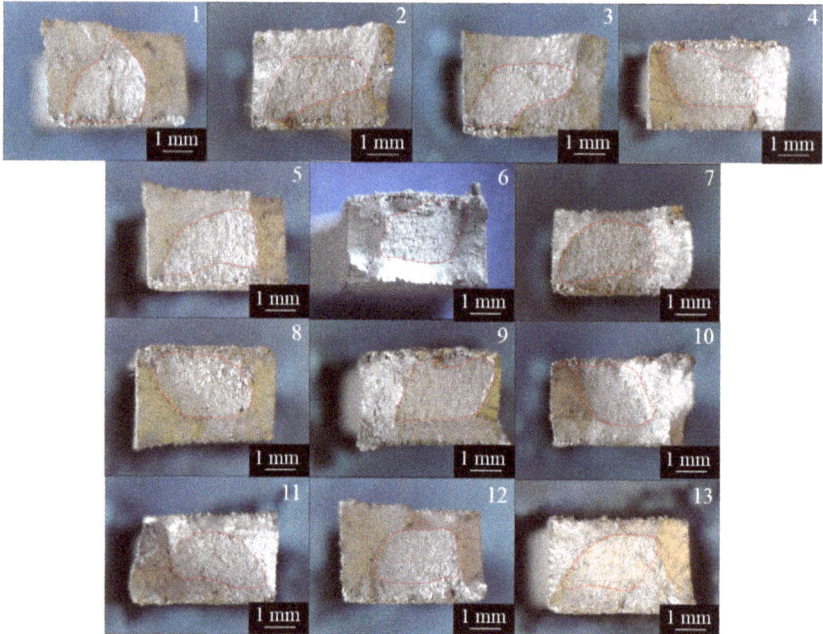

Figure 3. Surface fracture images for all the samples. The fracture propagation area is indicated with a red line.

Relative density

It should be observed that the material's relative density exhibits the same patterns as those required for tensile strength, leading to comparable properties in the three LED range values. The specimen's porosity is caused by a variety of phenomena. The influence of the scanning speed predominates for samples from ID1 to ID5, meaning that less active kinetics of the material in the molten state leads to a higher presence of trapped gas porosity.

The findings from the density measurement are presented in Figure 2. The influence of the kinetics of the material in the molten state, that determines the presence of gas porosity, is still seen in the trend for the samples ID6 to ID9. A stabilization of the material porosity is seen for samples ID10 to ID13. When the laser power is larger than 370 W, the formation of gas porosity is more related on the quantity of the melted material during the scanning phase rather than to the kinetics of the material in the molten state. As a result, the porosity level has stabilized.

Materials Research Forum LLC
https://doi.org/10.21741/9781644902714-29

Conclusions

Tensile tests, fracture surface analysis, and density measurements were performed in order to investigate the impact of the LED on the mechanical behavior of printed Ti6Al4V parts. The outcomes of this investigation can be used to identify the proper process parameters for the hybrid manufacturing process. The following is a summary of the key findings:

- For high scanning speeds, there is a refinement of the material structure that increases the mechanical strength at the expense of ductility, even when the laser power and the LED value increase.

- There is a process parameter transition zone in which material strengthening is linked to the activation of more severe heat fluxes that affect the martensite production mechanism.

- Increased laser power values stabilize the material-strengthening mechanism connected to heat fluxes, deactivating the refining phenomena caused by varying scanning speeds.

- The best compromise between material strength and ductility was obtained at the upper end of the Δ_{LED1} range, rather than at the optimum LED values for the printing process.

- The relative density of the material follows the same trends as those reported for the tensile strength. The specimen's porosity is also affected by several phenomena. For laser power values higher than 370 W, a stabilization of the material porosity occurs. The onset of gas porosity is more dependent on the amount of molten material than on the kinetics of the molten material.

References

[1] A. Schaub, B. Ahuja, L. Butzhammer, J. Osterziel, M. Schmidt, and M. Merklein, Additive manufacturing of functional elements on sheet metal, Phys. Procedia, 83 (2016) 797–807. https://doi.org/10.1016/j.phpro.2016.08.082

[2] M. D. Bambach, M. Bambach, A. Sviridov, and S. Weiss, New process chains involving additive manufacturing and metal forming - A chance for saving energy?, Procedia Eng., 207 (2017) 1176–1181. https://doi.org/10.1016/j.proeng.2017.10.1049

[3] J. M. Flynn, A. Shokrani, S. T. Newman, and V. Dhokia, Hybrid additive and subtractive machine tools - Research and industrial developments, Int. J. Mach. Tools Manuf., 101 (2016) 79–101. https://doi.org/10.1016/j.ijmachtools.2015.11.007

[4] M. Hirtler, A. Jedynak, B. Sydow, A. Sviridov, and M. Bambach, A study on the mechanical properties of hybrid parts manufactured by forging and wire arc additive manufacturing, Procedia Manuf., 47 (2020) 1141–1148. https://doi.org/10.1016/j.promfg.2020.04.136

[5] M. Merklein, R. Schulte, and T. Papke, An innovative process combination of additive manufacturing and sheet bulk metal forming for manufacturing a functional hybrid part, J. Mater. Process. Technol., 291 (2021) 117032. https://doi.org/10.1016/j.jmatprotec.2020.117032

[6] M. Bambach, I. Sizova, B. Sydow, S. Hemes, and F. Meiners, Hybrid manufacturing of components from Ti-6Al-4V by metal forming and wire-arc additive manufacturing, J. Mater. Process. Technol., 282 (2020) 116689. https://doi.org/10.1016/j.jmatprotec.2020.116689

[7] T. Pasang et al., Directionally-Dependent Mechanical Properties of Ti6Al4V, Materials (Basel)., 14 (2021) 13:3603, doi: https://doi.org/10.3390/ ma14133603

[8] D. Wang, W. Dou, and Y. Yang, Research on selective laser melting of Ti6Al4V: Surface morphologies, optimized processing zone, and ductility improvement mechanism, Metals (Basel)., 8 (2018) 7:471. https://doi.org/10.3390/met8070471

[9] A. Moridi, A. G. Demir, L. Caprio, A. J. Hart, B. Previtali, and B. M. Colosimo, Deformation and failure mechanisms of Ti–6Al–4V as built by selective laser melting, Mater. Sci. Eng. A, 768 (2019) 0–26. https://doi.org/10.1016/j.msea.2019.138456

[10] J. Liu et al., Achieving Ti6Al4V alloys with both high strength and ductility via selective laser melting, Mater. Sci. Eng. A, 766 (2019) 138319. https://doi.org/10.1016/j.msea.2019.138319

[11] D. Palmeri, G. Buffa, G. Pollara, and L. Fratini, The Effect of Building Direction on Microstructure and Microhardness during Selective Laser Melting of Ti6Al4V Titanium Alloy, J. Mater. Eng. Perform., (2021). https://doi.org/10.1007/s11665-021-06039-x

[12] D. Palmeri, G. Buffa, G. Pollara, and L. Fratini, Sample building orientation effect on porosity and mechanical properties in Selective Laser Melting of Ti6Al4V titanium alloy, Mater. Sci. Eng. A, 830 (2022) 142306. https://doi.org/10.1016/j.msea.2021.142306

Assembly, disassembly and circular economy

Italian Manufacturing Association Conference - XVI AITeM
Materials Research Proceedings 35 (2023) 250-257

Materials Research Forum LLC
https://doi.org/10.21741/9781644902714-30

Circular economy strategies at the manufacturing system scheduling level: the impacts on Makespan

Claudio Castiglione[1,a*], Arianna Alfieri[1,b] and Erica Pastore[1,c]

[1]Politecnico di Torino, Department of Management and Production Engineering, Corso Duca degli Abruzzi, 24, 10129 Torino, Torino, Italy

[a]claudio.castiglione@polito.it, [b]arianna.alfieri@polito.it, [c]erica.pastore@polito.it

Keywords: Production Planning, Scheduling, Circular Economy

Abstract. The use of end-of-life and end-of-use products to recover parts and raw materials can mitigate the severity of the increasing price of raw materials, the disruption of global supply chains for critical raw materials (e.g., chips and rare earth elements), and reduce the environmental impacts. Furthermore, circular economy strategies can improve scheduling by shortening the completion times of the components. This paper investigates the effects of implementing circular economy strategies (repair, reuse, and re-manufacturing) at the scheduling level in a manufacturing system involving disassembly, re-manufacturing, and assembly operations. A set of eight priority rules modify the job priority and the strategy implementation. The results show that including circular economy strategies through disassembly can reduce the makespan, but scheduling is pivotal to managing the frequent changes in the quality of end-of-life products and their volumes and the current production order mix.

Introduction

Industry 4.0 (I4.0) and Circular Economy (CE) paradigms have been leading the innovation in manufacturing companies and scholars' interests for at least a decade. Furthermore, current research highlights the enabling role of I4.0 technologies in implementing CE practices while addressing the manufacturing challenges of mass customisation macrotrend [1]. I4.0 provides enabling technologies for CE from a twofold point of view: (i) advanced manufacturing systems with a high degree of flexibility and reconfigurability, (ii) digitalisation and data-driven approaches to allow the design and management of more complex systems [2].

Adopting I4.0 technologies and moving to the CE paradigm is important for the international competitive advantage of manufacturing companies [3]. The supply of raw materials has become critical [4], especially for importing countries like Italy [5], because of the disruptions of global supply chains that began with the Covid19 pandemic and propagated due to the recent geopolitical conflicts [6]. At the same time, the advent of mass customisation and the transition towards sustainable development are increasing product varieties, fluctuations in product demand, and the need to increase product life-cycle through, for example, repairing and re-manufacturing [7].

The main three barriers to the effective implementation of I4.0 and its enabling role for the CE transition are (i) the interoperability among different processes, (ii) the modelling of the processes and their integration to optimise the system, and (iii) the coordination and management of the entire manufacturing system and the digital counterparts that support it. [1]. CE actions exponentially increase the severity of these barriers because of the many cycles and flows added to the manufacturing system through the 6Rs (Reduce, Reuse, Repair, Re-manufacture, Re-design, Recycle) [8]. This complexity increases the risk of using new tools and machines in an obsolete way [9]; for example, optimising the stand-alone processes can result inefficient from the point of view of the entire system.

A wider and more flexible implementation of CE strategies within manufacturing systems involves the introduction of disassembling operations [10]. Disassembly enhances the reuse and re-manufacturing of components from recovered end-of-life and end-of-use products, their repair, or their recycling [11]. Disassembly operations can make available many components and optional that can be bundled together to improve customer satisfaction while reducing lead times [12], or balance inventories [13]. However, disassembling operations may jeopardise manufacturing performance. For example, they lead to several flows of generally low-value items requiring space and generating holding costs, with different market demands for each disassembled part [14]. In this context, production planning and control approaches are crucial, especially in the many available strategies offered by the CE paradigm, combining technologies to optimise overall performance and overcoming challenges, also in system design [15]. Moreover, additional sources of uncertainty must be considered, such as quality issues, low value of recovered components, uncertainties in quality and volumes of provided products, and the fact that disassembly operations are mainly performed by workers rather than robots [16].

In the literature, scheduling problems involving disassembly, re-manufacturing, and reassembly operations focus on reducing tardiness and makespan [17] and on finding the minimal operation sequence to disassemble returned end-of-life or use products (namely, returns) [18]. However, scheduling problems with disassembly are NP-hard; small problems can be solved by finding optimal solutions, while industrial-scale problems require heuristic and approximated approaches [19]. Heuristics have been applied to families of products [20], multi-objective stochastic scheduling problems [21], and multi-product scheduling problems [22]. Priority rules are mainly used for scheduling problems because of ease of understanding and implementation and good performance [23], especially when calibrated on every single workstation [24].

From the production planning aspect, the literature is focused on the solution approach, while, from the disassembly point of view, the literature investigates the technical and economic performance. Instead, the literature neglects the intersection with CE strategies and approaches intertwined with sustainability that require a simultaneous multi-dimensional assessment [25].

This paper investigates the impact of including CE strategies in a scheduling problem for a system characterised by disassembly, re-manufacturing, and assembly workstations, including quality control and returns repair. The schedules are identified by applying eight priority rules derived from the literature and combined to deal with the CE strategies. Each priority rule is applied to a scenario with specific conditions of finished product demand and volumes of reparable and irreparable end-of-life and use products, which are exploited to recover parts and components.

Problem description

The inclusion of CE strategies and disassembly operations complicates the scheduling problem.

Apart from the standard scheduling decisions, the inclusion of CE strategies also includes decisions about the strategy each job (products, returns, or components) must follow. The impact of these further decisions on manufacturing performance is investigated in this paper by considering a realistic manufacturing system based on a structure diffused in the literature that assumes three main production areas: disassembly, processing, and reassembly. The scheduling problem includes the following further decisions: (i) allocating recovered products to repair or disassembly workstations, (ii) deciding which of the components recovered from the disassembly will be re-manufactured rather than reused as is.

Furthermore, the system must deal with increased system variability because the availability of the return depends on the quantity of end-of-life and use products disposed of by consumers. Also, the quality level of the returns can make them irreparable or particularly long and expensive to recover their components. Therefore, to consider the impact on manufacturing performance of these sources of uncertainty, a scenario analysis investigates (i) different combinations of production orders and (ii) quality of returns, and (iii) different quantities of returns compared to the total production orders. Finally, priority rules are investigated to manage synchronisation between production orders and returns disassembly in the different scenarios. In fact, disassembly operations may fast saturate buffers with components of low values required by production activities at different times and in different quantities.

Fig. 1 shows the studied system implementing the following CE strategies: repair, re-manufacturing, and reuse. It consists of the following five single-server stations, identified by grey circles in the figure: disassembly (D), manufacturing (M), assembly (A), quality inspection (Q), and repair (R).

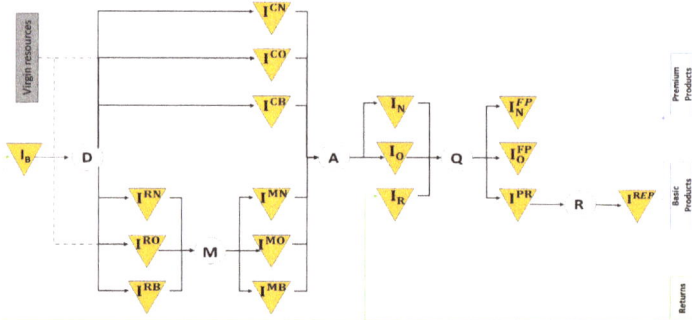

Fig. 1. Manufacturing system with five single-server workstations for the following operations: disassembly (D), manufacturing (M), assembly (A), quality inspection (Q), and repair (R).

The proposed system produces two types of finished products (FP): type N is a new top premium product, while type O is the basic version. Unlike FP-O, production orders for FP-N can be satisfied through repaired returns (REP). There are two types of returns (green arcs in Fig. 1): the first can be repaired (R), while the second (B) can be disassembled. The recovered products of type B are stored in a buffer (I_B) and provide components (C) and raw materials (R).

Components C are supplied by other companies (virgin resources) or recovered from the disassembled products and reused without any processing activity. At the same time, raw materials R are provided by other companies (virgin resources) or retrieved from the disassembled returns, but they are re-manufactured within the system to obtain manufactured components of the set M. The FP-N and FP-O are produced by assembling components from set C and manufactured parts from set M. Raw materials, components, and manufactured parts are clustered into three groups: (a) parts and components necessary for FP-N, i.e., RN, CN, and MN, respectively; (b) raw materials, parts, and components necessary for FP-O, i.e., RO, CO, and MO, respectively; c) raw materials, parts, and components necessary for both products, i.e., RB, CB, and MB, respectively. Assume that component CN can be reused for FP-N without processing, or it can be re-manufactured to become a part of MO.

FP-N, FP-O, and R are subjected to a quality inspection that: verifies the quality level for items FP-N and FP-O and identifies the necessary tasks for repairing items R.

Design of Experiment

Discrete Event Simulation is used to study the system in various scenarios. The scenarios are characterised by various proportions of the two types of returns and eight priority rules to find the sequence of jobs processed in each of the five workstations. The simulation model is developed and evaluated in Arena 16.2.

The experiment investigates the makespan because of its importance in manufacturing since it represents the time required to satisfy all the production orders. System processing times are proportionally reduced together with job arrival time to have a sufficient number of observations in a limited amount of time (1-2 work shifts of 8 hours) to improve the comprehension of CE strategies on makespan. Assembly and quality control are the most time-consuming activities because they include setups, packaging, and small reparations. At the same time, the full return disassembly is considered a destructive activity to quickly recover key components such as chipboards, metallic frames and bodies, and small electric motors.

In Table 1, the fixed parameters are reported: the processing times of workstations and the total number of production orders that must be satisfied.

Table 1. Fixed parameters of the simulation model.

Parameter	Value	Description
t_D	0.3 [min]	Processing time of disassembly workstation
t_M	0.44 [min]	Processing time of manufacturing workstation.
t_A	0.7 [min]	Processing time of assembly workstation.
t_{QI}	0.67 [min]	Processing time of the quality inspection workstation.
t_R	3 [min]	Processing time of repairing workstation.
N+O	600 [u]	The total number of production orders of both types.

The intertwined effects of variability and system characteristics can influence the makespan. Therefore, the following three factors have been considered:

- R/B. Ratio between the end-of-life and use of recovered products that can be repaired (R) out of those that can be disassembled (B): 1, 0.5, and 1.5.
- (DN+DO)/(R+B). Ratio of the demand of finished products DN and DO out of the recovered reparable and irreparable end-of-life and use products: 1, 0.5, and 1.5.
- DN/DO. Ratio between the demand for the premium level products of type FP^N (DN) of the basic products of type FP^O (DO): 1, 0.5, and 1.5.

For each combination of factors, eight priority rules are tested to model priority to one finished product or the other, the priority to reduce pressure on buffers, the priority in repairing strategy, or balance priority to all finished products and strategies:

Rule 1. Priority in all the workstations to the operations for FP-N and repairing activities to satisfy production orders.

Rule 2. Priority in all the workstations to the operations for FP-O and repairing activities to satisfy production orders.

Rule 3. Priority in manufacturing and assembly workstations to the operations that decrease buffer levels, while in quality inspection, FP-N has higher priority than FP-O, while repairing activities have the highest priority to satisfy production orders.

Rule 4. Priority in manufacturing and assembly workstations to the operations that decrease buffer levels, while in quality inspection, FP-O has higher priority than FP-N, while repairing activities have the highest priority to satisfy production orders.

Rule 5. Same priority in all the workstations to both production orders by serving the type with the maximum number of remaining orders. The highest priority is given to repairing activities.

Rule 6. Priority in manufacturing and assembly workstations to the operations that decrease buffer levels, balanced priority in quality inspection and maximum priority to repairing activities.

Rule 7. Same priority in all the workstations.

Rule 8. Priority in manufacturing and assembly workstations to the operations that decrease buffer levels, balanced priority in quality inspection.

Results and discussion

According to CE strategies, finished products can be repaired, reused as is or disassembled to recover components and raw materials, which, in turn, can be reused or re-manufactured.

Italian Manufacturing Association Conference - XVI AITeM   Materials Research Forum LLC
Materials Research Proceedings 35 (2023) 250-257   https://doi.org/10.21741/9781644902714-30

However, the impact of introducing these strategies on the required time to complete a set of given production orders (makespan) is unclear, and it also depends on other system characteristics (variability in production order types and in the volumes and the quality level of available returns). Fig. 2 shows the effects of the considered characteristics on the makespan. Specifically, from left to right: the proportion of returns devoted to the disassembly out of those repaired, the proportion of production orders covered by returns, the number of production orders that can be satisfied by repaired products out of the other production order type, and the eight priority policies. The number of returns can increase the makespan from around 480 minutes (one production shift) to 640 minutes (two production shifts).

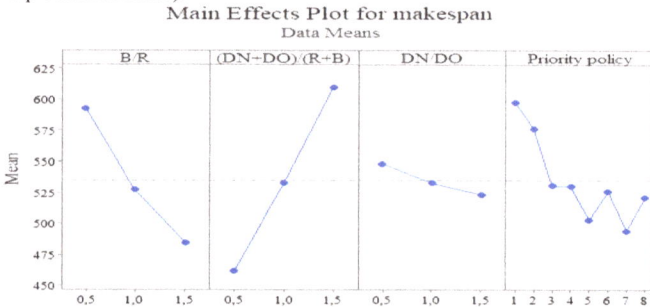

Fig. 2. The main effect graphs show the effects of the factors on the makespan.

Repairable returns have a short flow time within the system because they only require quality control and repair operations to satisfy production orders rather than disassembly, re-manufacturing, and reassembly. In fact, the presence of more production orders satisfiable through repaired returns reduces the makespan (DN/DO = 1.5). However, on average, the disassembly strategy, which provides components and resources for both product types, led to a lower makespan than the increasing repair strategy (B/R = 0.5). Priority rules that aim for a balanced satisfaction of production order types (5 and 7) lead to smaller makespan.

Fig. 3 shows the pairwise interactions on the makespan by highlighting that all the factors influence it since each level of the factor has a marker in a different level of the makespan. Also, there is an amplifying effect on the makespan between the proportion of return types and the number of returns (first frame in the top-left part of the figure). Therefore, in the case of many returns (1.5) and reparable returns larger than the others, the makespan increases (positive slope of the blue line).

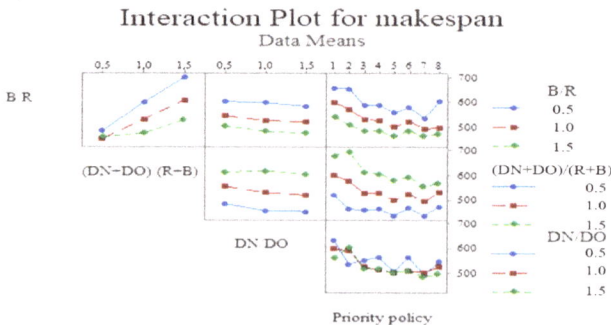

Fig. 3. The pairwise effects of the four factors influencing the makespan.

The pairwise interaction plot is not able to capture the real effectiveness of priority rules since it considers an average makespan of all the factors except for the two whose interaction is investigated. However, from the interaction between priority rules and production order mix (DN/DO, the last frame in the bottom), it can be seen that priority rules 1 (higher priority to FP-N) and 2 (high priority to FP-O) are not so bad how they appear in the other frames. Still, they are effective when applied during specific production order mixes. Conversely, priority rules 5 and 7 appear much more flexible since, on average, they led to a low makespan. In contrast, priority rules that focus on buffers (3, 4, 6, 8) mitigate the extreme effects of unbalanced rules towards one specific production order (3 and 4 lead to more robust and on average makespan than 1 and 2), but, when applied to the entire system they are widely influenced by return availability, returns quality, and production mix (6 and 8 lead to different result for different colour lines).

Conclusion

This paper investigates the impacts on the makespan of implementing some of the Circular Economy (CE) strategies (reuse, re-manufacturing, and repair) in a manufacturing system that exploits end-of-life and end-of-use products to recover components and satisfy the demand for new finished products. A scenario analysis evaluated through the Discrete Event Simulation model has been created to assess eight priority rules applied to the same system with different characteristics in production order types, types and numbers of returned end-of-life and use products.

Systems that include disassembly operations coupled with processing and reassembly are spreading because of the new laws and regulations regarding CE, the disruption of global supply chains, and the increasing lead times in the supply of critical raw materials. Therefore, the discussion of the results of the paper could provide technical and operational insights regarding the characteristics of manufacturing systems that foster or dampen the transition towards the manufacturing paradigm that includes CE strategies.

The results show that synchronising CE strategies (the numbers and the types of recovered products) with the production orders (numbers and types) reduces the makespan. However, CE strategies make short-term production planning more complex, and scheduling is important to improve technical performance. Furthermore, potential disequilibria between the types of returns and the types of production orders can be mitigated through priority rules. Finally, priority rules deeply affect the system performance, and they must follow the frequent changes in the system condition since they are not robust to the high uncertainty levels considered in this paper.

Future research will deepen the many critical issues related to the disassembly processes by intertwining them with the adoption of CE strategies. It will address other indicators, such as WIP and the total consumption of virgin materials. Also, other sustainable alternatives should be investigated, like industrial symbiosis.

A preliminary, introductive, and not peer-reviewed version of this paper is available at the EngrXiv database [26].

Acknowledgement

This study was carried out within the MICS (Made in Italy – Circular and Sustainable) Extended Partnership and received funding from the European Union Next-GenerationEU (PIANO NAZIONALE DI RIPRESA E RESILIENZA (PNRR) – MISSIONE 4 COMPONENTE 2, INVESTIMENTO 1.3 – D.D. 1551.11-10-2022, PE00000004). This manuscript reflects only the authors' views and opinions, neither the European Union nor the European Commission can be considered responsible for them.

References

[1] S. Rajput, S.P. Singh, Industry 4.0−Challenges to implement circular economy, Benchmarking: An International Journal 28(5) (2021) 1717-1739. https://doi.org/10.1108/BIJ-12-2018-0430

[2] P. Rosa, C. Sassanelli, A. Urbinati, D. Chiaroni, S. Terzi, Assessing relations between Circular Economy and Industry 4.0: a systematic literature review, International Journal of Production Research 58(6) (2020) 1662-1687. https://doi.org/10.1080/00207543.2019.1680896

[3] B. Ding, X. Ferras Hernandez, N. Agell Jane, Combining lean and agile manufacturing competitive advantages through Industry 4.0 technologies: an integrative approach, Production planning & control (2021) 1-17. https://doi.org/10.1080/09537287.2021.1934587

[4] J. Chen, H. Wang, R.Y. Zhong, A supply chain disruption recovery strategy considering product change under COVID-19, Journal of Manufacturing Systems 60 (2021) 920-927. https://doi.org/10.1016/j.jmsy.2021.04.004

[5] A. Coveri, C. Cozza, L. Nascia, A. Zanfei, Supply chain contagion and the role of industrial policy, Journal of Industrial and Business Economics 47 (2020) 467-482. https://doi.org/10.1007/s40812-020-00167-6

[6] S. Roscoe, E. Aktas, K.J. Petersen, H.D. Skipworth, R.B. Handfield, F. Habib, Redesigning global supply chains during compounding geopolitical disruptions: the role of supply chain logics, International Journal of Operations & Production Management (ahead-of-print) (2022). https://doi.org/10.1108/IJOPM-12-2021-0777

[7] M.C. Magnanini, W. Terkaj, T. Tolio, Robust optimisation of manufacturing systems flexibility, Procedia CIRP 96 (2021) 63-68. https://doi.org/10.1016/j.procir.2021.01.053

[8] C. Castiglione, A. Alfieri, Supply chain and eco-industrial park concurrent design, IFAC-PapersOnLine 52(13) (2019) 1313-1318. https://doi.org/10.1016/j.ifacol.2019.11.380

[9] C.O. Klingenberg, M.A. Viana Borges, J.A. Valle Antunes Jr, Industry 4.0 as a data-driven paradigm: a systematic literature review on technologies, Journal of Manufacturing Technology Management (2019). https://doi.org/10.1108/JMTM-09-2018-0325

[10] T. Tolio, A. Bernard, M. Colledani, S. Kara, G. Seliger, J. Duflou, ..., & S. Takata, Design, management and control of demanufacturing and re-manufacturing systems, CIRP Annals 66(2) (2017) 585-609. https://doi.org/10.1016/j.cirp.2017.05.001

[11] M. Colledani, O. Battaïa, A decision support system to manage the quality of End-of-Life products in disassembly systems, CIRP Annals 65(1) (2016) 41-44. https://doi.org/10.1016/j.cirp.2016.04.121

[12] A. Arianna, C. Castiglione, E. Pastore, A multi-objective tabu search algorithm for product portfolio selection: A case study in the automotive industry, Computers & Industrial Engineering 142 (2020) 106382. https://doi.org/10.1016/j.cie.2020.106382

[13] C. Castiglione, A. Alfieri, E. Pastore, Decision Support System to balance inventory in customer-driven demand, IFAC-PapersOnLine 51(11) (2018) 1499-1504. https://doi.org/10.1016/j.ifacol.2018.08.288

[14] E. Suzanne, N. Absi, V. Borodin, Towards circular economy in production planning: Challenges and opportunities, European Journal of Operational Research 287(1) (2020) 168-190. https://doi.org/10.1016/j.ejor.2020.04.043

[15] T.L. Olsen, B. Tomlin, Industry 4.0: Opportunities and challenges for operations management, Manufacturing & Service Operations Management 22(1) (2020) 113-122. https://doi.org/10.1287/msom.2019.0796

[16] A.J. Lambert, Disassembly sequencing: a survey, International Journal of Production Research, 41(16) (2003) 3721-3759. https://doi.org/10.1080/0020754031000120078

[17] W. Zhang, Y. Zheng, R. Ahmad, The integrated process planning and scheduling of flexible job-shop-type re-manufacturing systems using improved artificial bee colony algorithm, Journal of Intelligent Manufacturing (2022) 1-26. https://doi.org/10.1007/s10845-022-01969-2

[18] F. Ehm, A data-driven modeling approach for integrated disassembly planning and scheduling, Journal of Re-manufacturing, 9(2) (2019) 89-107. https://doi.org/10.1007/s13243-018-0058-6

[19] H.J. Kim, D.H. Lee, P. Xirouchakis, O.K. Kwon, A branch and bound algorithm for disassembly scheduling with assembly product structure, Journal of the Operational Research Society, 60(3) (2009) 419-430. https://doi.org/10.1057/palgrave.jors.2602568

[20] J.M. Yu, J.S. Kim, D.H. Lee, Scheduling algorithms to minimise the total family flow time for job shops with job families, International Journal of Production Research, 49(22) (2011) 6885-6903. https://doi.org/10.1080/00207543.2010.507609

[21] Y. Fu, M. Zhou, X. Guo, L. Qi, Stochastic multi-objective integrated disassembly-reprocessing-reassembly scheduling via fruit fly optimisation algorithm, Journal of Cleaner Production, 278 (2021) 123364. https://doi.org/10.1016/j.jclepro.2020.123364

[22] I. Ferretti, Multi-product economic lot scheduling problem with returns and sorting line, Systems, 8(2) (2020) 16. https://doi.org/10.3390/systems8020016

[23] V.D.R. Guide Jr, G.C. Souza, E. Van Der Laan, Performance of static priority rules for shared facilities in a re-manufacturing shop with disassembly and reassembly, European Journal of Operational Research, 164(2) (2005) 341-353. https://doi.org/10.1016/j.ejor.2003.12.015

[24] J.M. Kim, Y.D. Zhou, D.H. Lee, Priority scheduling to minimise the total tardiness for re-manufacturing systems with flow-shop-type reprocessing lines, The International Journal of Advanced Manufacturing Technology, 91(9) (2017) 3697-3708. https://doi.org/10.1007/s00170-017-0057-z

[25] C. Castiglione, E. Pastore, A. Alfieri, Technical, economic, and environmental performance assessment of manufacturing systems: the multi-layer enterprise input-output formalization method, Production Planning & Control (2022) 1-18. https://doi.org/10.1080/09537287.2022.2054743

[26] C. Castiglione, E. Pastore, A. Alfieri, Circular economy strategies at the manufacturing system scheduling level: impacts on Makespan, engrXiv, May 31, 2023. https://doi.org/10.31224/3027

Italian Manufacturing Association Conference - XVI AITeM
Materials Research Proceedings 35 (2023) 258-265

Materials Research Forum LLC
https://doi.org/10.21741/9781644902714-31

Laser welding with and without filler wire of aluminum sheets produced by rolling and additive manufacturing for e-mobility applications

Alessandro Ascari[1], Erica Liverani[1], Alessandro Fortunato[1],
Stefano Cattaneo[2], Marco Franzosi[2]

[1]University of Bologna, Viale Risorgimento 2, 40136 Bologna, Italy

[2]IPG Photonics Italy, Viale Kennedy 21, 20023 Cerro Maggiore (MI), Italy

Keywords: Laser Welding, Aluminum Alloys, E-Mobility

Abstract. One of the most critical factors to be taken into consideration in laser welding of aluminum alloys is the formation of pores in the fused zone, which depends strictly on the semi-finishing format of the parent sheets. According to these considerations the present paper deals with welding of AA6082 sheets with additively manufactured AlSi10 ones in a configuration that is typical for the production of casings for batteries for the e-mobility field. In order to understand the role of process strategies on weld bead quality, both autogenous welding and welding with filler wire are investigated and the eventual benefits of applying a wobbling beam shaping is also considered. For any of the above-mentioned strategies, the role of process parameters, such as laser power, welding speed, filler wire speed and wobbling, is underlined, with particular reference to the formation of pores and defects.

Introduction

Laser welding with filler wire has been studied for several years as a method for overcoming some of the most common autogenous welding drawbacks, such as difficult gab-bridging and impossibility to promote metallurgical modifications in the weld pool. The applications of filler wire in laser welding has always gone hand in hand with the development of the different beam generation technologies: starting from CO_2 lasers in late 90s and early 2000s [1,2], the attention moved to fiber delivered Nd:YAG and Disk ones [3-5], while the recent years have been characterized by a massive exploitation of modern fiber sources. In laser welding with filler wire applications, aluminum alloys play a very important role thanks to their specific applications especially in automotive and aerospace fields. In that direction Grunenwald et al. [3] pointed out the benefits of filler wire in gap bridging of AA5083 alloy. Vollertsen et al. [5] investigated the role of filler wire on crack formation in laser welding of AA6082 and AA6056 alloys. Pinto et al. [4] pointed out the importance of selecting the proper filler wire composition in laser welding of AA6xxx and AA5xxx alloys, both in similar and dissimilar configurations, with the aim of reducing both crack susceptibility and pores formation. Yu et al. [6] investigated the importance of wire-laser beam mutual position and distance for promoting a stable melt pool and a smooth wire melting. Schultz et al. [7] proposed the oscillation of a high brilliance laser beam as a mean for evenly distributing the energy on the tip of the filler wire instead of defocusing or using a low brilliance beam with a large spot. They also demonstrated that this technique can also have beneficial effects in gap bridging. Enz et al. [8] investigated the benefits of adding a filler wire in laser welding of AA7075 alloy for the production of tailored blanks. In a similar direction Adisa et al. [9] stressed on the possibility to apply pulsed laser sources in welding AA7020 alloy. In order to understand potential process productivity, Xu et al. [10] investigated high speed laser welding with filler wire of AA6xxx alloys. Li et al. [11] proposed the possibility to exploit hot wire laser welding of AA7075 alloy. Examilioti et al. [12] investigated the role of filler wire on AA2198

Italian Manufacturing Association Conference - XVI AITeM   Materials Research Forum LLC
Materials Research Proceedings 35 (2023) 258-265   https://doi.org/10.21741/9781644902714-31

alloy for aerospace applications and pointed out that it greatly reduces crack formation. Huang et al. [13] gave a comprehensive explanation of the role of filler wire positioning in laser welding of AA5xxx alloys: by selecting the proper angle and stand-off distance, melting of filler wire can be greatly optimized. Concerning welding of additively manufactured component, several studies demonstrated that laser techniques allow to achieve good results on steel [14] and stainless steel [15]. The above mentioned literature shows that, in aluminum alloy laser welding, the adoption of a filler wire implies three main benefits:

- Possibility to deal with gaps, misalignments, differences in thickness, etc.
- Reduction of pores formation.
- Reduction of cracks formation.

In the automotive field, the application of laser welded aluminum alloys has become very popular in the last years, both in structural (car-body) and electric applications (batteries) [16,17] and the search for optimized processes that guarantee good versatility and high weld quality is of utmost importance. In those fields the application of filler wire has proven to be very beneficial [18], since different alloys, thicknesses and joint configurations are involved. In the light of what has been underlined so far, the present paper reports an investigation concerning laser welding of rolled AA6082 sheets on Selective Laser Melting (SLM) printed AlSi10 sheets. The idea is to evaluate the possibility of welding different aluminum alloys in different semi-finishing conditions, guaranteeing the proper joint geometry and overall quality. According to this, laser welding with and without filler wire is proposed herein and wobbling application is also involved as a mean for achieving the proper joint characteristics.

Materials and Methods

The equipment exploited in the present investigation was composed by a IPG YLS-6000 fiber source equipped with a IPG D50 two axes scanning optics (see Table 1) for complete characteristics. The wire feed system was based on a Fronius KD7000 push-pull equipment with a maximum wire feeding speed of 10 m/min. The welding optics was mounted on a Yaskawa-Motoman HP-20 6 axes anthropomorphic robot (see Figure 1).

Figure 1: Welding system

The filler wire was a AA5356 one, with a diameter of 1.2 mm and a mixture of 15% He and 85% Ar was used as shielding gas with a flow rate of 30 l/min. The filler wire guiding nozzle was placed at an inclination of 30° with respect to the horizontal plane ant the welding process was

Italian Manufacturing Association Conference - XVI AITeM   Materials Research Forum LLC
Materials Research Proceedings 35 (2023) 258-265   https://doi.org/10.21741/9781644902714-31

carried out in the wire leading configuration (see Figure 2a). The sheet metals involved in this investigation were 50x50x1.5 mm rolled AA6082 and 50x50x5 mm SLM printed AlSi10. The welding configuration was a lap-joint one with AlSi10 at the bottom and AA6082 on top and the weld bead was realized with an inclination of the welding head of 45° (see Figure 2b).

Table 1: Laser source and optics characteristics

Maximum power	6 kW
BPP	4 mm·mrad
Collimation focal length	200 mm
Focalization focal length	300 mm
Magnification factor	1.5
Fiber core diameter	100 μm
Spot diameter	150 μm
Maximum scanning field	8x8 mm
Maximum wobbling frequency	350 Hz

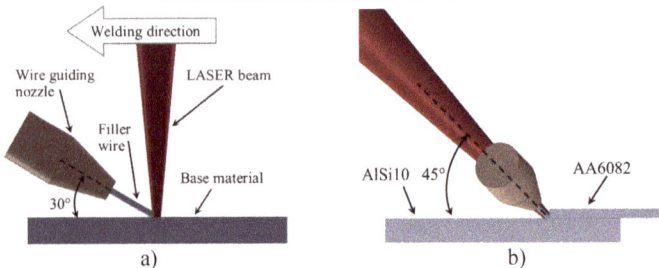

Figure 2: welding setup

The experimental campaign was carried out in three different configurations:
- Welding with filler wire, no wobbling. In this case the process parameters were selected as follows:
  - wire feed speed and welding speed 25 mm/s, Laser power 2.2, 2.4, 2.6, 2.8, 3.0 kW.
  - wire feed speed and welding speed 30 mm/s, Laser power 2.2, 2.4, 2.6, 2.8, 3.0 kW.
  - wire feed speed and welding speed 35 mm/s, Laser power 2.2, 2.4, 2.6, 2.8, 3.0 kW.
- Welding with filler wire and wobbling. In this case the process parameters were selected as follows:
  - wire feed speed and welding speed 25 mm/s, wobbling frequency 200 Hz, wobbling amplitude 1.2 mm, Laser power 2.2, 2.4, 2.6 2.8, 3.0 kW.
  - wire feed speed and welding speed 30 mm/s, wobbling frequency 200 Hz, wobbling amplitude 1.2 mm, Laser power 2.2, 2.4, 2.6 2.8, 3.0 kW.
  - wire feed speed and welding speed 35 mm/s, wobbling frequency 200 Hz, wobbling amplitude 1.2 mm, Laser power 2.2, 2.4, 2.6 2.8, 3.0 kW.
- Welding without filler wire and without wobbling. In this case the process parameters were selected as follows:
  - wire feed speed and welding speed 25 mm/s, Laser power 2.2, 2.4, 2.6 kW.
  - wire feed speed and welding speed 30 mm/s, Laser power 2.2, 2.4, 2.6 kW.
  - wire feed speed and welding speed 35 mm/s, Laser power 2.2, 2.4, 2.6 kW.

Italian Manufacturing Association Conference - XVI AITeM

Materials Research Forum LLC

Materials Research Proceedings 35 (2023) 258-265

https://doi.org/10.21741/9781644902714-31

The different values of process parameters selected were defined by means of previous preliminary tests aimed at defining plausible process windows suitable for comparing the three working conditions defined above. The welded samples were cross sectioned, resin mounted, polished and chemically etched, so that the main geometrical characteristics, such as weld bead width and penetration depth could be measured (see Figure 4). By means of the ImageJ image analysis software, average porosity was also measured.

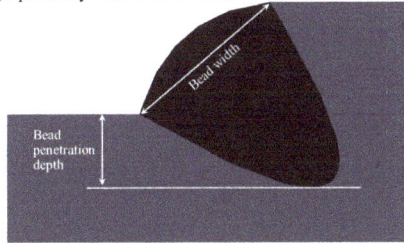

Figure 4: Weald bead measurement

Results and Discussion

Concerning welding with filler wire, the graphs in Figure 5 shows some results in terms of weld bead width, penetration depth and porosity.

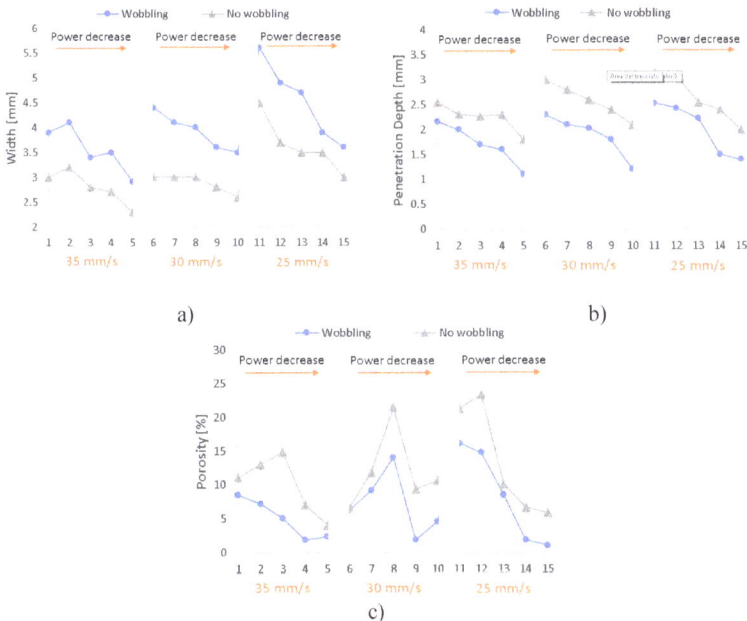

Figure 5: Welding with filler wire

Concerning weld bead width, the trend is exactly the same both with and without wobbling: this parameter tends to decrease if laser power decreases and if welding speed increases, as expected. The main difference between the two modes is that, when wobbling is enabled, bead

Italian Manufacturing Association Conference - XVI AITeM   Materials Research Forum LLC
Materials Research Proceedings 35 (2023) 258-265   https://doi.org/10.21741/9781644902714-31

width is larger, due to the "stirring" effect of beam oscillation, that tends to distribute the molten pool on a larger area. The trend characterizing bead penetration depth is very similar to the one analyzed above: an increase of welding speed leads to a decrease of penetration depth, while an increase of laser power leads to an increase of penetration depth.

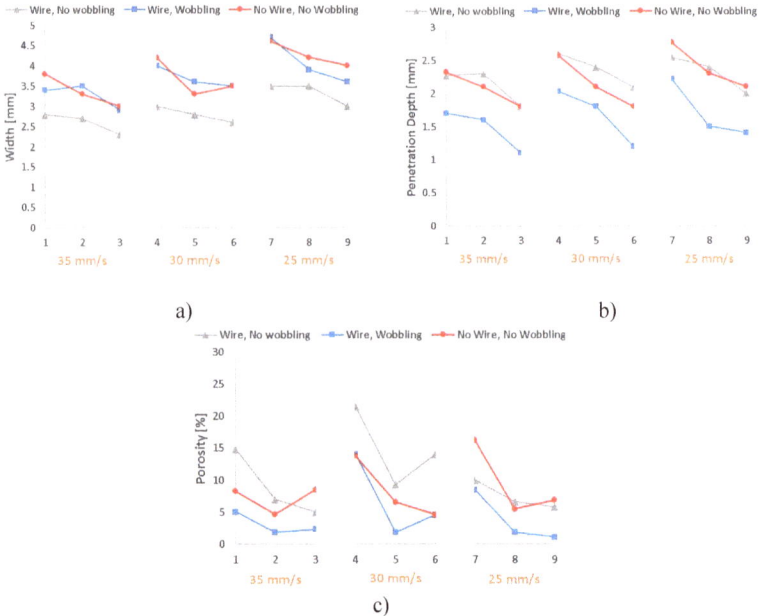

Figure 6: Overall comparison

The adoption of wobbling leads to a general decrease of penetration depth: this is due to the fact that oscillation of the beam enlarges the interaction zone between laser radiation and base material, determining a general decrease of energy density during the process and, thus, the formation of a larger molten pool. The effect of beam wobbling, in fact, is to determine a large apparent spot that leads to a general increase of melt pool dimension, while keeping the actual "instantaneous" spot small and promoting a good beam-material coupling especially in case of highly reflective materials. Concerning porosity, the trend is much different and the role of the main process parameters is not so clear. Analyzing laser power, it looks like there is not a definite trend: these results could be very much affected by the fact that porosity percentage was evaluated only in one specific section of every specimen and, thus, it is really far from being an average value of the whole bead. The investigation at the base of the present paper is, in fact, to give a preliminary idea of what to expect in the different welding modes analyzed: further investigations in this direction could involve tomographies, so that a more generalized porosity could be detected, and also a more comprehensive experimental campaign with repetitions, design of experiment and statistical analyses. Concerning welding speed, it looks like there is a more definite trend, since porosity increases as this parameter decreases: this is due to the fact that low welding speed determines a relatively high specific heat input and a high interaction time, that promote the formation of a large quantity of pores that fail to evolve from the melt pool. The most interesting

Italian Manufacturing Association Conference - XVI AITeM   Materials Research Forum LLC
Materials Research Proceedings 35 (2023) 258-265   https://doi.org/10.21741/9781644902714-31

result concerning porosity is related to which welding mode is applied: in case of wobbling mode, in fact, the average porosity always tends to be lower: the "stirring" effect of wobbling promotes a de-gassing of the molten pool given a certain set of process parameters. These results are also in accordance with the studies carried out by Ramiarison et al. [19]: the authors put in evidence that wobbling mode helps to mitigate several defects such as cracks and pores. Concerning welding without filler wire, the graphs in Figure 6 show a comprehensive comparison of the three welding modes investigated herein. The result that surely deserves more attention, in this case, is that the lowest porosity still remains a beneficial characteristic of welding with wobbling and filler wire, while welding without filler wire shows, generally, worse results. The number of trials in this case was lower because 3.0 kW and 2.8 kW without filler wire caused an excessive penetration in the lower sheet and, thus, they were excluded from the campaign. Figure 7 shows representative micrographs of the three investigated modes (Laser Power 2.2 kW, Welding speed 25 mm/s). The pictures confirm that welding with wobbling and filler wire gives the best results in terms of porosity.

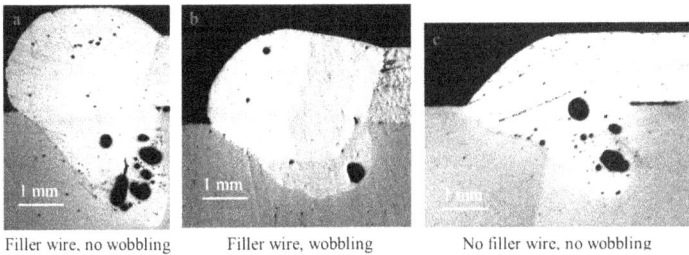

Filler wire, no wobbling   Filler wire, wobbling   No filler wire, no wobbling

Figure 7: Micrographs of representative weld beads

The presence of filler wire generates a higher reinforcement on the weld bead, that can be beneficial when gap bridging is needed. In particular, concerning weld bead reinforcement, the presence of wobbling and filler wire determines a shape of the top part of the weld bead that is more evenly distributed on the corner between the sheets. If no wobbling is adopted, the reinforcement is generally very prominent, and the transition between the sheets and the weld bead is abrupt: this situation very likely promotes a severe stress concentration in correspondence of the heat affected zone, causing a general decrease of mechanical performance of the joint. These results are also supported by the studies underlined by Zhao et al. [20], where it was put in evidence that the presence of wobbling has a beneficial effect in evenly distributing the molten pool during the process.

Summary

The present paper investigates welding with and without filler wire of different aluminum alloys in different semi-finished conditions. In particular a rolled sheet is welded on an additively manufactured one. The activity is a preliminary one and future developments will involve more thorough investigations of porosities and micro-structures and also tensile tests to assess the actual mechanical performance of the joints. The main outcomes resulting from the investigation are the following:

- Welding with filler wire and wobbling is the best solution for achieving a low porosity in the weld bead.

- The presence of filler wire determines a higher reinforcement in the weld bead and wobbling helps to distribute it more evenly.

Acknowledgements

The authors would like to thank IPG Photonics Italy for equipment and support supplied during the present investigation.

References

[1] [Salminen, A., Kujanpää, V. P. and Moisio, T. J. I. "Interactions between laser beam and filler metal." Welding Journal Vol. 75 (1996): pp. 9–13.

[2] Salminen, A. S. and Kujanpää, V. P. "Effect of wire feed position on laser welding with filler wire." Journal of Laser Applications Vol. 15 No. 1 (2003): pp. 2–10. https://doi.org/10.2351/1.1514220

[3] Grünenwald, S., Kujanpää, V. and Salminen, A. "Nd:YAG laser welding of 5083 aluminum alloy using filler wire." International Congress on Applications of Lasers & Electro-Optics 2007, Vol. 4831: pp. 562 – 569. 2007. Laser Institute of America, LIA. https://doi.org/10.2351/1.5060979

[4] Pinto, L. A., Quintino, L., Miranda, R. M. and Carr, P. "Laser Welding of Dissimilar Aluminium Alloys with Filler Materials." Welding in the world Vol. 54 (2010): pp. 333–341. https://doi.org/10.1007/BF03266747

[5] Vollertsen, F., Buschenhenke, F. and Seefeld, T. "Reduction of Hot Cracking in LaserWelding using Hypereutectic AlSi FillerWire." Welding in the world Vol. 52 (2008): pp. 3–8. https://doi.org/10.1007/BF03266635

[6] Yu, Y., Huang, W., Wang, G., Wang, J., Meng, X., Wang, C., Yan, F., Hu, X. and Yu, S. "Investigation of melting dynamics of filler wire during wire feed laser welding." Journal of Mechanical Science and Technology Vol. 27 (2013): pp.1097–1108. https://doi.org/10.1007/s12206-013-0218-4

[7] Schultz, V. "Process Stability during Laser Beam Welding with Beam Oscillation and Wire Feed." Journal of Manufacturing and Materials Processing Vol. 3 No. 1 (2019): pp. 1–16. https://doi.org/10.3390/jmmp3010017

[8] Enz, J., Riekehr, S., Ventzke, V., Sotirov, N. and Kashaev, N. "Laser Welding of High-strength Aluminium Alloys for the Sheet Metal Forming Process." Procedia CIRP Vol. 18 (2014): pp. 203–208. v10.1016/j.procir.2014.06.132

[9] Adisa, S. B., Loginova, I., Khalil, A. and Solonin, A. "Effect of Laser Welding Process Parameters and Filler Metals on the Weldability and the Mechanical Properties of AA7020 Aluminium Alloy." Journal of Manufacturing and Materials Processing Vol. 2 No. 2 (2018): pp. 1–10. https://doi.org/10.3390/jmmp2020033

[10] Xu, F., Chen, E. G., L. andHe and Guo, L. Y. "Laser welding 6A02 aluminum alloy with filler wire under high welding speed." IOP Conference Series: Materials Science and Engineering Vol. 504 No. 1 (2019): p. 012028. https://doi.org/10.1088/1757-899X/504/1/012028

[11] Li, S., Mo, B., Xu, W., Xiao, G. and Deng, X. "Research on nonlinear prediction model of weld forming quality during hot-wire laser welding." Optics & Laser Technology Vol. 131 (2020): p. 106436. https://doi.org/10.1016/j.optlastec.2020.106436

[12] Examilioti, T. N., Kashaev, N., Ventzke, V., Klusemann, B. and Alexopoulos, N. D. "Effect of filler wire and post weld heat treatment on the mechanical properties of laser beam welded AA2198." Materials Characterization Vol. 178 (2021): p. 111257. https://doi.org/10.1016/j.matchar.2021.111257

Italian Manufacturing Association Conference - XVI AITeM   Materials Research Forum LLC
Materials Research Proceedings 35 (2023) 258-265   https://doi.org/10.21741/9781644902714-31

[13] Huang, W., Chen, S., Xiao, J., Jiang, X. and Jia, Y. "Investigation of filler wire melting and transfer behaviors in laser welding with filler wire." Optics & Laser Technology Vol. 134 (2021): p. 106589. https://doi.org/10.1016/j.optlastec.2020.106589

[14] Feng, L., Gao, J., Liu, F., Liu, F., Huang, C., Zheng, Y. "Effect of grain orientation on microstructure and mechanical properties of laser welded joint of additive manufactured 300M steel". Materials Today Communications Vol. 35 (2023): p. 105497. https://doi.org/10.1016/j.mtcomm.2023.105497.

[15] Ascari, A., Fortunato, A., Liverani, E., Gamberoni, A., & Tomesani, L. "New Possibilities in the Fabrication of Hybrid Components with Big Dimensions by Means of Selective Laser Melting (SLM)". Physics Procedia Vol. 83 (2016): p. 839–846. https://doi.org/10.1016/j.phpro.2016.08.087

[16] Ceglarek, D., Colledani, M., Váncza, J., Kim, D.-Y., Marine, C., Kogel-Hollacher, M., Mistry, A. and Bolognese, L. "Rapid deployment of remote laser welding processes in automotive assembly systems." CIRP Annals Vol. 64 No. 1 (2015): pp. 389–394. https://doi.org/10.1016/j.cirp.2015.04.119

[17] Franciosa, P., Sun, T., Ceglarek, D., Gerbino, S. and Lanzotti, A. "Multi-wave light technology enabling closed-loop in-process quality control for automotive battery assembly with remote laser welding." Stella, Ettore (ed.). Multimodal Sensing: Technologies and Applications, Vol. 11059: p. 110590A. 2019. International Society for Optics and Photonics, SPIE. https://doi.org/10.1117/12.2526075

[18] Sun, T., Franciosa, P., Sokolov, M. and Ceglarek, D. "Challenges and opportunities in laser welding of 6xxx high strength aluminum extrusions in automotive battery tray construction." Procedia CIRP Vol. 94 (2020): pp. 565–570. https://doi.org/10.1016/j.procir.2020.09.076

[19] Ramiarison, H., Barka, N., Pilcher, C., Stiles, E., Larrimore, G., Amira S. "Weldability improvement by wobbling technique in high power density laser welding of two aluminum alloys: Al-5052 and Al-6061". Journal of Laser Applications 1 August 2021; 33 (3): 032015. https://doi.org/10.2351/7.0000353

[20] Zhao, J., Jiang, P., Geng, S., Guo, L., Wang, Y., Xu, B. "Experimental and numerical study on the effect of increasing frequency on the morphology and microstructure of aluminum alloy in laser wobbling welding". Journal of Materials Research and Technology Vol. 21 (2022): pp. 267-282. https://doi.org/10.1016/j.jmrt.2022.09.008

Italian Manufacturing Association Conference - XVI AITeM
Materials Research Proceedings 35 (2023) 266-274

Materials Research Forum LLC
https://doi.org/10.21741/9781644902714-32

Application of high voltage fragmentation to treat end-of-life wind blades

Marco Diani[1,a] *, Shravan Torvi[1,b] and Marcello Colledani[1,c]

[1]Politecnico di Milano, Department of Mechanical Engineering, via La Masa 1, 20156, Milan, Italy

[a]marco.diani@polimi.it, [b]shravan.torvi@polimi.it, [c]marcello.colledani@polimi.it

Keywords: Circular Economy, Composite Recycling, Mechanical Separation

Abstract. The use of composites is constantly increasing in several sectors, from wind energy to automotive, thanks to their mechanical properties, lightweight, and resistance to corrosion. Despite this, the recycling and reuse of these materials in high-added value applications is not yet performed at the industrial level. In particular, End-of-Life (EoL) products are sent to landfills (if possible), incinerated, or inserted in co-processing in cement plants. This work presents an experimental approach to treat End-of-Life wind blades based on High Voltage Fragmentation (HVF). This technology, based on the creation of electric spark channels, is able to generate localized shock waves at the interface between two different materials. The potential of its application has been shown in the literature, but an experimental campaign is needed to find the optimal parameters to obtain an output material with proper characteristics to feed specific output products, following a demand-driven approach.

Introduction

In the last years, the necessity to reduce carbon emissions and to become less dependent on fossil fuels has led to the wide exploitation of renewable energy, reaching a share of 22% of the European energy mix in 2021 [1]. Among them, wind energy is one of the most used. As an example, in the last decade wind power installations doubled in Europe with 236 GW of wind turbines (considering both onshore and offshore) [2]. To achieve the goal, defined in the European Green Deal, to become a climate-neutral continent by 2050, the EU has set a target of 32% of renewable energy by 2030 [3]. Wind Europe, the European Wind Energy Association, estimated in a realistic scenario the installation of more than 23 GW/y in the next five years [2]. However, as the typical life of a wind turbine is of 20/25 years, the issue of End-of-Life treatment of their components, especially the blades, has become increasingly pressing to satisfy future growth plans.

Wind turbine blades are made from composite materials that are difficult to recycle, leading to concerns about the environmental impact of wind turbine decommissioning. According to the European Union, around 80,000 wind turbines will need to be decommissioned in the EU alone by 2030, creating a significant waste management challenge [4]. Currently, most of the EoL wind blades are landfilled. This results not only in a massive environmental impact but also in a relevant loss of economic opportunities, as these products have a considerable residual value.

Composite materials can be recycled through mechanical, thermal, or chemical recycling [5]. Mechanical recycling allows to obtain granules composed of both fibers and resin that can be reprocessed for the manufacturing of new products with lower mechanical properties. Even if these processes are competitive in terms of cost, the obtained material cannot be reused for the same application, leading to a downgrade [6]. Thermal treatments exploit the temperature to volatilize the polymeric part leading to clean fibers and, using innovative processes, also liquid resin. Also in this case, due to the depolimerazion temperature, the fibers degrade, leading to a downcycling of the material, together with the relatively high cost of the process (mainly dependent on its

Italian Manufacturing Association Conference - XVI AITeM   Materials Research Forum LLC
Materials Research Proceedings 35 (2023) 266-274   https://doi.org/10.21741/9781644902714-32

energy-consuming nature) [7]. Chemical processes such as solvolysis are able to clean fibers with relatively low degradation but the costs to obtain them are high, mainly due to the solvents [8]. Independently from the adopted solutions, two different concepts have to be considered to enable robust and reliable recycling processes [9]. The first one is the so-called cross-sectorial approach. Materials obtained through the recycling of products in a specific sector have to be reused in another sector which requires lower mechanical properties, but with high-added value. In this way, recycled particles from wind energy can be reused in sectors such as automotive, avoiding their use as fillers or in the co-processing of cement. This can be enhanced by considering a second concept: the demand-driven approach. Traditionally, an EoL product is recycled and, only after that, an application for the obtained material is found. This, even if allows to have rigid but fast traditional processes, leads to the loss of the greatest part of the residual value. The demand-driven approach reverses the traditional chain. First of all, a product embedding recycled material is designed. As a consequence, the characteristics of the recycled material that enable to obtain the desired mechanical, physical, and aesthetical properties of the product are derived. On the basis of them, the recycling process is optimized to maximize the quantity of target material, minimizing, at the same time, the costs to obtain it.

Innovative technologies are studied at the research level. Among them, High Voltage Fragmentation (HVF) is one of the most promising. Exploiting the difference in electric conductivity of different materials, it is able to create local shockwaves to detach them at their interface. Due to the nature of composites, the application of HVF to their recycling is interesting to obtain homogeneous and, in the future, pure fractions of material. This work presents an experimental approach to preliminary investigate the potentiality of the innovative technology of HVF in the recycling of EoL wind blades, fostering the adoption of the demand-driven approach, to enable reliable and robust circular economy solutions for this kind of product.

Literature review

High Voltage Fragmentation (or electrodynamic fragmentation) was initially designed during the 1960s [10] for mineral extraction from ore. Since the technology exploits different electrical permittivity of material to separate them at the phase boundaries, it was traditionally used for obtaining high-value minerals, like gold, from its ore mineral extractions. Considering the shown high efficiency in treating rocks and ores, together with the possibility to control a wide range of parameters, the potentiality to apply HVF for other applications like the treatment of Municipal Solid Waste (MSW) ashes has been investigated. Examples of using HVF technology for the recovery of metals and immobilization of heavy metals from ashes generated by the thermal treatment of MSWs can be seen in [10]. After treatment, the ashes have been used as an aggregate to produce reinforced concrete.

Considering the similarity of minerals with concrete, made up of an aggregate structure, HVF technology has been applied to the recycling of building materials, primarily concrete and fiber-reinforced concrete [10] [11]. The HVF technology has been seen to be both very effective and energy efficient [11] in breaking down and separating its various components. Tests carried out on Ultra-High-Performance Fiber Reinforced Concrete (UHPFRC) have shown the potential of this technology. The results can be seen in [12], where concrete with different compression strengths was tested with HVF and it was seen the process was not only viable but the change in compression strength had little to no effect on the liberation.

HVF technology has recently been tested for the recycling of Printed Circuit Boards (PCBs) and photovoltaic panels [13] as well. For PCBs, HVF technology produces, in comparison with size reduction processes, a better liberation of metal fractions, but it was found to be significantly more energy-consuming [14]. HVF has proven to be a viable option also for recycling and dismantling of photovoltaic panels. It is comparatively better than mechanical recycling since it is less energy-consuming [13] and it produces specific size fractions of the target metallic materials.

Italian Manufacturing Association Conference - XVI AITeM   Materials Research Forum LLC
Materials Research Proceedings 35 (2023) 266-274   https://doi.org/10.21741/9781644902714-32

It also has benefits over chemical recycling methods since it reduces further contamination with chemicals and it is environmentally friendly due to the absence of solvents [15].

Considering fiber-reinforced plastics, it is possible to find some examples in the literature on the application of High Voltage Fragmentation to their recycling [16]. There have been researches conducted into the viability of the technology with thermoplastics with carbon fiber (CF) reinforcement [17]. Door hinges made from thermoplastics with CF reinforcement have been tested with HVF and the process was seen to be largely viable in terms of mechanical qualities of the recycled fibers. There was only a 17% reduction in the mechanical properties of the recycled fibers with respect to the virgin ones [17], and a new door hinge was manufactured with 100% recycled material with little to no post-treatment. Considering Glass Fiber Reinforced Plastics (GFRPs), the most present material in wind blades, some work can be found in the literature. High Voltage Fragmentation has been seen as a competitor of mechanical recycling [18] [19] of GFRP due to comparable comminution principles, but mechanical recycling by itself almost always results in short fibers with a lot of resin material, limiting the utilization of recycled particles as secondary raw materials in products. Also, the tensile strength and Young's Modulus of recycled particles obtained with HVF were slightly higher compared to mechanical recycling. It can also be seen that by increasing the number of pulses (discharges), the percentage of mixed particles (with fibers and resin still joined together) in the recycled material can be reduced [19]. However, the specific energy requirements for HVF recycling are considerably higher in comparison to mechanical recycling (at least 2,6 times higher) as seen in [19] [16].

Materials & methods

The experiments have been conducted using a High Voltage Fragmentation machine of SelFrag AG (laboratory-scale model) has been used. The machine works on the principle of the selective breakdown of material with the help of high-voltage electrical pulses. At the top of the working vessel, which contains the sample in a liquid (typically deionized water), there is an electrode, while at the bottom the counter-electrode is placed. The machine produces high-voltage electric discharges between the two electrodes in a nitrogen atmosphere. During the process, a plasma channel is created by an electrical discharge with high energy which leads to high temperatures, greater than 10^4 K, and high pressure, about 10^9-10^{10} Pa, which is able to pass through the sample to be treated, concentrating at the interface between two materials with two different electric permittivity. This generates stresses which usually exceed the strength of solid material. When the electrical pulse rises time goes down as low as 500 ns the ionization stops and the breakdown path in the plasma ends. All of these effects generate a localized shock wave, similar to that of a lightning strike, at the phase boundaries of the solid material leading to cracks. After multiple discharges, the cracks that have been created expand and reach the edges of the material that in the end causes fragmentation. Since the electrons are in the material boundaries, the fragmentation energy is situated along with the interface of the material [20]. This leads to high liberation and selective fragmentation of the product.

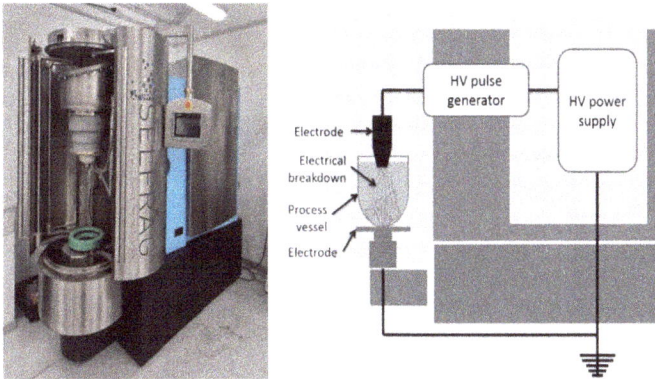

Fig. 1: High voltage fragmentation machine of SelFrag AG (functioning scheme on the right)

The process vessel in which the sample is processed can be of open or closed type. In the open type, a sieve inlay (grid) can be added to allow material that has reached the sieve size to fall into a rubber container. In the closed-type vessel, the material stays inside till the end of the process.

The different controllable parameters of this process are the voltage, which ranges between 90-200 KV, the distance between the electrodes, which can be between 10-40 mm, the pulse repetition rate, which ranges between 1-5 Hz, and the number of electric pulses, and a peak power consumption of 6 kW.

To analyze the material obtained with this process, a particle analyzer has been used, in particular Camsizer P4 of Microtrac, which uses a dual-camera imaging system to capture real-time images of material. The results obtained give information on the dimensional characteristics like the length and width of the material, together with morphological details like sphericity and elongation, with a wide dynamic range that extends from 20 microns to 30 millimeters. The software is able to elaborate the images, providing several dimensional parameters, like Feret Diameter (useful to evaluate fibers), and morphological parameters, such as the aspect ratio.

For the experiments, samples with dimensions of 50x50x15 mm with a weight of about 65 g obtained cutting an EoL wind blade have been used. They were composed of glass fibers in an epoxy resin with an intermediate layer made of polyurethane (PU) of about 10 mm.

Two different experimental campaigns have been conducted. The objective of the first one was to evaluate the potentiality of the HVF to liberate the PU. As a matter of fact, contamination of polyurethane can reduce the reuse possibilities of recycled material in high-added value products. In this case, the only parameter that has been changed is the number of discharges, with values of 50, 100, 250, 500, 1000, and 2000, while the voltage, the repetition rate, and the distance between the electrodes were fixed at, respectively, 200 kV, 5 Hz, and 20 mm. The second experimental campaign focused on the differences using a vessel with or without a grid. This is interesting not only in terms of final distribution but also to understand the potentiality of this technology to be used at an industrial scale. In this case, the considered factor was the presence of a 1 mm grid, while the voltage was of 200 kV, the repetition rate of 5 Hz, 1000 discharges, and the distance between the electrodes of 20 mm. Finally, the results obtained with the grid have been compared with historical data from a previous shredding campaign of the same material with a cutting mill of Retsch model SM300 with a 1 mm grate size.

Results and discussion

The obtained results will be presented and discussed in the following three subsections.

Italian Manufacturing Association Conference - XVI AITeM   Materials Research Forum LLC
Materials Research Proceedings 35 (2023) 266-274   https://doi.org/10.21741/9781644902714-32

Separation of polyurethane.

For this first set of experiments, the results have been evaluated through visual analysis as the PU is easily recognizable from GRFP. The obtained results are shown in Fig. 2. From these pictures, it is possible to say that high voltage fragmentation is able to completely liberate the polyurethane insert from the fiber matrix of the wind blade also with very few pulses (even 50). This effect can be explained by considering the physics of HVF. Since the machine works on the principle of selective fragmentation, which exploits the different electric permittivity of a material, the PU insert is distinctively different from the fiber-matrix component of the composite. However, with an increasing number of pulses the size of the PU reduces (mainly due to shockwave side effect that propagates in the medium), increasing the energy consumption and making its separation more difficult, requiring more complex technology with respect to, as an example, traditional optic separation. The high voltage fragmentation technology can hence be used for preliminary coarse process to separate different components, the target of which can be further processed through the same technology or through more traditional ones (as size reduction).

Fig. 2: Material obtained with (from the left) 50 pulses, 250 pulses, 1000 pulses.

Influence of the grid.

An interesting result was observed when the sample was processed with and without a sieve inlay. The grid has a unique effect on the process since all of the material that passes through it below is no longer processed. The distribution of the recycled particles is narrower around 1mm and the dimensions of the particles were in general smaller than processing the same sample without a sieve (see Fig. 3). This behavior can be explained by considering that since the material passes through the grid, less material remains inside the process vessel. As a consequence, treating less material with the same quantity of energy make the process more efficient as we have more energy per volume unit. Without the grid, all of the material remains inside the vessel till the end of the process. This disperses the energy in a more diffused manner and the material retains larger dimensions. In addition, when a particle is fragmented during the process, an higher number of smaller particles is generated, resulting in less efficient process at the subsequent step.

This behavior is of particular interest considering the already introduced demand-driven approach. Using a grid it seems to be possible to obtain narrower distribution, resulting in a larger quantity of material respecting the target characteristics to be reused in high-added value products. Different experiments will be conducted in the next future to better understand the repeatability of these results and the effect of grids with different dimensions.

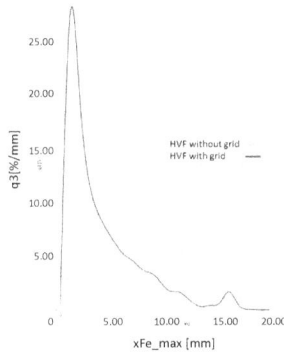

Fig. 3: Dimensional distributions obtained through HVF without (in blue) and with (in red) a 1 mm grid.

Comparison with a traditional mechanical process (shredding).
The material obtained with high voltage fragmentation with the sieve inlay (the most promising one) was also compared with the one obtained through a cutting mill in previous research, using also in this case a 1 mm grid. The results are reported in Fig. 4. The material processed with HVF is narrower around 1 mm, while the fibers mechanically recycled had a wider spread in terms of length, even if most of the particles are around 1 mm. In addition, considering the aspect ratio of the two samples, high voltage fragmentation seems to be able to obtain more elongated particles, while the ones obtained with shredding are more circular, suggesting an higher presence of powder. For sake of completeness, also at a visual inspection, the material obtained with shredding has a considerable quantity of powder material (even if cutting mill is one of the size reduction technologies that minimizes powder production), opposite with the one obtained with HVF, due to the physics behind the mechanical process, based both on impact and cutting. Also in this case, these results are not good or bad in an absolute way, as the shape of the material is also an important factor for the demand-driven recycling. The desired distribution depends on the reuse purposes (e.g. longer particles will have better mechanical properties while powder can substitute typical additives in composite formulation as calcium carbonate).

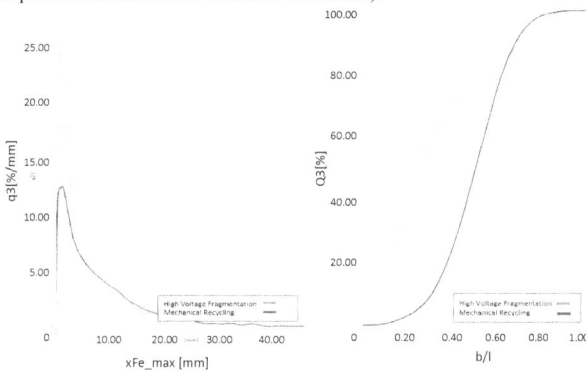

Fig. 4: Comparison of the results obtained with HVF (in blue) and with shredding (in red), both for dimensional distribution (on the left) and aspect ratio (on the right).

Conclusions

The recycling of End-of-Life wind blades is becoming urgent. The constant adoption of this technology to increase the share of green energy in the EU energy mix, together with their typical life of 20/25 years, is pushing the necessity to find reliable solutions to treat these products. Considering that most of the wind blades are made with glass fiber reinforced plastics, mechanical recycling is the current solution that allows obtaining a material competitive with virgin one. This can be reached only if both cross-sectorial and demand-driven approaches are applied. To enhance the beneficial effects of these approaches, innovative technologies can be exploited. Among them, one of the most promising is High Voltage Fragmentation. Using a fast-rising electric field with high voltage, it is able to act at the interface between two different materials, thanks to their difference in electric permittivity, liberating them. Composites, which have fibers in a resin matrix, seem to be the right candidate to be treated with this technology.

This work presented the results of two preliminary experimental campaigns to treat EoL wind blades. The first one was dedicated to the removal of undesired materials, while the objective of the second one was to investigate the potentiality of exploiting the presence of a grid to better select the target material. The results seem promising in both directions. The first experimental campaign showed that the polyurethane layer, which, if present in the recycled material, can reduce its reuse possibilities, can be completely liberated also after a few discharges, obtaining a purer material to be sent to subsequent processes. The second one suggested the improvement in obtaining material with specific target dimensions, in particular using a sieve inlay, with respect to shredding, fostering the demand-driven approach (increasing in addition possible options to treat the material).

Following the obtained results, future research will consider a full factorial design of experiments, including all the controllable parameters, with different levels. This will allow to better understand the effect of this technology on composite materials and to find the optimal parameters considering both the target characteristics of the material in output and the cost to obtain it. Also, in addition to the characterization of the material in terms of dimensions and morphology, the analysis with a Scanning Electron Microscope has to be done to see the cleanliness level of the fibers. Finally, the development and implementation of a model, physics or AI-based, of the machine can be interesting, in particular for possible future exploitation to optimize and control the process itself.

References

[1] European Environment Agency, Share of energy consumption from renewable sources in Europe, www.eea.europa.eu/ims/share-of-energy-consumption-from, Last accessed 06/03/2023

[2] Komusanac, I.; Brindley, G.; Fraile, D.; Ramirez, L. Wind Energy in Europe-2021 Statistics and the Outlook for 2022-2026; WindEurope: Brussels, Belgium, 2022; p. 37.

[3] Directive (EU) 2018/2001 of the European Parliament and of the Council of 11 December 2018 on the promotion of the use of energy from renewable sources, https://eur-lex.europa.eu/legal-content/EN/TXT/?uri=uriserv:OJ.L_.2018.328.01.0082.01.ENG&toc=OJ:L:2018:328:TOC, Last accessed 06/06/2023

[4] Beauson, J., & Brøndsted, P. (2016). Wind turbine blades: an end of life perspective. MARE-WINT: New Materials and Reliability in Offshore Wind Turbine Technology, 421-432. https://doi.org/10.1007/978-3-319-39095-6_23

[5] Oliveux, G., Dandy, L. O., & Leeke, G. A. (2015). Current status of recycling of fibre reinforced polymers: Review of technologies, reuse and resulting properties. Prog. in mat. Sc., 72, 61-99. https://doi.org/10.1016/j.pmatsci.2015.01.004

[6] Diani, M., Picone, N., & Colledani, M. (2023). Smart Composite Mechanical Demanufacturing Processes. In Systemic Circular Economy Solutions for Fiber Reinforced Composites (pp. 61-80). Cham: Springer International Publishing. https://doi.org/10.1007/978-3-031-22352-5_4

[7] García-Arrieta, S., Sarlin, E., Calle, A. D. L., Dimiccoli, A., Saviano, L., & Elizetxea, C. (2023). Thermal Demanufacturing Processes for Long Fibers Recovery. In Systemic Circular Economy Solutions for Fiber Reinforced Composites (pp. 81-97). Cham: Springer International Publishing. https://doi.org/10.1007/978-3-031-22352-5_5

[8] Oliveux, G., Bailleul, J. L., Gillet, A., Mantaux, O., & Leeke, G. A. (2017). Recovery and reuse of discontinuous carbon fibres by solvolysis: Realignment and properties of remanufactured materials. Composites Science and Technology, 139, 99-108. https://doi.org/10.1016/j.compscitech.2016.11.001

[9] Colledani, M., Turri, S., & Diani, M. (2023). The FiberEUse Demand-Driven, Cross-Sectorial, Circular Economy Approach. In Systemic Circular Economy Solutions for Fiber Reinforced Composites (pp. 17-35). Cham: Springer International Publishing. https://doi.org/10.1007/978-3-031-22352-5_2

[10] Bluhm, H., Frey, W., Giese, H., Hoppe, P., Schultheiss, C., & Strassner, R. (2000). Application of pulsed HV discharges to material fragmentation and recycling. IEEE Transactions on Dielectrics and Electrical Insulation, 7(5), 625-636. https://doi.org/10.1109/94.879358

[11] Ménard, Y., Bru, K., Touzé, S., Lemoign, A., Poirier, J. E., Ruffie, G., ... & von Der Weid, F. (2013). Innovative process routes for a high-quality concrete recycling. Waste management, 33(6), 1561-1565. https://doi.org/10.1016/j.wasman.2013.02.006

[12] Bru, K., Touzé, S., Auger, P., Dobrusky, S., Tierrie, J., & Parvaz, D. B. (2018). Investigation of lab and pilot scale electric-pulse fragmentation systems for the recycling of ultra-high performance fibre-reinforced concrete. Minerals engineering, 128, 187-194. https://doi.org/10.1016/j.mineng.2018.08.040

[13] Song, B. P., Zhang, M. Y., Fan, Y., Jiang, L., Kang, J., Gou, T. T., ... & Zhou, X. (2020). Recycling experimental investigation on end of life photovoltaic panels by application of high voltage fragmentation. Waste Management, 101, 180-187. https://doi.org/10.1016/j.wasman.2019.10.015

[14] Zherlitsyn, A. A., Alexeenko, V. M., Kumpyak, E. V., & Kondratiev, S. S. (2022). Fragmentation of printed circuit boards by sub-microsecond and microsecond high-voltage pulses. Minerals Engineering, 176, 107340. https://doi.org/10.1016/j.mineng.2021.107340

[15] Song, B. P., Zhang, M. Y., Fan, Y., Jiang, L., Kang, J., Gou, T. T., ... & Zhou, X. (2020). End-of-life management of bifacial solar panels using high-voltage fragmentation as pretreatment approach. Journal of Cleaner Production, 276, 124212. https://doi.org/10.1016/j.jclepro.2020.124212

[16] Leißner, T., Hamann, D., Wuschke, L., Jäckel, H. G., & Peuker, U. A. (2018). High voltage fragmentation of composites from secondary raw materials-Potential and limitations. Waste management, 74, 123-134. https://doi.org/10.1016/j.wasman.2017.12.031

[17] Roux, M. A. X. I. M. E., Eguemann, N., Giger, L. I. A. N., & Dransfeld, C. L. E. M. E. N. S. (2013, March). High performance thermoplastic composite processing and recycling: from cradle to cradle. In Proceedings of the SAMPE 34th International Technical Conference, Paris, France (pp. 11-12).

[18] Rouholamin, D., Shyng, Y. T., Savage, L., & Ghita, O. (2014, June). A comparative study into mechanical performance of glass fibres recovered through mechanical grinding and high voltage pulse power fragmentation. In Proceedings of the ECCM16-16th European Conference on Composite Materials, Seville, Spain (pp. 22-26).

[19] Mativenga, P. T., Shuaib, N. A., Howarth, J., Pestalozzi, F., & Woidasky, J. (2016). High voltage fragmentation and mechanical recycling of glass fibre thermoset composite. CIRP annals, 65(1), 45-48. https://doi.org/10.1016/j.cirp.2016.04.107

[20] SELFRAG CFRP (High Voltage Pulse Fragmentation Technology to recycle fibre-reinforced composites), Final Report Summary, https://cordis.europa.eu/project/id/323454/reporting, last accessed 06/03/2023.

Process and system simulation, optimization and digital manufacturing

Italian Manufacturing Association Conference - XVI AITeM
Materials Research Proceedings 35 (2023) 276-285

Materials Research Forum LLC
https://doi.org/10.21741/9781644902714-33

The ranking-aggregation problem in manufacturing: potential, pitfalls, and good practices

Fiorenzo Franceschini[1,a], Domenico Augusto Maisano[1,b] * and
Luca Mastrogiacomo[1,c]

[1] Dept. of Management and Production Engineering (DIGEP), Politecnico di Torino, Italy

[a] fiorenzo.franceschini@polito.it, [b] domenico.maisano@polito.it, [c] luca.mastrogiacomo@polito.it

Keywords: Decision Making, Performance Indicators, Ranking-Aggregation Problem

Abstract. A number of *experts*, who individually rank a set of *objects* based on a certain attribute, and the need to aggregate the resulting (subjective) *rankings* into a *collective judgement*: these are the "ingredients" of the ranking-aggregation problem, which is typical of *social choice*, *psychometrics* and *economics*. This paper shows that the problem has many interesting applications even in *manufacturing* and must be approached with care, in order to avoid misleading results. Through a real-world case study concerning cobot-assisted manual (dis)assembly, the paper illustrates (i) a methodology to tackle the problem in a practical and effective way and (ii) various useful tools (e.g., for estimating the degree of *concordance* among experts, the *consistency* and *robustness* of collective judgment, etc.). The article is addressed to scientists and practitioners in the manufacturing field.

Introduction

Ranking aggregation is an ancient problem with three characteristic elements: (i) a set of *objects* to be prioritised according to a certain *subjective attribute*, (ii) a set of *experts* (equally important or with a hierarchy of importance), who formulate *preference rankings* of the objects of interest, and (iii) a *collective judgement* concerning the objects, resulting from the aggregation of expert rankings through a suitable *aggregation technique* [1-3].

Due to the great generality, disciplinary transversality, and multiplicity of potential applications, the ranking-aggregation problem is of interest to many scientific disciplines and operational contexts, including *manufacturing* [1, 3-6]. Some of the many possible manufacturing applications are:

*   *Conceptual design*, regarding the opinions of different designers about alternative design concepts, from the perspective of a specific attribute [7];
*   *Production management*, regarding the selection of the most appropriate production system on the basis of productivity, flexibility or another performance attribute [8-9];
*   *Quality control*, regarding the prioritization of defects on manufactured parts, aggregating (subjective) expert judgments by visual inspection [10].

The analyst's attention is often directed to the aggregation technique, which can be interpreted as a "black box" transforming *input* data (i.e., experts' rankings and importance hierarchy) into *output* data (i.e., collective judgement and related data) [6]. However, this may lead to overlooking other important aspects that characterize the ranking-aggregation problem, e.g., preliminary assessment of the degree of *concordance* among experts, verification of the *consistency* and *robustness* of output data, etc. The above considerations can turn into the research question: "*What methodological approach should be adopted to address the problem of interest with full awareness?*". Despite the variety of applications to specific cases of interest to individual authors,

Italian Manufacturing Association Conference - XVI AITeM   Materials Research Forum LLC
Materials Research Proceedings 35 (2023) 276-285   https://doi.org/10.21741/9781644902714-33

the scientific literature in the manufacturing field lacks general guidelines and a collection of good practices for addressing the ranking-aggregation problem.

Aimed at scientists and manufacturing professionals, this work is intended to increase their awareness of the complexity of the ranking-aggregation problem, while providing a set of useful tools to tackle it in a practical and effective manner. Following a pedagogical approach, the description is accompanied by a case-study application concerning cobot-assisted manual (dis)assembly.

Case study

A company reconditions different types of automotive components, mainly starters and alternators. Because of the wide variety of components and the complexity of (dis)assembly and repair operations, the company has been supporting human operators with *collaborative robots*, or simply *cobots*. Cobots are useful for assisting operators in manual operations that require great precision, dexterity and strength [11]. They are extremely versatile for multiple tasks, such as (i) picking up, clamping, handing the tools and parts to be machined/assembled, (ii) supporting dimensional inspection, online quality control, etc., and (iii) guiding less experienced operators, like virtual tutors.

The current market includes a relatively wide range of cobot models, which could be adapted to the operational context of interest. The company management decided to identify the most appropriate cobot model depending on *programming practicality*; in fact, this aspect is crucial in making task preparation faster and easier, while reducing the level of technical skills required by operators [11]. Having previously selected five cobot models from those at the cutting edge of the market, the company relies on the evaluation of a panel of eight experts, including technicians, engineers and external consultants with relatively in-depth and complementary expertise.

Methodology

The flowchart in Fig. 1 summarizes the proposed methodology, which can be divided into three operational phases, illustrated in the following subsections. The multiple feedback loops denote the iterative nature of the proposed procedure, which includes several intermediate verifications, with possible in-progress corrections and adjustments.

Problem definition

First, the specific problem and its characteristics should be identified clearly and unambiguously.

Based on the above case study, a specific ranking-aggregation problem can be formulated: the $n = 5$ *objects* (o_1 to o_5) are the cobot models that will be evaluated in terms of programming practicality, i.e., the *attribute* of interest, which is inherently *subjective*. The $m = 8$ *experts* are technicians (e_1 to e_8), engineers and external consultants who formulate their individual preference rankings of the cobot models.

In selecting experts, (at least) two aspects must be taken into account:

1. The greater the number of experts formulating their individual rankings, the higher the statistical relevance of the problem output [12, 13]. Pragmatically, it would be desirable for m to be no less than 5-6, in order for the results of the study to be relevant [6].
2. It may sometimes be appropriate to have a hierarchy of importance of experts, for instance by discriminating those with greater technical expertise. This hierarchy can be constructed in different ways, typically by associating each expert with a *weight* or defining an *importance ranking* [12]. For simplicity, in the case study all experts are considered as equally important.

Next, the type of expert rankings can be determined depending on several factors, such as the *goal* of the problem (e.g., identifying the best/worst object(s), drawing up a complete ranking, etc.), the *data-collection* strategy (e.g., through focus groups, personal telephone/street interviews, online forms, etc.), etc.

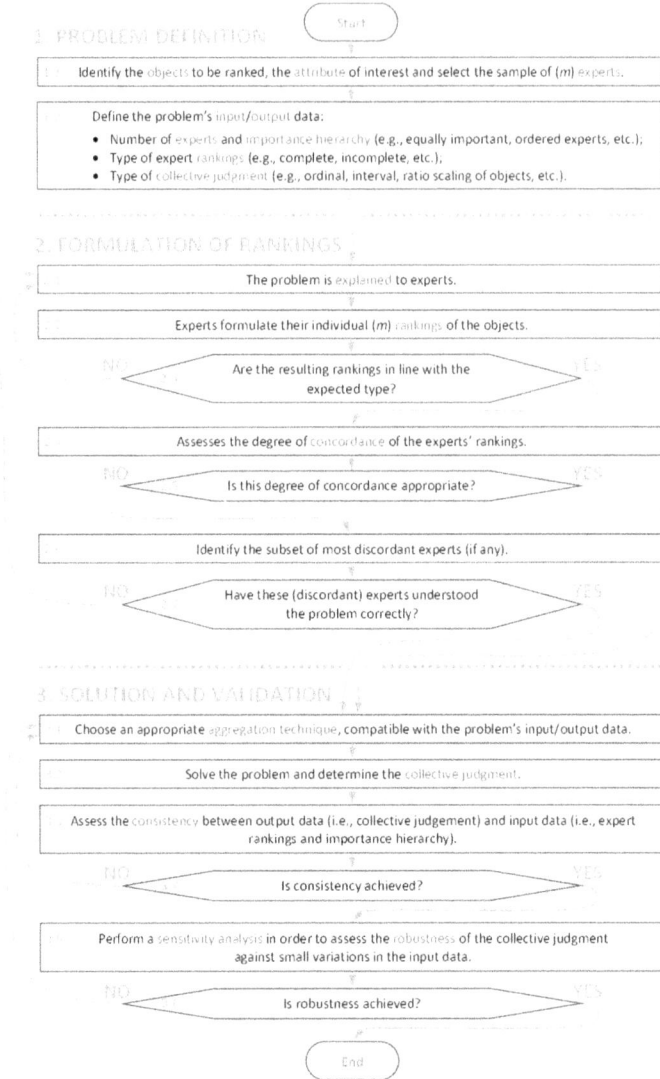

Fig. 1. Flow chart summarizing the proposed operational methodology.

Complete rankings – i.e., rankings in which experts order all objects by linking them with *strict preference* ("$o_i \succ o_j$") or *indifference* relationships ("$o_i \sim o_j$") – represent a classic scenario, although their formulation requires some effort, especially if the number of objects is large [12]. On the other hand, *incomplete* rankings are more "digestible" for experts, because they can take into account possible hesitations or doubts. E.g., incomplete are those rankings in which only a small number of top or bottom objects are included (e.g., the three most/least preferred), or in which the expert decides to omit an object from his/her ranking (e.g., since he/she is not familiar with), or even rankings with *incomparability* relationships between objects ("$o_i \parallel o_j$") [6]. Given the relatively small number of objects, in the present case experts formulate complete rankings of all five objects.

Subsequently, the type of collective judgment has to be defined depending on the properties that are "desirable" for the specific problem; there is a wide range of possibilities: *rankings*, *scalings* on different scale types (e.g., *interval*, *ratio*), *clusterings*, *scorings*, or collective judgments designating only the winner/loser object, etc. [6]. For the sake of simplicity, in the case study the expected collective judgment is represented by a complete ranking.

Formulation of rankings
This stage begins with a detailed explanation of the problem to experts, who need to understand exactly which objects are to be evaluated, the attribute against which the evaluation is to be made, and how to formulate individual rankings. As seen before, in the case study each expert is required to formulate a complete ranking of all objects; Fig. 2(a) reports the resulting expert rankings, which include relationships of strict preference ("$o_i \succ o_j$") and indifference ("$o_i \sim o_j$") between objects. At this stage, it must be ensured that the expert rankings are formulated consistently with the expected type; if necessary, the formulation must be corrected/revised (see feedback loop from block 2.3 in Fig. 1).

Evaluating the *concordance* among expert rankings is a preliminary check of the plausibility of input data, which is useful to prevent difficulties, such as excessive heterogeneity in the selection of experts, poor understanding of the problem, errors in the formulation of rankings, or other potential obstacles to achieving consensus. The scientific literature includes various statistical indicators, which can be used depending on the problem characteristics [12, 14]. Since the present case is characterized by complete expert rankings with equally-important experts, the Kendall's W and Spearman's ρ can be used [6].

W, known as *coefficient of concordance*, is a *multivariate* statistic that applies at the level of expert rankings and is related to the dispersion of the ranks associated with each object [6, 15]. This measure belongs to [0, 1], with 1 indicating perfect concordance and 0 indicating independence [12]. Returning to the case study, each ranking can be translated into a set of ranks – that is, permutations of the integers {1, 2, 3, 4, 5} – which are then organized into a so-called *rank table*, i.e., a bidirectional matrix of size $m \times n$, with row and column labels designating experts and objects (see Fig. 2(b)). In the case of *tied* objects (i.e., pairs of objects with *indifference* relationships, e.g., "$o_i \sim o_j$"), we conventionally use the *average ranks* that each set of bound objects would occupy if a preference could be expressed [12]; for example, in a ranking where objects o_1 and o_3 are tied for 3rd and 4th place (e.g., see the ranking by e_6 in Fig. 2(a)), the average rank of $(3+4)/2 = 3.5$ would be assigned to both.

W is defined as:

$$W = \frac{\sum_{j=1}^{n}(R_j - \bar{R})^2}{[m^2 \cdot n \cdot (n^2-1) - m \cdot \sum_{i=1}^{m} T_i]/12},\tag{1}$$

being
n the number of objects;
m the number of experts;
R_j the column total related to the j-th column of the rank table;
$\bar{R} = m \cdot (n+1)/2$ the average column total (i.e., 24 in the present case);
$T_i = \sum_{k=1}^{g_i}(t_k^3 - t_k)$ a correction factor for ties, in which t_k is the number of tied ranks in the k-th group of tied ranks (where a group is a set of values having constant tied rank) and g_i is the number of groups of ties in the set of ranks (ranging from 1 to n) for expert i.

Italian Manufacturing Association Conference - XVI AITeM
Materials Research Proceedings 35 (2023) 276-285
Materials Research Forum LLC
https://doi.org/10.21741/9781644902714-33

	(a) Rankings						(b) Rank table					
						o_1	o_2	o_3	o_4	o_5	Row totals	T_i
e_1	$o_3 \succ (o_1 \sim o_5) \succ o_2 \succ o_4$					2.5	4.0	1.0	5.0	2.5	15	6
e_2	$o_5 \succ o_1 \succ (o_2 \sim o_3 \sim o_4)$					2.0	4.0	4.0	4.0	1.0	15	24
e_3	$(o_1 \sim o_3) \succ o_2 \succ (o_4 \sim o_5)$					1.5	3.0	1.5	4.5	4.5	15	12
e_4	$o_1 \succ o_5 \succ o_3 \succ o_4 \succ o_2$					1.0	5.0	3.0	4.0	2.0	15	-
e_5	$o_4 \succ (o_1 \sim o_2 \sim o_5) \succ o_3$					3.0	3.0	5.0	1.0	3.0	15	24
e_6	$o_5 \succ o_4 \succ (o_1 \sim o_3) \succ o_2$					3.5	5.0	3.5	2.0	1.0	15	6
e_7	$o_3 \succ o_5 \succ o_4 \succ o_2 \succ o_1$					5.0	4.0	1.0	3.0	2.0	15	-
e_8	$(o_3 \sim o_5) \succ (o_1 \sim o_2) \succ o_4$					3.5	3.5	1.5	5.0	1.5	15	12
	Col. totals (R_j)		22.0	31.5	21.5	28.5	17.5				120	

Fig. 2. (a) Complete rankings of n = 5 objects, formulated by m = 8 experts; (b) corresponding rank table. T_i is a correction factor for ties (cf. Eq. 1).

With reference to the case study, it is obtained $W = 23.1\%$, denoting a relatively low level of concordance. To further investigate the reasons for this low inter-expert concordance, the *bivariate* perspective of Spearman's correlation coefficient (ρ) related to each possible pair of rankings can be considered. Tab. 1 contains the ρ coefficients between all the possible pairs of expert rankings under consideration [15].

Tab. 1. Spearman's ρ correlation table for the expert rankings in Fig. 2(a). The most pronounced negative correlations (i.e., $\rho < -0.4$) are bolded.

Ranking	e_1	e_2	e_3	e_4	e_5	e_6	e_7	e_8
e_1	1							
e_2	0.287	1						
e_3	0.649	-0.177	1					
e_4	0.564	0.783	0.316	1				
e_5	**-0.918**	0.000	**-0.707**	-0.224	1			
e_6	-0.026	0.574	**-0.649**	0.410	0.344	1		
e_7	0.462	-0.112	-0.158	-0.100	**-0.447**	0.410	1	
e_8	0.865	0.412	0.250	0.369	**-0.825**	0.162	0.632	1

Rather pronounced negative correlations (i.e., $\rho < -0.4$) between certain pairs of expert rankings stand out. Curiously, they (almost) always involve the ranking by e_5, denoting a sort of "countertrend" with respect to the other rankings. Upon brief investigation, it turns out that e_5 expert misunderstood the ranking construction, formulating it in the sense of reverse preference; therefore, the correct ranking should be "$o_3 \succ (o_1 \sim o_2 \sim o_5) \succ o_4$" instead of "$o_4 \succ (o_1 \sim o_2 \sim o_5) \succ o_3$" (see feedback loop from block 2.7 in Fig. 1). After this correction, the W value is significantly higher than before (i.e., $W = 39.6\%$ versus 23.1%). Simultaneously, the relatively large negative ρ values for e_5 are "reabsorbed".

As exemplified, the concordance analysis can be useful in pointing out possible anomalies and "pitfalls" in the formulation of expert rankings [6].

Solution and validation

At this point, it is needed to solve the ranking-aggregation problem utilizing an appropriate aggregation technique and, subsequently, verifying the plausibility of the resulting output. Unfortunately, presenting an exhaustive overview of the state-of-art techniques would require an encyclopaedic analysis. Far from this ambition, Tab. 2 simply recalls some possible aspects to be taken into account while selecting the aggregation technique [6]. For an overview of the aggregation techniques in the "mare magnum" of the scientific literature, we refer the reader to relevant surveys and extensive reviews [2, 6, 16].

Tab. 2. Aspects to consider when selecting a ranking-aggregation technique [6].

(a) Input-data characteristics	(b) Aggregation mechanism	(c) Output-data characteristics
• Problem size: - Number of objects (n); - Number of expert rankings (m). • Type of expert rankings • Type of expert hierarchy	• Rule-based. • Optimization-based; • Distribution-based.	• Designation of a unique winner/loser; • Complete/incomplete *ranking*; • *Classification* in categories; • Collective *scoring* of objects; • Collective *scaling* of objects. ...

A relatively simple aggregation technique is applied for the problem of interest: the so-called *Borda Count* (BC), according to which, for each expert ranking, the first object accumulates one point, the second two points, and so on [3, 17]. The collective score of one object can be calculated by cumulating the scores related to each ranking.

The application of the BC technique to the expert rankings (after the correction of the ranking by e_5) leads to the following collective scoring: $o_1 = 22.0$, $o_2 = 31.5$, $o_3 = 16.5$, $o_4 = 32.5$, $o_5 = 17.5$ (cf. also Tab. 5(b-i)), from which the collective ranking $o_3 > o_5 > o_1 > o_2 > o_4$ is deduced.

Every aggregation technique provides a collective judgment; but how does one know whether it is plausible? Certainly, the rationale of the aggregation technique represents a conceptual guarantee that it is capable of producing reasonable results. However, the aggregation technique that most consistently reflects expert rankings cannot be assessed *ex ante*, but only *ex post* and on a case-by-case basis [4, 18].

Studies have focused on the concept of *consistency of the collective judgment with respect to input data*, defined as "*the ability of a collective judgment to reflect the rankings of experts, while taking the importance hierarchy into account*" (i.e., giving priority to the more important experts) [6].

Among the available tools to assess the degree of consistency of the solution to a certain ranking-aggregation problem, p-indicators are very versatile, as they can be adapted to a variety of contexts, such as those in which expert rankings are (i) not necessarily complete, (ii) equally important, or (iii) characterized by an importance hierarchy [6]. In general, p-indicators can be divided into two families:

- p_j, indicators of *local consistency*, which are based on the comparison of each j-th expert's ranking with the collective judgement. A preliminary operation for determining p_j is constructing "*a paired-comparison table*" in which each ranking (i.e., those from experts and that one deduced from the collective judgment) is transformed into sets of paired-comparison relationships (see symbols ">" and "~" in Tab. 3(a)). Next, a "*consistency table*" – which turns the paired-comparison relationships of each expert into scores, according to the following scoring system is constructed:

  1. *Full consistency*, i.e., identical relationship of strict preference (">") or indifference ("~") $\Rightarrow$ score 1;
  2. *Weak consistency*, i.e., consistency with respect to a weak preference relationship only (">" or ~" and "<" or ~", i.e., strict preference or indifference); e.g., when comparing the relationship $o_1 > o_2$ with $o_1 \sim o_2 \Rightarrow$ score 0.5;
  3. *Inconsistency* (with respect to both strict and weak preference relationships); e.g., when comparing the relationship $o_1 > o_2$ with $o_2 > o_1 \Rightarrow$ score 0.

The conventional assignment of 0.5 points in the case of *weak consistency* is justified by the fact that this is the intermediate case between that of *full consistency* (with score 1) and that of *inconsistency* (with score 0) [6]. The consistency table also reports the sum of the scores (x_j)

obtained by each j-th expert ranking. Tab. 3(b) exemplifies the consistency table related to the case study of interest.

Next, for each j-th expert, the portion of "consistent" paired-comparisons can be calculated as:

$$p_j = \frac{x_j}{\binom{n}{2}} = \frac{x_j}{10}, \tag{2}$$

being:

    x_j the total score related to the j-th expert;

    $\binom{n}{2} = \frac{n \cdot (n-1)}{2}$ the overall number of paired comparisons (i.e., 10 in the present case).

- p, i.e., indicator of *global consistency*. In the case of equally-important experts, the p_j values are aggregated through the arithmetic average [6]:

$$p = \frac{1}{m} \cdot \sum_{j=1}^{m} p_j, p \in [0,1]. \tag{3}$$

In this specific case, the aggregation technique results into $p = 73.8\%$ (see Tab. 3(c)), denoting a relatively good consistency [6, 10].

Tab. 3. (a) Paired-comparison table, (b) consistency table, and (c) p-indicators related to the BC technique.

(a) Paired-comparison table										(b) Consistency table							
Paired			Experts						Collective					Scores			
comparison	e_1	e_2	e_3	e_4	e_5	e_6	e_7	e_8	judgment	e_1	e_2	e_3	e_4	e_5	e_6	e_7	e_8
1 o_1, o_2	>	>	>	>	~	>	<	~	>	1	1	1	1	0.5	1	0	0.5
2 o_1, o_3	<	>	~	>	<	~	<	<	<	1	0	0.5	0	1	0.5	1	1
3 o_1, o_4	>	>	>	>	>	<	<	>	>	1	1	1	1	1	0	0	1
4 o_1, o_5	~	<	>	>	~	<	<	<	<	0.5	1	0	0	0.5	1	1	1
5 o_2, o_3	<	~	<	<	<	<	<	<	<	1	0.5	1	1	1	1	1	1
6 o_2, o_4	>	~	>	<	>	<	<	>	>	1	0.5	1	0	1	0	0	1
7 o_2, o_5	<	>	>	<	~	<	<	<	<	1	1	0	1	0.5	1	1	1
8 o_3, o_4	>	~	>	>	>	<	>	>	>	1	0.5	1	1	1	0	1	1
9 o_3, o_5	>	<	>	>	>	<	>	~	>	1	0	1	0	1	0	1	0.5
10 o_4, o_5	<	<	~	<	<	<	<	<	<	1	1	0.5	1	1	1	1	1

| | | | | | | | | | x_j | 9.5 | 6.5 | 7 | 6 | 8.5 | 5.5 | 7 | 9 |

(c) *p*-indicators p_j **95%**   **65%**   **70%**   **60%**   **85%**   **55%**   **70%**   **90%** $\Rightarrow p = 73.8\%$

The formulation of rankings is often affected by inherent variability, which can "propagate" onto the variability of the output [19]. In general, it may be useful to perform a *sensitivity analysis* to assess the robustness of the solution against small variations in the input data [19]. An example of sensitivity analysis follows.

Tab. 4 contains three sets of expert rankings: the initial one and two additional ones, obtained by applying small distortions (e.g., some "rank-reversal") to the initial one. For each set, the collective scoring/ranking was determined by applying the BC aggregation technique (see results in Tab. 5). Next, the average dispersion in the rank position of individual objects was used as a proxy for the robustness of the resulting collective rankings (see Tab. 5(c)). In this specific case, BC seems to provide a somewhat robust result (i.e., mean standard deviation of 0.44). Thus, no revision of the aggregation technique adopted is necessary (cf., feedback loop from block 3.6 of Fig. 1).

Discussion and general remarks

This paper focused on the ranking-aggregation problem, due to the variety of potential applications in manufacturing. Through a pedagogical approach based on a case study, the paper illustrated a sequential and iterative operational methodology to tackle the problem of interest at multiple levels:

Tab. 4. Set of rankings used for sensitivity analysis.

Experts	(i) Initial set of rankings	(ii) 1st additional set	(iii) 2nd additional set
e_1	$o_3 \succ (o_1 \sim o_5) \succ o_2 \succ o_4$	$(o_3 \sim o_1) \succ o_5 \succ (o_2 \sim o_4)$	$o_5 \succ o_3 \succ o_1 \succ o_4 \succ o_2$
e_2	$o_5 \succ o_1 \succ (o_2 \sim o_3 \sim o_4)$	$o_1 \succ o_5 \succ (o_2 \sim o_3) \succ o_4$	$(o_5 \sim o_1 \sim o_2) \succ o_4 \succ o_3$
e_3	$(o_1 \sim o_3) \succ o_2 \succ (o_4 \sim o_5)$	$(o_1 \sim o_3) \succ o_5 \succ (o_2 \sim o_4)$	$(o_1 \sim o_3 \sim o_2) \succ o_4 \succ o_5$
e_4	$o_1 \succ o_5 \succ o_3 \succ o_4 \succ o_2$	$o_1 \succ (o_5 \sim o_3) \succ o_4 \succ o_2$	$o_5 \succ o_1 \succ (o_3 \sim o_4 \sim o_2)$
e_5	$o_3 \succ (o_1 \sim o_2 \sim o_5) \succ o_4$	$o_3 \succ (o_1 \sim o_2) \succ o_5 \succ o_4$	$o_3 \succ o_2 \succ o_1 \succ (o_5 \sim o_4)$
e_6	$o_5 \succ o_4 \succ (o_1 \sim o_3) \succ o_2$	$o_5 \succ o_4 \succ (o_1 \sim o_2) \succ o_3$	$o_4 \succ o_5 \succ (o_1 \sim o_3 \sim o_2)$
e_7	$o_3 \succ o_5 \succ o_4 \succ o_2 \succ o_1$	$o_5 \succ o_3 \succ o_2 \succ o_4 \succ o_1$	$o_3 \succ o_4 \succ o_5 \succ (o_2 \sim o_1)$
e_8	$(o_3 \sim o_5) \succ (o_1 \sim o_2) \succ o_4$	$(o_3 \sim o_5) \succ o_1 \succ o_2 \succ o_4$	$o_1 \succ (o_3 \sim o_5) \succ (o_2 \sim o_4)$

Tab. 5. Rank tables and collective scorings/rankings resulting from sensitivity analysis.

	(a) Rank table								(b) Collect. scoring (rank)		(c) Rank dispersion
	e_1	e_2	e_3	e_4	e_5	e_6	e_7	e_8			

(i) Initial set of rankings

	e_1	e_2	e_3	e_4	e_5	e_6	e_7	e_8	Collect. scoring (rank)	
o_1	2.5	2	1.5	1	3	3.5	5	3.5	22.0	(3.0)
o_2	4	4	3	5	3	5	4	3.5	31.5	(4.0)
o_3	1	4	1.5	3	1	3.5	1	1.5	16.5	(1.0)
o_4	5	4	4.5	4	5	2	3	5	32.5	(5.0)
o_5	2.5	1	4.5	2	3	1	2	1.5	17.5	(2.0)

Collective ranking: $o_3 \succ o_5 \succ o_1 \succ o_2 \succ o_4$

(ii) 1st additional set

	e_1	e_2	e_3	e_4	e_5	e_6	e_7	e_8	Collect. scoring (rank)		Object	St. dev.
o_1	1.5	1	1.5	1	2.5	3.5	5	3	19.0	(3.0)	o_1	0.3
o_2	4.5	3.5	4.5	5	2.5	5	3	4	30.5	(4.0)	o_2	0.3
o_3	1.5	3.5	1.5	2.5	1	5	2	1.5	18.5	(2.0)	o_3	0.8
o_4	4.5	5	4.5	4	5	2	4	5	34.0	(5.0)	o_4	0.3
o_5	3	2	3	2.5	4	1	1	1.5	18.0	(1.0)	o_5	0.6

Collective ranking: $o_5 \succ o_3 \succ o_1 \succ o_2 \succ o_4$ Mean st.dev. **0.44**

(iii) 2nd additional set

	e_1	e_2	e_3	e_4	e_5	e_6	e_7	e_8	Collect. scoring (rank)	
o_1	3	2	2	2	3	4	4.5	1	21.5	(2.5)
o_2	5	2	2	4	2	4.5	4.5	4.5	28.0	(4.5)
o_3	2	5	2	4	1	4	1	2.5	21.5	(2.5)
o_4	4	4	4	4	4.5	1	2	4.5	28.0	(4.5)
o_5	1	2	5	1	4.5	2	3	2.5	21.0	(1.0)

Collective ranking: $o_5 \succ (o_1 \sim o_3) \succ (o_2 \sim o_4)$

- Checking the plausibility of expert rankings in terms of *concordance*, through *multivariate* and *bivariate* statistical indicators;
- Guiding the aggregation-technique selection, depending on the desired types of input and output data;
- Evaluating the *consistency* and *robustness* of the resulting collective judgment.

Interestingly, the application of the aggregation technique is only one of several steps in the proposed methodology. This study gives greater awareness of the complexity of the ranking-aggregation problem, providing some practical tools for dealing with it in a structured and effective way. The results of this study may be useful for scientists and practitioners in manufacturing, who are facing various kinds of decision-making problems that can be linked to that of ranking aggregation. Although the case study focused on a few specific practical tools (e.g. ρ, W, and p-indicators), the proposed methodology is open to the use of other similar tools.

Italian Manufacturing Association Conference - XVI AITeM   Materials Research Forum LLC
Materials Research Proceedings 35 (2023) 276-285   https://doi.org/10.21741/9781644902714-33

Regarding the future, it is planned to define an in-depth taxonomy of the aggregation techniques and analytical tools, so as to facilitate their selection for a specific problem of interest.

References

[1] Spohn, W. (2009). A survey of ranking theory. In Degrees of belief (pp. 185-228). Springer, Dordrecht. https://doi.org/10.1007/978-1-4020-9198-8_8

[2] Reich, Y. (2010). My method is better!, Research in engineering design 21(3): 137-142. https://doi.org/10.1007/s00163-010-0092-3

[3] Saari D.G. (2011). Decision and elections, Cambridge: Cambridge University Press.

[4] Arrow K.J. (2012). Social choice and individual values, 3rd edn. Yale University Press, New Haven.

[5] Köksalan, M., Wallenius, J., Zionts, S. (2013). An early history of multiple criteria decision making. Journal of Multi-Criteria Decision Analysis, 20(1-2): 87-94. https://doi.org/10.1002/mcda.1481

[6] Franceschini, F., Maisano, D., Mastrogiacomo, L. (2022) Rankings and Decisions in Engineering: Conceptual and Practical Insights. International Series in Operations Research & Management Science Series, Vol. 319, Springer International Publishing, Cham (Switzerland), ISSN: 0884-8289. https://doi.org/10.1007/978-3-030-89865-6

[7] Franceschini, F., Maisano, D. (2019). Design decisions: concordance of designers and effects of the Arrow's theorem on the collective preference ranking. Research in Engineering Design, 30(3), 425-434. https://doi.org/10.1007/s00163-019-00313-9

[8] Chatterjee, P., Chakraborty, S. (2014). Flexible manufacturing system selection using preference ranking methods: A comparative study. International Journal of Industrial Engineering Computations, 5(2), 315-338. https://doi.org/10.5267/j.ijiec.2013.10.002

[9] Qin, Y., Qi, Q., Scott, P.J., Jiang, X. (2020). An additive manufacturing process selection approach based on fuzzy Archimedean weighted power Bonferroni aggregation operators. Robotics and Computer-Integrated Manufacturing, 64, 101926. https://doi.org/10.1016/j.rcim.2019.101926

[10] Franceschini, F., Maisano, D. (2018). Classification of objects into quality categories in the presence of hierarchical decision-making agents. Accreditation and Quality Assurance, 23(1), 5-17. https://doi.org/10.1007/s00769-017-1291-7

[11] Gervasi, R., Mastrogiacomo, L., Maisano, D. A., Antonelli, D., Franceschini, F. (2022). A structured methodology to support human-robot collaboration configuration choice. Production Engineering, 16: 435-451. https://doi.org/10.1007/s11740-021-01088-6

[12] Gibbons, J.D., Chakraborti, S. (2010). Nonparametric statistical inference (5th ed.). CRC Press, Boca Raton, ISBN 978-1420077612. https://doi.org/10.1201/9781439896129

[13] Kendall, M. G. (1962). Ranks and measures. Biometrika, 49(1/2), 133-137. https://doi.org/10.1093/biomet/49.1-2.133

[14] Agresti, A. (2010). Analysis of Ordinal Categorical Data (2nd ed.), New York: John Wiley & Sons. https://doi.org/10.1002/9780470594001

[15] Ross, S.M. (2009). Introduction to probability and statistics for engineers and scientists. Academic Press. https://doi.org/10.1016/B978-0-12-370483-2.00006-0

[16] Figueira, J., Greco, S., Ehrgott, M. (2005) Multiple criteria decision analysis: state of the art surveys. Springer, New York. https://doi.org/10.1007/b100605

[17] Borda, J.C. (1781). Mémoire sur les élections au scrutin, Comptes Rendus de l'Académie des Sciences. Translated by Alfred de Grazia as Mathematical derivation of an election system, Isis, 44:42-51. https://doi.org/10.1086/348187

[18] McComb C., Goucher-Lambert K., Cagan J. (2017), Impossible by design? Fairness, strategy and Arrow's impossibility theorem. Design Science, 3:1-26. https://doi.org/10.1017/dsj.2017.1

[19] Saltelli, A., Ratto, M., Tarantola, S., & Campolongo, F. (2006). Sensitivity analysis practices: Strategies for model-based inference. Reliability engineering & system safety, 91(10-11), 1109-1125. https://doi.org/10.1016/j.ress.2005.11.014

Italian Manufacturing Association Conference - XVI AITeM   Materials Research Forum LLC
Materials Research Proceedings 35 (2023) 286-294   https://doi.org/10.21741/9781644902714-34

Digital upgrade of a bandsaw machine through an innovative guidance system based on the digital shadow concept

Federico Scalzo[1,a *], Davide Bortoluzzi[1,b], Giovanni Totis[1,c], Marco Sortino[1,d]

[1]Polytechnic Department of Engineering and Architecture, University of Udine, Via Delle Scienze 206, 33100 Udine, Italy

[a]federico.scalzo@uniud.it, [b]davide.bortoluzzi@uniud.it, [c]giovanni.totis@uniud.it, [d]marco.sortino@uniud.it

Keywords: Man-Machine System, Digital Shadow, Sensors

Abstract. Nowadays, there is an increasing trend towards advanced CNC machine tools having a high level of automation. Nevertheless, manually operated equipment is still playing an important role in many industrial workshops. Operators' experience is still essential in the perspective of increasing productivity, enhancing product quality, reducing manufacturing costs related to tool wear, waste and maintenance. Thus, even manual operations that are apparently less important in terms of product added value may deserve attention and need to be improved according to the principles of the digital transformation era. This paper introduces a structured approach for design, development and implementation of an operator guidance system for a manual bandsaw machine, based on the digital shadow concept and additional feedback sensors. This provides an actual example of how the digital transformation of a small-scale equipment may improve the manufacturing performance and ergonomics as well.

Introduction

Advances in saw technology and blade materials have significantly improved the effectiveness, versatility, and cost-efficiency of sawing. In the past, sawing was regarded as a secondary step in the machining process and saws were mainly utilized for cutting bar stock in preparation for further machining operations. However, thanks to the recent technological advancements, bandsaws are now being used as the primary tool for shaping many metal components. The benefits of using band sawing include minimal material loss (low kerf loss) and process efficiency [1].

Observing modern manufacturing shopfloors, the growing trend towards highly automated advanced CNC machine tools is obvious. Nevertheless, manually operated equipment still holds significant importance in many industrial settings. In such cases, the expertise of operators is still crucial for increasing productivity, improving product quality, and reducing costs associated with tool wear, waste, and maintenance. The application of the principles of the digital transformation era can therefore give a significant contribution, providing support to technical operators, improving the performance of the production process.

The fusion of Cyber-Physical Systems (CPS) and the Internet of Things favored the evolution of smart manufacturing in the context of Industry 4.0. Nowadays at the core of CPS is Digital Twin (DT) technology, which is capable of accurately sensing and reflecting the behavior and real-time state of the production system, thereby enabling the analysis, simulation, prediction, and optimization of processes [2]. In particular, the ability to perform data-driven simulations executed on real-time gathered data, thanks to the availability of adequate hardware and software tools, should be highlighted. Numerous studies have tackled the challenge of providing a comprehensive categorical overview of the existing literature on Digital Twin, as demonstrated by the works of Kritzinger et al. [3] and,Jones et al. [4]. The plethora of review works underscores the growing interest of the research community in this subject matter. Nevertheless, most of the literature

Italian Manufacturing Association Conference - XVI AITeM   Materials Research Forum LLC
Materials Research Proceedings 35 (2023) 286-294   https://doi.org/10.21741/9781644902714-34

focuses mainly on theoretical foundation and conceptual aspects pertaining to DT, while concrete implementation and case studies are still at an early stage. One of the first examples of manufacturing digital twin applied at the shop-floor level was proposed by SIEMENS in 2015 [5]. A Cyber-Physical Production System (CPPS) comprised of four Cyber-Physical Production units was under investigation: a Robotic Cell for handling loading and unloading operations, a Drilling Machine, a Milling Machine, and a Transport System, all were managed by a Manufacturing Execution System (MES). However, the model proposed was merely an illustrative example, and only preliminary work was conducted to implement it. Further implementations of DT can be observed in the aerospace [6], additive manufacturing [7], injection molding [8] and end-of-life [9]. DT was also applied to machining process design. A new method for process optimization driven by a DT for fast reconfiguration of automated manufacturing systems was proposed by Liu et al. [10]. A novel model-based methodology for tactical and strategical decisions in manufacturing systems, applied to a real case study of o a manufacturing company producing axles for the railway sector was proposed in [11]. Digital twins for operation and maintenance are especially beneficial for managing complex equipment such as aircraft, ships, and wind turbines that are susceptible to performance degradation from environmental factors. Some applications of DT in operation and maintenance were developed for prognostic and health management of wind turbines [12], damage accumulation and remaining useful life [13]. DT applications were developed also in the manufacturing sector. An active fixture prototype was developed to detect and mitigate the level of chatter vibrations in general rough-milling operations [14]. A survey of the state of the art and of relevant standards of Digital Twins in Industrial IoT can be found in [15]. Chatter vibrations arising during machining operations are particularly detrimental when appropriate strategies cannot be applied [16], leading to high production cost and extended pre-production runs. In order to optimize the thin-walled part manufacturing process, a Digital Twin-driven framework for thin-walled part manufacturing was proposed in [17]. DT of dynamic mechanical systems could be used for experimental setup design and optimization purposes, since as demonstrated in [18] the influence of external disturbances has to be taken into account to enhance dynamic characterization accuracy.

Since this study involves a one-way flow of data from the physical counterpart to the digital object, the attention is focused on the Digital Shadow characteristics. On the basis of the level of data integration between the physical and digital counterpart, a comprehensive classification of Digital Model (DM), Digital Shadow (DS) and Digital Twin is drawn by Kritzinger et al. [3]. In a manufacturing environment a Digital Shadow requires data harvesting from each machine, product or component. The data flow includes operation and condition data, process data, and other information. The Digital Shadow is continuously linked to the manufacturing system and generates a database. Hence, a real-time, accurate representation of the production system is available for optimization purposes. Data collected can be analyzed, labeled and linked to their appropriate context. The Digital Shadow is a prerequisite for application of methods and models of data analysis and evaluation in a manufacturing environment, and it serves as the foundation for additional applications that utilize the aggregated data for smart manufacturing applications.

As previously mentioned, the literature review has demonstrated that the majority of research papers on digital twin and digital shadow is theoretical, with only a few examining applications or actual case studies. To address this gap, this study introduces a structured approach for design, development, and implementation of an operator guidance system (OGS) for a manual band saw machine. The initial section examines the selection and installation of sensors required for generating the digital shadow, followed by the analysis of the infrastructure created for collecting, analyzing, and visualizing the gathered data. Finally, results of cutting tests performed with and without the operator guidance system are presented, demonstrating that the OGS allows for a quick evaluation and control of the actual process efficiency.

Italian Manufacturing Association Conference - XVI AITeM   Materials Research Forum LLC
Materials Research Proceedings 35 (2023) 286-294   https://doi.org/10.21741/9781644902714-34

This case study provides an actual example of how the digital transformation of a small-scale equipment may be beneficial for ergonomics and manufacturing performance improvement.

Machining efficiency evaluation in band sawing

The cutting process efficiency can be evaluated considering several factors such as tool wear, cutting forces, material removal rate (MRR), chip ratio, chip characteristics and temperature. Surface finish, sound and vibrations can also indicate the efficiency of machining. Nevertheless, none of these indicators are suitable for a quantitative measurement of the machining efficiency when different combinations of machining process, tool and workpiece are under investigation. A more effective way of quantitatively measuring the workpiece machinability is the specific cutting energy u_c [1]. Assuming that the material properties of the blank to be cut are almost constant, the initial machining performance corresponds to the cutting behavior of a brand-new blade, then u_c tends to increase with tooth wear. The wear rate can be determined by the slope of the u_c curve plotted against the blade's life, expressed in terms of the number of workpieces cut. The specific cutting energy is the energy required to remove a specific volume of material. Therefore, it can be calculated as the ratio between the cutting power P_c to the material removal rate MRR, as follows:

$$u_c = P_c/MRR \quad [\text{J/mm}^3] \tag{1}$$

Another useful parameter to characterize the cutting process is the average, uncut chip thickness h. This parameter is the average chip thickness perceived by each cutting edge when it is engaged with the workpiece. Specifically, for a band saw the average, uncut chip thickness can be calculated as follows:

$$h = \frac{v_f p}{1000 \, v_c} \quad [\text{mm/tooth}] \tag{2}$$

where v_f [mm/min] is the instantaneous feed speed perpendicular to the band saw, v_c [m/min] is the cutting speed parallel to the bandsaw and p is the average pitch between subsequent teeth (which are generally not equally spaced along the bandsaw). It has to be highlighted that this parameter depends on the instantaneous feed speed, which does further depend on the feed control strategy. Ideally, the chip thickness should be almost constant – within a given range recommended by the bandsaw manufacturer – to reduce the risk of a rapid tool wear or cutting edge chipping. In the current case (manual band saw), in order to keep the uncut chip thickness constant, the force applied by the operator should be adapted throughout the cutting process.

Materials and experimental setup

The machine on which the operator guidance system based on the digital shadow concept was implemented is a BIANCO 370 M horizontal pivot style manual bandsaw installed at the Laboratory for Advanced Mechatronics (LAMA FVG) of the University of Udine. The machine is powered by a SIEMENS three-phase electric motor. The use of star and delta connections enables the selection of two distinct rotational speeds. The machine is equipped with a bimetal band saw blade with variable raker tooth setting (6/10 teeth/in, alternating right-left-straight teeth). The blade's edge is made of HSS AISI M42 alloy (8% Co) while the body is made of spring steel, ensuring an optimal balance between flexibility and resistance to fatigue, tensile and torsional stresses. The material selected to perform the cutting tests is a 40 mm diameter Al 7075 aluminum alloy rod.

The BIANCO 370 M band saw is not equipped with sensors for process monitoring. In order to analyze the cutting process, some low-cost sensors were chosen and installed. The design choice to limit the cost of the sensors is appropriate to the type of machine being studied. The quantities monitored are the force exerted by the operator who manually controls the feed, the power absorbed by the three-phase electric motor, the cutting speed and the blade angle.

Italian Manufacturing Association Conference - XVI AITeM   Materials Research Forum LLC
Materials Research Proceedings 35 (2023) 286-294   https://doi.org/10.21741/9781644902714-34

The force exerted by the operator is calculated starting from the data acquired from two active HBM 120LY41 120 Ω strain gage elements glued to the actuating lever, positioned where the maximum deflection is expected. The first one mounted in the direction of axial strain on the top side of the lever and the other mounted in the direction of axial strain on the bottom side (bending configuration). The half bridge is connected to a National Instruments (NI) Data Acquisition System (DAQ) NI-9986 bridge adapter, connected in turn with a NI-9218 that allows the analog signal acquisition. Calibration was performed using known masses fixed at the end of the actuation lever. Through this procedure it was possible to correlate the sensors' output in volts with the force applied to the operating lever. Hence, the calibration coefficients were obtained by linear regression.

The power absorbed by the electric motor is measured with a Montronix PS200-DGM power sensor. The amount of current in each phase conductor is measured by means of three Hall effect sensors. Via current and voltage, the PS200-DGM determines the effective power and creates a corresponding analog voltage output signal. Future developments will include replacement with an economy class industrial power sensor.

The cutting speed is calculated starting from the data acquired from an OMRON E2E-S05S12-WC-B1 proximity sensor that picks up the transit of the spokes of the rotating driven pulley. A tailor-made 3D printed pulley inspection cover was designed in order to facilitate the proximity sensor installation. The passage of a spoke triggers the sensor that outputs a "high" logic state. Knowing the geometry of the driven pulley and the number of "high" pulses from the proximity sensor in a certain period of time, it is possible to calculate the rotation speed and therefore the cutting speed.

The blade angle is calculated using a 10 kΩ VISHAY potentiometer mounted close to the rotation axis of the cutting head. To increase the resolution of the potentiometer, the coupling with the axis of rotation was achieved by means of a pair of straight-toothed gear wheels with a 5:1 gear ratio. As the blade angle changes, the potentiometer acts as a voltage divider, outputting a variable voltage proportional to the angle of rotation. Similarly to the case of the strain gage, the calibration was performed through linear regression using the data acquired from the potentiometer and comparing them with those obtained through direct measurements of the pivot angle.

The potentiometer and the proximity sensors are powered by a TRACO POWER TPP 15-109-J 9V stabilized power supply, while the power sensor is powered by a KERT 24V stabilized power supply. The NI-9218 provides the 2V excitation voltage to the half bridge strain gage. The analog voltage outputs of power sensor, proximity sensor and potentiometer are acquired by means of a NI-9215 four channel voltage input module. Fig. 1A shows the upgraded band saw, evidencing the sensors installed to acquire the cutting process data.

The digital signals are acquired, displayed, recorded, and analyzed using a tailor-made system created using MATLAB and NI LabVIEW. In order to determine the volume of chips removed during a specific time frame, two essential data have to be known: the starting blank's geometry and the blade angle.

A MATLAB-based application was developed to calculate the volume of chip removed during the cutting process. This application necessitates the loading of the STL 3D model of the workpiece to be cut. Hence, the blank needs to be positioned in the virtual space using the appropriate keys to perform translations and rotations, see Fig. 2A. The software allows to clearly view the blank and its position relative to the band saw with a three-dimensional, side and frontal views. The virtual cutting plane is indicated by the blue transparent plane with a red outline. Selecting from the top menu the "Slice & Cut" tab it is possible to perform the virtual cutting operation and calculate some process parameters, according to the angular position of the blade (e.g. number of engaged teeth, contact length and cut area), see Fig. 2B.

Italian Manufacturing Association Conference - XVI AITeM
Materials Research Proceedings 35 (2023) 286-294

Materials Research Forum LLC
https://doi.org/10.21741/9781644902714-34

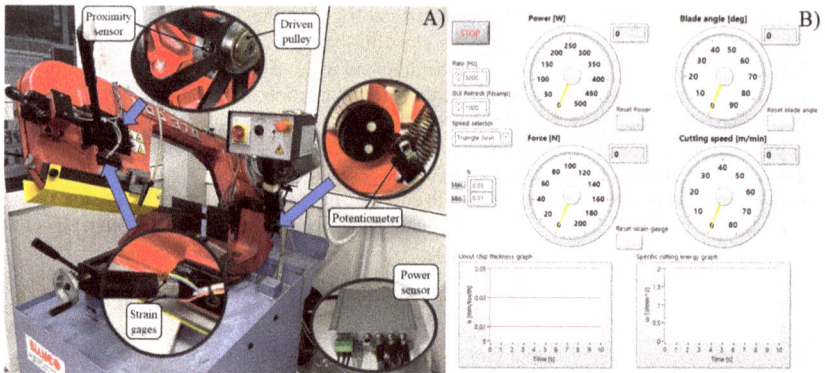

Figure 1 Upgraded band saw and dashboard for data acquisition and visualization (OGS).

The blue outline indicates the edge of the areas involved in the cutting process, while the red circles represent the engaged teeth. The volume of the removed material is easily calculated multiplying the cut area (expressed as a function of the blade angle) by the mean blade thickness. The data table generated by the virtual cutting process is then stored in a spreadsheet, which is subsequently imported into the LabVIEW application to allow real-time u_c and h calculations. The data visualization and control dashboard of the operator guidance system is shown in Fig. 1B. Fig. 3A shows a diagram that summarizes the tasks performed by the MATLAB app. On the left side, the dark blue boxes highlight the virtual 3D environment initialization and the blank's STL file import and positioning. The light blue boxes in the center show the cutting process slicing and simulation stages. These phases allow working teeth number, contact length and cut area estimation. On the right side a spreadsheet containing the calculated parameters is generated (green box), ready for import into NI LabView (yellow box). Similarly, Fig. 3B shows a diagram that summarizes the tasks performed by the LabVIEW VI. The dark blue boxes on the left side highlight the preliminary operations to be performed prior to the main loop start. Local variables are initialized and analog input channels, sample clock and shift registers are configured. Hence, the cutting process data calculated by the MATLAB app are imported by means of the aforementioned spreadsheet. The main loop starts (light blue boxes), analog input data acquisition, filtering and elaboration is performed allowing real-time cutting process data calculation and visualization. Therefore, acquired data are stored in a LabVIEW measurement file (green boxes).

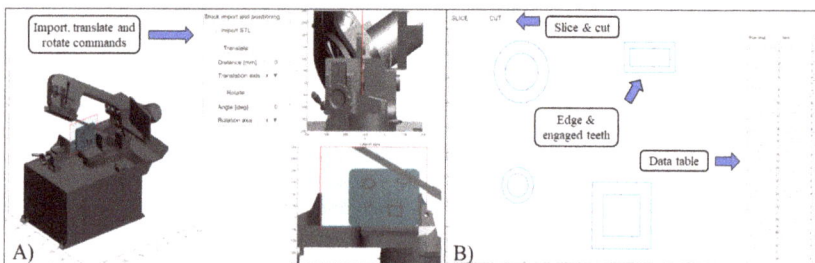

Figure 2 MATLAB application for virtual cut data calculation.

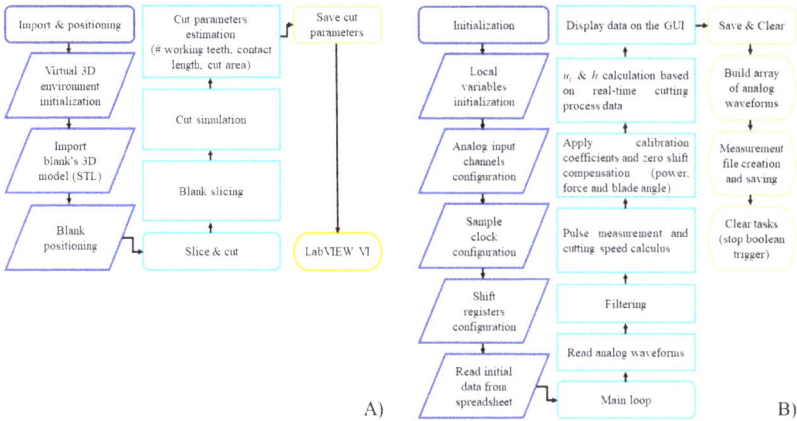

Figure 3 MATLAB app and LabVIEW VI operation diagrams.

Results and discussion

The conceived operator guidance system provides the operator with the cutting process data required for a quick evaluation and control of the actual process efficiency, see OGS of Fig. 1B. The program allows to set the sampling frequency, the refresh rate of the GUI, the cutting speed and the uncut chip thickness max.\min. values. Furthermore, the absorbed power, the blade angle, the force applied by the operator and the actual cutting speed are shown for in-process evaluation. The power is calculated taking into account the existence of a contribution due to friction and machine efficiency (no-load operation). The reset keys next to the power, blade angle and force gauges allow the sensors zeroing to eliminate any offsets due to electrical noise or changes in environmental conditions. Finally, two graphs allow the in-process evaluation of cutting process efficiency by means of u_c and h. To evaluate the effectiveness of the OGS, cutting tests were performed on a 40 mm Al 7075 rod. The tests were repeated 3 times, carrying out the measurements with and without OGS, for a total of 6 data sets. Comparing the results obtained with and without OGS, see Fig. 5 and Fig. 6, it can be observed that the developed system allows to keep the specific cutting energy almost constant, at the same time the uncut chip thickness is kept within the optimal range, avoiding rubbing phenomena and excessive tooth load. This helps extend the tool life and improve the cutting process efficiency. These results are confirmed observing the u_c standard deviation (Fig. 4) and the uncut chip thickness average values (Fig. 5).

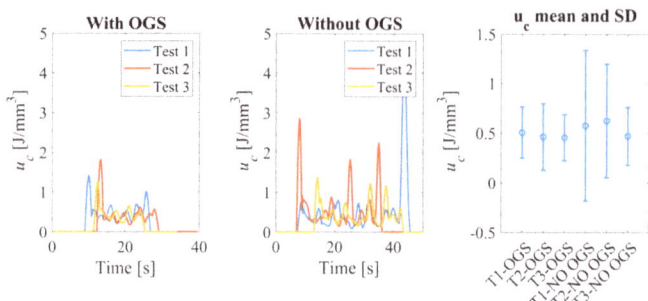

Figure 4 Measured specific cutting energy, with and without OGS.

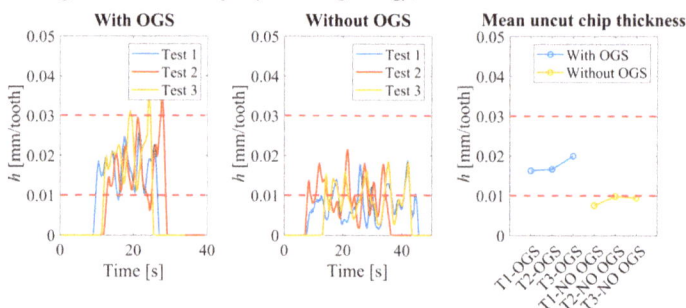

Figure 5 Measured uncut chip thickness, with and without OGS.

Conclusions

The operator guidance system (OGS) based on the digital shadow of a bandsaw machine presented in this paper demonstrated the effectiveness of digital innovation for machining processes performance improvement. The cutting process parameters measurement was made possible through the design and on-board deployment of a tailor-made measurement system made mostly by low-cost sensors. In-process data acquisition, visualization and elaboration were performed by means of NI DAQ devices and a specifically devised NI LabVIEW application. Preliminary geometrical data required to calculate the real-time specific cutting energy u_c were computed by means of a MATLAB app that performed the cut simulation. The combination of these systems made it possible to develop an OGS which allows the operator to view real-time process parameters such as power, cutting speed, blade angle and applied force. These parameters are then processed to obtain the specific cutting energy and the uncut chip thickness, allowing for real time cutting process optimization, avoiding rubbing and excessive tooth load. To evaluate the effectiveness of the OGS, a series of cutting tests were performed on a 40 mm Al 7075 rod with and without the support of the OGS. These tests demonstrated the effectiveness of the OGS, allowing the operator to keep the u_c stable and maintaining the uncut chip thickness within optimal limits throughout the cutting operation. The OGS utilizes high-performance DAQ and power sensor. By using industrial PLC and I/O, it is possible to contain costs within 15% of the market price of the bandsaw. A further development of the system can be achieved by implementing an automatic feed rate control, effectively obtaining a digital twin.

Acknowledgements

This study was carried out within the Interconnected Nord-Est Innovation Ecosystem (iNEST) and received funding from the European Union Next-GenerationEU (PIANO NAZIONALE DI

Italian Manufacturing Association Conference - XVI AITeM   Materials Research Forum LLC
Materials Research Proceedings 35 (2023) 286-294   https://doi.org/10.21741/9781644902714-34

RIPRESA E RESILIENZA (PNRR) – MISSIONE 4 COMPONENTE 2, INVESTIMENTO 1.5 – D.D. 1058 23/06/2022, ECS00000043). This manuscript reflects only the authors' views and opinions, neither the European Union nor the European Commission can be considered responsible for them.

References

[1] Sarwar, M., Persson, M., Hellbergh, H., & Haider, J., Measurement of specific cutting energy for evaluating the efficiency of bandsawing different workpiece materials. International Journal of Machine Tools and Manufacture 49 12-13 (2009), pp. 958-965. https://doi.org/10.1016/j.ijmachtools.2009.06.008

[2] Lu, Y., Liu, C., Kevin, I., Wang, K., Huang, H., & Xu, X., Digital Twin-driven smart manufacturing: Connotation, reference model, applications and research issues. Robotics and Computer-Integrated Manufacturing 61 (2020), 101837. https://doi.org/10.1016/j.rcim.2019.101837

[3] Kritzinger, W., Karner, M., Traar, G., Henjes, J., & Sihn, W., Digital Twin in manufacturing: A categorical literature review and classification. Ifac-PapersOnline 51 11 (2018), pp. 1016-1022. https://doi.org/10.1016/j.ifacol.2018.08.474

[4] Jones, D., Snider, C., Nassehi, A., Yon, J., & Hicks, B., Characterising the Digital Twin: A systematic literature review. CIRP Journal of Manufacturing Science and Technology 29 (2020), pp. 36-52. https://doi.org/10.1016/j.cirpj.2020.02.002

[5] Lattanzi, L., Raffaeli, R., Peruzzini, M., & Pellicciari, M., Digital twin for smart manufacturing: A review of concepts towards a practical industrial implementation. International Journal of Computer Integrated Manufacturing 34(6) (2021), pp. 567-597. https://doi.org/10.1080/0951192X.2021.1911003

[6] Li, C., Mahadevan, S., Ling, Y., Choze, S., & Wang, L., Dynamic Bayesian network for aircraft wing health monitoring digital twin. Aiaa Journal 55(3) (2017), pp. 930-941. https://doi.org/10.2514/1.J055201

[7] Knapp, G. L., Mukherjee, T., Zuback, J. S., Wei, H. L., Palmer, T. A., De, A., & DebRoy, T. J. A. M. Building blocks for a digital twin of additive manufacturing. Acta Materialia 135 (2017), pp. 390-399. https://doi.org/10.1016/j.actamat.2017.06.039

[8] Liau, Y., Lee, H., & Ryu, K., Digital Twin concept for smart injection molding. In: IOP Conference Series: Materials Science and Engineering 324(1), IOP Publishing, 2018. https://doi.org/10.1088/1757-899X/324/1/012077

[9] Wang, X. V., & Wang, L., Digital twin-based WEEE recycling, recovery and remanufacturing in the background of Industry 4.0. International Journal of Production Research 57(12) (2019), pp. 3892-3902. https://doi.org/10.1080/00207543.2018.1497819

[10] Liu, J., Zhou, H., Tian, G., Liu, X., & Jing, X., Digital twin-based process reuse and evaluation approach for smart process planning. The International Journal of Advanced Manufacturing Technology 100 (2019), pp. 1619-1634. https://doi.org/10.1007/s00170-018-2748-5

[11] Magnanini, M. C., & Tolio, T. A. M., A model-based Digital Twin to support responsive manufacturing systems. CIRP Annals 70.1 (2021), pp. 353-356. https://doi.org/10.1016/j.cirp.2021.04.043

[12] Tao, F., Zhang, M., Liu, Y., & Nee, A. Y., Digital twin driven prognostics and health management for complex equipment. Cirp Annals 67(1) (2018), pp. 169-172. https://doi.org/10.1016/j.cirp.2018.04.055

[13] Sivalingam, K., Sepulveda, M., Spring, M., & Davies, P., A review and methodology development for remaining useful life prediction of offshore fixed and floating wind turbine power converter with digital twin technology perspective. In: 2018 2nd international conference on green energy and applications (ICGEA), IEEE, 2018, pp. 197-204. https://doi.org/10.1109/ICGEA.2018.8356292

[14] Sallese, L., Tsahalis, J., Grossi, N., Scippa, A., Campatelli, G. & Tsahalis, H. Case Study 1.3: Auto-adaptive Vibrations and Instabilities Suppression in General Milling Operations. In: Intelligent Fixtures for the Manufacturing of Low Rigidity Components. Lecture Notes in Production Engineering. Springer, Cham., 2018. https://doi.org/10.1007/978-3-319-45291-3_3

[15] Vuković, M., Mazzei, D., Chessa, S., & Fantoni, G. Digital Twins in Industrial IoT: a survey of the state of the art and of relevant standards. 2021 IEEE International Conference on Communications Workshops (ICC Workshops). IEEE, 2021. https://doi.org/10.1109/ICCWorkshops50388.2021.9473889

[16] Scalzo F, Totis G, Vaglio E & Sortino M. Passive Chatter Suppression of Thin-Walled Parts by Means of High-Damping Lattice Structures Obtained from Selective Laser Melting. Journal of Manufacturing and Materials Processing 4(4):117 (2020). https://doi.org/10.3390/jmmp4040117

[17] Zhu, Z., Xi, X., Xu, X., & Cai, Y. Digital Twin-driven machining process for thin-walled part manufacturing. Journal of Manufacturing Systems 59 (2021), pp. 453-466. https://doi.org/10.1016/j.jmsy.2021.03.015

[18] Scalzo, F., Totis, G. & Sortino, M. Influence of the Experimental Setup on the Damping Properties of SLM Lattice Structures. Exp Mech 63 (2023), pp. 15-28. https://doi.org/10.1007/s11340-022-00898-8

Italian Manufacturing Association Conference - XVI AITeM
Materials Research Proceedings 35 (2023) 295-301

Materials Research Forum LLC
https://doi.org/10.21741/9781644902714-35

The influence of material properties and process parameters on energy consumption in the single-screw extrusion of PVC tubes

Andrea Pieressa[1,a] *, Enrico Bovo[1,b], Marco Sorgato[1,c], Giovanni Lucchetta[1,d]

[1]University of Padova, Department of Industrial Engineering, Via Venezia 1, Padova 35131, Italy

[a]andrea.pieressa@phd.unipd.it, [b]enrico.bovo.1@phd.unipd.it, [c]marco.sorgato@unipd.it, [d]giovanni.lucchetta@unipd.it

Keywords: Energy Efficiency, Polymers, Single-Screw Extruder

Abstract. Extrusion is one of the most widely used but energy-intensive processes for shaping polymers. It is thus fundamental to understand the process conditions that enable a cost-efficient operation of the extruders. The energy required by the process is highly affected by system variables and material properties in a complex manner. The aim of this work is thus to investigate the correlation between the process settings, the material properties, and the extrusion energy consumption. An extensive experimental campaign was conducted, testing different flexible PVCs. The power was recorded for both the motor of the extruder and the entire machine during continuous single-screw extrusion for various process conditions. A regression model was developed to correlate the specific energy consumption with the material properties, thus providing a valuable tool for estimating the cost of the final manufactured products.

Introduction

Extrusion is one of the most popular processes for shaping polymers because it is used to manufacture various products such as pipes, films, and sheets. Additionally, most plastic parts are subjected to this processing stage at least once in their realization. Therefore, improving the process's efficiency is crucial to contribute to global energy savings and reduce environmental emissions [1].

Extrusion allows the conversion of raw polymeric material (in the form of pellets) using three main steps: melting, forming, and cooling. Among these, the first one represents the most energy-intensive [2].

The power consumption is typically determined by the processing parameters that the operator selects, the rheology of the polymer, and material properties. However, optimizing the extrusion process is still challenging due to the complex relationship between these factors and energy usage. Many efforts have been devoted to studying the influence of processing parameters and rheology. However, a complete correlation between energy consumption and material characteristics still needs to be found.

A strategy to minimize power usage was presented by Rauwendaal [3]. Rasid et al. analyzed the effect of the variation of barrel temperatures on energy consumption, noting that the heating element positioned close to the feeding area prevails over the others in terms of energy [4]. In 2016 Abeykoon et al. confirmed the decreasing dependence of an extruder's specific energy consumption on the rotational speed using PS, LDPE, and LLDPE [5]. In addition, they obtained contrasting results with those already present in the literature regarding the barrel temperature's effect [6].

Abeykoon et al. found possible correlations between specific energy usage and polymer viscosity using PMMA, PS, and LDPE. Furthermore, polymer rheology shows some links with melt temperatures, torque, power factor, and active power fluctuations [7].

This work investigates the unclear correlation between extrusion energy consumption and material properties by varying them within the specific family of flexible PVCs. Moreover, for estimating process costs, a regression model capable of linking the characteristics of the materials with specific energy consumption was developed. It could represent a valuable tool for process optimization, cost reduction, and sustainable manufacturing.

As anticipated, there are few works in the literature related to this issue, such as those of Sikora et al., in which they analyzed the effect of granulometric properties and bulk pellet density on extruder energy demand [8,9].

Materials and method

The experimental tests were conducted using a lab-scale 19 mm single screw extruder (Rheomex), manufactured by Haake Company, Germay, and a rod die with 1.5 mm diameter x 25 mm lenght. Ten flexible PVC types were tested among the most frequently used ones for tube manufacturing. The properties and compositions of these materials, provided by the supplier datasheets, are reported in Table 1. The materials refer to the nomenclature used internally by the company involved in the study.

Table 1 *Main properties of the PVCs used in the experiments*

PVC type	Hardness, [ShA]	Specific weight, $[\frac{g}{cm^3}]$	DOTP Plasticizer, [%]	Suspension PVC K70, [%]	$CaCO_3$ (inorganic filler), [%]	Recommended Barrel Temperature, [°C]
20_02000	71	1.07	37.3	59.0	/	155-165
20_02784	71	1.22	32.0	48.6	16.3	155-165
20_03162	75	1.26	18.7	46.6	21.0	155-165
20_03161	71	1.05	37.3	59.7	/	155-165
20_01609	80	1.35	24.9	43.0	28.8	155-165
20_02748	80	1.22	28.8	52.9	15.3	155-165
20_00043	85	1.08	29.1	67.6	/	155-165
20_01970	85	1.31	22.3	49.0	26.4	155-165
20_02868	80	1.08	32.2	63.1	/	155-165
20_02992	78	1.41	23.2	34.2	39.3	155-165

The viscosity curves for each material were obtained by capillary rheometry (Ceast, Rheologic 500) operating at 160°C. All the materials exhibited a shear-thinning behavior. Therefore the viscosity was modeled through a power-law equation:

$$\eta = m(T)\dot{\gamma}^{n-1} \tag{1}$$

Italian Manufacturing Association Conference - XVI AITeM   Materials Research Forum LLC
Materials Research Proceedings 35 (2023) 295-301   https://doi.org/10.21741/9781644902714-35

where n is the shear-thinning exponent of the material, m(T) is the consistency index which allows modeling the temperature dependence of the viscosity according to the WLF equation [10].

For the acquisition of energy data, two DIRIS A-10 power meters were installed; they allow the measurement of active, reactive, and apparent power in a three-phase network. To distinguish the consumption due to the motor alone from that of the entire extruder, the devices were installed upstream of the inverter and upstream of the whole machine, respectively.

Since the effect of temperature has already been extensively discussed in the literature, the experiments were carried out varying only the material and the screw speed. In particular the extruder temperatures were fixed for all the experiments at the following values: T_{zone1}=150°C, T_{zone2}=155°C and T_{zone3}=160°C, from the hopper to the die, and T_{die}=155°C. The temperature of the melt (~160°C) was thus in the middle of the range suggested by the material spreadsheet reported in Table 1. At the same time, the screw speed was varied from 5 rpm to 170 rpm through 11 levels to replicate the operating conditions of an extrusion line in an industrial framework. The extruder was initially heated-up, then, for each material, the screw speed was progressively increased. For each screw speed level, once the process had reached a steady-state condition, the extrudate was collected after two minutes and weighted using a digital scale to calculate the mass flow rate. The energy meters data were recorded with a sampling time of three seconds.

Results and discussion

The motor's specific energy consumption (SEC) is shown in Figure 1 as a function of screw rotational speed. In agreement with the results reported in the literature, the SEC decreases with the increasing rotational speed of the screw.

One of the most relevant aspects that emerged from this study is the fact that there is a relationship between the motor's active power and the hardness of the material, as shown in Figure 2. From the plot, it is possible to note that the materials analyzed can be clustered into three groups:

1. 20_03161, 20_02784, 20_02000, 20_03162
2. 20_02992, 20_01609, 20_02868, 20_02748
3. 20_00043, 20_01970

Extending the results to motor SEC is possible since it is defined as power usage over the volumetric flow rate.

Further analyses have shown that within the clusters, the motor SEC increases as the viscosity of the material increases. Figure 3 shows in a 3D plot the correlation between hardness, viscosity, and motor SEC.

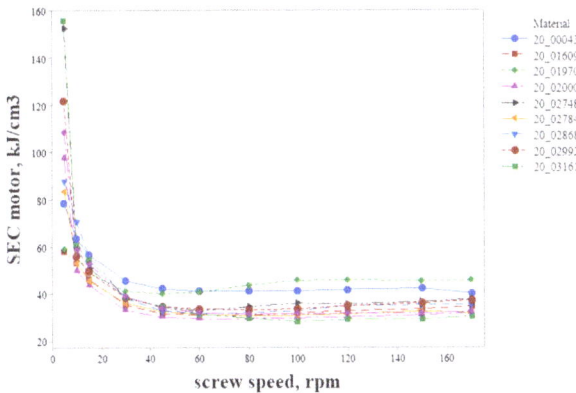

Fig. 1 *Motor specific energy consumption as a function of the screw speed and material*

Italian Manufacturing Association Conference - XVI AITeM
Materials Research Proceedings 35 (2023) 295-301

Materials Research Forum LLC
https://doi.org/10.21741/9781644902714-35

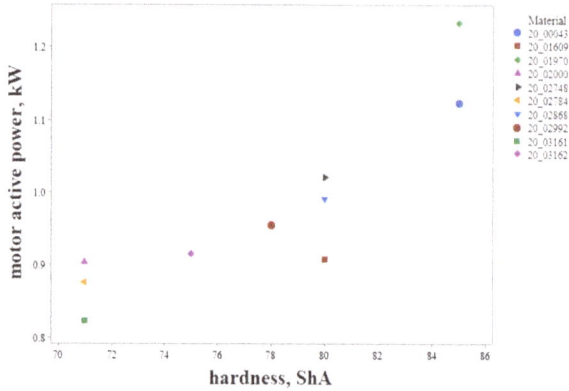

Fig. 2 *Relationship between motor active power and hardness of the materials at 120 rpm*

A linear regression model, added to the plot, can approximate the trend of the points with a coefficient of determination of 0.93. Therefore, this simple model can be used as a cost estimation tool since it can provide a value linked to energy demand depending on viscosity and material hardness.

Viscosity is a measure of a polymer's resistance to flow and deformation, affecting the force required to move the polymer through the extrusion die. Therefore, polymers with higher viscosity need higher screw torque to melt, leading to higher energy consumption. Similar results were found in [8] working with very different materials.

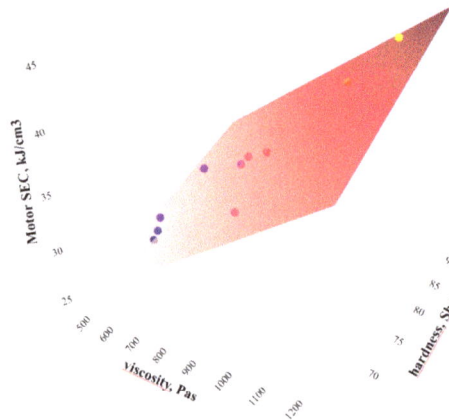

Fig. 3 *Correlation between viscosity, hardness, and specific energy consumption*

The hardness of a polymer depends on its molecular structure, particularly the size, shape, and rigidity of its polymer chains, as well as the degree of intermolecular forces and crosslinking [11].

Italian Manufacturing Association Conference - XVI AITeM   Materials Research Forum LLC
Materials Research Proceedings 35 (2023) 295-301   https://doi.org/10.21741/9781644902714-35

Since hardness is the resistance to deformation and scratching, polymers characterized by a higher hardness need more energy to be extruded due to the greater mechanical work required [3]. Therefore the increase in the motor SEC can be explained by the greater demand for power to implement the contiguous solid melting process.

Another important aspect is that the materials exhibit the wall slip phenomenon at various degrees, especially at high screw speeds. Wall slip phenomena can be caused by an adhesive failure at the interface between the polymer and the barrel surface (direct detachment of adsorbed polymer chains from the wall), or by a sudden disentanglement of the chains in the bulk from those adsorbed/attached to the wall. [12,13]

This is evident from Figure 4 as the volumetric flow rate dependence on the screw speed is less than proportional, especially for the materials 20_01970 and 20_01609.

Table 2 reports the deviation value from linearity for each material at 150 rpm.

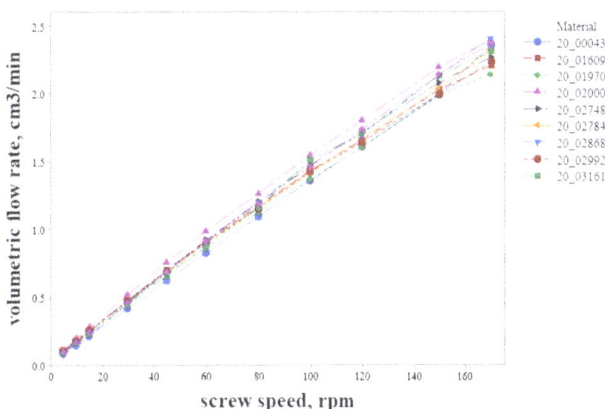

Fig. 4 *Volumetric flow rate dependence on screw speed and material*

Table 2 *Deviation from linearity for each material at 150 rpm*

	20 02000	20 02784	20 03162	20 03161	20 01609	20 02748	20 00043	20 01970	20 02868	20 02992
Deviation from linearity [cm³/min]	0.178	0.031	0.075	0.056	0.206	0.171	0.019	0.241	0.101	0.178

In Figure 5, these two high-slipping materials are compared with those that show less slippage (e.g. 20_00043, and 20_02784, respectively). The motor active power curves are almost parallel, indicating a constant motor torque value (the slope of the curves in Figures 5a and 5b). However, the volumetric flow rate curves for the same conditions (Figures 5c and 5d) show a lower dependence of flow rate on screw speed for materials 20_01970 and 20_01609, which can be attributed to the wall slip phenomenon.

The observed behaviors resulted in distinct trends in the motor's specific energy consumption plots. The materials 20_01970 and 20_01609 have a lower SEC at low screw speeds than 20_00043 and 20_02784, respectively, and vice versa at high screw speeds.

Italian Manufacturing Association Conference - XVI AITeM   Materials Research Forum LLC
Materials Research Proceedings 35 (2023) 295-301   https://doi.org/10.21741/9781644902714-35

This is due to the reduction of the volumetric flow rate progressively caused by wall slip at increasing values of the screw speed. Therefore, it is essential to consider wall slip phenomena when predicting energy consumption in an extrusion line, as they can generate a global minimum point in the SEC curve.

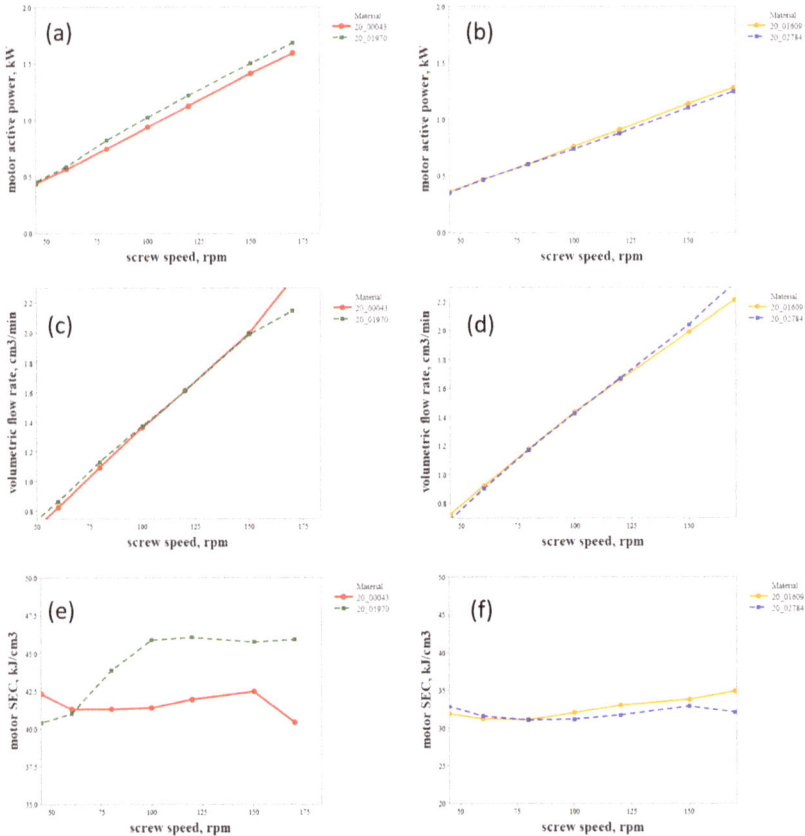

Fig. 5 *Comparison between materials with different wall slip behaviour in terms of motor active power (a and b), volumetric flow rate (c and d), and motor SEC (e and f)*

Conclusion

A study to investigate the unclear correlation between specific energy consumption and material properties in single screw extrusion was conducted. The experimental campaign was run by varying material properties within the specific family of flexible PVCs. The results highlighted a relationship between the motor's active power and the material's hardness, which was also possible to extend to the motor's specific energy consumption.

A linear regression model was used to approximate the motor SEC dependence on material hardness and viscosity with a coefficient of determination of 0.93, thus providing a cost estimation tool for process optimization, cost reduction, and sustainable manufacturing.

Italian Manufacturing Association Conference - XVI AITeM   Materials Research Forum LLC
Materials Research Proceedings 35 (2023) 295-301   https://doi.org/10.21741/9781644902714-35

The results obtained suggest that wall slip can significantly affect the specific energy consumption in the extrusion process since it progressively reduces the volumetric flow rate at increasing values of the screw speed. This phenomenon can generate a global minimum in the SEC curve, which can be exploited for optimizing the process.

Further studies are required taking into consideration other types of material and carrying out a more in-depth analysis of issues related to wall slip.

References

[1] J. Vlachopoulos and D. Strutt, Polymer processing, Materials Science and Technology, 2003 vol. 19, no. 9, pp. 1161–1169. https://doi.org/10.1179/026708303225004738

[2] Kruder GA and Nunn RE, Optimizing energy utilization in extrusion processing, SPE ANTEC technical papers, Ed. 1981, pp. 648–652.

[3] C. Rauwendaal, Polymer Extrusion, 4th ed. München: Carl Hanser Verlag GmbH & Co. KG, 2014. https://doi.org/10.3139/9781569905395.fm

[4] R. Rasid and A. K. Wood, Effect of process variables on melt temperature profiles in extrusion process using single screw plastics extruder, Plastics, Rubber and Composites, vol. 32, no. 5, pp. 187–192. https://doi.org/10.1179/146580103225002731

[5] C. Abeykoon, A. L. Kelly, E. C. Brown, and P. D. Coates, The effect of materials, process settings and screw geometry on energy consumption and melt temperature in single screw extrusion, Appl Energy, 2016, vol. 180, pp. 880–894. https://doi.org/10.1016/j.apenergy.2016.07.014

[6] J. Deng, K. Li, E. Harkin-Jones, M. Price, N. Karnachi, and M. Fei, Energy Consumption Analysis for a Single Screw Extruder, 2013, pp. 533–540. https://doi.org/10.1007/978-3-642-37105-9_59

[7] C. Abeykoon, P. Pérez, and A. L. Kelly, The effect of materials' rheology on process energy consumption and melt thermal quality in polymer extrusion, Polym Eng Sci, 2020, vol. 60, no. 6, pp. 1244–1265. https://doi.org/10.1002/pen.25377

[8] J. W. Sikora, Feeding an Extruder of a Modified Feed Zone Design with Poly(vinyl chloride) Pellets of Variable Geometric Properties, International Polymer Processing, 2014, vol. 29, no. 3, pp. 412–418. https://doi.org/10.3139/217.2860

[9] B. Samujło and J. W. Sikora, The impact of selected granulometric properties of poly(vinyl chloride) on the effectiveness of the extrusion process, Journal of Polymer Engineering, 2013, vol. 33, no. 1, pp. 77–85. https://doi.org/10.1515/polyeng-2012-0100

[10] T. A. Osswald, Understanding polymer processing : processes and governing equations. Hanser Publishers, 2011. https://doi.org/10.3139/9783446446038.fm

[11] Agassant, J. F., Avenas, P., Carreau, P. J., Vergnes, B., & Vincent, M. Polymer processing: principles and modeling. Carl Hanser Verlag GmbH Co KG, 2017. https://doi.org/10.3139/9781569906064.fm

[12] M. M. Denn, Extrusion instabilities and wall slip, Annu Rev Fluid Mech, 2001, vol. 33, no. 1, pp. 265–287. https://doi.org/10.1146/annurev.fluid.33.1.265

[13] S. G. Hatzikiriakos, Wall slip of molten polymers, Prog Polym Sci, 2012, vol. 37, no. 4, pp. 624–643. https://doi.org/10.1016/j.progpolymsci.2011.09.004

Italian Manufacturing Association Conference - XVI AITeM   Materials Research Forum LLC
Materials Research Proceedings 35 (2023) 302-309   https://doi.org/10.21741/9781644902714-36

Numerical evidence of submerged arc welding at changing of the main process parameters

Francesco Raffaele Battista[1,a] *, Romina Conte[1,b] , David Izquierdo Rodriguez[1,c],
Francesco Gagliardi[1,d], Giuseppina Ambrogio[1,e] and Luigino Filice[1,f]

[1]Department of mechanical, energy and management engineering, University of Calabria, Rende (CS), Italy

[a]francesco.battista@unical.it, [b]romina.conte@unical.it, [c]david.rodriguez@unical.it, [d]francesco.gagliardi@unical.it, [e]giuseppina.ambrogio@unical.it, [f]luigino.filice@unical.it

Keywords: Assembly and Joining Processes, Arc Welding, FEM

Abstract. In the arc welding processes, the parts to be connected are subjected to thermal cycles that can generate distortions. Furthermore, owing to clamping constraints, residual stresses can arise affecting the mechanical properties of the welding bead. In the work proposed, multi-pass submerged arc welding (SAW) process was investigated by a numerical analysis, properly set by experimental evidence. Indeed, the welding bead of a reference sample, was section and etched before performing micrograph inspections to achieve the contours of both welded material and affected zone for each welding pass. Arc current, voltage and welding speed were the investigated process parameters. Their values were changed proportionally to maintain constant the generated specific heat of the process. Furthermore, the effect of different clamping set up was considered. As a result of the analysis, process configurations, according to the employed process parameters, able to reduce distortions and to minimize residual stress, were identified.

Introduction

Welding technologies are one of the most widespread joining processes in several industrial fields thanks to their flexibility, due to the many process alternatives currently available, and because they can be successfully applied to any material of engineering interest [1,2,3]. A fast localized heating of the material to melting temperature occurs during electric arc welding processes, followed by a rapid cooling phase. The main macroscopic effect of such thermal cycle results in an uneven distribution of displacements and thus in a deformed state of the welded joint [1,4,5]. The magnitude and distribution of welding-induced distortions negatively affect the aesthetic quality and dimensional accuracy of joints as well as may lead to additional post-weld processing [2,4,5,6]. Furthermore, the mechanical constraint action during process execution generates post-weld residual stresses. These can be critical because they adversely affect the fatigue strength of components and increase the risk of brittle fracture [2,4]. Both distortions and residual stresses in welded joints are influenced by numerous factors other than thermal cycle and clamping. Among these, the thermo-physical properties of the materials involved, the plate thickness, the welding method and parameters used, the sequence of passes, the joint configuration and the preheat and interpass temperatures deserve to be cited [3,4,5,7].

The mechanical restraint condition is one of the most significant factors affecting both the values and the distributions of welding deformations and residual stresses [1,5]. Indeed, numerous numerical-experimental studies attest that the sequence of constraints' application, as well as their location and dimension, results in changes in the displacement and tension fields [4,5]. In this regard, the degree of stiffness imposed on the components to be connected has a mirror effect on the post-weld strain and tension values [2]. In fact, in absence of clamping the welded parts have more possibility of movement and therefore high deformation and low internal post-welding stresses occur. Therefore, when planning a welding process, it is essential to identify the optimal

Italian Manufacturing Association Conference - XVI AITeM   Materials Research Forum LLC
Materials Research Proceedings 35 (2023) 302-309   https://doi.org/10.21741/9781644902714-36

clamping condition that provides the right trade-off between deformations and internal stresses after cooling. Several FEM-based numerical simulation techniques are widely applied in order to optimize strain and tension fields in welded joints [1,2,4,5]. The strong interest in numerical simulation of welding-induced distortions and stresses has led to the development of several approaches for FEM modeling, from complex thermo-metallurgical-mechanical and thermo-mechanical models to simpler purely mechanical ones [1,6,8]. Welding processes numerical analysis enables fast, flexible, and cost-effective assessment of the effects of different process conditions and allows time and cost advantages related to the development and execution of additional experimental tests [2,9]. Indeed, through modern welding process numerical simulation techniques, it is possible to predict the final shape of the parts and then act at the planning stage of the process to match the required dimensional tolerances through compensation operations. Furthermore, numerical weld residual stresses assessment enables to carry out component service life estimates and/or predict areas where fatigue cracks can occur. Presently, among the most widely used FEM-based software for welding process simulation are ANSYS, ABAQUS and SYSWELD [1,4,5,9]. In particular, SYSWELD is a specially designed software for the simulation of welding processes and heat treatments and provides a wide range of tools and computational methods [1,3,7,9,10].

This paper deals with the numerical simulation of the submerged arc welding process on plates that differ in thickness and number of passes. The process numerical modeling and validation were carried out within SYSWELD software based on experimental tests. The main purpose of the work is to evaluate the effect produced on displacements and residual stress distributions by the different process configurations both in terms of plate thickness and process variables. Furthermore, the influence of experimental clamping on the investigated results was assessed compared with a process condition in the absence of clamping.

Material and Method

The experimental welding evidence was obtained joining ASTM A516 Gr. 70 steel plates. This is a C-Mn steel with excellent mechanical properties of notch toughness and is the standard specification for the fabrication of components subject to high pressure. As far as filler material is concerned, the low-carbon steel AWS 5.17 grade EH 14 was selected. This is a high Mn content, copper-coated wire commonly used in submerged arc welding. Two different sizes were used for the filler wire diameter, i.e., 2.4 mm for the root pass (pass 1) and 4 mm for all the others. The mechanical properties of the materials involved are reported in Table 1. As for the covering flux, ESAB OK Flux 10.61 was used.

Table 1 - Main mechanical properties of base material and filler metal.

	Tensile strength [MPa]	Yield strength (min) [MPa]	Elongation in 200 mm (min), %	Elongation in 50 mm (min), %
ASTM A516 Gr. 70	486 – 620	260	17	21
AWS 5.17 Gr. EH14	550	470	30	-

The experimental evidence involved multi-pass submerged arc welding of butt joints with different thicknesses. In particular, two different configurations were tested. The first experimental test concerned the joining of plates of 100 mm width, 400 mm length and, 10 and 20 mm thickness, respectively. While, the second trial involved plates of the same length and width as the previous ones but with thicknesses of 10 and 25 mm, respectively. In both tests, the plates were rigidly fixed at the edges along their whole length and thickness, over a width of 50 mm. The plates were neither subjected to initial pre-heating nor post-welding heat treatment, but an interpass temperature of 250°C was maintained between two consecutive passes.

Italian Manufacturing Association Conference - XVI AITeM   Materials Research Forum LLC
Materials Research Proceedings 35 (2023) 302-309   https://doi.org/10.21741/9781644902714-36

Both experimental tests were performed using LINCOLN ELECTRIC's Place Power Wave® AC/DC 1000 SD CE welding machine and were carried out with reverse polarity DC. The first experimental test was carried out by a first root pass (pass 1) from the lower side of the plates, then two weld passes (pass 2 and 3) from the opposite side were enough to fill the whole milled channel previously made. Both passes 2 and 3 were performed employing the same values of voltage, current and welding speed. The second experimental test was carried out in the same manner and with the same process parameters as the first, to evaluate the effect of different plate thickness. Next, two more welding passes were performed to evaluate the effect of additional passes to strength the weld bead. Table 2 highlights the followed process sequences.

Table 2 - Main welding process parameters employed in the experimental tests.

Pass number	Voltage [V]	Current [A]	Welding speed [mm/min]
1	27	440	320
2	25.4	540	384
3	25.4	540	384
4	34	675	768
5	34	675	768

The experimental work also involved the obtainment of specimens from the welded samples to conduct macrographic inspections. The specimens were achieved through the steps of cutting, cold embedding, grinding, polishing and chemical etching. Then, the macrographs of the welds were achieved, through which the melted and heat-affected zones were identified for each pass. Fig. 1 shows the macrographs of the weld seams with the number of each pass highlighted.

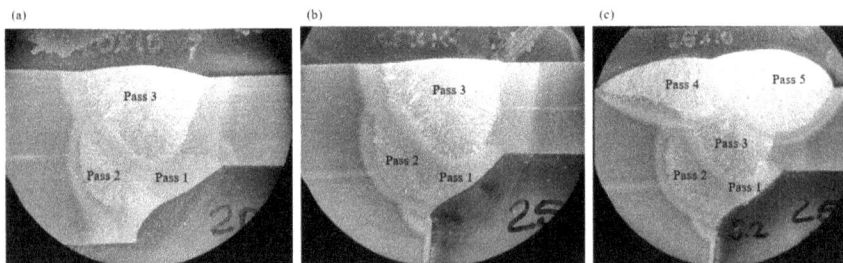

Fig. 1 - (a) Cross-sectional macrograph of the weld bead obtained from the first experimental test, (b) cross-sectional macrograph of the weld bead achieved from the second experimental test performed through three passes and (c) five passes.

Numerical modeling
A FEM model within SYSWELD software was developed based on the experimental evidence. Then an uncoupled sequential thermo-metallurgical mechanical analysis aimed at evaluating the welding-induced displacement and stress fields was conducted.

The first phase of numerical modeling involved the creation of 2D joint models. The FEM models of the tested specimens were developed in the Visual-mesh environment of the SYSWELD package. Due to computational time reasons, all FEM models were developed with a 100 mm length, while the other dimensions reflected those of the experimental specimens. Near the melted and heat affected zone, a finer mesh was made to obtain more accurate results. The FEM models of the three simulated process conditions are shown in Fig. 2.

Italian Manufacturing Association Conference - XVI AITeM   Materials Research Forum LLC
Materials Research Proceedings 35 (2023) 302-309   https://doi.org/10.21741/9781644902714-36

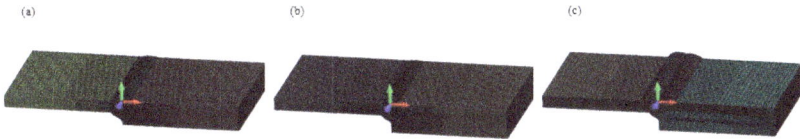

Fig. 2 – (a) 3D FEM model of the specimen related to the first experimental test, (b) 3D FEM model of the specimen related to the second experimental test performed with three passes and (c) five passes.

A transient thermal analysis was first carried out in the Visual-Weld environment through the moving heat source method already implemented within the software [11]. This kind of analysis is mainly affected by the availability of the physical materials properties at changing temperatures and the modeling of the heat source that simulates the arc movement [5,9,11]. Regarding temperature-dependent thermo-mechanical materials parameters such as thermal conductivity and yield stress were already implemented in the SYSWELD database. As an example, in Fig. 4 the trends of these properties at varying temperature and metallurgical phases for the base material extracted from the software have been shown [12]. The 3D double ellipsoid model proposed by Goldak [13] was employed for heat source modeling because it provides a reliable description of the heat distribution produced by the electric arc during welding [1,3]. Heat source design was performed for each welding pass based on the values of specific heat input, welding speed, process efficiency and the single pass sizes. The latter were obtained from the macrographic specimen analysis achieved by the experimental tests.

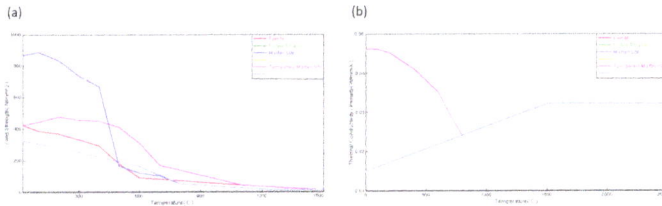

Fig. 3 - Trends of yield strength (a) and thermal conductivity (b) of ASME A516 Gr.70 steel as a function of temperature and metallurgical phases from the SYSWELD materials database [12].

To solve the thermal analysis, the cooling conditions were set in a way that the models' external surfaces exchanged heat by convection and radiation to the environment. For that, a room temperature of 20 °C was set. Since the subsequent mechanical analysis took as input the temperature field obtained as a result of the thermal analysis, the FEM models required validation by experimental data. Hence, heat source calibration was performed with the geometric parameters describing Goldak's model [13] that were iteratively changed for each pass until the predicted melted zone was in good agreement with that obtained from the macrographs. Fig. 4 shows the result of heat source calibration for each of the three simulated process conditions, while Table 3 shows the final numerical values of the parameters of Goldak's model that were subject to optimization. The geometric parameters are referred to the model proposed in the study [13], in addiction, it is specified that the root pass (pass1) was not simulated in the present investigation.

305

Italian Manufacturing Association Conference - XVI AITeM   Materials Research Forum LLC
Materials Research Proceedings 35 (2023) 302-309   https://doi.org/10.21741/9781644902714-36

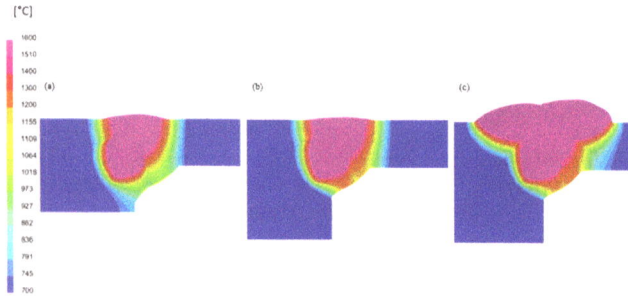

Fig. 4 - Melted and heat affected zone after calibration step for process conditions related to: (a) first experimental test, (b) second experimental test with three passes and (c) five passes.

Table 3 - Final values of the geometric parameters of Goldak's model according to the nomenclature of the study [13].

Pass Number	Test 1				Test 2			
	a [mm]	c1 [mm]	c2 [mm]	b [mm]	a [mm]	c1 [mm]	c2 [mm]	b [mm]
2	5.5	6.667	13.33	7	1.25	3.333	6.667	7
3	1.25	3.333	6.667	7.2	5.5	6.667	13.33	7
4					7	3.333	6.667	2
5					6	3.333	6.667	2

Based on the thermal analysis results carried out on the calibrated models, mechanical analysis was performed to predict post-cooling distortions and stresses. For this purpose, an elastic-plastic material behavior with isotropic hardening and small displacement theory was assumed. In this analysis the FEM models were rigidly constrained in the same manner as the experimental tests. Thereafter, a clamping-free condition was evaluated in view of analyzing its effects on the displacement and tension fields.

Results and discussion
The results of numerical simulations aimed to analyze the effect of different plate thickness and passes number on the displacements and residual stress distributions. Furthermore, the effect produced on the same output through the experimental clamping compared with a case without clamping was assessed.

The normal displacements and residual stresses after the cooling phase from the two experimental tests both performed with three passes were compared to evaluate the effect of different plate thicknesses. Figs 5 and 6 show the normal displacement and residual stresses at the end of welding for the same number of passes. As for the displacement field this appeared to be in good agreement between the two models, the only observable difference being lower values at the root pass in the case of the 10x25-thickness model (Fig. 5). The stress field between the two cases was also comparable. In the case of the thicker model, bead sections with slightly lower stress values were observed. Lower values for that model were also detected at the lower edge of the thicker plate (Fig. 6).

Italian Manufacturing Association Conference - XVI AITeM   Materials Research Forum LLC
Materials Research Proceedings 35 (2023) 302-309   https://doi.org/10.21741/9781644902714-36

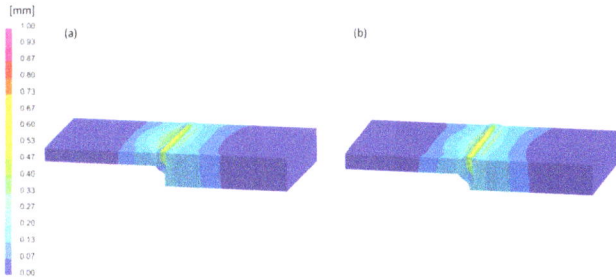

Fig. 5 - Post-weld displacement field for 10x20 mm (a) and 10x25 mm (b) thickness models.

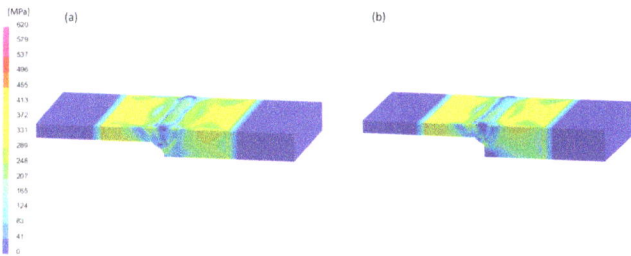

Fig. 6 - Post-weld stress field for 10x20 mm (a) and 10x25 mm (b) thickness models.

As for the influence of the different process parameters employed is concerned, the distortions and residual stresses of the second test performed with three and five passes, respectively, were compared. Fig. 7 shows the displacement fields for the second experimental test carried out with three and five passes, while Fig. 8 shows the post-welding residual stresses. The model performed with five passes compared with that using three produced an enlargement of the areas with higher value of the displacement field at the bead (Fig. 7). As for residual stresses, the weld bead produced by five passes exhibited slightly lower values at the areas adjacent to the bead not subject to the constraint. Low stress values were also observed at the surface portions of the bead, particularly for the last pass (Fig. 8).

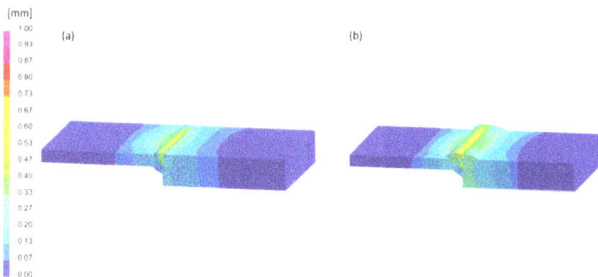

Fig. 7 - Post-weld normal displacements for the 10x25-thickness model performed with three (a) and five passes (b).

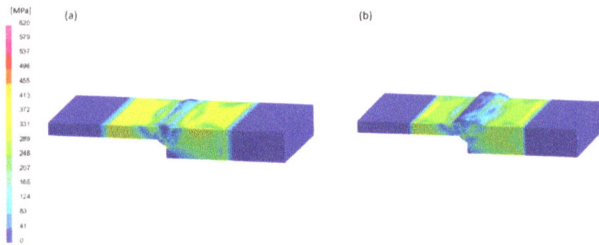

Fig. 8 - Post-weld residual stresses for the 10x25-thickness model performed with three (a) and five passes (b).

Finally, the FEM model related to the first experimental trial was considered in order to evaluate the effect of experimental clamping on the mechanical outputs investigated. In particular, the post-welding displacement and stresses were compared in the case of constrained sample as in the experimental tests and under free conditions. The latter condition was obtained by blocking model displacements on a lower edge of the thicker plate in the three directions of space. This prevents rigid motion of the body during numerical simulation and allows the possibility of free thermal expansion of the material. Fig. 9 show the post-welding normal displacement and residual stress fields for the examined model in the case of experimental clamping and without clamping. As can be observed from Fig. 9, much greater distortion is generated in the absence of clamping. The post-weld stress distribution also changes; a significant reduction in the extensions of the areas of greatest stress can be noticed.

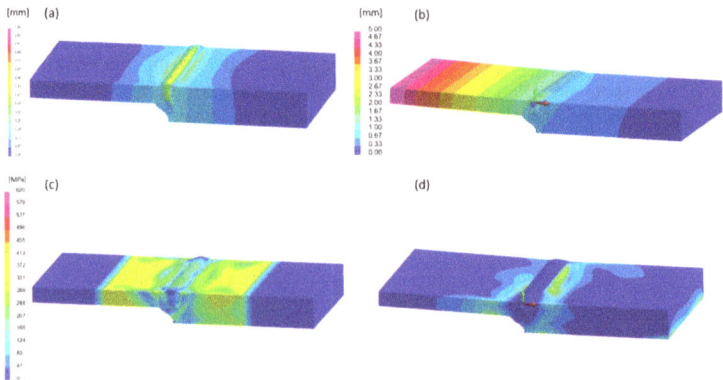

Fig. 9 – Post-weld normal displacements and residual stresses in the cases of experimental clamping (a) - (c) and without clamping (b) - (d).

Summary
This paper deals with the numerical simulation of the submerged arc welding process on plates with different thicknesses performed under different number of passes. The process numerical modeling and validation were carried out based on experimental tests. The study is aimed at evaluating the effect produced on the displacements and residual stress distributions by the different process configurations employed both in terms of plate thickness and process parameters. Moreover, the effect of the clamping condition applied in the experimental tests was assessed in

Italian Manufacturing Association Conference - XVI AITeM   Materials Research Forum LLC
Materials Research Proceedings 35 (2023) 302-309   https://doi.org/10.21741/9781644902714-36

comparison with a case without clamping. Analysis of the results showed that the effect of the five-mm thickness difference between the plates involved is negligible in terms of both displacement range and post-welding residual stresses. Regarding the effect of process parameters, an enlargement around the bead area of the higher-value displacement field was observed in the case of welding by five passes. In addition, the effect of a higher number of passes is reflected by a slight reduction in the post-weld stress values in the areas adjacent to the bead free of constraints. Finally, it was observed that in the absence of clamping a much more pronounced deformation state is obtained, with maximum values approximately five times higher than in the experimental clamping case for the chosen investigated plates' dimensions. Furthermore, a significant reduction in high-stress areas adjacent to the bead was observed without clamping. The obtained evidence are in agreement with the experimental process knowing confirming the reliability of the employed numerical model.

References

[1] E. S.V. Marques, F. J.G. Silva, A. B. Pereira, Comparison of finite element methods in fusion welding processes - A review, Metals 10.1 (2020) 75. https://doi.org/10.3390/met10010075

[2] M. Slováček, T. Kik, Use of Welding Process Numerical Analyses as Technical Support in Industry. Part 1: Introduction to Welding Process Numerical Simulations. Biuletyn Instytutu Spawalnictwa. 4 (2015) 25-31. https://doi.org/10.17729/ebis.2015.4/3

[3] T. Kik, Computational techniques in numerical simulations of arc and laser welding processes. Materials 13.3 (2020) 608. https://doi.org/10.3390/ma13030608

[4] Y. Li et al, Effect of structural restraint caused by the stiffener on welding residual stress and deformation in thick-plate T-joints. J. of Mater. Res. and Tech. 21 (2022) 3397-3411. https://doi.org/10.1016/j.jmrt.2022.10.127

[5] D. Venkatkumar, D. Ravindran, Effect of boundary conditions on residual stresses and distortion in 316 stainless steel butt welded plate. High Temp. Mater. Proc. 38 (2019) 827-836. https://doi.org/10.1515/htmp-2019-0048

[6] D. Tikhomirov et al. Computing welding distortion: comparison of different industrially applicable methods. Advanced mater. research. Vol. 6. Trans Tech Pub. Ltd (2005) 195-202. https://doi.org/10.4028/www.scientific.net/AMR.6-8.195

[7] T. Kik, J. Górka, Numerical simulations of laser and hybrid S700MC T-joint welding. Materials 12.3 (2019) 516. https://doi.org/10.3390/ma12030516

[8] Z. Barsoum, M. Ghanadi, S. Balawi, Managing welding induced distortion-comparison of different computational approaches. Procedia Engineering 114 (2015) 70-77. https://doi.org/10.1016/j.proeng.2015.08.043

[9] A. A. Deshpande et al, Combined butt joint welding and post weld heat treatment simulation using SYSWELD and ABAQUS. Proceedings of the Institution of Mechanical Engineers, Part L: J. of Mater.: Design and Applications 225.1 (2011) 1-10. https://doi.org/10.1177/14644207JMDA349

[10] T. Kik, M. Slováček, M. Vaněk, Use of Welding Process Numerical Analyses as Technical Support in Industry. Part 2: Methodology and Validation. Biuletyn Instytutu Spawalnictwa 5 (2015) 25-32. https://doi.org/10.17729/ebis.2015.5/4

[11] L. Iorio (2021) ESI Welding simulation solution, ESI Group internal presentation.

[12] Welding Simulation User Guide, Sysweld Manual ESI Group.

[13] J. Goldak, A. Chakravarti, M. Bibby, A new finite element model for welding heat sources. Metallurgical transactions B 15 (1984) 299-305. https://doi.org/10.1007/BF02667333

Materials processing technology

Italian Manufacturing Association Conference - XVI AITeM
Materials Research Proceedings 35 (2023) 311-317

Materials Research Forum LLC
https://doi.org/10.21741/9781644902714-37

Augmented multi-scale instrumented indentation test characterization of complex multi-layered coatings for tribological application

Giacomo Maculotti[1,a] *, Gianfranco Genta[1,b] and Maurizio Galetto[1,c]

[1]Department of Management and Production Engineering, Politecnico di Torino, Corso Duca degli Abruzzi 24, 10129 Turin, Italy

[a]giacomo.maculotti@polito.it, [b]gianfranco.genta@polito.it, [c]maurizio.galetto@polito.it

Keywords: Coating, Tribology, Instrumented Indentation Test

Abstract. Multi-layer coatings for steel bushings consisting of an innermost layer of sintered bronze and an outermost composite layer of lead-reinforced polytetrafluoroethylene (PTFE+Pb) have been used in several power transmission applications. The PTFE+Pb layer provides lubrication by material transfer on the counter-body reducing friction and smoothing motion. The mechanical characterization of such a complex system is challenging and essential to provide input data necessary to design and predict the service life of the components. This work innovatively mechanically characterizes the coating by augmented multi-scale Instrumented Indentation Test (IIT). Nano-IIT will evaluate the uniformity of the Pb particles' dispersion. Dynamic nano-IIT will investigate the damping properties of the material as a function of load frequency. Micro-IIT will tackle the layer thickness evaluation and the gradient of mechanical properties through the layers, by continuous multi-cycle and by data augmentation provided by electric contact resistance.

Introduction

Metal-polymer coatings exploiting polytetrafluoroethylene (PTFE)-based materials have high lubricity and are an efficient alternative to conventional low-friction metallic materials, e.g., bronze and lead, for sliding bearing elements. In fact, these coatings rely on the superior tribological performance of PTFE, which has a friction coefficient lower than traditional materials. Several power transmission and automation applications that need gentle driving of sliding and rotating mechanical components under moderate load with very low frictional losses have adopted metal-polymer PTFE-based coatings. Due to its self-lubricating properties, PTFE can support the interfacial sliding motion and bear contact loads even without adding any lubricating material. This is crucial for applications in controlled environments where a spill of fat contaminants must be prevented at all costs, such as surgery machines and the food industry [1].

Using the symbiotic tribological mechanism between PTFE and a low-friction tin-bronze porous structure, metal-polymer composite coatings with reinforcing low-friction metal structure have proven to be the most successful solution for the specific application to plain bearings and bushings. Such a combination takes advantage of the polytetrafluoroethylene's self-lubricating qualities and the tin-bronze's mechanical properties, which bears some of the contact load preventing the PTFE from wearing out too soon. Metal-polymer coatings are usually manufactured as strips of steel sheet reinforcement on which the porous bronze matrix is sintered with a thickness of (0.2~0.4) mm. The manufacturing of guide bushings for linear pneumatic actuators is among the uses for these composite coatings that are of interest. Their incorporation with servo systems for highly automated Industry 4.0 production lines is a vital feature that has lately allowed them to broaden their area of use.

The literature reports some research works where component testing was carried out to investigate the wear process of the guide bushings of pneumatic actuators [2]. However, despite the wide application of this class of coatings, a thorough characterization of the mechanical behaviour of PTFE-based metal-polymer composite coatings with reinforcing low-friction metal

Italian Manufacturing Association Conference - XVI AITeM   Materials Research Forum LLC
Materials Research Proceedings 35 (2023) 311-317   https://doi.org/10.21741/9781644902714-37

structures is still missing. Some research papers have been published on this topic, but only application-oriented tribological results were presented [3,4].

The objective of this work is to thoroughly characterize the surface mechanical characteristics of the PTFE/Sintered Bronze composite coatings for simple bearings and bushings. Instrumented Indentation Tests (IIT) will be exploited to characterize the mechanical response of the material, which, to the best of the author's knowledge, is unreported and will complement tribological characterization already available in the literature. Innovatively, in-situ electric contact resistance measurement (ECR) will augment the mechanical characterization to highlight the relationship between the electromechanical response and the material structure, which can be henceforth non-destructively investigated. With a multi-scale methodology, the characterization will also address the material's elastic and viscoelastic properties. The paper introduces a methodology to characterize multi-layer composite coating. The methodology is tested on a commercial grade PTFE+Pb/Bronze coating.

Materials and Methods

This work considers a commercial grade of PTFE+Pb/Bronze composite coating manufactured by GGB Bearing Technology Inc (Thorofare, New Jersey, USA). The nominal composition of the coating presents a steel lamina as backing, coated with a low-friction lining consisting of two superimposed layers. The innermost layer (nominally of 280 µm) is a sintered porous bronze structure impregnated with a polymer compound of PTFE filled with Pb. The outermost layer is a 20 µm thick lining out of the same PTFE+Pb compound (see Figure 1). While the PTFE-based polymer composite offers self-lubrication, the porous bronze structure reinforces the metal-polymer coating, ultimately extending the service life. In fact, during early operation, running-in occurs, and part of the PTFE-based material is transferred to the opposite surfaces, forming a protective third layer that insulates the metallic counterpart from direct contact with the coating.

The thickness of the layers has been measured with a commercial optical metallographic microscope, taking 15 images and, per each of them, measuring the layers' thickness by manual marking, as in Figure 1. Results in terms of average and measurement uncertainty are reported showing a PTFE+Pb thickness of (50 ± 20) µm and for the Sintered porous Bronze of (278 ± 10) µm. Measurement uncertainty has been evaluated at a 95% confidence level, combining the resolution and the reproducibility according to the law of uncertainty propagation [5]. The resolution of the objective pixel size (0.5 µm) was propagated with a uniform resolution as a type B contribution. The reproducibility was estimated from the empirical standard deviation as a type A contribution. Surface roughness was measured on ten different locations with a white light interferometer, showing Sa of (0.95 ± 0.224) µm, Sq (1.47 ± 0.32) µm, and Sz of (37.38 ± 23.88) µm, propagating the sole contribution of reproducibility at 95% confidence level.

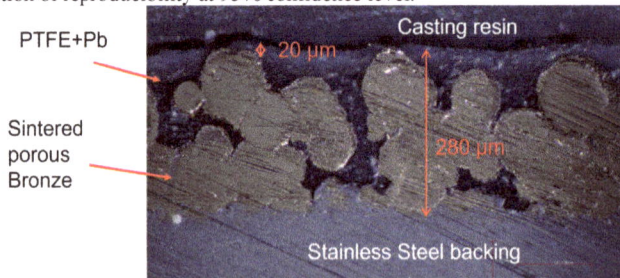

Figure 1 Section view of the DU® coating from GGB Inc.

Italian Manufacturing Association Conference - XVI AITeM   Materials Research Forum LLC
Materials Research Proceedings 35 (2023) 311-317   https://doi.org/10.21741/9781644902714-37

Instrumented Indentation Test for mechanical characterization. Instrumented Indentation Test (IIT) is a depth-sensing nonconventional hardness measurement technique that can characterize materials in terms of indentation hardness H_{IT}, Young modulus estimate, i.e. the indentation modulus E_{IT}, creep and relaxation from macro- to nano-scale. IIT applies a loading-holding-unloading force-controlled cycle with an indenter of known and calibrated shape to a test specimen [6]. During the indentation cycle, the applied force F and indenter penetration depth h in the material are measured, and the analysis of the measured quantities allows characterization relying on fundamental equations which are standardized [7]. IIT allows characterizing composite materials [8], coating [9] and surface treatments [10], while distinguishing and quantitatively characterizing different phases and structure variations of materials [6].

In-situ electrical contact resistance (ECR) augments IIT to evaluate the electromechanical response, highlighting differences in the microstructure due to a different electrical response [11], and estimating critical stress states for material phase transformation [12]. ECR relies on a doped-diamond conductive indenter and applies a controlled current at the contact between the indenter and the test sample surface. While the indentation cycle is performed, the voltage is measured, and changes in the contact resistance are appreciated, allowing electromechanical characterization [13,14].

An IIT variation, namely, continuous multi cycle (CMC) applies in the same location several indentations with increasing maximum test force [15]. CMC allows characterizing material properties in depth: it determines the mechanical properties as a function of the (increasing) maximum penetration depth $h_{c.max}$, for each successive unloading, resulting in a mechanical characterization, which can be related to the maximum penetration depth.

More recently, dynamic indentations have been implemented, by superimposing a sinusoidal trend to a conventional indentation cycle. Dynamic indentation allows evaluating viscoelastic properties of the material in terms of the loss modulus $E^{''}$ and the storage modulus $E^{'}$, whose evaluation is essential to provide a thorough evaluation of polymeric materials also apt for damping [16,17]. Instrumented indentation test will be performed with an Anton Paar STeP6 platform equipped with the measuring heads NHT[3] and MCT[3] hosted in the metrological room at the MInd4Lab laboratory of the Politecnico di Torino.

PTFE+Pb characterization. The layer of PTFE+Pb was mechanically characterized by Anton Paar NHT[3]. Quasistatic indentation cycles (loading and unloading of 30 s, holding of 10 s) at 10 mN mapped an equally spaced grid of 7×7 indentations with a step of 90 μm. Characterization is reported in terms of H_{IT} and E_{IT} and investigates the homogeneity of the PTFE+Pb layer.

The viscoelastic properties of the PTFE+Pb layer are obtained by dynamic indentation through a set of 15 indentations (distance of 90 μm) at a maximum load of 10 mN. A typical oscillation frequency for linear pneumatic actuators applications of 2 Hz is considered to superimpose the sinusoidal trend on the holding part of the indentation cycle with the amplitude of oscillation set at 5% of the maximum load [16,17]. The replication of tests on bulk PTFE aims at comparing the response with respect to the base material.

Additionally, a full factorial design with 15 replications and factors the material (bulk base material, i.e. PTFE, and reinforced PTFE+Pb) and loading frequency, i.e. (1, 2, 3, 5) Hz, (for 120 tests overall) was deployed to test the factors' effect on the mechanical response.

Micro-scale characterization. Micro-IIT is considered to evaluate the material's mechanical response as a function of penetration depth by CMC, which avoids cross-sectioning the sample. Data augmentation by ECR of micro-IIT is performed to obtain insights on the composition of the multi-layer coating through the electromechanical characterization. In fact, the PTFE+Pb has a resistivity greater than the sintered bronze layer. A set of 36 CMC indentations were performed with Anton Paar MCT[3], each featuring 15 cycles with a quadratic increase of the maximum load from 0.5 N to 30 N and with the loading, holding and unloading each lasting 30 s. A further set of

Italian Manufacturing Association Conference - XVI AITeM   Materials Research Forum LLC
Materials Research Proceedings 35 (2023) 311-317   https://doi.org/10.21741/9781644902714-37

18 CMC indentations ranging from 0.01 N to 0.5 N (other parameters were left the same) was performed to get data at shallow depths. ECR was set up in controlled current at 10 mA with a maximum voltage of 6 V to avoid sparks.

Results and Discussion

PTFE+Pb characterization. Nanoindentation resulted in mapping shown in Figure 2, in terms of E_{IT}. Both the indentation modulus and indentation hardness did not show significant deviations from a normal distribution, according to the Anderson-Darling test, thus supporting the hypothesis of homogeneous PTFE+Pb composition. The higher precision of the E_{IT} identifies outliers (see Figure 2), which can be ascribed to surfacing bronze particles (see Figure 1).

Viscoelastic response of the material was performed, and Table 1 summarizes the characterization results, reporting expanded uncertainty with a confidence level of 95%. Hypothesis tests on the average mechanical characterization, at a confidence level of 95%, show that the Pb fine powder stiffens the base material, e.g. induces an increase of the Young modulus.

ANOVA on the full factorial design showed, with a risk of error of 5%, that both the material and the loading frequency introduce a systematic effect on the response. In particular, an increase is appreciated due to the PTFE+Pb, with respect to base material, and at increasing frequency, this being an apparent effect due to the reduced time for elastic recovery of the material.

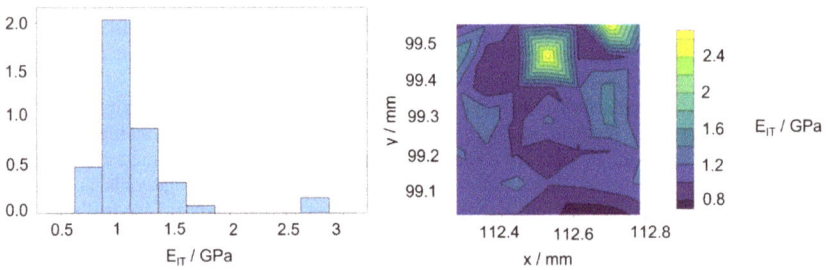

Figure 2 Histogram and spatial mapping of PTFE+Pb E_{IT}.

Table 1 Nano-IIT mechanical characterization in terms of Indentation Modulus E_{IT}, storage modulus E', loss modulus E'' and indentation hardness H_{IT}. Notice the increase of mechanical properties induced by the Pb reinforcement.

Material	E_{IT} / GPa	E' / GPa	E'' / GPa	H_{IT} / MPa
PTFE+Pb	1.1±0.07	2.1±1	0.15±0.01	46.2±2.4
PTFE	0.7±0.05	0.8±0.2	0.10±0.02	40.4±1.9

Micro-scale characterization. The experimental plan based on CMC indentations aimed at characterizing the gradient of mechanical properties as a function of depth was implemented and Figure 3(a-b) shows the trend of the mechanical characterization in terms of E_{IT} and H_{IT}, respectively. Consistently with the material structure appreciated with optical microscopy (see Figure 1), the mechanical response presents a trend depending on the indenter penetration. Specifically, an increase of the mechanical response is measured when moving from the PTFE+Pb to the sintered porous bronze, with a gradient in between.

The data augmentation through ECR provides insights. In fact, the transition can be seen in Figure 3(a-b) onsetting at 10 μm and terminating at 50 μm from the surface. In particular, ECR highlights a highly insulating material up to 20 μm, i.e. the PTFE+Pb. Electrical resistance decreases, as shown in Figure 3(c), at deeper penetrations up to 50 μm, where both electrical

Italian Manufacturing Association Conference - XVI AITeM   Materials Research Forum LLC
Materials Research Proceedings 35 (2023) 311-317   https://doi.org/10.21741/9781644902714-37

resistance and mechanical response tend to constant values, indicating the end of the transition to the sintered porous bronze. Furthermore, Figure 3(c) shows a highly insulating material, i.e. PTFE+Pb, even farther from the surface, e.g. values of 600 Ω at depths greater than 40 μm. This is in agreement with the optical analysis and suggests the presence of sacks of PTFE+Pb in the bronze matrix, consistently with the manufacturing procedure of the material. Such insights are further highlighted in the CMC indentation curve shown in Figure 4. In fact, the resistance starts to drop at about 45 μm, indicating the end of pure contact with PTFE+Pb, and then further decreases at about 50 μm indicating that the contact takes place majorly with bronze.

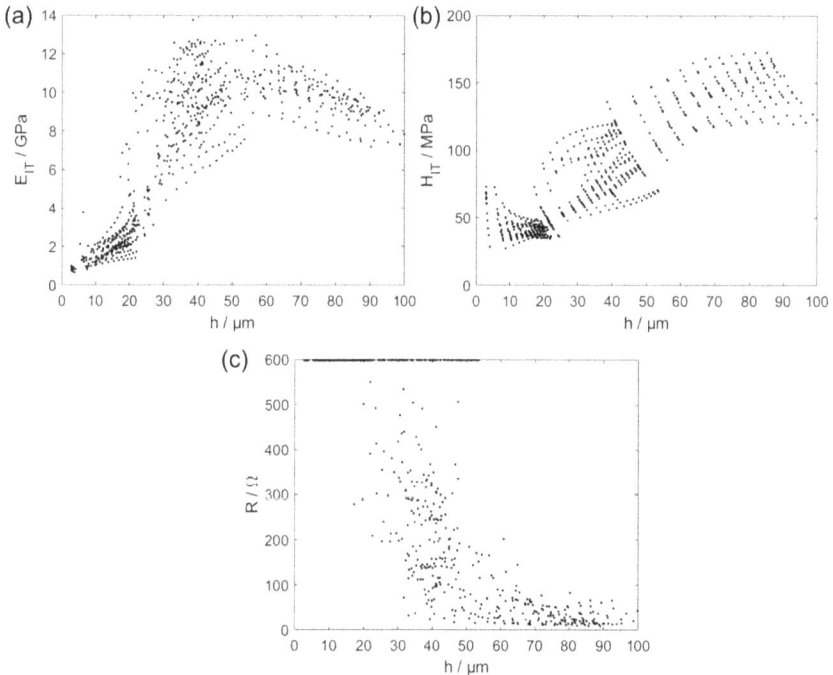

Figure 3 Results of the micro-scale characterization with augmented CMC indentation. Scatter plot of (a) E_{IT}, (b) H_{IT} and (c) R as a function of the penetration depth.

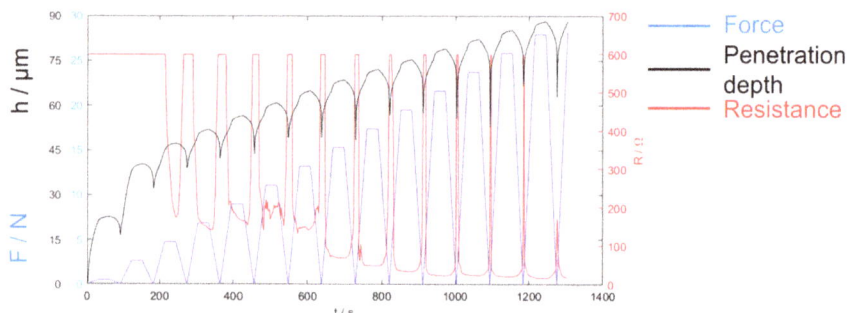

Figure 4 Augmented CMC indentation results: force (blue), penetration (black) and resistance (red) as a function of time.

Conclusions

This work presented a thorough mechanical and tribological characterization of a complex composite coating for plain bearings. Although this kind of coatings is widespread in many industrial applications, the detailed knowledge of their mechanical properties was lacking in the scientific literature.

This work has introduced an empirical methodology to achieve such a characterization. The characterization methodology was based on instrumented indentation test augmented by in-situ contact resistance measurements. The methodology has been applied, and demonstrated effective, on a commercially available grade of composite multi-layer coating (PTFE+Pb/Bronze). The approach proved effective in characterizing the material while obtaining information through the data augmentation otherwise achievable only through destructive cross-sectioning.

Quasistatic and dynamic nano-scale analysis performed on the PTFE+Pb layer evaluated its mechanical characteristic in terms of Young modulus (1.1±0.07) GPa, indentation hardness (46.2±2.4) MPa and dynamic loss modulus (0.15±0.01) GPa. Outliers in the surface mapping, otherwise homogeneous, were related to surfacing bronze particles inherent in the manufacturing of the coating, and their presence was confirmed by optical metallographic microscopy.

Microscale characterization with continuous multi cycle evaluated the evolution of mechanical properties as a function of depth. The mathematical description of the evolution of the material property as a function of depth will allow more accurate numerical modelling of the material behaviour, only roughly approximated in previous literature, even if essential in designing its application.

Data augmentation by ECR showed that transition and systematic differences in the material properties are consistent with the measured layer thickness and allowed insights into the data dispersion from the mechanical characterization.

The obtained results will allow future work to investigate traceable and metrologically trustworthy contact virtual experiments to support the design of components and applications.

References

[1] Rondinella A, Andreatta F, Turrin D, Fedrizzi L Degradation Mechanisms Occurring in PTFE-Based Coatings Employed in Food-Processing Applications Coatings 2021 11:1419. https://doi.org/10.3390/coatings11111419

[2] Ambu R, Bertetto A M, Mazza L Re-design of a guide bearing for pneumatic actuators and life tests comparison Tribol Int 2016 96:317-25. https://doi.org/10.1016/j.triboint.2015.12.043

[3] Güngör K, Demirer A Effects of impregnating PTFE filled with added graphite on wear behavior of sintered bronze plain bearings Int J Mater Res 2021 112:623-35. https://doi.org/10.1515/ijmr-2020-8067

[4] Goti E, Mazza L, Manuello Bertetto A Wear tests on PTFE+pb linings for linear pneumatic actuator guide bushings Int J Mech Control 2020 21:155 - 162.

[5] JCGM100: Evaluation of measurement data - Guide to the expression of uncertainty in measurement (GUM) 2008:Sèvres, France.

[6] Lucca D A, Herrmann K, Klopfstein M J Nanoindentation: Measuring methods and applications CIRP Ann - Manuf Technol 2010 59:803-19. https://doi.org/10.1016/j.cirp.2010.05.009

[7] ISO 14577-1:2015 Metallic materials-Instrumented indentation test for hardness and materials parameters - Part 1: Test method. ISO, Genève.

[8] Engqvist H, Wiklund U Mapping of mechanical properties of WC-Co using nanoindentation Tribol Lett 2000 8:147-52. https://doi.org/10.1023/A:1019143419984

[9] Maculotti G, Senin N, Oyelola O, Galetto M, Clare A, Leach R Multi-sensor data fusion for the characterization of laser cladded cermet coatings. Eur. Soc. Precis. Eng. Nanotechnology, Conf. Proc. - 19th Int. Conf. Exhib. EUSPEN 2019, 2019.

[10] Maculotti G, Genta G, Lorusso M, Galetto M Assessment of heat treatment effect on AlSi10Mg by selective laser melting through indentation testing Key Eng Mater 2019 813:171-7. https://doi.org/10.4028/www.scientific.net/KEM.813.171

[11] Yunus E M, McBride J W, Spearing S M The relationship between contact resistance and contact force on Au-coated carbon nanotube surfaces under low force conditions IEEE Trans Components Packag Technol 2009 32:650-7. https://doi.org/10.1109/TCAPT.2009.2014964

[12] Pharr G ., Oliver W C, Cook R F, Kirchner P D, Kroll M C, Dinger T R, et al. Electrical resistance of metallic contacts on silicon and germanium during indentation J Mater Res 1992 7:961-72. https://doi.org/10.1557/JMR.1992.0961

[13] Sprouster D J, Ruffell S, Bradby J E, Stauffer D D, Major R C, Warren O L, et al. Quantitative electromechanical characterization of materials using conductive ceramic tips Acta Mater 2014 71:153-63. https://doi.org/10.1016/j.actamat.2014.02.028

[14] Galetto M, Kholkhujaev J, Maculotti G Improvement of Instrumented Indentation Test accuracy by data augmentation with Electrical Contact Resistance CIRP Ann - Manuf Technol 2023 72: 10.1016/j.cirp.2023.03.034. https://doi.org/10.1016/j.cirp.2023.03.034

[15] Oliver W C, Pharr G M Measurement of hardness and elastic modulus by instrumented indentation: Advances in understanding and refinements to methodology J Mater Res 2004 19:3-20. https://doi.org/10.1557/jmr.2004.19.1.3

[16] Herbert E G, Oliver W C, Pharr G M Nanoindentation and the dynamic characterization of viscoelastic solids J Phys D Appl Phys 2008 41:74021. https://doi.org/10.1088/0022-3727/41/7/074021

[17] Hay J, Crawford B Measuring substrate-independent modulus of thin films J Mater Res 2011 26:727-38. https://doi.org/10.1557/jmr.2011.8

Italian Manufacturing Association Conference - XVI AITeM   Materials Research Forum LLC
Materials Research Proceedings 35 (2023) 318-325   https://doi.org/10.21741/9781644902714-38

Effect of sub-zero deformation temperatures and rolling directions on the formability of AISI 316 stainless steel sheets

Rachele Bertolini[1,a] *, Enrico Simonetto[1,b] Andrea Ghiotti[1,c] and Stefania Bruschi[1,d]

[1]Dept. of Industrial Engineering, University of Padova, Via Venezia 1, 35131, Padova, Italy

[a]rachele.bertolini@unipd.it, [b]enrico.simonetto.1@unipd.it, [c]andrea.ghiotti@unipd.it, [d]stefania.bruschi@unipd.it

Keywords: Stainless Steel, Tensile Strength, Cryogenic Forming

Abstract. Sheet forming carried out at sub-zero temperatures is gaining more and more interest for deforming metal sheets as it can represent an alternative to room temperature forming to increase the sheet metal formability and to high-temperature forming to avoid the need for subsequent heat treatments aimed at restoring the alloy microstructural characteristics. In this framework, the present work deals with mechanical testing of AISI 316 stainless steel sheets in a wide range of temperatures, namely from -100°C to 700°C, and rolling directions. After mechanical testing, the strain at ultimate tensile strength, identified as an indicator of steel formability, was measured at varying process conditions. Then, the samples were analyzed in terms of microstructural features, micro-hardness, and martensite formation. The obtained results showed that deforming at -50°C induced a substantial increase in formability preserving the hardness, compared to testing at higher temperatures.

Introduction

Austenitic stainless steels are used in several industrial sectors thanks to some characteristics they offer, such as high corrosion resistance, good ductility, and reasonable cost, even if their mechanical strength is quite low due to the face-centered cubic (FCC) austenite phase they present. During cold deformation processes, the phenomenon called strain-induced martensitic transformation (SIMT) may occur, which leads to the steel mechanical resistance increase but also reduces its ductility and corrosion resistance. When forming corrosion-resistant parts, SIMT needs to be decreased, which can be fulfilled through forming in the warm temperature range, which includes temperatures below the steel recrystallization one. For example, in [1], it was proved that forming AISI 304 stainless steel sheets at temperatures lower than 150 °C can help in SIMT suppression and formability enhancement. However, it is well known that warm forming decreases the work hardening and, therefore, the part's final strength. To this regard, in very recent years, sheet forming processes at temperatures below room one have been investigated with two main aims, namely i) to delay the necking onset in view of enhancing the sheet uniform elongation characteristics, and ii) to avoid possible heat treatments that can become mandatory when forming at elevated temperatures in order to recover the sheet initial mechanical and microstructural characteristics.

Literature records are available describing the application of these forming processes to mainly aluminum alloy sheets. AA7075 aluminium alloy sheets, deformed in [2] at varying stress triaxiality and temperature ranging from -100° and 400 °C, showing that the mechanical characteristics at necking were improved when deforming below the room temperature compared to high temperature testing, as a consequence of the formation of a higher amount of precipitates and suppression of dynamic recovery. AA7075 sheets were deformed in [3] to analyze the damage and fracture behaviour at cryogenic temperatures making use of a phenomenological fracture

Italian Manufacturing Association Conference - XVI AITeM   Materials Research Forum LLC
Materials Research Proceedings 35 (2023) 318-325   https://doi.org/10.21741/9781644902714-38

model; the validation of the latter through experimental trials carried out on U-shaped parts proved lower damage values in the part when cryogenic formed.

On the contrary, the performances of sheets other than those made of aluminum alloys under cryogenic temperatures have been scarcely investigated. In addition, just metals with a FCC structure have been addressed, pointing out a significant sensitivity of this crystal lattice to sub-zero temperature forming.

In this view, to enlarge the knowledge about the metal sheet response to sub-zero temperatures, the paper focuses on the effect of sub-zero deformation temperatures on the mechanical behaviour of austenitic stainless steel sheets. In particular, the AISI 316 sheets were strained under uni-axial tensile testing conditions in a wide range of temperatures, spanning from -100 °C to 700 °C and their response in terms of stress and deformation at necking as well at fracture was evaluated together with the post-deformation micro-hardness. The effect of the rolling direction was taken into account as well. The sheet microstructural features were analyzed and correlated to the mechanical response in order to explain the different behavior at varying testing temperatures.

Experimental procedures

Material

The material under investigation was the AISI 316 stainless steel supplied in form of sheets of 1 mm thickness. Its nominal chemical composition in the as-received condition is reported in Table 1.

Table 1. Chemical composition of the AISI 316 sheets (weight %) [4].

Fe	Cr	Ni	Mo	Mn	Si	C
67.69	16.63	10.85	2.42	0.38	1.28	0.018

Uniaxial tensile tests

The universal test frame (MTS^{TM} 322) was used to perform uniaxial tensile tests at different temperatures. Mechanical tests were conducted at temperatures ranging from 700 to −100 °C (including 25, -50, and 300 °C) at a fixed strain rate of 0.1 s^{-1}.

The sub-zero temperature tests were carried out by using an environmental chamber connected to a liquid nitrogen (LN$_2$) tank, as visible in Fig. 1a. Fig. 1b offers a magnified view of the dog-bone sample tightly fixed to the gauges. Liquid nitrogen was used as a refrigerant and its flow rate was digitally controlled to maintain the testing temperature with an accuracy of ± 2 °C. Uniaxial tests at elevated temperatures were performed with the same testing apparatus using a resistance heating system to heat the sample to the testing temperature.

The tensile tests were carried out using dog-bone-shaped samples, characterized by a nominal gauge length of 65 ± 0.1 mm and a nominal gauge width of 12 ± 0.05 mm, as prescribed in the ISO 6892 standard [5]. The samples were water-jet cut from the rolled sheets along the sheet rolling direction and the orthogonal one. The former samples will be hereinafter called *0deg* samples whereas the latter *90deg* samples.

In Fig. 1c the rolling direction (RD) refers to the direction of motion of the rolling plate, the transverse direction (TD) refers to the direction that parallel the rolling plate, and the normal direction (ND) refers to the direction that is perpendicular to the rolling plate. Given that, under uniaxial tensile testing, RD and TD represent the resistant section of the *0deg* and *90deg* samples, respectively.

At the industrial level, uniform elongation is the usual indicator of metal sheet formability. For this reason, for each testing condition, two sets of experiments were carried out, namely one at fracture to provide the whole engineering stress-strain curve and, thus, identify the ultimate tensile strength (UTS) for each testing temperature, and the other at UTS to assess the material characteristics at the necking occurrence.

Italian Manufacturing Association Conference - XVI AITeM   Materials Research Forum LLC
Materials Research Proceedings 35 (2023) 318-325   https://doi.org/10.21741/9781644902714-38

To have insights into the overall sheet ductility, the fracture surfaces were inspected using FEI™ QUANTA 450 Scanning Electron Microscope (SEM) with the Secondary Electron (SE) probe. An image at 100X magnification was acquired for each sample to evaluate the deformation at fracture.

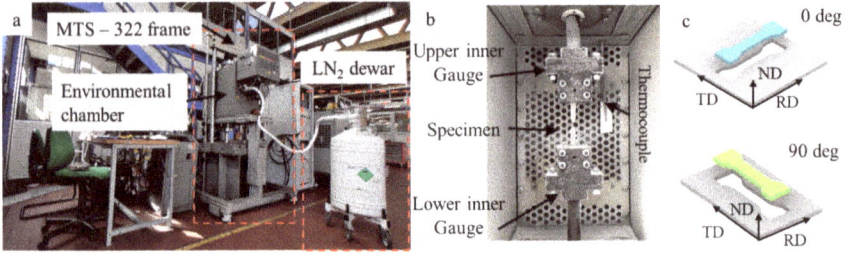

Figure 1. (a) Uni-axial tensile test equipment for sub-zero and elevated temperature testing, (b) zoom of the specimen fixed on the machine gauges, and (c) geometries of the samples.

Characterization methods after tensile testing
The samples strained at UTS were further characterized to gain more information about the formability of the AISI 316 at varying temperatures.

Firstly, the samples were cut, hot-mounted, and then ground with silicon carbide abrasive papers P320, P500, P800, P1200, P2400, P4000. Afterward, a finishing polishing pass was performed with a napped cloth using a diamond slurry. The polished samples were then etched twice to evidence: (i) the austenite grain boundaries, and ii) the possible presence of martensite. The first etching was electrolytic with a solution of 65% nitric acid (HNO_3) in water at 1 V for 10 s while the second one used Behara's solution (100 ml H_2O, 20 ml HCl, 1 g $K_2S_2O_5$) for 10 s.

The etched samples were observed through an optical microscope (Leica DMRE™) and their grain size was determined using the linear intercept method according to ASTM E112 [6].

Later on, an X-ray diffractometer (Siemens™ D500), equipped with a monochromator on the detector side and a Cu radiation tube, was used to quantify the martensitic phase percentage in the samples strained at UTS. The $2\Theta = 40$–100 ° angular range was investigated with a step scan of 0.04 ° and a counting time of 8 s.

The Vickers micro-hardness of the samples strained at UTS was measured using a Leitz Durimet™ micro-hardness tester with a load of 50 gr for 30 s; five values for each section were recorded and then the average value was calculated.

Results & Discussion
Mechanical properties
The engineering tensile curves of the AISI 316 samples are shown in Fig. 2 at varying temperatures and rolling directions. All the curves exhibited the same trend, except for the -100 °C curve presenting an S-shaped, more evident in the case of the 90 deg sample. Such peculiar behavior can be attributed to the SIMT effect that occurs at sub-zero temperatures.

A parameter conventionally used to predict the stability of the austenite phase with respect to martensite one is M_{d30}, namely the temperature at which 50 % of martensite is formed at a true strain of 0.3 [7]. The calculated M_{d30} of the AISI 316 settles to -83 °C, meaning that a fully austenitic structure cannot be kept at -100 °C during plastic deformation, which is consistent with the S-shape of the AISI 316 curves at -100 °C.

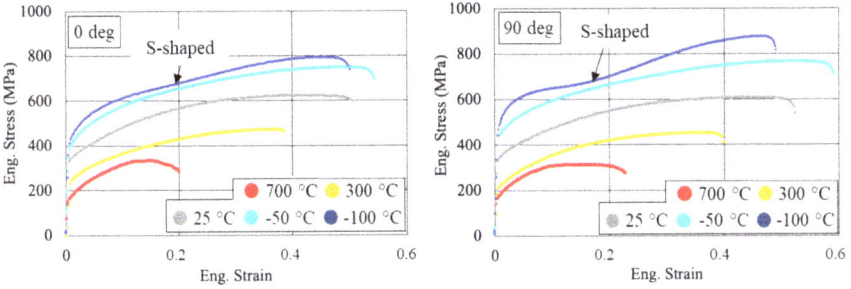

Figure 2. Tensile curves at varying temperatures and rolling directions.

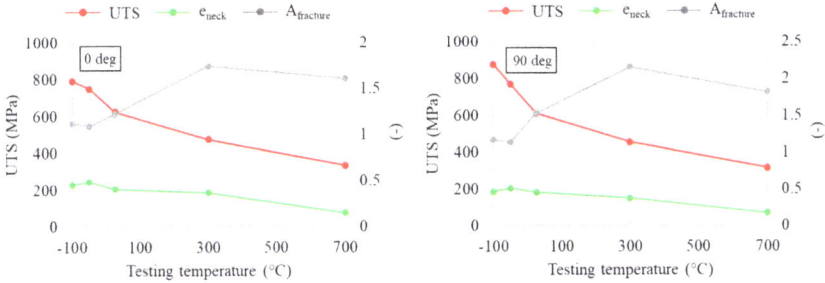

Figure 3. UTS, e_{neck} and $e_{fracture}$ at varying temperatures and rolling directions.

Fig. 3 shows the UTS, e_{neck}, and $A_{fracture}$ at varying temperatures and rolling directions Compared to room temperature, UTS increased on average by 23 % and 35 % for the samples tested at -50 °C and -100 °C, respectively. Interestingly, such increase in strength did not affect the uniform elongation. In fact, the e_{neck} of the sample tested at sub-zero temperatures is higher than that at room temperature, increasing on average of 16 % at -50 °C and of 7 % at -100 °C. On the contrary, $A_{fracture}$ decreases on average of 18 % at -50 °C and of 16 % at -100 °C, respectively.

It is worth noting that the highest values of e_{neck} as well as the lowest value of $A_{fracture}$ were registered when testing at -50 °C.

A completely different scenario is observed when considering testing at temperatures over the room one. In fact, both UTS and e_{neck} decrease while $A_{fracture}$ drastically increases. A 48 % and 25 % reduction in UTS was registered for samples tested at 700 °C and 300 °C, respectively. For the same testing temperatures, a reduction of 63 % and 14 % was achieved in e_{neck} while an increase of 25 % and 42 % in $A_{fracture}$.

The uniform elongation increase at decreasing testing temperature can be attributed to the face centered cubic (FCC) crystalline structure that characterizes stainless steels with quite elevated stacking fault energy (SFE). The SFE effectively affects the deformation mechanisms and the mechanical behavior of metals through its influence on the dislocations mobility: the higher the SFE the easier the dislocation movement as well as the dynamic recovery. On the other hand, the SFE is drastically influenced by the temperature, namely as the temperature increases the SFE increases as well. The application of sub-zero temperatures leads to the effective suppression of dynamic recovery, like thermal-induced cross slip and climb movements of dislocations,

Italian Manufacturing Association Conference - XVI AITeM   Materials Research Forum LLC
Materials Research Proceedings 35 (2023) 318-325   https://doi.org/10.21741/9781644902714-38

increasing the work hardening and postponing the formation of geometric instabilities to higher values of strain.

The influence of the rolling direction is notable only when considering the strain at fracture, which was always higher for the *90deg* samples, regardless of the testing temperature.

Characteristics after tensile testing

Fig. 4 shows the microstructure of the samples strained at UTS at varying temperature and rolling direction. The microstructures of the *0deg* and *90deg* samples deformed at 25 °C and 300 °C are very similar, indicating no thermally-induced microstructural phenomena. The same was found for the *0deg* and *90deg* samples deformed at -50 °C and -100 °C, here not reported for sake of brevity.

On the contrary, the *0deg* and *90deg* samples deformed at 700 °C present partially recrystallized austenite grains with annealing twins, with an average size slightly lower than the one of the samples deformed at lower temperatures.

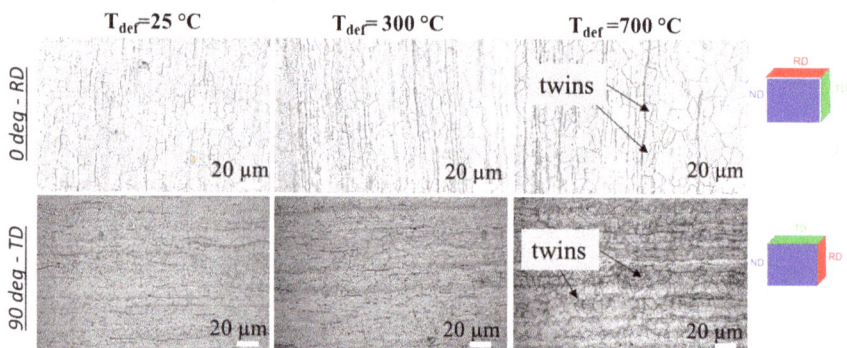

Figure 4. Microstructure of the samples strained at UTS at room and elevated temperatures.

The Beraha's etchant was then used on the same samples as indicated in the description of the experimental methods. Fig. 5 shows the microstructure of the *0deg* and *90deg* samples strained at UTS at room and sub-zero temperatures and varying rolling directions.

The deformation behavior of the samples tested at -50 °C and -100 °C is predominantly governed by SIMT, involving the deformation-induced transformation of austenite into martensite. On the contrary, the samples deformed at 25 °C, 300 °C and 700 °C (here not reported for sake of brevity) barely show the presence of the martensite phase.

SIMT of austenitic stainless steels deformed at cryogenic temperatures is a well-known phenomenon [8, 9]. The SIMT phenomenon is induced by both mechanical and thermodynamic driving forces. Specifically, the lower the temperature, the higher the thermodynamic driving force. This is due to the fact that lower temperatures impair the atomic diffusion ability suppressing the dynamic recovery. At the same time, sub-zero deformation temperatures produce a large number of defects and dislocations, which is conducive to martensite nucleation and growth.

Italian Manufacturing Association Conference - XVI AITeM   Materials Research Forum LLC
Materials Research Proceedings 35 (2023) 318-325   https://doi.org/10.21741/9781644902714-38

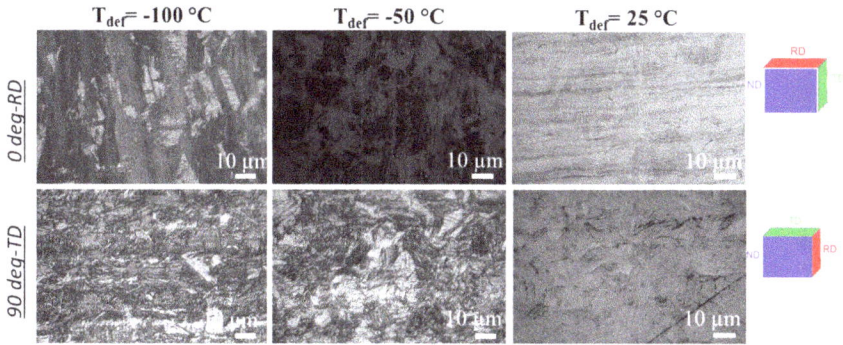

Figure 5. Microstructure of the samples strained at UTS at room and sub-zero temperatures.

XRD analyses were performed on the samples strained at UTS to investigate the microstructural evolution of the material, and the results are shown in Fig. 6. It can be seen that only the peaks related to the face-centred cubic (FCC) γ-austenite phase are visible at 25°C, regardless of the rolling directions. With the temperature decrease, peaks related to the body-centred cubic (BCC) martensite phase are evident at both -50 °C and -100 °C.

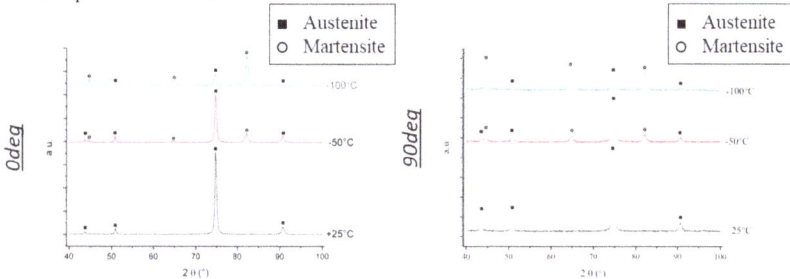

Figure 6. XRD spectra of the samples strained at UTS at room and sub-zero temperatures.

The XRD spectra give the chance to quantify the martensite percentages that are reported in Fig. 7a. At -100 °C, 78 % and 72 % of martensite formed in the *0deg* and *90deg* samples deformed at UTS. These results are in accordance with the previous considerations on M_{d30} temperature. At -50 °C, the martensite content decreased to 25 % and to 35 % in the *0deg* and *90deg* samples deformed at UTS, respectively. On the contrary, a fully austenitic structure was detected in the samples tested at 25 °C.

Fig. 7b reports the hardness values at varying testing temperatures and rolling directions calculated with respect to the baseline value obtained at room temperature. The highest hardness increase, namely 15 % on average, compared to the room temperature baseline, was found in the samples deformed at the lowest temperature. These outcomes are in accordance with the microstructures reported in Fig. 5, which shows the highest amount of martensite content when the material was deformed at -100 °C, and, therefore, the highest hardness.

On the contrary, a drastic decrease in hardness was visible for the samples tested at temperatures higher than room one. Specifically, the hardness of the samples deformed at 300 °C ad 700 °C decreased by 14 % and 25 %, respectively, with respect to one of the samples deformed at room temperature.

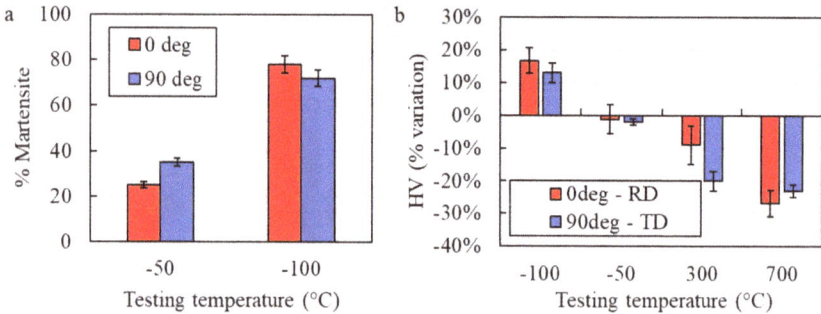

Figure 7. (a) Percentage of martensite calculated from the XRD spectra of Fig. 6 and (b) percentage variation of hardness at varying testing temperature and rolling direction.

Summary

In this paper, AISI 316 stainless steel samples were subjected to uniaxial tensile testing in a wide range of temperatures, from -100 °C to 700 °C. Their strain at UTS, namely the maximum strain for uniform elongation, was considered as a measure of the steel ductility at varying temperatures and rolling directions. The samples strained at UTS were also analyzed in terms of microstructural features, phase constituents and micro-hardness.

The main following conclusions can be drawn:

- All the tensile curves exhibited an approximately parabolic shape, except for the -100 °C curves that were S-shaped, especially in case of the 90 deg sample. Such peculiar behavior can be attributed to the SIMT effect that occurs at sub-zero temperatures.
- Regardless of the rolling direction, UTS monotonically decreased at increasing temperature. The same trend was registered for e_{neck}, which, however, showed the maximum value at T_{def}=-50 °C. Finally, $A_{fracture}$ increased at increasing temperature, presenting the minimum value at T_{def} =-50 °C.
- Regardless of the rolling direction, partially recrystallized austenite grains with annealing twins were evidenced only in the samples deformed at 700 °C. On the contrary, a certain amount of martensite was registered in the samples deformed at sub-zero temperatures. The outcomes of the microstructural analysis were also confirmed by the XRD measurements.
- Regardless of the considered section with respect to the rolling direction, the hardness of the sample strained at UTS at -100 °C was the highest, as a consequence of the most significant SIMT effect, with a less significant influence in the case of the sample strained at UTS at -50 °C. The opposite situation was registered for the samples deformed within the warm forming range, whose hardness was reduced as expected.

Acknowledgements

This research was developed in the framework of the project "Sub-zero temperature forming processes for lightweight components - StepLight" ref. BIRD200305-2020 funded by the University of Padova, Department of Industrial Engineering.

Italian Manufacturing Association Conference - XVI AITeM   Materials Research Forum LLC
Materials Research Proceedings 35 (2023) 318-325   https://doi.org/10.21741/9781644902714-38

References

[1] H. Takuda, K. Mori, T. Masachika, E. Yamazaki, Y. Watanabe, Finite element analysis of the formability of an austenitic stainless steel sheet in warm deep drawing. J. Mat. Proc. Tech. 143-144 (2003) 242-248. https://doi.org/10.1016/S0924-0136(03)00348-0

[2] E. Simonetto, R. Bertolini, A. Ghiotti, S. Bruschi, Mechanical and microstructural behaviour of AA7075 aluminium alloy for sub-zero temperature sheet stamping process. Int. J. Mech. Sci. 187 (2020) 105919. https://doi.org/10.1016/j.ijmecsci.2020.105919

[3] W. Liu, H. Hao, Damage and fracture prediction of 7075 high-strength aluminum alloy during cryogenic stamping process. Mech. Mater. 163 (2021) 104080. https://doi.org/10.1016/j.mechmat.2021.104080

[4] S.M. Hussaini, S.K. Singh, A.K. Gupta, Formability and fracture studies of austenitic stainless steel 316 at different temperatures. J King Saud Univ-Eng Sci 26(2) (2014)184-190. https://doi.org/10.1016/j.jksues.2013.05.001

[5] J. Aegerter, H.J. Kühn, H. Frenz, C. Weißmüller. EN ISO 6892-1: 2009 tensile testing: Initial experience from the practical implementation of the new standard. Materials Testing (2011) https://doi.org/10.3139/120.110269

[6] ASTM E 112-96 Standard Test Methods for Determining Average Grain Size.

[7] K. Couturier, S. Sgobba, Phase stability of high manganese austenitic steels for cryogenic applications (No. CERN-EST-2000-006-SM) (2000).

[8] G.M. Karthi, E.S. Kim, P. Sathiyamoorthi, A. Zargaran, S.G. Jeong, R. Xiong, H.S. Kim, Delayed deformation-induced martensite transformation and enhanced cryogenic tensile properties in laser additive manufactured 316L austenitic stainless steel. Additive Manufacturing, 47 (2021) 102314. https://doi.org/10.1016/j.addma.2021.102314

[9] X. Li, Z. Wei, X. Wang, L. Yang, X. Hao, M. Wang, J. Guo, Effect of cryogenic temperatures on the mechanical behavior and deformation mechanism of AISI 316H stainless steel. J. Mater. Res. Technol. 22 (2023) 3375-86. https://doi.org/10.1016/j.jmrt.2022.12.190

Italian Manufacturing Association Conference - XVI AITeM   Materials Research Forum LLC
Materials Research Proceedings 35 (2023) 326-333   https://doi.org/10.21741/9781644902714-39

A novel application of cryogenics in dieless sheet metal piercing

Paolo Albertelli[1,a] *, Valerio Mussi[2,b] and Michele Monno[1,a]

[1]Mechanical Engineering Department, Politecnico di Milano, via la Masa 1, 20156 Milan, Italy

[2]Consorzio MUSP via Callegari, 29122 Piacenza, Italy

[a]paolo.albertelli@polimi.it, [b]valerio.mussi@musp.net, [c]monno.michele@polimi.it

Keywords: Piercing, Cryogenic, Constitutive Models

Abstract. In tube punching, if the internal die is necessary to properly pierce the tube avoiding its collapse. it also represents a bottleneck to a rapid change of the punching set. In this research an innovative dieless tube punching approach has been conceived and studied. The use of a cryogenic fluid to force the material ductile-brittle transition is a way to limit the sheet deformation during the piercing process. The analysis of the innovative cryogenic punching was carried out both adopting numerical and experimental methodologies. A finite element FE model of the cryogenic punching was developed and updated in two stages. First, experimental tensile tests, performed at cryogenic temperatures, were used to characterize some material properties. Secondly, some piercing tests in cryogenic conditions were performed at different velocities and temperatures to fine update the model. A validation session was carried out to assess the model and the process feasibility. It was found that the FE model reproduced the experimental results within a maximum estimation error of 10% on both the punching force and tube deflection. Results showed that both the increment of the punching velocity and especially the decrement of the punching temperature could be the only viable solution for making the tube dieless punching industrially feasible.

Introduction

The use of cryogenic fluids in manufacturing processes represents a rather new research branch. Jawahir et al. in [1] critically reviewed the main applications in which cryogenics are used in manufacturing and they outlined the most relevant opportunities and challenges in the field. Some aluminum alloys show a high deformability at cryogenic temperatures [2] that is used is in several processes like deep drawing or extrusion to produce more complex parts. Conversely, although the ductile-brittle transition of some materials (e.g., body-centered cube bcc materials) at low temperatures is well known, this material feature has not yet been exploited in manufacturing processes. To bridge this gap, in this study it has been investigated the possibility to exploit the ductile-brittle transition that occurs at cryogenic temperatures for making the dieless punching feasible from the product quality perspective. The novelty of this research is associated to the development of a cryogenic punching model that can be used to robustly estimate the required punching force and the tube deformation according to the specific adopted process parameters. The paper is structure as follows. In the material and methods session, a first process description was provided. The modelling aspects, both considering the FE modelling and the material rheological behavior were reported. The strategy to update the model and the design of experiments were also presented. The results and discussion section presents the main experimental findings and a proper validation of the conceived modelling approach.

Materials and Methods

In this paper section, a description of the innovative dieless cryogenic tube piercing process was provided. A conceptual comparison between the regular tube punching process and the cryogenic dieless punching was reported in Fig. 1 a) and b). Since some materials show a ductile-brittle transition at cryogenic temperatures, the possibility to pierce a tube without the internal die but, at the same time, limiting the sheet deformation has been investigated, Fig. 1 c) and d). The first implementation of the cryogenic punching process was depicted in Fig. 1 e). It is worth of noting that the tube is cooled by an inner flow on liquid nitrogen LN. The tube is hold by the bottom part of the external die while the punch and the upper part of the die can move vertically to allow the tube feed after the piercing execution. In this specific cryogenic punching release, a 30mmx30mm tube of S235JR steel with a thickness of 2.45mm was adopted. The punch diameter is equal to 10mm. Additional information on the experimental set-up were provided in the Design of Experimental session. In the following subsections details regarding the model development, calibration and validation are provided.

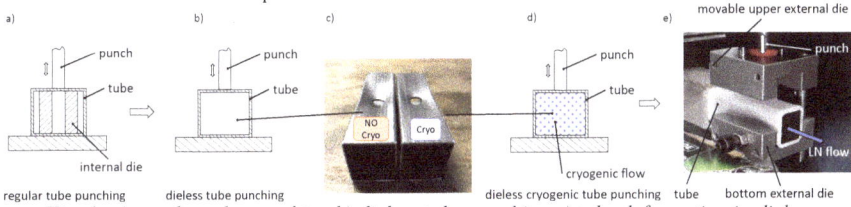

Fig. 1: a) regular tube punching b) dieless tube punching c) tube deformation in dieless punching (cryogenic vs no cryogenic) d) cryogenic punching e) cryogenic experimental set up.

Finite Element modelling
The FE model was developed in with ForgeNxT2.1. Three-dimensional finite elements (tetrahedrons with five nodes and linear interpolation) were used for meshing the punch and the tube. The implemented formulation allows considering both the deformation mechanics and thermal equations. An implicit integration scheme, with the Newton-Raphson method was used. An overview of the dieless punching process model, its main settings, the main approach adopted for its updating and validation are reported in Fig. 2 a) e c). Additional details will be provided in the model updating section. Four mesh regions were defined. The mesh size was refined in three specific regions as reported in Fig. 2 b). It was decided to model the punch as a rigid body since the deformation is mostly related to the tube. Even the tube supports (external equipment) were modelled as rigid bodies.

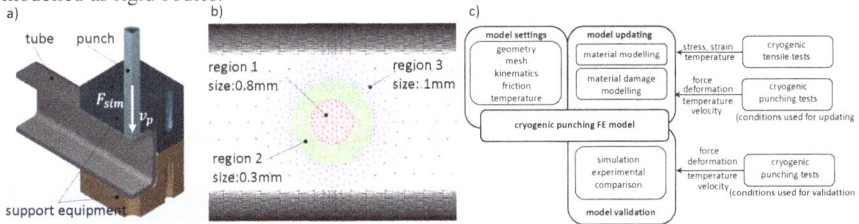

Fig. 2: a) Dieless tube punching modeling b) mesh regions c) model settings and updating

In accordance with the experimental set-up, the punch stroke was set equal to 8mm. The punch velocity v_p was considered constant during the punching. The range of 4-140mm/s for the punching velocity was considered for the simulations. An upper threshold for the punching force F_{sim} was imposed to reproduce the pressure limitations associated to the hydraulic units used for driving the punch, see the Design of Experiments session. Indeed, if the resulting punching force overcomes the threshold the simulation is quitted. Regarding the punching temperature T_p, according to the described implementation of the cryogenic process (Fig. 1), it was set uniformly into the whole tube while the punch temperature was set equal to the environmental temperature (20°C). The heat exchange between the tube and the punch/environment was limited since a steady state punching temperature was assured by a calibrated flow of liquid nitrogen inside the tube. For what concerns the punch-tube contact, the chosen friction model was a stick-slip Coulomb-Tresca model (Eq. 1). μ is the friction coefficient, m_f the friction factor, τ the shear stress, σ_n the normal pressure and τ_{lim} the yield shear stress which depends on the material flow stress σ according to Von-Mises.

$$\tau = \mu \cdot \sigma_n \qquad\qquad \text{when } \mu \cdot \sigma_n < m_f \cdot \tau_{lim}$$

$$\tau = m_f \cdot \tau_{lim} \qquad\qquad \text{when } \mu \cdot \sigma_n \geq m_f \cdot \tau_{lim}$$

(1)

In the developed model, the following values were adopted $\mu = 0.4; m_f = 0.8$.

Material modelling

It was decided to adopt the basic Johnson-Cook (JC) formulation [3], described as follows $\sigma = (A + B\varepsilon^n)\big(a + C \cdot ln(\dot{\varepsilon}/\varepsilon_0)\big)(1 - ((T - T_0)/T_m - T_0)^m)$. σ is the flow stress, $\dot{\varepsilon}$ the strain rate, T_m and T_0 are respectively the melting temperature and a reference temperature. A is the initial yield stress, B allows considering the material hardening phenomenon, n is the index associated to the strain sensitivity, C is the index that considers the dependency due to the strain rate, $\dot{\varepsilon}_0$ is the reference value for the strain rate and m is the coefficient that considers the temperature dependency. The innovative aspect of the present study is related to the characterization of the material behavior at the cryogenic temperatures. Thus, a strategy for the model parameter identification was conceived. For what concerns the damage modelling, it was decided to adopt the Latham-Cockcroft LC formulation $D = \int_0^{\varepsilon_f} \sigma_I d\varepsilon$ that is the most established approach for modelling fracture after large strain deformation. D is the threshold value that is used for eliminating the damaged elements during the simulations, σ_I is the highest principal component of the stress tensor, and ε_f is the limit fracture strain. More specifically, the critical damage value is calculated for each element under deformation at each time-step. Once the damage value in an element reaches the critical one, a crack is initiated in two steps: (i) this element is deleted with all the parameters related to it, including element connectivity definition, the strain and stress values; (ii) the rough boundary produced by element deletion is smoothed by cutting out the considered rough angle and adding new points.

Model Simulations and settings

The developed model was used to simulate the dieless punching process. Specifically, the tube deformation d_t (including the maximum value $d_{t,max,sim}$, the stress and the strain distribution and the punching force F_{sim} (including its maximum value $F_{sim,max}$) can be estimated. For sake of example, some simulation outputs were shown in Fig. 3 a) and b) and c).

Model Updating Strategy

The model updating procedure is summarized in Fig. 3 d). For what concerns the Johnson-Cook model parameters, the following strategy was adopted. A, B, n were identified from quasi-static test (tensile tests) at the reference temperature (-80°C). The tensile tests were executed considering a reference strain rate ($\dot{\varepsilon}_0 = 0.002/s$) to neglect the strain rate effect. The adopted conditions also allow neglecting the temperature related term. Three specimens were used, and the average nominal stress-strain curve was computed for the identification process. The approach suggested in Schwab and Harter [4] was adopted to estimate the true stress strain curve $\sigma = f(\varepsilon)$ from the corresponding nominal quantities (s and e). The identification of the parameters was twofold. First, the A parameter was identified as the initial yielding stress.

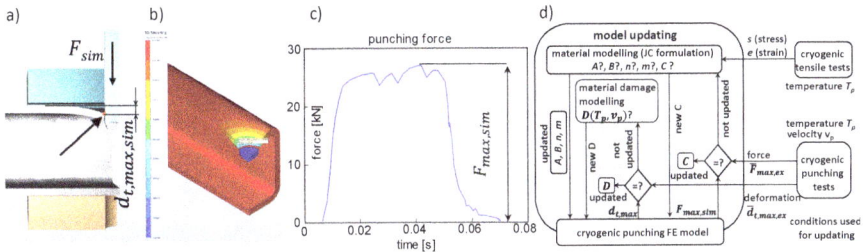

Fig. 3: Example of FE model simulations a) maximum tube deflection estimation $d_{t,max,sim}$ b) deformation map c) punching force $F_{max,sim}$ d) model updating procedure

In the second step, a regression procedure performed on the tensile test data corresponding to the homogeneous plastic deformation region, thus excluding both the elastic and necking region, was carried out for the identification of B and n. m is calculated using the results of tests (tensile tests) at different temperatures. (-80°C, -60°C, -40°C). For each temperature T, the stress σ data associated to different strain values ($\varepsilon_1 = 0.1, \varepsilon_2 = 0.15, \varepsilon_3 = 0.2$) were used. Starting from the JC formulation and introducing $K = (A + B\varepsilon^n)$, the following relationships can be obtained $ln(1 - \sigma/K) = m \cdot ln((T - T_0)/(T_m - T_0))$. As a result, the identification of the m parameter was done through a linear regression of the experimental data. See Results and discussion session session, Table 3. C is updated comparing the results (simulations and experimental data) of the punching tests in terms of cutting forces (F_{max}). In Fig. 3 c) an example of the simulated piercing force was reported. This approach was used to characterize the contribution of the strain rate in realistic conditions that are not achievable through tensile tests since the feed velocity of the tensile machine crossbar is generally rather limited. Additional details on the experimental force measurements were provided in the next paper session. Regarding the LC model parameter D, in this work, it was decided to characterize the material damaging considering the combined effect of temperature and strain rate. To accomplish it, a procedure based on experimental punching tests performed at different punching temperatures T_p and different punching velocities v_p was adopted. As a result, the formulation of the LC model can be resumed as $D = D(v_p, T_p)$. According to what reported in Fig. 3 d), for each tested condition, the FE model was updated, properly tuning the D parameter, to get the experimental maximum tube deflection $d_{t,max}$. Additional details on experimental measurement of the tube deflection were provided in the next session.

Design of Experiments and equipment description

Tensile tests

As previously described in the Model Updating Strategy, the tensile tests were carried out at five T_t temperatures (-80°C, -60°C, -40°C, 0°C and 20°C). For each temperature, three specimens were tested. According to ISO 6892-3 [5], a flat tensile test specimen with shoulders was used, Fig. 4 c). The specimen was characterized by a nominal cross-sectional area (6mm (width) x 2.45mm (thickness)) and a calibrated length equal to 25mm. The specimens were manufactured through abrasive water jet technology starting from the same tube batch. In order to take into account the process accuracy and variability, the cross-sectional dimensions of each single specimen were measured. The MTS Alliance RF/l50 tensile test machine, equipped with the MTS 651-environmental chamber, was used, Fig. 4. The machine has a thermal room able to reach different temperatures. In these tests the room was fed by nitrogen until the defined temperature is reached. The specimens to be tested at this temperature were positioned inside the chamber to reach the thermal equilibrium.

Italian Manufacturing Association Conference - XVI AITeM   Materials Research Forum LLC
Materials Research Proceedings 35 (2023) 326-333   https://doi.org/10.21741/9781644902714-39

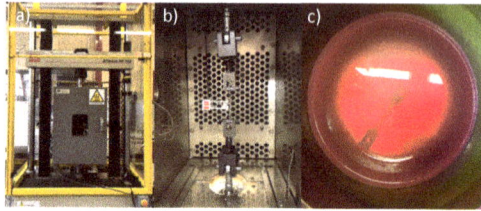

Fig. 4: a) Tensile test machine b) tensile test c) cryogenic chamber and specimen

The velocity of the crosshead that holds the specimens was equal to $4\,mm/min$, that corresponds to $\dot{\varepsilon}_0$. During the tensile tests, the crosshead position, the extensometer data (specimen extension) and the applied load were acquired with a time span of 0.1s. The data were used to compute the nominal s and e.

Punching tests

According to the model updating approach conceived for the identification of both the C parameter (JC) and the damage material behaviour $D = D(v_p, T_p)$, it was necessary to perform piercing tests in different conditions both in terms of temperatures and punching velocities. For this purpose, a 2-level factorial approach with a center point (intermediate conditions) was adopted. For each punching conditions three repetitions were executed. The punching tests were executed in randomized order. Table 1 resumes the tested conditions. The conditions used for updating the model and the ones used for the model validation were indicated, last column of Table 1.

Table 1: design of experiments – punching tests

Test condition ID	hydraulic unit HU	punching velocity $v_p[mm/s]$	punching temperature $T_p[°C]$	updating parameter (measured quantity)
ID1	HU1	$v_{p,high}$	$T_{p,high}$	updating D($d_{t,max,ex}$), C($F_{max,ex}$)
ID2	HU1	$v_{p,high}$	$T_{p,mid}$	updating C($F_{max,ex}$)
ID3	HU1	$v_{p,high}$	$T_{p,low}$	updating D($d_{t,max,ex}$)
ID4	HU2	$v_{p,mid}$	$T_{p,high}$	validation
ID5	HU2	$v_{p,mid}$	$T_{p,mid}$	updating D($d_{t,max,ex}$)
ID6	HU2	$v_{p,mid}$	$T_{p,low}$	validation
ID7	HU3	$v_{p,low}$	$T_{p,high}$	updating D($d_{t,max,ex}$)
ID8	HU3	$v_{p,low}$	$T_{p,mid}$	validation
ID9	HU3	$v_{p,low}$	$T_{p,low}$	updating D($d_{t,max,ex}$)

More specifically, in accordance with the explanation provided in the previous session and schematized in Fig. 3 d), Table 1 describes the experimental data, reported in brackets, used for tuning the specific model parameter. The FE model parameters tuning was considered achieved when the experimental/simulated matching was limited to a percentage error of about 3%. The punching tests were carried out with the equipment described in Fig. 5. The actuator cylinder Fig. 5 a) and b) directly drives the punch through the pressurized oil provided by the hydraulic unit HU Fig. 5 a). The punching temperature was set making the LN flows inside the tube, see Fig. 1 b) and Fig. 5 c). The LN flow rate was tuned, through the regulating valve, to set the desired tube temperature T_p. A thermocouple (see Fig. 5 c)) was used to monitor the temperature. Three temperature levels were considered $T_{p,low} = -80°C$, $T_{p,mid} = -40°C$ and $T_{p,high} = 20°C$. To change the punching velocity v_p, three different hydraulic units were used, see Table 1. Since no detailed information about the hydraulic units were available, an experimental characterization

Italian Manufacturing Association Conference - XVI AITeM   Materials Research Forum LLC
Materials Research Proceedings 35 (2023) 326-333   https://doi.org/10.21741/9781644902714-39

was carried out. Several air punching tests (without the tube) were executed to measure the average punching velocity. This was done exploiting an eddy current sensor that was installed to detect the punch motion.

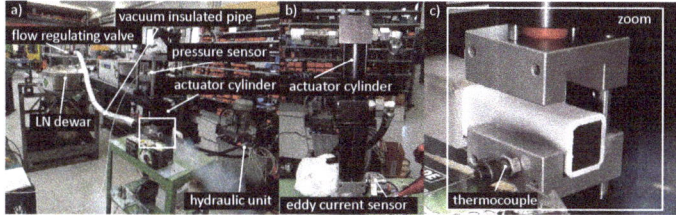

Fig. 5: a) experimental set up and sensors, b) actuator cylinder, c) process details and thermocouple for temperature measurements

From this experimental characterization it was found that respectively HU1 allowed to set a nominal punching velocity $v_{p,high} = 731.4mm/s$, HU2 a punching velocity of $v_{p,mid} = 156.6mm/s$ and the HU3 a punching velocity of $v_{p,low} = 44.5mm/s$. Since each hydraulic unit has its own control unit and its dynamic properties, the real average punching velocities were also computed exploiting the experimental piercing data. It is worth of noting that the calculated real velocities and the maximum achievable pressures were properly set in the simulation FE model. As previously described, both for model updating and for validation, the maximum punching force F_{max} and the maximum tube deflection $d_{t,max}$ were exploited. The punching force $F_{max,ex}$ was estimated measuring the cylinder inner pressure (pressure sensor, Fig. 5 a) and knowing the cylinder area that is equal to 1256mm^2. The tube deflection was measured scanning the upper tube area with Alicona infinite focus. For each tube, the $d_{t,max,ex}$ were measured. Since three repetitions were carried out for each piercing condition, the average values were computed $(\bar{d}_{t,max,ex}, \bar{F}_{max,ex})$.

Results and discussion
In this section the main results in terms of material modelling development and cryogenic punching modelling were reported. As described in the Model updating strategy section, most of the JC parameters (A, B, n, m) were identified from the tensile test data. All the identified parameters were reported in Table 2. Fig. 6 a) shows the results of experimental tensile tests performed at different temperatures while Fig. 6 b) shows the effect of the temperature on the identified JC model. Regarding the Lathan Cockcroft model, the following formulation (Eq. 2) was developed through a regression procedure that shows a $R^2 = 0.98$ and an adjusted value of $R^2_{adj} = 0.91$.

$$D = D(v_p, T_p) = 0.831 - 0.000154 \cdot V_p + 0.003278 \cdot T_p + 2.761 \cdot 10^{-5} \cdot V_p \cdot T_p \qquad (2)$$

Table 2: JC identified model parameters (reference temperature -80°C)

A [MPa]	B [MPa]	n	C	m	$\varepsilon_0 [s^{-1}]$
460	150	0.1	0.045	0.53	0.002

Italian Manufacturing Association Conference - XVI AITeM   Materials Research Forum LLC
Materials Research Proceedings 35 (2023) 326-333   https://doi.org/10.21741/9781644902714-39

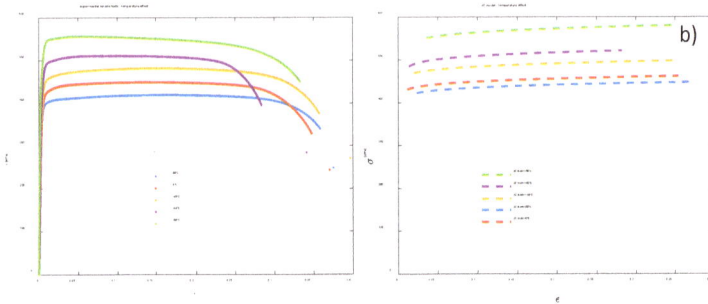

Fig. 6: effect of temperature: a) experimental tensile test data (average on three specimens) b) identified JC model and effect of temperature.

Both the formulations (refer to Table 2 and Eq. 2) were implemented in Forge to perform the punching simulations. As anticipated in Table 1, some of the test conditions were used to calibrate the model according to the experimentally measured tube deflection and punching force while some other test conditions were used to validate developed FE model. Among brackets were reported the updated numerical quantities. Table 3 resumes the main achieved results in the tested conditions. The percentage errors in both the maximum force $e_F = 100(F_{max,sim} - \bar{F}_{max,ex})/\bar{F}_{max,ex}$ and tube deflection estimation $e_d = 100(d_{t,max,sim} - \bar{d}_{t,max,ex})/\bar{d}_{t,max,ex}$ were reported. In addition, the real average punching velocities were also shown. As anticipated, they were computed for each HU and considering each specific punching conditions (ID). They were estimated analysing the experimental pressure profiles. It was found that the real punching velocities are much lower than the corresponding nominal ones (measured during air punching experiments) due to the dynamic behaviour of the HU and of its control unit under the effect of the punching load.

Table 3: Experimental results and validation of the model.

ID	HU	v_p [mm/s]	T_p [°C]	$\bar{F}_{max,ex}$ [kN]	$F_{max,sim}$ [kN]	$\bar{d}_{t,max,ex}$ [mm]	$d_{t,max,sim}$ [mm]	e_F [%]	e_d [%]
ID1	HU1	137	20	23	(23.2)	3.95	(3.82)	(0.9)	(-3.3)
ID2	HU1	105	-40	25.1	(25.1)	3.48	3.38	(0)	-2.9
ID3	HU1	85	-80	28.6	27.4	2.79	(2.74)	-4.2	(-1.8)
ID4	HU2	20	20	22.2	22.23	3.92	(3.79)	0.1	(-3.3)
ID5	HU2	10	-40	25.3	24.4	3.89	3.53	-3.6	-9.3
ID6	HU2	4.5	-80	28.3	25.43	3.53	(3.44)	-10.1	(-2.5)
ID7	HU3	20	20	22.5	22.23	3.9	(3.79)	-1.2	(-2.8)
ID8	HU3	20	-40	25.7	24.9	3.55	(3.49)	-3.1	(-1.7)
ID9	HU3	20	-80	28	26.3	3.16	3.24	-6.1	2.5

It can be observed that the model reproduces the experimental data with a maximum error of 10% both on the punching force and tube deflection. This demonstrates that the developed material models, suitable for describing its behaviour at cryogenic temperatures, and the FE model can be considered rather accurate and can be used for predicting the sheet deformation and the punching force involved in the process. The developed model can be rather useful to design a new dieless punching system especially in the definition of its specifications in term of maximum punching force, punching velocity and punching temperature to be adopted. Referring to the specific analyzed case, considerations on the effect of the punching velocity and temperature can be carried

Italian Manufacturing Association Conference - XVI AITeM   Materials Research Forum LLC
Materials Research Proceedings 35 (2023) 326-333   https://doi.org/10.21741/9781644902714-39

out. Specifically, the temperature has a strong effect on both the punching force and on the maximum tube deflection. It was found that averagely the punching force increases of about 20% decreasing the T_p from 20°C to -80°C while the tube deflection decreases of about 29%. The effect of the punching velocity is limited with respect to the one associated to the temperature. This is more evident on the tube maximum deflection than on the punching force. Moreover, it seems that the punching velocity shows its effect mainly at cryogenic temperatures (-80°C). Indeed, in such conditions the deflection decreases of about 20% if the punching velocity is increased from 4.5mm/s to 85mm/s. At -40°C the tube deflection decreases of about 12% increasing the punching velocity from 10mm/s to 105mm/s while at 20°C the effect of the punching velocity seems negligible.

Conclusions

In this research, the use of cryogenics for dieless tube punching was explored. Both experimental cryogenic punching tests and numerical simulations of the process were carried out. Specifically, a FE model was developed and properly updated in terms of rheological material behavior. For this purpose, both punching tests, executed at different temperatures and velocities, and tensile tests carried out at different velocities were done. The experimental data (punching force and tube deformation) were exploited to identify the parameters of both the JC and LT models that feed the FE model. The estimating capabilities of the FE model were assessed considering some validation test conditions. The model showed quite good performances. Indeed, estimation errors lower than 10% on the maximum punching force and maximum tube deflection were observed. It was noted that the punching velocity seems affecting the tube deformation only at cryogenic temperatures. Although the execution of the dieless punching at cryogenic temperatures (-80°C) with the maximum achievable punching velocity (considering the equipment available in this research) allowed reducing the tube deformation, the result seems still much higher than the ones obtained with the regular tube punching process. According to the described results, to make the cryogenic punching a feasible solution for piercing the tubes without the internal die, it would be necessary to investigate the performance even setting lower punching temperatures and especially imposing a much higher velocity to the punch. For this purpose, a specific punching control unit should be developed. An electric drive seems more adequate for the implementation of this innovative process.

References

[1] I.S. Jawahir, H. Attia, D. Biermann, J. Duflou, F. Klocke, D. Meyer, S.T.T. Newman, F. Pusavec, M. Putz, J. Rech, V. Schulze, D. Umbrello, Cryogenic manufacturing processes, CIRP Ann Manuf Technol. 65 (2016) 713–736. https://doi.org/10.1016/j.cirp.2016.06.007

[2] X. Wang, X. Fan, X. Chen, S. Yuan, Forming limit of 6061 aluminum alloy tube at cryogenic temperatures, J Mater Process Technol. 306 (2022) 117649. https://doi.org/10.1016/J.JMATPROTEC.2022.117649

[3] P. Albertelli, M. Strano, M. Monno, Simulation of the effects of cryogenic liquid nitrogen jets in Ti6Al4V milling, J Manuf Process. 85 (2023) 323–344. https://doi.org/10.1016/J.JMAPRO.2022.11.053

[4] R. Schwab, A. Harter, Extracting true stresses and strains from nominal stresses and strains in tensile testing, Strain. 57 (2021). https://doi.org/10.1111/STR.12396

[5] ISO 6892-3, Metallic Materials - Tensile testing - part 3, (2015).

Italian Manufacturing Association Conference - XVI AITeM
Materials Research Proceedings 35 (2023) 334-341

Materials Research Forum LLC
https://doi.org/10.21741/9781644902714-40

UV picosecond laser processing for microfluidic applications

Vincenzina Siciliani[1,a*], Alice Betti[2,b], Claudio Ongaro[2,c], Leonardo Orazi[1,d], Barbara Zardin[2,e], Barbara Reggiani[1,f]

[1] DISMI - University of Modena and Reggio Emilia, via Amendola 2, 42122 Reggio Emilia, Italy

[2] DIEF – University of Modena and Reggio Emilia, via Vivarelli 10, 41125 Modena, Italy

[a]vincenzina.siciliani@unimore.it, [b]alice.betti@unimore.it, [c]claudio.ongaro@unimore.it, [d]leonardo.orazi@unimore.it, [e]barbara.zardin@unimore.it, [f]barbara.reggiani@unimore.it

Keywords: Laser Processes, Laser Machining, Micro Machining

Abstract. In recent years, the fields of nanomedicine and nanopharmaceuticals have seen extensive use of microfluidic technologies. A microfluidic system transports fluids through micrometer-sized channels usually generated by replication on silicones. However, using laser technology and glass material together reduces manufacturing time, while maintaining high accuracy, and has the greatest benefit of being flexible. In this frame, the work aims to evaluate the potential of an ultrafast laser source for rapid and precise prototyping of a glass micromixer device. The study involves scanning electron microscopy to analyze the morphology, and confocal microscopy to investigate the topography of the sample. In addition, the study investigates the main process strategy to gain optimization of the processing time. Finally, the functionality of the manufactured devices is assessed, through a mixing test of two fluids with different pH.

Introduction

Nanomedicine and nanopharmaceuticals are two fields where microfluidics has potential as an emerging technology. A microfluidic system can carry fluids through channels with a micrometer dimension and can translate the distinctive properties of nanomaterials into therapeutic products. Different materials, including silicon, glass, hydrogel, paper, and polymers, and different technologies can be used to create these devices.

While polydimethylsiloxane (PDMS) silicone is the most used material in laboratory research, due to its ease of production and excellent resolution, glass has advantages over PDMS for microfluidic applications which requires resistance to various chemical and organic solvents, and good thermal and mechanical stability. Particularly, the "replica molding" process involves curing PDMS through photolithography on a mold that creates a negative image of the circuit to be made. Afterward, a glass layer is plasma-bonded onto the PDMS to seal the circuit. This implies creating a mold for each specific circuit [1, 2]. So, thinking about the device process chain, producing micro-molds required for PDMS-based systems is a time-consuming process that affects development and prototyping.

For applications where a batch may consist of several hundred pieces and where prototyping is a crucial step, the use of glass in combination with laser capability can be particularly advantageous. Examples of these applications include the production of diagnostic kits or drug delivery systems, and the creation of micromixers for the precipitation of nanoparticles for active ingredients [3], respectively. Particularly, the variety of applications of a micromixer device range from the synthesis of several types of nanoparticles to cell separation and microfiltration, and others that are continuously expanding.

Ultrashort laser ablation is considered an ideal method for the rapid and precise prototyping of microdevices since only a minimal amount of material is removed during the process: several recent works have used this technology to engrave channels on glass in different ways [3–10].

Italian Manufacturing Association Conference - XVI AITeM   Materials Research Forum LLC
Materials Research Proceedings 35 (2023) 334-341   https://doi.org/10.21741/9781644902714-40

Three different processes of 3D laser processing are proposed in [9], including laser ablation, laser reduction, and laser-induced surface nano-engineering. While the work [4] reviews the ultrafast laser-based process to fabricate nanofluidic systems, such as additive laser or subtractive laser, ablation or assisted etching. They present the challenges for issues concerning channel sizes and fluid dynamics, and some integrated solutions of micro/nano devices. These studies follow the approach of performing in-bulk processes, within the material volume, often assisted by chemical etching.

A more consolidated approach is to create the 2.5D structures on the surface of the glass and seal it with a silicon cover, to create a closed channel [8, 11]. By allowing the depth and width of each channel to vary, this approach offers a straightforward method to increase the precision of the generated geometries [12]. An aspect to consider is that the microchannel surface roughness can be altered by laser irradiation. Compared to longer laser pulse durations, the issue for ultrashort laser pulses is significantly reduced, and in some cases, may not require post-treatment. Ultrashort laser processing characterized by the removal of material without significant heat transmission to surrounding areas is also interesting for good quality microdrilling [8, 13], where the thermal load is critical for successful processing.

Therefore, in this work, the aim is to create a hybrid microfluidic mixer using laser ablation to create microchannels on glass, followed by plasma bonding of a silicone layer with punched holes for the device inlet and outlet. In addition, the possibility of laser drilling the holes directly in the glass will be demonstrated, keeping all conclusions about the functionality of the device valid. Once the channel geometry has been selected to allow mixing between the fluids fed into the system, the accuracy that the laser can achieve will be demonstrated. Then, the mixing efficiency of the device will be qualitatively evaluated by pH measurement of the outlet flow of two fluids with different inlet pH.

Materials and Methods
The material used in this study is a commercially available float glass of 76x26x1 mm. The same material was used in [12], where chemical and physical characteristics are given.

The geometry chosen for the micromixer is adapted from previous works [14–16]. It has a double "Y" shape with a pair of inlet branches, a pair of outlet branches, and a central main channel, as depicted in Figure 1 and Table 1. Each branch has an inlet or outlet seat, while holes replace the seats in the case of inlets and outlets drilled into the glass. Each pair of branches joins at opposite ends of the main channel in which the mixing unit, a series of grooves, is realized. These grooves, or so-called "herringbones", are asymmetrically V-shaped and are periodically reversed to form different cycles.

Table 1: Geometric parameters

Channel depth H [μm]	80
Grove depth h [μm]	30
Groove step s [μm]	90
Groove width w [μm]	45
Slope Θ [°]	60
Distance d [μm]	140
Number of grooves per cycle	5

Italian Manufacturing Association Conference - XVI AITeM   Materials Research Forum LLC

Materials Research Proceedings 35 (2023) 334-341   https://doi.org/10.21741/9781644902714-40

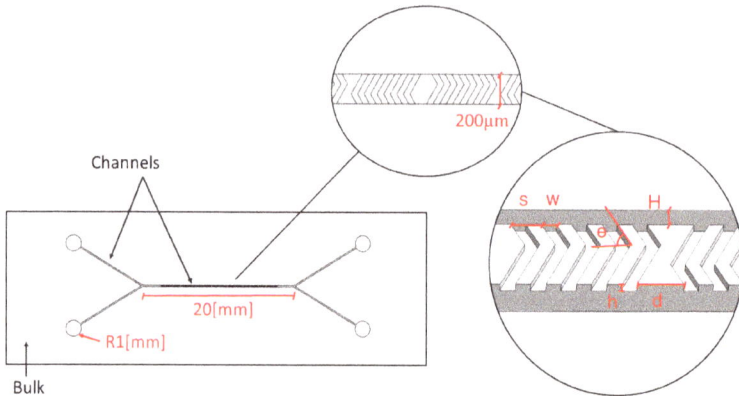

Figure 1: Perspective view of the device with detail on herringbones

This design exploits passive microfluidics by generating flow chaoticity over a wide range of Reynolds numbers to lead to mixing, see Figure 2. The geometry previously studied by Stroock et al. [16] was used as a starting point. The other dimensions were chosen taking into account the overall dimensions of the feeding accessories and the slide on which the system was derived [17].

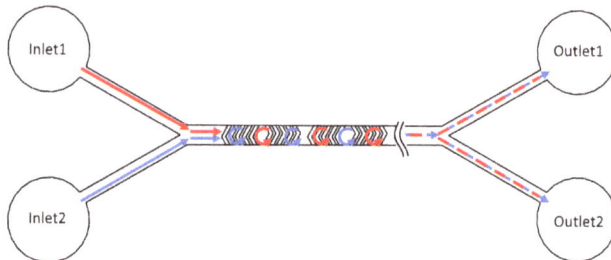

Figure 2: Fluid dynamics and mixing process

Channels have been etched onto a glass surface using a picosecond laser. The experimental set-up of the ultrafast laser system used consists of an EKSPLA Atlantic 50 laser source, emitting at the 355 nm wavelength, a Galvanometric scanner and a f-theta lens that focuses the beam to about 10 μm in diameter. The glass sample was processed by placing the ends on fixtures, suspending the lower surface in air during the ablation, to avoid back reflections.

The scan strategy was mainly a hatching of the geometry to be created. During the processing the galvanometer mirrors are kept in motion while working on laser emission, turning it on or off in creating the different geometries. In this way, micrometer-sized geometries can be processed with a scan speed of millimeters per second, reaching time resolutions of microseconds, removing the effects of accelerations, and optimizing the processing time. The laser parameters were chosen based on our previous work [11], optimizing them to achieve the accuracy and depths imposed by the device design. For the main channels the laser parameters chosen are 100 kHz of repetition rate, a pulse energy of 7 μJ, a scan strategy made of parallel longitudinal lines 3 μm spaced, and a scan speed of 300 mm/s to equally distribute the energy along and between the scanlines, all repeated for 11 passes. For the herringbones the same parameters were chosen, but with 6 passes.

Italian Manufacturing Association Conference - XVI AITeM   Materials Research Forum LLC
Materials Research Proceedings 35 (2023) 334-341   https://doi.org/10.21741/9781644902714-40

The total process time is about 5 minutes: the seats are the largest and slowest feature to manufacture. In this context, the drilled holes can be made on the 1 mm thick glass using a 5 μm spaced spiral scanned at 2500 mm/s, 300 kHz and 16 μJ in about 1.5 minutes per drill. The automated procedure consists of slowly lowering the scan head while performing the spiral to change the focal height and achieve breakage. Furthermore, in this way, all the processing steps are made on a single glass slide.

Morphological analysis of the samples was obtained using the secondary electron detector (ETD) of a Scanning Electron Microscope (SEM) and then compared with the topography obtained from confocal microscopy. The confocal microscope setup used in this study was a Leica inverted research microscope with motorized XYZ, 40x HCX objective at Oil immersion, white laser light (WWL) set at a wavelength of 532nm.

After ablating the channels using the laser, the device was assembled by bonding the glass with a commercial silicone layer to seal the channels.

A pH qualitative test was performed to evaluate the mixing functionality. An RWD R462 double syringe pump was connected to the two inlets of the device through Tygon® hosing with a 1.6 mm internal diameter, as in Figure 3. One syringe was filled with acetic acid, the other with distilled water.

The pump flow rate was set for a flow in the main channel of the device with a Reynolds number equal to 10, to have a perfectly laminar flow in the channel [15]. The pH of the two outlets of the device was checked using litmus paper.

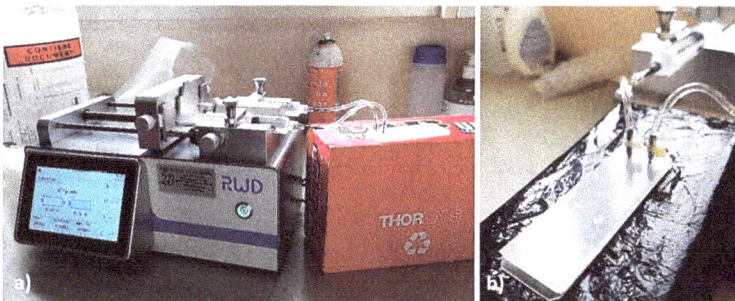

Figure 3: a) Feeding unit; syringe pump R462 RWD; b) Micromixer with detail on inlets and outlets.

Results and Discussion

Morphological analysis

Figure 4 shows SEM images of the main channel with its surface aspects. The topography looks chaotic and random. It is not easy to obtain a reliable measure of the areal parameters, but the resulting structures are at the micrometric and sub-micrometric scales, Figure 4– b).

Figure 5 shows a tilted image of the different details of the micromixer: a) the seat, b) the Y channel and the herringbones at two different magnifications, c) and d). Sharp edge delimiting the geometry, quasi-vertical sidewalls and the two-level depth are evidently achieved.

Figure 4: SEM images showing the microstructure of the channel obtained by laser ablation at two different magnifications, a) and b).

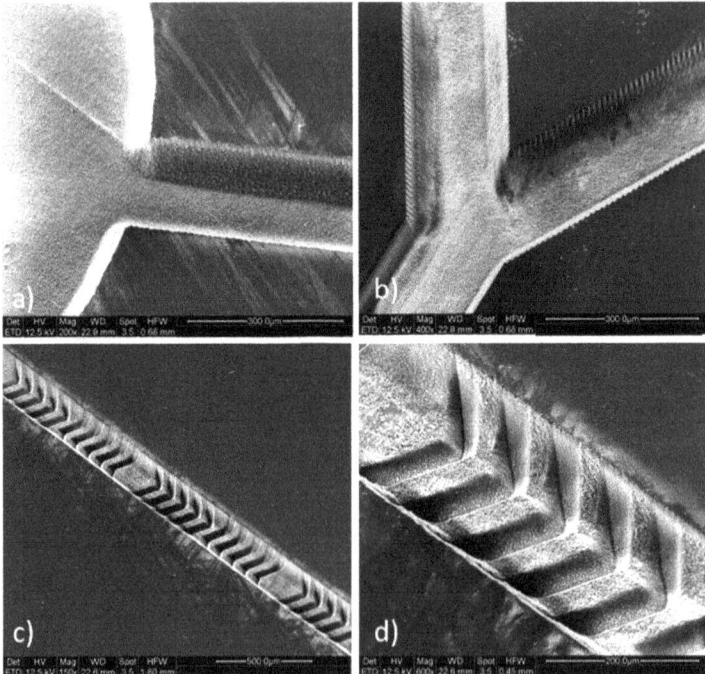

Figure 5: Tilted image of the different details of the micromixer: a) the seat, b) the Y channel and the herringbones at 2 different magnifications, c) and d).

The drilled seats, namely the holes alternative to the seats of Figure 5 - a), showed very good quality. The slight ellipticity in Figure 6-a) is probably due to the effects of linear polarization, beam ellipticity and astigmatism. It should be mentioned also that the scan strategy adopted for the holes was a spiral. The inlet flow port is gradual and smooth, Figure 6- b), with a weak conicity irrelevant to fluid entry and the device functionality.

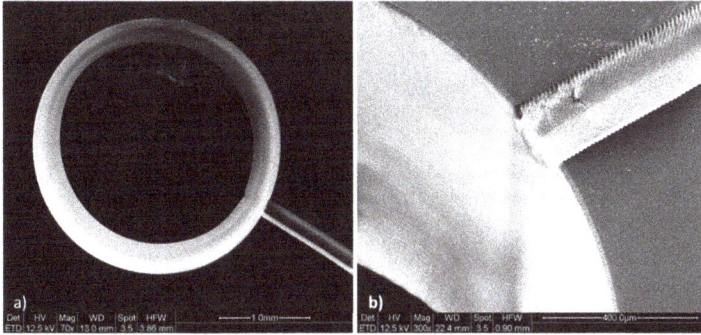

Figure 6: SEM images showing the quality of laser-drilled holes: a) frontal image b) focus on the fluid entry into the channel.

Topographic analysis

To evaluate the topography of the samples, images were acquired with a confocal microscope at different heights with a Z stack selection and a step size of 1 μm. The results can be seen in the 2D profiles with scale bars in Figure 7. The actual depth reached is about 80 μm for the main channel and 35 μm for the herringbone channel. The noise signals in Figure 7 may have been caused by artifacts generated by glass reflections, microscope data acquisition or data conversion software error. The inclination of the walls is constantly about 25°. It depends on reflections of oblique laser light at the vertical wall and plasma formation in the cavity. Consequently, the walls of the herringbones are almost V-shaped because of their narrow width.

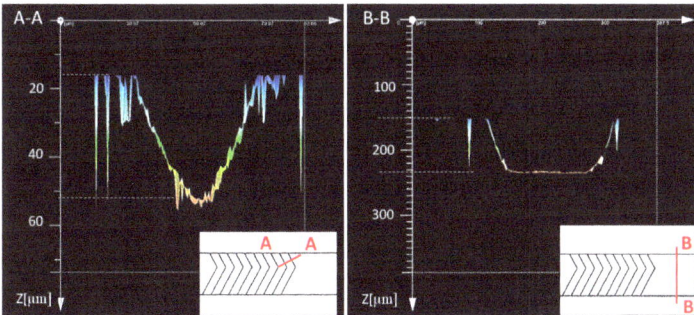

Figure 7: Cross-sectional profile reconstruction by confocal microscope of a single herringbone groove (A-A) and the section of the main channel (B-B).

Qualitative mixing test

The pH test results are reported in Figure 8, where "A" identifies the side of the micromixer branch where the acetic acid was introduced, "B" identifies the side of distilled water branch.

Comparing the colors obtained by dipping the litmus papers into the fluids at the outlets of the micromixer device without herringbone grooves, Figure 8– 1, and with herringbones grooves, Figure 8– 2, the pH test shows different results.

At the outlets of the device without grooves (1), the pH of the two fluids is easily distinguishable, the side "A" litmus paper has a deep red color that is associated with a pH ≅ 2, the side "B" litmus paper has a light orange corresponding to a pH ≅ 5. On the contrary, for the

Italian Manufacturing Association Conference - XVI AITeM

Materials Research Proceedings 35 (2023) 334-341

Materials Research Forum LLC

https://doi.org/10.21741/9781644902714-40

device with herringbones (2), the two litmus papers, "A" and "B", have an intermediate color to the previous two, between the pH =3 and pH = 4, showing the successful mixing of the two flows.

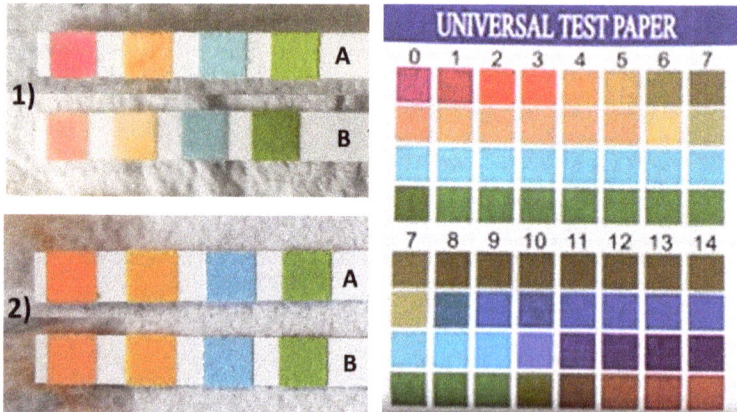

Figure 8: Litmus paper test where A = acid-side outlet channel and B=water-side outlet channel. Results at device outlets without (1) and with herringbones (2).

Conclusions

Ultrashort laser ablation offers a direct manufacturing approach for quick and accurate prototyping of glass microfluidic devices. The major advantages that this technology can offer are high accuracy, low manufacturing lead time and unprecedented flexibility. The beam can follow a wide variety of geometries easily generated by scanning software and modified as needed once functional tests have been performed, without previously creating a mold for each model, and that's a great advantage.

In this study, a complex channel geometry was selected to allow the mixing of two fluids, with dimensions ranging from millimeters to micrometers. Nevertheless, by choosing the optimal scanning parameters, the obtained geometry was faithful in shape. Certainly, with an appropriate set-up of laser processing parameters, i.e., the number of passes or the laser power, tight tolerance can be achieved.

Mixing between the two fluids has been shown to occur correctly, using a simple but effective qualitative test with litmus paper.

Additionally, the inlet and outlet holes, which are commonly made in the silicone layer, were laser drilled into the same glass slide on which the channels were made, resulting in excellent hole quality and a reduced device manufacturing time.

As a potential avenue for future work, it may be worthwhile to:
- seal the processed glass slide directly with another glass slide, i.e., by diffusion bonding, to obtain an entire glass device.
- quantitatively evaluate the mixing efficiency by introducing two different fluids of known composition into the device's inlets and measuring the resulting composition of the fluids at the outlets; this would provide a more accurate assessment of the degree of mixing.

In this way, ultrafast laser technology can be a viable alternative to fabricate highly functional biochips integrated with almost arbitrary shapes.

References

[1] Nguyen N-T, Wereley ST (2006) Fundamentals and applications of microfluidics, 2. ed. . Artech House, Boston, Mass.

[2] Bhushan B (ed) (2010) Springer Handbook of Nanotechnology. . Springer Berlin Heidelberg, Berlin, Heidelberg.

[3] Erfle P et al (2019) Stabilized Production of Lipid Nanoparticles of Tunable Size in Taylor Flow Glass Devices with High-Surface-Quality 3D Microchannels. Micromachines 10, 4 220. https://doi.org/10.3390/mi10040220

[4] Sima F, Sugioka K (2021) Ultrafast laser manufacturing of nanofluidic systems. Nanophotonics 10, 9 2389-2406. https://doi.org/10.1515/nanoph-2021-0159

[5] Kim J et al (2021) Two-step hybrid process of movable part inside glass substrate using ultrafast laser. Micro and Nano Syst Lett 9, 1 16. https://doi.org/10.1186/s40486-021-00142-3

[6] Qi J et al (2020) A Microfluidic Mixer of High Throughput Fabricated in Glass Using Femtosecond Laser Micromachining Combined with Glass Bonding. Micromachines 11, 2 213. https://doi.org/10.3390/mi11020213

[7] Li W et al (2020) A three-dimensional microfluidic mixer of a homogeneous mixing efficiency fabricated by ultrafast laser internal processing of glass. Appl. Phys. A 126, 10 816. https://doi.org/10.1007/s00339-020-04000-8

[8] Wlodarczyk K et al (2018) Rapid Laser Manufacturing of Microfluidic Devices from Glass Substrates. Micromachines 9, 8 409. https://doi.org/10.3390/mi9080409

[9] Yu Y et al (2018) Ultra-Short Pulsed Laser Manufacturing and Surface Processing of Microdevices. Engineering 4, 6 779-786. https://doi.org/10.1016/j.eng.2018.10.004

[10] Li Y et al (2011) Fabrication of microfluidic devices in silica glass by water-assisted ablation with femtosecond laser pulses. J. Micromech. Microeng. 21, 7 075008. https://doi.org/10.1088/0960-1317/21/7/075008

[11] Ongaro C et al (2022) An Alternative Solution for Microfluidic Chip Fabrication. J. Phys.: Conf. Ser. 2385, 1 012029. https://doi.org/10.1088/1742-6596/2385/1/012029

[12] Orazi L et al (2022) Ultrafast laser micromanufacturing of microfluidic devices. Procedia CIRP 110 122-127. https://doi.org/10.1016/j.procir.2022.06.023

[13] Orazi L et al (2021) Ultrafast laser manufacturing: from physics to industrial applications. CIRP Annals 70, 2 543-566. https://doi.org/10.1016/j.cirp.2021.05.007

[14] Du Y et al (2010) A simplified design of the staggered herringbone micromixer for practical applications. Biomicrofluidics 4, 2. https://doi.org/10.1063/1.3427240

[15] Forbes TP, Kralj JG (2012) Engineering and analysis of surface interactions in a microfluidic herringbone micromixer. Lab on a Chip 12, 15 2634-2637. https://doi.org/10.1039/c2lc40356k

[16] Hadjigeorgiou AG et al (2021) Thorough computational analysis of the staggered herringbone micromixer reveals transport mechanisms and enables mixing efficiency-based improved design. Chemical Engineering Journal 414. https://doi.org/10.1016/j.cej.2021.128775

[17] Ahmed F et al (2021) Design and validation of microfluidic parameters of a microfluidic chip using fluid dynamics. AIP Advances 11, 7. https://doi.org/10.1063/5.0056597

Italian Manufacturing Association Conference - XVI AITeM

Materials Research Proceedings 35 (2023) 342-349

Materials Research Forum LLC

https://doi.org/10.21741/9781644902714-41

Effect of carbon nanotube content on the mechanical behaviour of CFRP composite materials

Marina Andreozzi[1,a], Iacopo Bianchi[1,b], Archimede Forcellese[1,c],
Serena Gentili[1,d*], Luciano Greco[1,e], Tommaso Mancia[1,f], Chiara Mignanelli[1,g],
Michela Simoncini[1,h], Alessio Vita[1,i]

[1]Università Politecnica delle Marche, Dipartimento di Ingegneria Industriale e Scienze Matematiche, Via Brecce Bianche 12, 60131, Ancona, AN, Italy

[a]m.andreozzi@pm.univpm.it, [b]i.bianchi@pm.univpm.it, [c]a.forcellese@staff.univpm.it,
[d]s.gentili@pm.univpm.it, [e]l.greco@pm.univpm.it, [f]t.mancia@pm.univpm.it,
[g]c.mignanelli@pm.univpm.it, [h]m.simoncini@staff.univpm.it, [i]alessio.vita@staff.univpm.it

Keywords: Composite Materials, CFRP, Carbon Nanotubes, Mechanical Performances

Abstract. The present investigation aims at studying the effect of carbon nanotubes (CNTs) content on the mechanical performances of carbon fiber reinforced polymer (CFRP) composite materials. To this purpose, percentages of carbon nanotubes, ranging from 0.5 to 4%, were dispersed in the epoxy resin used to impregnate carbon fibers. Tensile and flexural tests were performed in order to evaluate the effect of CNTs content on the tensile and flexural performances of CFRP composites reinforced using different percentages of carbon nanotubes. Furthermore, a scanning electron microscopy analysis was carried out in order to analyze the dispersion of the CNTs in the composite laminates. The results showed that the addition of nanofillers up to the value of 3% leads to an improvement in the tensile and flexural performances of CFRP composites.

Introduction

Carbon Fiber Reinforced Polymer (CFRP) are increasingly used in the field of advanced materials for a wide range of sectors due to their high specific mechanical properties [1-4]. The most common CFRPs are obtained by combining thermosetting matrix with continuous carbon fibres. The properties of these materials can be further enhanced by the addition of nanoparticles dispersed within the CFRP polymer matrix [5]. Among others, carbon nanotubes (CNTs) are receiving most of the attention due to their remarkable reinforcement properties. Nanotubes can be single-walled (SWCNT) or multi-walled (MWCNT) [6]; the latter are easier and less expensive to produce than the former; furthermore MWCNTs have excellent properties in terms of stiffness, strength, thermal and electrical conductivity [7, 8].

Several studies have highlighted how the dispersion of CNTs within thermosetting matrix significantly improves the matrix properties in terms of strength, modulus and interlaminar shear strength [9-11]. In addition, the use of CNTs can improve the fatigue resistance of CFRP composites, making them stronger and more durable. For these reasons, the addition of CNTs to CFRP composite materials can offer several benefits and innovations in various industrial fields, including aerospace, automotive and sport equipment. CFRP composites with CNTs can be lighter than traditional composites and can lead to more efficient and sustainable components in service. Costs could also be lowered by dispersing carbon nanotubes within CFRP composites, as the improved mechanical behaviour of the materials [12] can reduce the amount of material required to achieve the desired performance. However, the high potential of CNT reinforced polymer composites can be exploited through the definition of the proper CNTs content; furthermore, the agglomeration of CNTs in the polymer matrix must be avoided since the agglomerates can lead to stress concentration that weakened the interface between CNT and matrix [13]. Several methods

Italian Manufacturing Association Conference - XVI AITeM
Materials Research Proceedings 35 (2023) 342-349

Materials Research Forum LLC
https://doi.org/10.21741/9781644902714-41

can be used to obtain a high quality dispersion, such as solution-assisted dispersion, tip/bath sonication, three-roll milling and thermal agitation [14].

Another problem presented by CNTs is the filtering effect that occurs during the manufacturing process of CFRPs, as it exploits long resin flows. This happens because the reinforcement retains the nanofiller during the matrix flow, preventing uniform dispersion. Then, the dispersion of CNTs in the matrix enhances its thixotropic behaviour, making the impregnation and the curing processes more complicated [13]. Therefore, the CNTs content must be carefully defined to improve mechanical performances and to guarantee uniform mechanical properties of the component, without promoting the agglomerates formation and avoiding the CNTs filtering effect. For this reason, even though some studies have been already presented on this topic, it is of extreme important to enrich the literature with a comprehensive study of the reinforcement effect of CNTs on composite laminates, especially at high dispersion concentrations. To do this, composite panels have been realized exploiting the liquid infusion process and varying the CNTs volume fraction in an epoxy matrix from 0 to 4%, with steps of 0.5%. This allows to evaluate both the filtering and the agglomerating processes of CNTs and how these affect the mechanical properties of laminates.

The characteristic of the CFRP composite laminates have been evaluated in terms of tensile and flexural behaviours and fractured surfaces were analysed by means of scanning electron microscopy. In addition, void contents in each composite panel have been measured in order to investigate the processability of unfilled and nanofilled matrices.

Materials and Experimental Procedures
Materials
The CNTs used in the present research are industrial-grade -OH functionalized multi-walled carbon nanotubes supplied by Nanoamor. The reinforcement used in the composite material is a TWILL-type carbon fiber, realized with an aerial weight of 200 g/m2. The thermosetting matrix is a high performing and low viscosity two component epoxy resin (EC157) and W152 amine hardener provided by Elantas. This matrix is widely used in the high-performance composites industry due to its outstanding mechanical properties that make it suitable for racing and aerospace applications. The CNTs were dispersed in the liquid resin through a patented three-roll milling process [15] developed by Nanotech S.p.A; different weight contents, ranging from 0.5 to 4% were investigated. After the dispersion, the liquid nanofilled resin was collected and mixed with the hardener in a mixing ratio of 100:30, as reported in the technical datasheet.

Manufacturing Process of Composite Laminates
The manufacturing processes used to fabricate the composite laminates is the liquid infusion, by which the matrix is heated to decrease the viscosity and, by means of a vacuum pump, is flowed inside the fibrous reinforcement, suitably disposed over a mold and closed in a vacuum bag. In order to facilitate the impregnation and to accelerate the process, a grid with high permeability is laid above the laminate, in this way it uniforms and speeds up the flow of resin, limiting the possibility of the formation of voids or dry areas.

Once the impregnation process is completed, the composite material is placed in an autoclave to allow the resin solidification under pressure. The process, according to the matrix data sheet, was carried out at 120°C for 6 hours at a pressure of 4 bar. Once the autoclave cycle was completed, a post-curing process was carried out at 120°C for 6 hours in an oven to achieve complete cross-linking. Figure 1 schematically shows the steps of the manufacturing process of the composite laminates reinforced by CNTs. This manufacturing process has been repeated to obtain composite laminates with different CNTs dispersion.

Italian Manufacturing Association Conference - XVI AITeM   Materials Research Forum LLC
Materials Research Proceedings 35 (2023) 342-349   https://doi.org/10.21741/9781644902714-41

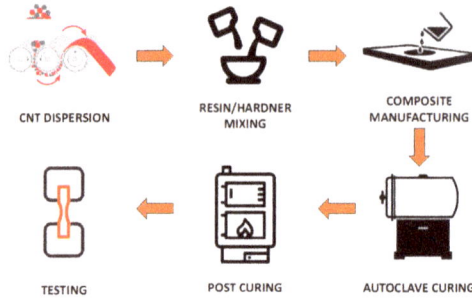

Fig. 1 Scheme of the manufacturing process of composite laminates with carbon nanotubes.

Void quantity analysis

The presence of voids in composite materials, that strongly reduces the material strength [16], was detected by performing the resin digestion test, according to the ASTM D3171 standard - procedure B "matrix digestion using sulfuric acid/hydrogen peroxide". This procedure allows the quantification of the constituent materials of the composite (matrix and fiber). This test, in addition to allowing the determination of the fiber-to-matrix ratio, allows the quantification of the void volume percentage, a key parameter in identifying the quality of a composite and its manufacturing process. In fact, void percentages greater than 2% are typically considered unacceptable in high demanding sectors, thus leading to laminate nonconformity [17].

Mechanical tests

In order to evaluate the effect of carbon nanotubes content on the mechanical performances of CFRP composite materials, tensile and flexural tests were performed at room temperature, according to the ASTM D3039 and ASTM D7264 – Procedure A standards, respectively. Such tests were carried out on the MTS 810® servo-hydraulic testing machine, with a load cell of 250 kN. The specimens were extracted from cured laminates, at different CNTs contents, by means of waterjet cutting operations. The tensile samples were characterized by a nominal length, width and thickness equal to 250, 15 and 1 mm, respectively. In order to reduce stress concentrations that can be caused by the clamping system of testing machine and to avoid premature failure, tabs in composite material, with a thickness of 1.5 mm, were bonded to each end of the tensile specimens using two components epoxy adhesive. Tensile tests were carried out at a loading rate of 2 mm/min. During tests, the load and nominal strain along the loading direction were acquired using the load cell and an extensometer; such results allowed to plot tensile stress vs. tensile strain curves, by which the maximum value of tensile strength was obtained.

In order to investigate the effect of CNTs content on the flexural behavior of the CFRP composite laminates, flexural tests were performed by three-point bending tests, carried out at a constant crosshead motion of 1 mm/min. To this purpose, rectangular specimens, 154 mm in length, 13 mm in width and 4 mm in thickness, were cut by laminates at different values in percentage of CNTs contents. Tests were carried out until fracture using an equipment consisting in a loading nose and two supports, characterized by cylindrical contact surfaces, finely ground surfaces free of indentation and burrs, with all sharp edges relieved. The diameter of the loading nose and supports span are respectively 5.0 ± 0.1 mm and 57.5 mm. The force applied to the specimen and resulting specimen deflection at the center of span was measured and recorded until the sample failure occurs on the outer surface. The experimental results were plotted as flexural stress vs. flexural strain curves, derived after the acquisition of punch load and punch stroke (ps) during experimental tests.

Italian Manufacturing Association Conference - XVI AITeM   Materials Research Forum LLC
Materials Research Proceedings 35 (2023) 342-349   https://doi.org/10.21741/9781644902714-41

To guarantee the repeatability of the results, five tensile and flexural tests were performed for each experimental condition investigated. The results obtained by tensile and flexural tests on samples obtained using different CNTs contents were compared to the mechanical behavior of the CFRP sample obtained using the net resin, in order to evaluate the influence of the CNTs content in percentage.

To evaluate more precisely the results obtained at different CNTs contents, a normalization process with respect to the fiber volume fraction was performed. As a matter of fact, the variation of fiber content in the composite laminate leads to significant changes in the mechanical properties of the laminates. Therefore, the results obtained from the mechanical tests were normalized to a quantity of fiber volume fraction equal to 60%, by taking into account the results of the resin digestion, as reported in composite materials [18].

Scanning Electron Microscopy
The fracture surfaces of tensile specimens in CFRP composite material, at different CNTs contents, were analyzed using the scanning electron microscope FESEM ZEISS SUPRA TM40, with compact GEMINI® objective lens, in order to acquire high magnification three-dimensional topography of materials and to evaluate the effect of the dispersion of the CNTs in the composite laminates. The samples for SEM investigation were coated by means of a metallization process in order to make it conductive for the analysis.

Results and Discussion
The presence of voids in laminates significantly reduces the strength of the material, degrading the physical and chemical properties of the fibers due to the consequent moisture absorption and crack propagation. As far as the evaluation of the void contents is concerned, typical results of resin digestion test, conducted according to ASTM D3171- Procedure B, are shown in Table 1, in which the results of the resin digestion test are reported.

Table 1: Results of the resin digestion

MEASURE	SAMPLES								
	0% CNT	0.5% CNT	1.0% CNT	1.5% CNT	2.0% CNT	2.5% CNT	3.0% CNT	3.5% CNT	4.0% CNT
Composite density [g/cm³]	1.506	1.492	1.477	1.457	1.449	1.462	1.451	1.516	1.467
Fiber volume fraction Vf [%]	50.85	50.17	47.92	46.35	43.20	47.22	46.21	52.12	46.16
Matrix volume fraction Vm [%]	48.78	49.06	51.25	51.97	56.16	52.17	51.97	46.85	53.11
Void volume fraction [%]	0.37	0.77	0.83	1.68	0.64	0.61	1.82	1.03	0.73

It can be observed that, irrespective of the CNTs content, the value of the void volume fraction in nano-reinforced composite materials is always higher than the unreinforced composite. Such a result can be attributed to the higher viscosity of the matrix, which makes the impregnation process more difficult. As the viscosity increases, resin flow between the fibers becomes more challenging, resulting in air trapping and void formation in the cured composite. In addition, the consolidation process that occurs in the autoclave is less effective when using high viscosity resins because their flow towards the breather and bleeder is lower. However, it should be noted that all measured

Italian Manufacturing Association Conference - XVI AITeM
Materials Research Proceedings 35 (2023) 342-349

Materials Research Forum LLC
https://doi.org/10.21741/9781644902714-41

values are below 2%, which is considered the maximum acceptable value for high-performance applications.

In Fig. 2, typical tensile stress vs. tensile strain curves and flexural stress vs. flexural strain curves are shown. As can be seen, the increase of CNTs content leads to an increase in both strength and stiffness of the laminates. This effect is more marked as the quantity of CNTs is the highest, whilst at values of 0.5 and 2% the mechanical performances improvement is limited.

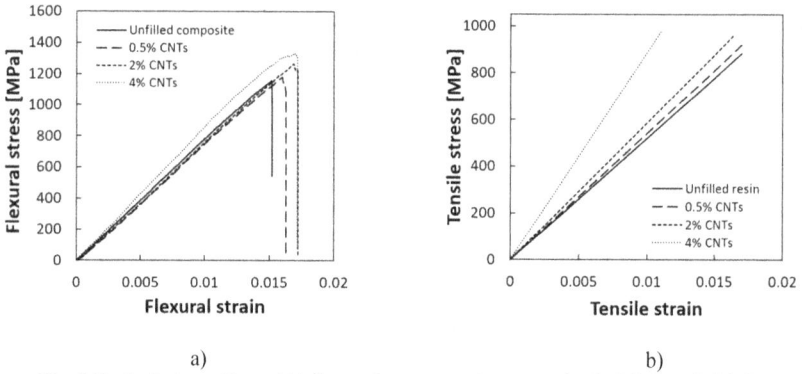

a)   b)

Fig. 2 Typical a) tensile and b) flexural stress strain curves for 0, 0.5, 2 and 4% CNTs concentration

Fig. 3a shows that, irrespective of the CNTs content, the ultimate tensile strength is higher than the one of the unfilled composites. Such increase can be attributed to the reinforcement effect of carbon nanotubes on the matrix, which results both in a higher strength at the matrix fiber interface which leads to an improved load transfer mechanism, and in a stiffening effect on the matrix. Furthermore, the elongation to failure tends to decrease with increasing CNTs content, resulting in a reduction of about 50% with respect to the unfilled composite at the highest CNTs content investigated. In the literature, it has been shown that for low CNTs content agglomerate formation is not observed; as an example, Forcellese et al. [19] demonstrates that samples reinforced using a multiwalled carbon nanotube content of 1% did not exhibit evident agglomerates. However, for higher CNTs content dispersed, conflicting results are reported. In one hand, some studies, such as the one carried out by Hong et al. [15], show an increase in mechanical performances, while in the other hand other research show a decrease due to the formation of agglomerates as the content of CNTs increases [20].

In this study, as far as the slope of the UTS vs. % CNTs curve is concerned, a decrease with increasing carbon nanotubes content is reported; such behavior is more evident as the CNTs content is higher than 2%. This effect can be related to the formation of agglomerates due to the occurrence of CNTs filtering effect. In fact, as the CNTs content increases, the strengthening effect produced by CNTs is mitigated by the weakening effect caused by the CNTs agglomerates.

a)   b)

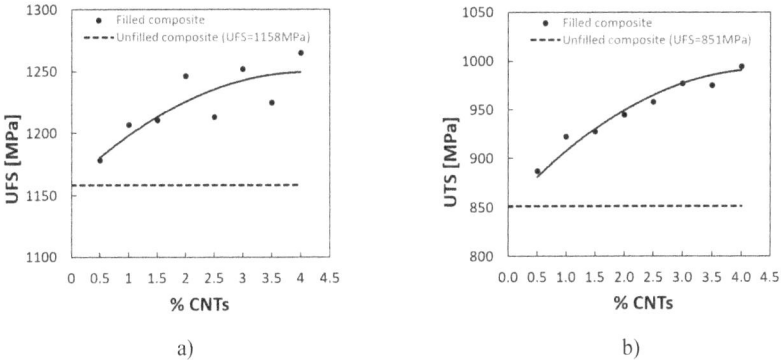

Fig. 3 (a) Maximum tensile stress values of CFRP composite material as a function of CNTs, (b) Maximum flexural stress of CFRP composite material as a function of CNTs

As far as the ultimate flexural strength (UFS) is concerned, a trend similar to the one exhibited by UTS is shown. As a matter of fact, Fig. 3b shows that UFS values of nanofilled composite are higher than the unfilled one; in addition, the UFS vs. % CNTs curve is characterized by a decrease in the slope with increasing CNTs content.

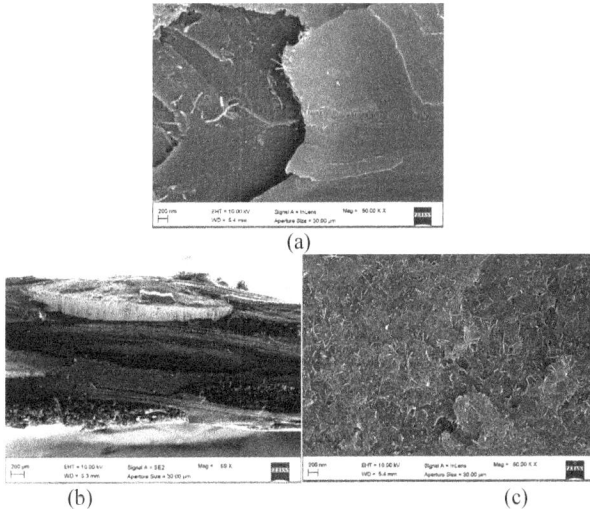

(a)

(b)   (c)

Fig. 5 SEM of fractured tensile specimen characterized by a CNTs content of: (a) 0.5, (b) 2 and (c) 4%.

Fig. 5 shows the SEM microscopies of the fracture surface of filled laminates at different CNTs contents. It can be observed that the CNTs tend to form agglomerates with the increase in CNTs content (Fig.s 5b and 5c). Such agglomerates lead to stress concentration and weakened CNT-matrix interfacial adhesion, resulting in a reduction of mechanical properties of the filled CFRP laminate. Furthermore, the matrix results in rough and jagged surface, denoting an increasingly

Italian Manufacturing Association Conference - XVI AITeM   Materials Research Forum LLC
Materials Research Proceedings 35 (2023) 342-349   https://doi.org/10.21741/9781644902714-41

ductile behavior as the CNTs content increases. On the other hand, at the lowest CNT content investigated, the surface appears smooth, indicating a more brittle failure of matrix (Fig. 5a).

Conclusion

The present study aimed at investigating the effect of carbon nanotube content on the mechanical properties of carbon fiber reinforced composites. Carbon nanotubes were dispersed in an epoxy resin using a patented process, and carbon fiber composites were realized by using resin infusion with an autoclave curing process. Laminates characterized by different CNTs weight contents (ranging from 0.5% to 4% in steps of 0.5%) were tested to evaluate tensile and flexural behavior and to measure void content. Fracture surfaces were analyzed using SEM images, to analyze the effect of CNTs content on failure mechanisms.

The main results can be summarized as follows:

- the increase in CNTs content leads to an increase in void quantities, due to the increase in resin viscosity, and a consequent poor fiber - matrix impregnation;
- the CNTs dispersion in composite laminates results in an improvement in tensile and flexural properties as compared to the unfilled composite;
- the ultimate tensile and flexural strength increase with CNTs content with a decreasing rate; such behavior tends to be more marked when a CNTs content of 2% is exceeded;
- SEM images showed that high CNTs content resulted in the formation of agglomerates, which caused stress concentration and weakened CNT-matrix interfacial adhesion.

Future research will focus on optimizing CNTs dispersion through appropriate processing techniques to minimize void formation and optimize the composite's mechanical properties.

References

[1] S. J. Jitha, A. Saritha, W. Runcy, J. Kuruvilla, G. Gejo, O. Kristiina, An introduction to fiber reinforced composite materials, Fiber Reinforced Composites: constituents, compatibility, perspectives and applications, Woodhead Publishing, (2021),1-24. https://doi.org/10.1016/B978-0-12-821090-1.00025-9

[2] M. Sreejith, R.S. Rajeev, Fiber reinforced composites for aerospace and sports applications, Fiber Reinforced Composites: constituents, compatibility, perspectives and applications, Woodhead Publishing, (2021), 821-859. https://doi.org/10.1016/B978-0-12-821090-1.00023-5

[3] A. Oludaisi, M. Thokozani, Industrial and biomedical applications of fiber reinforced composites, Fiber Reinforced Composites: constituents, compatibility, perspectives and applications, Woodhead Publishing, (2021), 753-783. https://doi.org/10.1016/B978-0-12-821090-1.00004-1

[4] I. Bianchi, A. Forcellese, M. Simoncini, A. Vita, Life cycle impact assessment of safety shoes toe caps realized with reclaimed composite materials, J Clean Prod, vol. 347, (2022) https://doi.org/10.1016/j.jclepro.2022.131321

[5] I. Bianchi, S. Gentili, L. Greco, M. Simoncini, Effect of graphene oxide reinforcement on the flexural behavior of an epoxy resin, Procedia CIRP, vol. 112, (2022) pp. 602-606. https://doi.org/10.1016/j.procir.2022.09.057

[6] M. F. Yu, O. Lourie, M. J. Dyer, K. Moloni, T. F. Kelly, R. S. Ruoff, Strength and breaking mechanism of multiwalled carbon nanotubes under tensile load, Science (1979), 287 no. 5453 (2000) 637-640. https://doi.org/10.1126/science.287.5453.637

[7] E. W. Wong, P. E. Sheehan, C. M. Lieber, Nanobeam mechanics: Elasticity, strength, and toughness of nanorods and nanotubes, Science (1979), 277 no. 5334 (1997), 1971-1975. https://doi.org/10.1126/science.277.5334.1971

[8] M. M. J. Treacy, T. W. Ebbesen, J. M. Gibson, Exceptionally high Young's modulus observed for individual carbon nanotubes, Nature, 381 (Jun. 1996) 678-680. https://doi.org/10.1038/381678a0

[9] A. Pantano, Mechanical Properties of CNT/Polymer, in: R. Rafiee (Eds.), Carbon Nanotube-Reinforced Polymers: From Nanoscale to Macroscale, Elsevier Inc., (2017), pp. 201-230. https://doi.org/10.1016/B978-0-323-48221-9.00009-1

[10] S. Imani Yengejeh, S. A. Kazemi, A. Öchsner, Carbon nanotubes as reinforcement in composites: A review of the analytical, numerical and experimental approaches, Comput Mater Sci, 136 (Aug. 2017) 85-101. https://doi.org/10.1016/j.commatsci.2017.04.023

[11] T. Glaskova-Kuzmina, A. Aniskevich, M. Zarrelli, A. Martone, M. Giordano, Effect of filler on the creep characteristics of epoxy and epoxy-based CFRPs containing multi-walled carbon nanotubes, Compos Sci Technol, vol. 100, (Aug. 2014) pp. 198-203. https://doi.org/10.1016/j.compscitech.2014.06.011

[12] A. Forcellese, M. Simoncini, A. Vita, A. Giovannelli, L. Leonardi, Performance analysis of MWCNT/Epoxy composites produced by CRTM, J Mater Process Technol, (2020). https://doi.org/10.1016/j.jmatprotec.2020.116839

[13] A. Forcellese, A. Nardinocchi, M. Simoncini, A. Vita, Effect of Carbon Nanotubes Dispersion on the Microhardness of CFRP Composites, Key Eng Mater, vol. 813, (2019). https://doi.org/10.4028/www.scientific.net/KEM.813.370

[14] S. N. Bhattacharya, M. R. Kamal, R. K. Gupta, Polymeric nanocomposites: Theory and practice, Polymeric Nanocomposites: Theory and Practice, pp. 20061-383.

[15] S. K. Hong, D. Kim, S. Lee, B. W. Kim, P. Theilmann, and S. H. Park, Enhanced thermal and mechanical properties of carbon nanotube composites through the use of functionalized CNT-reactive polymer linkages and three-roll milling, Compos Part A Appl Sci Manuf, vol. 77, (Oct. 2015), pp. 142-146. https://doi.org/10.1016/j.compositesa.2015.05.035

[16] P. Olivier, J. P. Cottu, and B. Ferret, Effects of cure cycle pressure and voids on some mechanical properties of carbon/epoxy laminates, Composites, vol. 26, no. 7, (1995). https://doi.org/10.1016/0010-4361(95)96808-J

[17] K. Naresh, K.A. Khan, R. Umer, W.J. Cantwell, The use of X-ray computed tomography for design and process modeling of aerospace composites: A review, Mater. Des, (2020). https://doi.org/10.1016/j.matdes.2020.108553

[18] Department of defense, handbook composite materials handbook vol. 1, Polymer matrix A.

[19] A. Forcellese, M. Simoncini, A. Vita, A. Giovannelli, and L. Leonardi, Performance analysis of MWCNT/Epoxy composites produced by CRTM, J Mater Process Technol, vol. 286, Dec. 2020, p. 116839. https://doi.org/10.1016/j.jmatprotec.2020.116839

[20] M. Tarfaoui, K. Lafdi, and A. El Moumen, "Mechanical properties of carbon nanotubes based polymer composites," Compos B Eng, vol. 103, (Oct. 2016), pp. 113-121. https://doi.org/10.1016/j.compositesb.2016.08.016

Italian Manufacturing Association Conference - XVI AITeM Materials Research Forum LLC
Materials Research Proceedings 35 (2023) 350-358 https://doi.org/10.21741/9781644902714-42

Static indentation properties of basalt fiber reinforced composites for naval applications

Chiara Borsellino[1,a], Mohamed Chairi[2,b], Jalal El Bahaoui[2,c], Federica Favaloro[1,d], Fabia Galantini[3,e] and Guido Di Bella[1,f] *

[1]Department of Engineering, Messina University, Messina 98166, Italy

[2]Department of Physics, Abdelmalek Essaadi University, Tetouan, Morocco

[3]Technical Department, Intermarine Spa, Sarzana 19038, Italy

[a]chiara.borsellino@unime.it, [b]mchairi@uae.ac.ma, [c]jelbahaoui@uae.ac.ma, [d]federica.favaloro@unime.it, [e]f.galantini@intermarine.it, [f]guido.dibella@unime.it

Keywords: Delamination and Debonding, Naval, Composites

Abstract. In recent years, the attention toward the use of basalt fiber reinforced composite in shipbuilding is significantly grown. Basalt is a green and environmentally friendly high-tech fiber made without environmental pollution. Among the natural fibers that can be used as reinforcement, it represents one of the most interesting due to its excellent mechanical properties. The goal of this research is to mechanically characterize some laminates used by Intermarine to make several structural or non-structural parts (i.e., hulls, deck), where the glass fibers are substituted with basalt ones at varying the manufacturing process (i.e., hand-lay-up and vacuum infusion). Specifically, static indentation tests were performed with different pin diameters (i.e., 17 mm, and 20 mm) and speeds (i.e., 1.25 mm/min, and 2.50 mm/min) to study the difference between glass and basalt in terms of resistance and failure modes.

Introduction

In shipbuilding, steel has been widely used because of its mechanical properties such as high tensile strength, yield strength, resilience, hardness, and weldability. However, steel structures are characterized by several problems such as the corrosive phenomena induced by the aggressive marine environment, and the pollution issues connected to the weight of the vessels and the consequent high amount of fuel required for navigation [1]. Recently, the attention has been focused on the concept of sustainable and efficient shipbuilding with the aim of reducing greenhouse gas emissions produced by navigation, seeking materials and techniques to increase sustainability, including the construction of lighter ship weights to reduce fuel consumption [2]. This has led to the need to use multi-material systems to make structural and non-structural parts of vessels, and the most suitable materials for this purpose are composites characterized by high corrosion resistance, tensile and shear strength, impact resistance, and low weight. Basalt is a natural material that is found in volcanic rocks. It is mainly used (as crushed rock) in construction, industrial and highway engineering [3]. Due to its good properties such as chemical stability, non-toxicity, non-combustibility, corrosion and high temperature resistance, thermal and acoustic insulation, low moisture absorption and better mechanical properties than those of E-glass ones [4-5], basalt fibres began to be used as a new reinforcing material for concrete [6-8] and polymer composites as well as for hybrid composite laminates in marine applications [9-11]. The main advantages that characterise these materials include the capability to create complex geometries and the lightness. However, they are susceptible to low velocity transverse impact damage, which can be occurred during construction (i.e., accidental impacts, "falling gear") or during the use (i.e., common impact events are collisions with floating debris, other vessels, docks, groundings, strandings, all of which are low velocity impacts) [12-13]. This fact is critical, taking in account that in the typical glass-polyester based laminates the failure can occur for low incident energies [14].

In this study some resin-based laminated composites with glass fibres were considered, actually these products are used by Intermarine at Sarzana shipyard [La Spezia (SP), Italy] in the military ships.

Three kinds of composites, employed in different parts of the ship, were considered. The first kind is used for structural and non-structural parts (hull, decks, bulkheads, ship tanks), the second is used for internal non-structural bulkheads, the third is a new proposal to be assigned as a function of its performances. The different composites are therefore realized with different lamination sequences, employing two different technologies (i.e., hand lay-up and vacuum infusion). More details are reported in the next paragraphs.

The study describes the results of the static indentation tests, developed in the context of a research project, where a complete mechanical characterisation of these products is required, to evaluate the possibility to replace the glass fibres with the basalt ones, without losing performances of the products.

The static indentation tests have been performed both with two different pin diameters and two pin speeds. The resistance of the laminate composites has been evaluated in terms of nominal maximum stress and the mode of fracture have been determined to compare the behaviour of the different materials.

Materials and method
Laminates panels preparation
The tested laminated panels were produced by Intermarine at Sarzana shipyard [La Spezia (SP), Italy]. These last were made with a size of 1x1 m, using two different resins (i.e., a polyester resin and a vinyl ester one) through two production technologies (i.e., hand lay-up and vacuum infusion). The raw materials used in the production of the composite materials are listed in Table 1, while Table 2 shows the configurations of the laminated panels.

Table 1: Raw materials.

Polyester resin SYNOLITE 288: 2 [kg]
Vinylester Resin ATLAC 580 AC 300: 2 [kg]
Catalyst TRIGONOX 61: 0.1 [kg]
Catalyst NOROX: 0.1 [kg]
Accelerator CP 12 PERGAQUICK: 40 [g]
Balsa 1" Diab: 0.5 [m²]
Fiber glass reinforcement 1100 [g/m²]- Jerago: 0.5 [m²]
Basalt fiber reinforcement: 0.5 [m²]

An identification code was used, i.e., IT "resin type"_"production type"_"fibre type"_"fibre orientation"_"indenter speed"_"indenter diameter"; where:
- IT means "item",
- "resin type" is P or V respectively for polyester or vinylester,
- "production type" is M or I respectively for manual lay-up or vacuum infusion,
- "fibre type" is G or B for glass or basalt,
- "fibre orientation" is "or" when the fibre are oriented,
- "indenter speed" is V1 for 1.25 mm/min and V2 for 2.5 mm/min,
- "indenter diameter" is IND1 for 17 mm and IND2 for 20 mm.

For example, the test with D=17 mm V=1.25 mm/min on the first kind of panel (realized with polyester resin by manual lay-up) is coded as: IT P_M_V1_IND1.

Table 2: Characteristics of laminates.

Panel ID	Thickness [mm]	Skin	Manufacturing Process	Resin	Lamination Sequence
IT_P_M_G	10	Single-skin	Manual	Polyester	6 layers glass
IT_V_I_G	6	Single-skin	Infusion	Vinylester	7 layers glass
IT_P_M_G_or	12	Single-skin (0°,45°)	Manual	Polyester	Orientation +45°/-45°/0°/+45°/-45°/90°
IT_P_M_B	8	Single-skin	Manual	Polyester	6 layers basalt
IT_V_I_B_or	6	Single-skin (0°,45°)	Infusion	Vinylester	6 layers basalt
IT_P_M_B_or	8	Single-skin (0°,45°)	Manual	Polyester	Orientation +45°/-45°/0°/+45°/-45°/90°

Experimental tests

Static indentation tests were carried out using a Zwick Roell Z600 tensile machine, in accordance with ASTM D6264 standard [15]. Each item was tested using two different indenter's diameters (i.e., 17 and 20mm) with two different test's speeds (i.e., 1.25 and 2.5 mm/min). Figure 1 shows the test set-up.

Fig. 1: Static Indentation Test set-up.

Results and discussion

Static indentation test

A typical load – displacement curve is reported in Figure 2 for each kind of sample by varying both speed and indenter diameter.

The curves show similar trends and similar maximum load values by considering the samples with glass fibre (top) and the ones with basalt fibers (bottom image). After a first quasi-linear elastic trend, a series of drops in the load are present when the fibres start to break.

With the aim to better evidence both the effect of this substitution and the effect of the speed and indenter diameters, the maximum stress was evaluated as the ratio between the standard force generated by the indenter on the surface of the specimen:

$$\sigma_{max} = \frac{F}{S_0} [MPa] \tag{1}$$

Therefore, the stress values at the maximum load of all laminates, subjected to the action of the two indenters and the two speeds, are summarized in Figure 3.

In Table 3 the results of ANOVA, applied on the maxium stress, are reported.

The results allow to affirm that both the kind of material and the diameter are significant factors affecting the resistance of the material (p-value < 0.05), while the effect of the speed of the indenter is not significant (p-value > 0.05). The effect of the diameter is to slightly reduce the resistance of the materials. To evidence how the resistance changes at varying the material, the data were evaluated by using the Tukey test that compares all the couple of materials. The result of such test is reported in Figure 4.

If the interval of the differences of the SigmaMax means values includes the "0" than the two materials compared cannot be considered significantly different.

Italian Manufacturing Association Conference - XVI AITeM   Materials Research Forum LLC
Materials Research Proceedings 35 (2023) 350-358   https://doi.org/10.21741/9781644902714-42

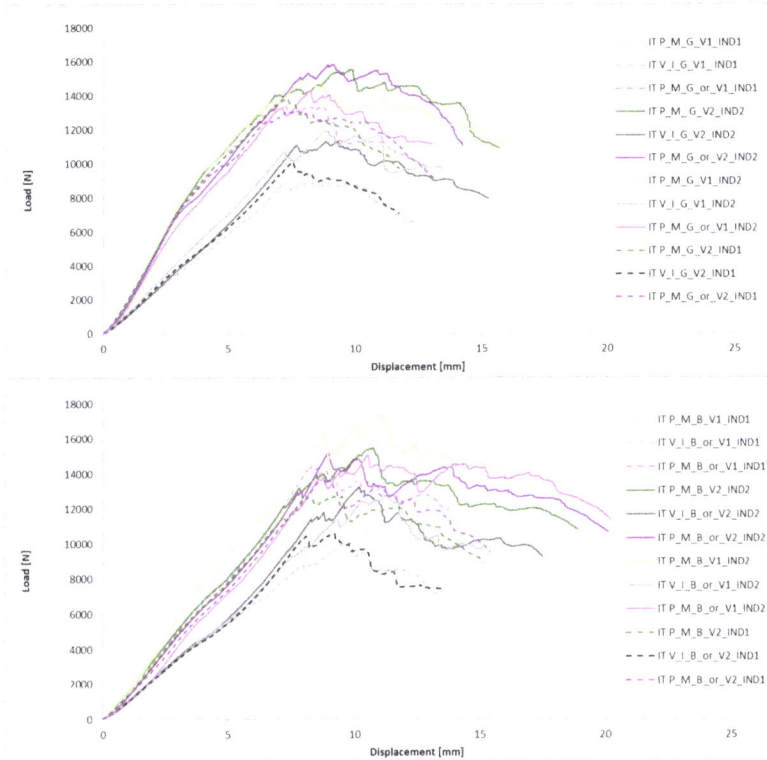

Fig.2: Typical load and laminate displacement curves for glass fiber and basalt samples at indenter speeds V1 and V2 with diameter D1 and D2.

By examining the position of the intervals for all the couples "basalt-glass" (i.e: IT P_M_G vs IT_P_M_B, IT V_I_G vs IT_V_I_B_or, IT P_M_G_or vs IT P_M-B_or) is evident that there is no significant difference in substituting the basalt with the glass, maintaining the other factors, for all kind of laminates.

Moreover, to evaluate the effect of the changing in orientation of the fibres the samples IT P_M_G vs IT P_M_G_or, and IT P_M_B vs IT P_M_B_or are compared. Also in this case, it is evident that the change has no effect on the resistance of the laminates.

To evaluate the effect of changing the production technology, the couples IT P_M_G vs IT V_I_G, and IT P_M_B vs IT V_I_B_or are compared. In this case the change causes the reduction of the laminate resistance, more in the ones with glass than in the basalt.

Finally, the comparison of IT P_M_G and IT V_I_B_or, allows to determine that replacing the glass with the basalt and changing the technology from manual lay-up to infusion leads to a significant reduction in the laminate resistance.

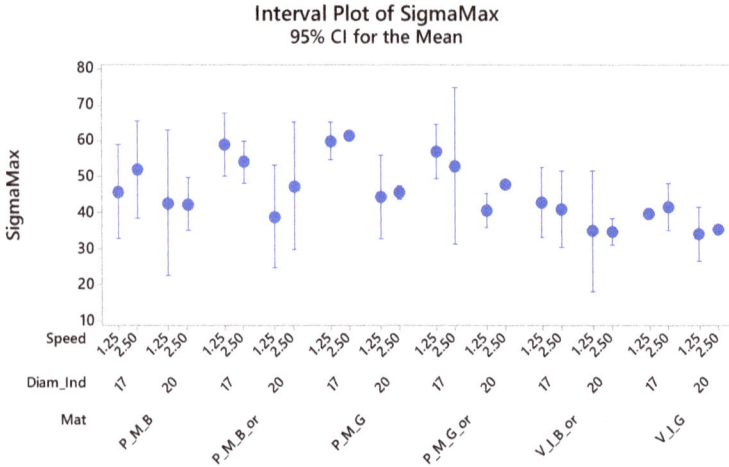

Individual standard deviations were used to calculate the intervals.

Fig.3: Interval plot for Sigma-max 95% CI for the Mean.

Table 3: ANOVA SigmaMax versus Mat; Speed; Diam_Ind.

```
Factor Information
Factor     Levels  Values
Mat          6      P_M_B; P_M_B_or; P_M_G; P_M_G_or; V_I_B_or; V_I_G
Speed        2      1.25; 2.50
Diam_Ind     2      17; 20

Analysis of Variance
Source                    DF    Adj SS    Adj MS   F-Value   P-Value
Model                     23   4376.22    190.27      9.70     0.000
  Linear                   7   3805.14    543.59     27.72     0.000
    Mat                    5   2263.18    452.64     23.08     0.000
    Speed                  1     31.92     31.92      1.63     0.208
    Diam_Ind               1   1688.14   1688.14     86.08     0.000
  2-Way Interactions      11    290.56     26.41      1.35     0.231
    Mat*Speed              5     24.95      4.99      0.25     0.935
    Mat*Diam_Ind           5    221.81     44.36      2.26     0.064
    Speed*Diam_Ind         1     40.88     40.88      2.08     0.156
  3-Way Interactions       5    213.96     42.79      2.18     0.072
    Mat*Speed*Diam_Ind     5    213.96     42.79      2.18     0.072
Error                     46    902.09     19.61
Total                     69   5278.31

Model Summary
      S     R-sq   R-sq(adj)   R-sq(pred)
4.42839   82.91%     74.36%       61.55%
```

Italian Manufacturing Association Conference - XVI AITeM   Materials Research Forum LLC
Materials Research Proceedings 35 (2023) 350-358   https://doi.org/10.21741/9781644902714-42

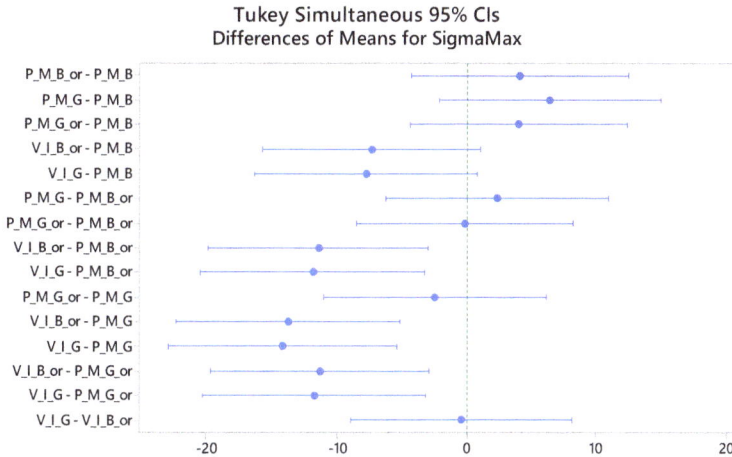

Tukey Simultaneous 95% CIs
Differences of Means for SigmaMax

If an interval does not contain zero, the corresponding means are significantly different.

Fig.4: Tukey simultaneous test on the Means for SigmaMax.

Fracture modes

The failure modes that occur on the specimens are shown in Figure 5. In all tested samples, the fracture propagates throughout the thickness of the specimens, and it has been noted that the sizes of the damage depend mainly on the diameter of the indenter. In fact, it is possible to observe that the increase in the indenter diameter increases the extension of the damage.

As it was expected, the changing in technology, from manual lay-up to vacuum infusion reducing the presence of air in the product, causes that the delamination is almost absent, and the fracture is more concentrated around the area of the indenter. In this case the fracture is due to both large crack propagation and breakage of the fibres.

The employ of an oriented structure of the fibre (substitution of mat with 0°/45° skin) induces a preferential direction of the delamination, thus qualitatively, the fracture appears more extended in the "or" samples.

The substitution of the glass with basalt does not determine changing in the kind of fracture, even though the fracture is more extended in the laminates with basalt fibres. It occurs for delamination coupled with splits.

Finally, is possible to observe, by analysing the corresponding load/displacement curves (Figure 2), that the occurrence of fracture in the glass-based laminates is more abrupt than in the basalt one (i.e. the fracture of the fibres is completed for lower values of displacement in the grass samples than in the basalt ones, of about 20%).

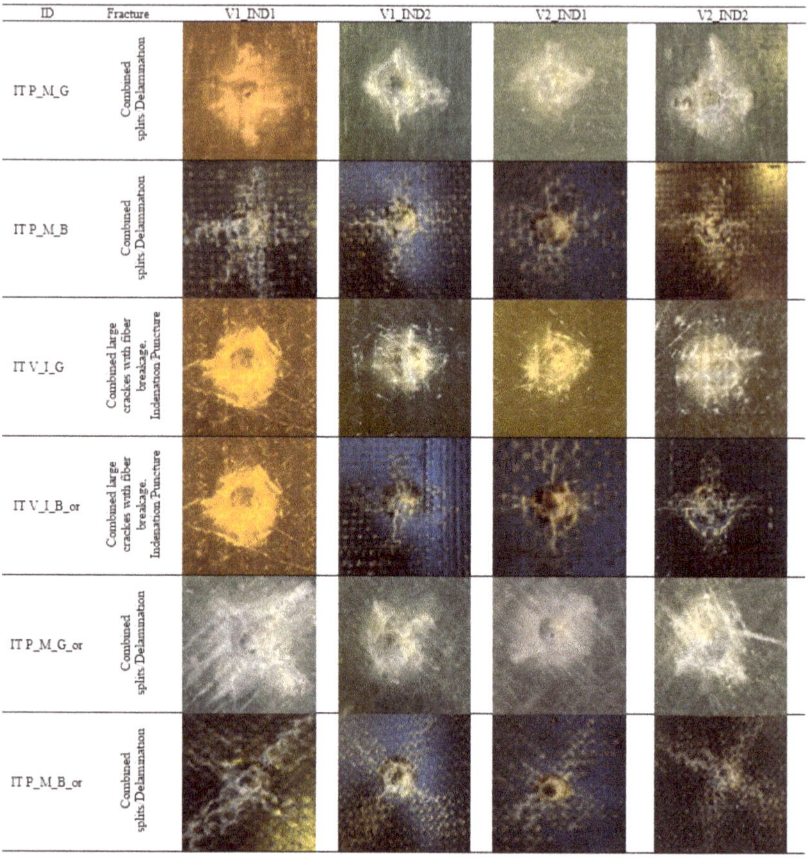

Fig. 5: Failure Mode Comparison of laminate samples tested with IND1 and IND2 diameters and at V1 and V2 velocities.

Conclusion

Static indentation tests, conducted on laminated materials made with vinylester and polyester resins, produced through hand lay-up and vacuum infusion processes, have shown that glass fibre composite materials exhibit similar loads than those made from basalt fibres.

The ANOVA on the Sigma max data allow to conclude the following.

- among the analysed factors, both the materials and the indenter diameters influence the final resistance (particularly the increase in diameter from 17 to 20 mm, induced a reduction in the resistance from 10% to 25% depending on the kind of sample),
- the changing in speed, in the considered range, is not significant.

The Tukey test allows comparing the means of samples taken in pairs. The result of this last tests allows to draw out that:

- replacing glass fibre with basalt fibre does not lead to significant changing in the resistance of any kind of sample,

Italian Manufacturing Association Conference - XVI AITeM   Materials Research Forum LLC
Materials Research Proceedings 35 (2023) 350-358   https://doi.org/10.21741/9781644902714-42

- the Vacuum Infusion technology has led to a worsening in mechanical resistance under indentation. By employing vacuum infusion technology, it is possible to halve the thickness with respect to the manual lay-up obtaining a poorer product (the Sigma max reduction ranges from 17% to 27% depending on the kind of sample).

By analysing the failure modes, it can be noticed that the failure is influenced by the technology, kind, and orientation of fibres.

The samples produced by Vacuum and the presence of orientation in the fibres limits the delamination phenomena.

The glass fibres determine a more abrupt fracture of the samples with respect to the basalt ones.

The results of this experimentation confirm that basalt fibre-reinforced composites are generally a suitable alternative to the use of glass fibre-reinforced composites and thus can be considered for marine applications.

This campaign is being finalized with other tests, where flexural and drop tests are expected to complete the mechanical behaviour of the laminates.

Funding
This research was funded by Ministry of Economic Development on the resources provided by the Decree 5 March 2018 Chapter III, as part of the project "Development of Ahead Systems and Processes for Highly AdvaNced TechnOlogies for low Magnetic Signature and HIghly eFFicient Electromagnetic shielded eco-friendly vessel – DAS PHANTOMSHIFFE", grant number F/190001/01/X44.

References
[1] I. Zivkovic, C. Fragassa, A. Pavlovic, T. Brugo, Influence of moisture absorption on the impact properties of flax, basalt and hybrid flax/basalt fiber reinforced green composites, Compos Part B 111 (2017) 148-164. https://doi.org/10.1016/j.compositesb.2016.12.018

[2] Information on https://cordis.europa.eu/article/id/422007-major-step-towards-sustainable-and-efficient-ship-construction.

[3] K. Van de Velde, P. Kiekens, L. Van Langenhove, Basalt fibres as reinforcement for composites, In: Proceedings of 10th International Conference on Composites / Nano Engineering ICCE/10, International Community for Composites Engineering and College of Engineering, University of New Orleans (2003).

[4] Z. Li, J. Ma, H. Ma, X. Xu, Properties and Applications of Basalt Fiber and Its Composites. In IOP Conference Series: Earth and Environmental Science, IOP Publishing: Banda Aceh, Indonesia, 186 (2018). https://doi.org/10.1088/1755-1315/186/2/012052

[5] V. Dhand, G. Mittal, K.Y. Rhee, S-J. Park, D. Hui. A short review on basalt fiber reinforced polymer composites, Comp B 73 (2015) 166-180. https://doi.org/10.1016/j.compositesb.2014.12.011

[6] C. Jiang, K. Fan, F. Wu, D. Chen, Experimental study on the mechanical properties and microstructure of chopped basalt fibre reinforced concrete, Mater Des, 58 (2014) 187-193. https://doi.org/10.1016/j.matdes.2014.01.056

[7] C.H. Jiang, T.J. McCarthy, D. Chen, Q.Q. Dong, Influence of basalt fibre on performance of cement mortar, Key Eng Mater, 426-427 (2010) 93-96. https://doi.org/10.4028/www.scientific.net/KEM.426-427.93

[8] D.P. Dias, C. Thaumaturgo, Fracture toughness of geopolymeric concretes reinforced with basalt fibres, Cem Concr Compos, 27 (2005) 49-54. https://doi.org/10.1016/j.cemconcomp.2004.02.044

Italian Manufacturing Association Conference - XVI AITeM   Materials Research Forum LLC

Materials Research Proceedings 35 (2023) 350-358   https://doi.org/10.21741/9781644902714-42

[9] V. Fiore, T. Scalici, G. Di Bella, A. Valenza, A review on basalt fibre and its composites, Compos Part B, 74 (2015) 74-94. https://doi.org/10.1016/j.compositesb.2014.12.034

[10] V. Fiore, G. Di Bella, A. Valenza, Glass-basalt/epoxy hybrid composites for marine applications, Mater Des 32 (2011) 2091-2099. https://doi.org/10.1016/j.matdes.2010.11.043

[11] P. Davies, W. Verbouwe, Evaluation of Basalt Fibre Composites for Marine Applications, Appl Compos Mater 25 (2018) 299-308. https://doi.org/10.1007/s10443-017-9619-3

[12] L.S. Sutherland, C. Guedes Soares, Contact indentation of marine composites, Comp Struct 70 (2005) 287-294. https://doi.org/10.1016/j.compstruct.2004.08.035

[13] L.S. Sutherland, A review of impact testing on marine composite materials: Part I - Marine impacts on marine composites, Comp Struct 188 (2018) 197-208. https://doi.org/10.1016/j.compstruct.2017.12.073

[14] L.S. Sutherland, C. Guedes Soares, Impact characterisation of low fibre-volume glass reinforced polyester circular laminated plates, Int J Impact Eng 31 (2005) 1-23 https://doi.org/10.1016/j.ijimpeng.2003.11.006

[15] C. Borsellino, L. Calabrese, G. Di Bella, Windsurf board sandwich panels under static indentation, Appl Compos Mater 15 (2008) 75- 86. https://doi.org/10.1007/s10443-008-9058-2

Materials Research Forum LLC
https://doi.org/10.21741/9781644902714-43

Surface micro – texturing of tapping tools with complex geometry

Manuel Mazzonetto[1,a]*, Vincenzina Siciliani[1,b], Riccardo Pelaccia[1,c],
Leonardo Orazi[1,2,d]

[1]DISMI – University of Modena and Reggio Emilia, Via Amendola 2, Reggio Emilia 42122, Italy

[2]EN&TECH – University of Modena and Reggio Emilia, Piazzale Europa 1, Reggio Emilia 42124, Italy

[a]manuel.mazzonetto@unimore.it, [b]vincenzina.siciliani@unimore.it,
[c]riccardo.pelaccia@unimore.it, [d]leonardo.orazi@unimore.it

Keywords: Laser Machining, Texture, Cutting Tool

Abstract. Lubrication control and tribological aspects affect all the machining operations: minimizing the friction coefficient, reducing the heat generated during machining or decreasing stresses during cutting operations are just some sought goals during machining. A fundamental aspect is also related to the drop of the amount of cutting fluid required during manufacturing processes. In fact, this can respond to issues related to the process sustainability thus allowing machining under MQL (Minimum Quantity Lubrication) conditions. The aim of this work is to propose an effective methodological procedure to texture cutting tools with complex geometry. A picosecond laser system operating at infrared frequency was used to structure the whole cutting tool surface. Rectangular dimples were selected as texture pattern to ensure the evaluation of the dimensions achieved by the dimples. Textured surfaces topography will allow the evaluation of the effectiveness of the developed process.

Introduction

In recent few years, the development of new coatings, applied to cutting tools, has made it possible to reach significant technological results such as the achievement of high performance in terms of friction reduction, the mitigation of adhesion, or the decrease of forces and torques during machining [1]. Moreover, these aspects are also reflected to the sustainability of mechanical processes. In fact, some coated tools have contributed to minimum quantity lubrication (MQL) thus reducing environmental pollution related to the disposal of cutting fluids [1]–[3]. Despite that, the development of new coating materials is getting harder [1].

In view to meet these needs, a solution can be represented by the application of a combination of micro-textured surface and surface coating. Several studies have emphasized the benefits seen from this method [4], [5].

Dimensions and texture geometry of dimples have a lot of influence in determining the tribological characteristics of the machining process. Kawasegi et al. [5] performed turning of aluminum with textured tools by varying the parallel dimple width. They showed how dimple textures parallel to the cutting edge contribute better in terms of forces reduction in comparison to dimples perpendicular to the cutting edge and cross patterned textures. Similar results were presented by Sugihara and Enomoto [6] who applied micro texture and generated shallow dimples on a diamond-like carbon (DLC) coated cutting tool by femtosecond laser technology. They investigated aluminum alloy cutting performance and stated a decrease in adhesion in wet cutting using micro-textured cemented carbide tools. They also found improvement in the lubrication of the cutting tool surface. As a subsequent result, they studied [7] the influence of the texture dimension on the anti-adhesiveness by varying the parallel dimple width of the functionalized tools.

Italian Manufacturing Association Conference - XVI AITeM   Materials Research Forum LLC
Materials Research Proceedings 35 (2023) 359-366   https://doi.org/10.21741/9781644902714-43

Obikawa et al. [1] investigated the effects of four different micro surface textures on the lubrication conditions at the tool rake in machining aluminum alloy A6061-T6. Their findings showed how parallel and square-dot type of micro-textures effectively improved the lubrication conditions by decreasing of coefficient of friction in orthogonal cutting of aluminum alloy A6061-T6. A similar study was recently made by Xing et al. [8] in which three different types of dimples (rectangular, circular and linear) were generated on the rake face of cemented carbide tools. During orthogonal dry cutting tests of A6061 aluminum alloy tubes, they studied the effects in terms of friction reduction, anti-adhesiveness, wear resistance and cutting force reduction of these textured geometries compared to the results achieved by a conventional tool.

Several studies, due to the positive effects of surface textures related to anti-adhesion properties [7], [9], [10], heat reduction [11], [12], anti-friction [6], have demonstrated a lot of fields of applications of these types of micro- and nano-machining, not only referred to the significant impact on tribological performance of cutting tool during machining [8] but also related to bearings [13], seal rings [14] and engine cylinder liners [15].

Textured tools surfaces can also lead to positive results in terms of processing sustainability via reaching minimum quantity lubrication as stated in [4], [16]. Nevertheless, if textures on cutting tools aren't correctly used, they can deteriorate the mechanical strength of the tool or worsen the cutting conditions of metals [17].

Throughout this study, the thread of a commercial tapping tool has been textured. During machining, this kind of tool needs a large supply of lubricant, especially when high depth must be reached. Rectangular shape dimples were chosen to better allow the characterization of the entire geometric development of the generated dimples in terms of dimensions, depth and inclination evaluations. These dimples, during machining, could operate as lubricant retention zone to enhance lubrication. Preliminary tests were conducted on flat surfaces in order to optimize process and laser parameters with the aim to obtain the specified dimples. Then, these findings were extended to complex geometry represented by the thread of a tapping tool. To do this, an ad-hoc kinematic procedure has been implemented, also via system integration, to optimize the entire process. Picosecond laser processing with infrared (IR) beamline was used for structuring. Morphological analyses via optical (OM) and scanning electron microscopes (SEM) were also performed to confirm the effectiveness of the proposed procedure.

Materials and methods
Workpiece and material
The tools used for texturing are commercially supplied by Seco Tools under the MF-M10x1.50-ISO-6HX-XC-V055 codification. These are cold forming TiN coated tapping tools used for steels, hardened steels with <62 HRC and non-ferrous metals. Their peculiarity refers to the presence of a conical inlet aimed to facilitate the entry of the various threads into the pre-hole. This zone involves the first three threads and is characterized by an inclination with respect to the longitudinal development of the tool of $\beta = 7.5°$. Overtaken this area, the threat develops its behavior regularly conforming the DIN 2174 standard of M10x1.5. The inlet conical profile of several SECO M10x1.5 tapping tools was analyzed via an optical microscope (Mod. Nikon LV100ND). Subsequently, by using the image processing program "ImageJ", the value of β angle as stated in the datasheet provided by SECO was confirmed with a tolerance of $\pm 0.5°$. Fig. 1 shows the geometry of the considered tool.

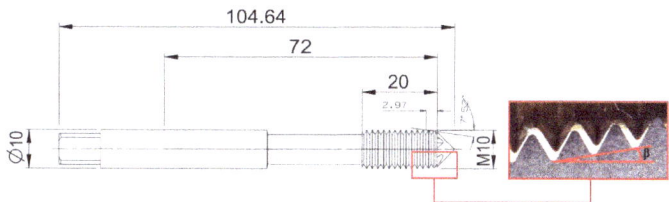

Fig. 1 SECO tapping tool used for the experiments (left); magnification in the conical inlet zone to measure and verify β angle (right). (All dimensions are in mm).

Laser texturing

Picosecond laser technology was adopted to perform textures on the thread. This technology is very promising compared to all other micro-surface structuring methods. Indeed, it operates with ultrashort pulses in picosecond duration which can produce, due to the high speed that can be reached, regular dimples on the workpiece. The necessity of post-processing is eliminated and no special tooling is required [4]. As the major advantage, pulsed laser systems, in the ultrashort regime, can structure surfaces of workpieces without limitation in terms of hardness of the material to be treated [4]. The laser amplifier EKSPLA Atlantic 50 operated at the IR wavelength of 1064 nm. The focused beam diameter was $\varnothing \sim 10$ μm evaluated at $1/e^2$ intensity. The same laser processing parameters, as summarized in Table 1, were used to texture both the flat surface and the entire tool.

Table 1 Parameters used for texturing experiments.

Wavelength λ [nm]	1064
Average output power P [W]	0.82
Pulse frequency f [kHz]	400
Pulse energy E [μJ]	4.1
Pulse duration τ [ps]	~ 10
Pulse fluence F [J/cm^2]	5.22
Line spacing s [μm]	5
Cumulated dose d [J/cm^2]	10.44
Marking speed v_s [m/s]	1
Number of passes on each dimple p	2

The Raylase Superscan IV galvanometric scanner coupled with an 80 mm F-theta lens assures a working area of 39 x 39 mm^2 through movements in X and Y of two mirrors mounted inside the scanner. The remaining displacements, outside the above scan area, are guaranteed by the translation of the Z axis, which allows positioning at different focal heights, and by the X- and Y-translation of the workpiece table. Rotation (both clockwise and counterclockwise) is permitted by a rotator, mounted on the automated X-Y table. Therefore, the entire system exhibits four global degrees of freedom.

Laser source, roto-translator system for workpiece and head manipulation, optical scanning system, beam shaping and sensors for process control are totally well integrated via C# libraries and LabVIEW to allow movement in the three translation directions and to permit rotation thus following the helical path of the thread. The setup used for laser texturing is illustrated in Fig. 2.

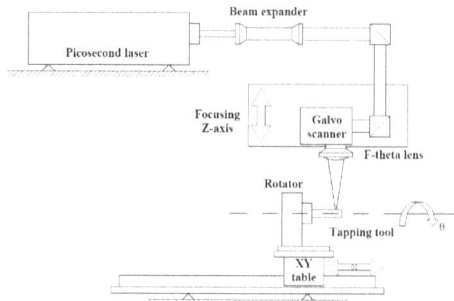

Fig. 2 Laser system used for the experiments.

Micro-texture fabrication

The entire textured area of the tapping tool had a width of 10 mm and a length of 20 mm.

The designed dimples were rectangular in shape with a maximum size of 30x60 μm and 10 μm as expected maximum depth. This geometry has remained unchanged for both flat and inclined machining. Dimples have been distributed along the entire circumferential and longitudinal development of the tapping tool with the aim of achieving a 15% coverage of the entire development of the thread surface. In both cases (flat and inclined surface) were performed three different dimples in terms of the trend of the depth variation of the dimples: flat bottom, positive- and negative-inclination with respect to the chip flow direction. These trends have been obtained by overlapping 7 rectangles with a constant height of 30 μm and gradually decreasing widths (60, 50, 40, 35, 30, 25, 20 μm). Every single rectangle was filled with an equally spaced horizontal hatch 5 μm in step to generate the desired maximum depth of the dimples of ~ 10 μm.

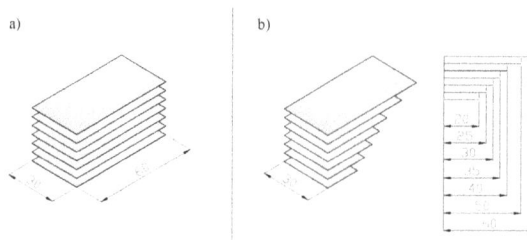

Fig. 3 Dimples generation criteria. a) flat bottom b) wedge dimples. Fillings of each rectangle are omitted to ease reading. Decreasing in width proceeds from top to bottom (All dimensions are in μm).

Planar samples were firstly textured to investigate the response of the material to infrared interaction with the aim to obtain a pulse overlap of 5 μm corresponding to a laser spot overlap of 50% of the whole surface. These tests were conducted by means of variation of the average output power, the number of passes on each dimple and the marking speed. The adopted optimized values settle as stated in Table 1 thus guaranteeing precise regular dimples formation, a gradual increase in depth up to 10 μm (as maximum) of the dimples preventing the formation of recast materials and a homogeneous distribution of energy on the treated area. During these tests, dimples with flat bottom, positive and negative inclination were generated, as mentioned above. To switch from planar to inclined surface an evaluation of the allowable depth of focus is required.

Italian Manufacturing Association Conference - XVI AITeM   Materials Research Forum LLC
Materials Research Proceedings 35 (2023) 359-366   https://doi.org/10.21741/9781644902714-43

By considering a beam quality factor M^2 of 1.5, subsequent calculations made it possible to evaluate the diameter of the laser beam waist D_f which settles at 11.6 μm and the Rayleigh length z_f of about 100 μm. This evaluation indicates the possibility to be robust in obtaining the inclined thread flank results comparable to what was achieved on preliminary planar tests. Then, experiments on the thread of the tool, which can be treated as a strongly inclined surface, were performed. Considering these surfaces, the maximum size of the designed dimples of 30x60 μm must be treated as the maximum dimension with respect to the surface normal. This series of experiments, as well as the preliminary tests, has consisted of using the three different depth variations, as previously mentioned. Positive and negative inclination in the depth of the dimples is visible in the inset of Fig. 4.

The distribution of the dimples on the cutting tool was very crucial. The laser beam must follow the thread in its entirety. The implemented handling system allowed following the winding direction of the thread helix through the movement of the X and the Z axis and by rotation of the workpiece. This aspect was essential to operate at the correct focal height. As a direct subsequent of the adopted criteria, a simplified method was developed for the programming and the conduction of laser texturing part program of complex surfaces under study.

The automatic process of creating dimples on tapping tools consists of two stages. First, laser beam performs the progressive creation of the dimples by using a multi-quote Z approach for a fixed value of rotation. Thus, the laser initially focuses on the crest and creates textures on it. After completing this task, the system, through a movement of the Z axis, moves progressively to the next lower level until it reaches the thread root. This allows to calculate a single set of $\{X,Z\}$ position of the laser vectors. The system can texture only one angular position in their entire height, from the crest to the root of the thread. The heights involved in the process were $z^* = 0$, -0.06, -0.19, -0.38, -0.57, -0.76, -0.89, -0.99, -1.17 mm with the origin for the z^* dimensions located at the outer radius of the tread. Then, by automatically rotating the tool with an angular step of 4.5° and shifting the X position to follow the helical path, the process restarts as mentioned above.

The parameters of the laser beam that can be calculated starting from the value of M^2 (e.g. D_f and z_f) allow for process optimization. In fact, it is possible to texturize in the same placement for a given value of z^*, three contiguous lines of dimples with an angular distance dθ of 1.5° as can be seen from the dotted rectangles in Fig. 4. The progressive rotations can thus be increased to θ = 4.5° (as mentioned above) resulting in a reduction of the number of mechanical placements to perform and in the reduction of the execution times thanks to the speed of the galvanometric optical scanner.

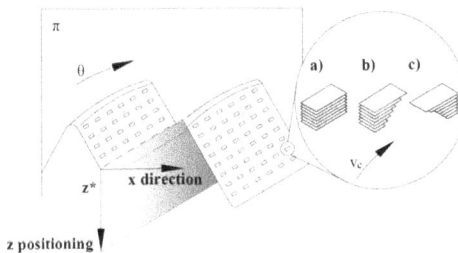

Fig. 4 Schematic diagram of textures on a circular sector of the tapped tool. In the enlarged figure are shown the projected views of: a) flat bottom dimples; b) - c) positively (and negatively) inclined dimples with respect to cutting speed v_c.

Characterization and measurements

The textured tools were cleaned in an ultrasonic bath. A Nikon LV100ND optical microscope was used to qualitatively measure the depth of the dimples by changing the depth of focus. Then, the textured surfaces were characterized via a Scanning Electron Microscope (Mod. Nova NanoSEM 450 - FEI) to measure the two dimensions of the dimples.

Results and discussions

The comparison of the images obtained via SEM analysis (Figs. 5-6) confirms the possibility to get regular dimples with specific dimensions on the thread of the tool. As first result, SEM images of the textured dimples on the planar surface confirmed the chosen laser parameters. In fact, dimples were unchanged in dimensions in the passage from the designed to the actually realized geometry. The maximum depth is evaluated to be about 10 μm.

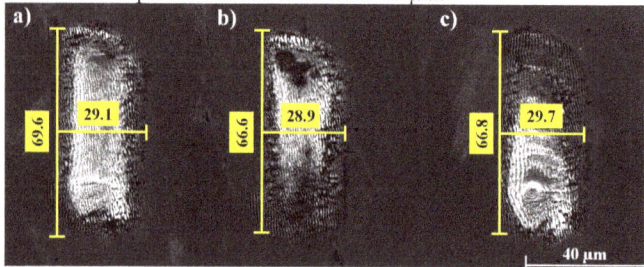

Fig. 5 SEM images of preliminary tests on planar surface: a) flat bottom; b) and c) shows the positive and negative increasing in depth.

Extending the study to the inclined surface similar results are achieved. The maximum depth reached by the dimples settles around 8 μm.

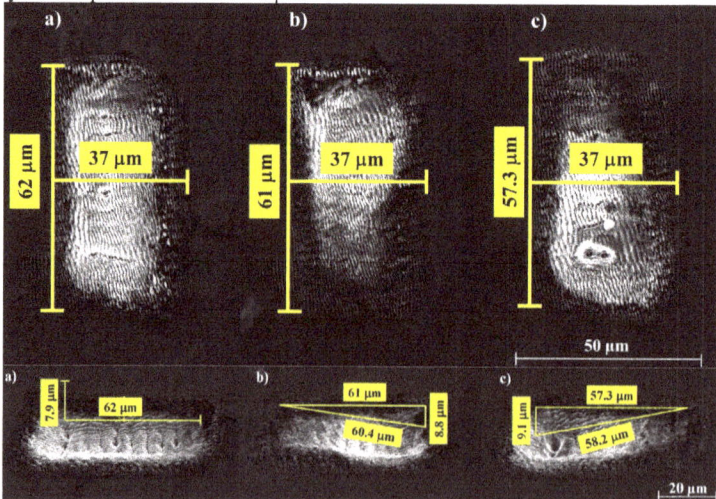

Fig. 6 SEM images of tests on 60 degrees inclined surface: a) flat bottom; b) and c) shows the positive and negative increasing in depth. (All dimensions are in μm).

Italian Manufacturing Association Conference - XVI AITeM   Materials Research Forum LLC
Materials Research Proceedings 35 (2023) 359-366   https://doi.org/10.21741/9781644902714-43

SEM images also confirm the adopted procedure showing the achievement of a homogeneous distribution of the dimples, especially in the conical inlet zone as visible in Fig. 7.

Fig. 7 SEM image of the treated inlet conical zone.

Conclusions

In this study, the possibility to implement a procedure to generate regular dimples on steeply inclined surface is analyzed. It has been shown how picosecond laser sources are suitable technology for metals micro-texturing. The procedure adopted allowed to achieve precise and regular dimples both on planar and inclined surface which was represented by the thread of the commercial tapped tool under investigation. A fully automated method to efficiently texture all the complex geometry of a real tapping tool was moreover presented.

The experimental results obtained were in accordance with the predicted geometry. In fact, maximum dimension of the dimples remained constant from the designed to the realized processing while they have changed due to trigonometry in the case of an inclined surface. The depth reached by the dimples of about 8 μm were interesting to probe in the next future the possibility to work under MQL conditions during tapping operations.

As subsequent studies, starting from the presented results, cutting tests will be performed to compare the thrust force from machining using textured and conventional SECO tapped tool.

Acknowledgments

This research was partially funded under the National Recovery and Resilience Plan (NRRP), Mission 04 Component 2 Investment 1.5 – NextGenerationEU, Call for tender n. 3277 dated 30/12/2021. Award Number: 0001052 dated 23/06/2022.

References

[1] T. Obikawa, A. Kamio, H. Takaoka, and A. Osada, "Micro-texture at the coated tool face for high performance cutting," *Int J Mach Tools Manuf*, vol. 51, no. 12, pp. 966–972, Dec. 2011. https://doi.org/10.1016/j.ijmachtools.2011.08.013

[2] H. Hanyu, S. Kamiya, Y. Murakami, and M. Saka, "Dry and semi-dry machining using finely crystallized diamond coating cutting tools," *Surf Coat Technol*, vol. 173, pp. 992–995, 2003. https://doi.org/10.1016/S0257-8972Ž03.00688-1

[3] T. Obikawa, Y. Asano, and Y. Kamata, "Computer fluid dynamics analysis for efficient spraying of oil mist in finish-turning of Inconel 718," *Int J Mach Tools Manuf*, vol. 49, no. 12–13, pp. 971–978, Oct. 2009. https://doi.org/10.1016/j.ijmachtools.2009.06.002

[4] T. Özel, D. Biermann, T. Enomoto, and P. Mativenga, "Structured and textured cutting tool surfaces for machining applications," *CIRP Annals*, vol. 70, no. 2, pp. 495–518, Jan. 2021. https://doi.org/10.1016/j.cirp.2021.05.006

[5] N. Kawasegi, H. Sugimori, H. Morimoto, N. Morita, and I. Hori, "Development of cutting tools with microscale and nanoscale textures to improve frictional behavior," *Precis Eng*, vol. 33, no. 3, pp. 248–254, Jul. 2009. https://doi.org/10.1016/j.precisioneng.2008.07.005

[6] T. Sugihara and T. Enomoto, "Development of a cutting tool with a nano/micro-textured surface-Improvement of anti-adhesive effect by considering the texture patterns," *Precis Eng*, vol. 33, no. 4, pp. 425–429, Oct. 2009. https://doi.org/10.1016/j.precisioneng.2008.11.004

[7] T. Sugihara and T. Enomoto, "Improving anti-adhesion in aluminum alloy cutting by micro stripe texture," *Precis Eng*, vol. 36, no. 2, pp. 229–237, Apr. 2012. https://doi.org/10.1016/j.precisioneng.2011.10.002

[8] Y. Xing, J. Deng, X. Wang, K. Ehmann, and J. Cao, "Experimental assessment of laser textured cutting tools in dry cutting of aluminum alloys," *Journal of Manufacturing Science and Engineering, Transactions of the ASME*, vol. 138, no. 7, Jul. 2016. https://doi.org/10.1115/1.4032263

[9] S. Lei, S. Devarajan, and Z. Chang, "A study of micropool lubricated cutting tool in machining of mild steel," *J Mater Process Technol*, vol. 209, no. 3, pp. 1612–1620, Feb. 2009. https://doi.org/10.1016/j.jmatprotec.2008.04.024

[10] J. Kümmel *et al.*, "Study on micro texturing of uncoated cemented carbide cutting tools for wear improvement and built-up edge stabilisation," *J Mater Process Technol*, vol. 215, pp. 62–70, 2015. https://doi.org/10.1016/j.jmatprotec.2014.07.032

[11] Y. Lian, J. Deng, G. Yan, H. Cheng, and J. Zhao, "Preparation of tungsten disulfide (WS2) soft-coated nano-textured self-lubricating tool and its cutting performance," *International Journal of Advanced Manufacturing Technology*, vol. 68, no. 9–12, pp. 2033–2042, Oct. 2013. https://doi.org/10.1007/s00170-013-4827-y

[12] W. Ze, D. Jianxin, C. Yang, X. Youqiang, and Z. Jun, "Performance of the self-lubricating textured tools in dry cutting of Ti-6Al-4V," *International Journal of Advanced Manufacturing Technology*, vol. 62, no. 9–12, pp. 943–951, Oct. 2012. https://doi.org/10.1007/s00170-011-3853-x

[13] I. Etsion, G. Halperin, V. Brizmer, and Y. Kligerman, "Experimental investigation of laser surface textured parallel thrust bearings."

[14] S. Bai, X. Peng, Y. Li, and S. Sheng, "A hydrodynamic laser surface-textured gas mechanical face seal," *Tribol Lett*, vol. 38, no. 2, pp. 187–194, May 2010. https://doi.org/10.1007/s11249-010-9589-1

[15] W. Grabon, W. Koszela, P. Pawlus, and S. Ochwat, "Improving tribological behaviour of piston ring-cylinder liner frictional pair by liner surface texturing," *Tribol Int*, vol. 61, pp. 102–108, 2013. https://doi.org/10.1016/j.triboint.2012.11.027

[16] R. Sasi, S. Kanmani Subbu, and I. A. Palani, "Performance of laser surface textured high speed steel cutting tool in machining of Al7075-T6 aerospace alloy," *Surf Coat Technol*, vol. 313, pp. 337–346, Mar. 2017. https://doi.org/10.1016/j.surfcoat.2017.01.118

[17] S. Durairaj, J. Guo, A. Aramcharoen, and S. Castagne, "An experimental study into the effect of micro-textures on the performance of cutting tool," *International Journal of Advanced Manufacturing Technology*, vol. 98, no. 1–4, pp. 1011–1030, Sep. 2018. https://doi.org/10.1007/s00170-018-2309-y

Italian Manufacturing Association Conference - XVI AITeM
Materials Research Proceedings 35 (2023) 367-375

Materials Research Forum LLC
https://doi.org/10.21741/9781644902714-44

Influence of silica aerogel filler on strength-to-weight ratio of carbon/epoxy composite made by vacuum resin infusion

Luigi Benvenuto[1,a*], Enrico Lertora[1,b], Chiara Mandolfino[1,c], Matteo Benvenuto[1,d], Marco Pizzorni[1,e]

[1]Department of Mechanical Engineering, Polytechnic School of University of Genoa, Via Opera Pia 15, 16145 Genoa, Italy

[a]luigi.benvenuto@edu.unige.it, [b]enrico.lertora@unige.it, [c]chiara.mandolfino@unige.it, [d]matteo.benvenuto@edu.unige.it, [e]marco.pizzorni@unige.it

Keywords: Composites, Resin Infusion, Mechanical Strength

Abstract. Thermoset matrix composites are a competitive solution in high-performance applications due to their superior characteristics. Several previous studies have shown that the addition of particles to thermosetting resins - particularly epoxy resins, which are currently the most widely used - can improve many of the physical properties of composites. In this context, this study aims to investigate the effect of silica aerogel content on the mechanical and weight properties of composites made via vacuum infusion. The results have shown that the strength-to-weight ratio increases with increasing filler percentage. However, the experiments also made it possible to recognise process limits, which occur when the percentage of aerogel in the resin exceeds 2%.

Introduction

Over the years, the transport industry has tended increasingly to replace conventional materials with composite materials, as in the case of hybrid composites[1], mainly because of the gain in lightness without sacrificing mechanical performance. Furthermore, the use of this type of material allows their mechanical, thermal, electrical, acoustic, etc. properties to be modified in a very flexible manner by inserting additives into the matrix. Commonly used fillers include carbon nanotubes (CNTs)[2,3], graphene, metal nanoparticles[4,5], or flame-retardant additives in cases where reinforcements are of organic origin[6] and silica aerogels[7,8]. Recent studies have shown that the introduction of the latter within the polymer matrix increases the mechanical, thermal and soundproofing properties of the composite[9,10]. For instance, Mazlan et al.[11] found that the addition of silica aerogel in a concentration of 2 wt% increased the flexural strength and modulus by 8 and 11%, respectively.

Impact resistance also improves with the introduction of silica aerogel. For example, Riahipour et al.[12] estimated that a 10 vol% aerogel percentage increased impact strength by about 100%. This trend, however, tends to reverse when the volume percentage of aerogels increases above 10%.

However, the use of nanoscale fillers presents certain problems including the lack of a strong interfacial bond between the filler particles and the polymer chains[13]. By using porous fillers with larger particle sizes, not only is the interfacial interaction between the matrix and the filler improved due to the infiltration of the resin within the inorganic scaffold of the aerogel[14], but there is also a benefit in terms of weight since more porous fillers are also less dense.

The aim of this work is to study the influence of silica aerogel within an epoxy matrix composite reinforced with carbon-fibre fabrics produced by vacuum resin infusion. In particular, using a porous filler, the aim of the work is to establish how the strength-to-weight ratio of the finished product varies with the amount of filler.

Italian Manufacturing Association Conference - XVI AITeM   Materials Research Forum LLC
Materials Research Proceedings 35 (2023) 367-375   https://doi.org/10.21741/9781644902714-44

Experimental details

Material and processing. An epoxy resin (SX8 EVO by Mates Italiana srl, Segrate, MI, Italy) and a carbon fiber fabric of 220 g/m² (provided by Industria Tessuti Tecnici srl, Lesmo, MB, Italy), with densities of 1.11 g/cm³ and 1.8 g/cm³ respectively, were used to produce the composite laminates.

The test campaign involved the production of four panels. For each laminate, the weight ratio of fibre to total mass was kept constant at 40%, while the percentage by weight between resin and silica aerogel particles with a density of 0.074 g/cm³, varied according to three different ratios 100:2, 100:5 and 100:10. In addition, a reference composite type with no aerogel was also fabricated for the sake of comparison.

Table 1 summarizes some of the characteristics of the samples produced, in particular the percentage weight content of each material.

Table 1 – Denomination and composition of the different laminated panels

Sample denomination	Fibre content [%]	Epoxy content [%]	Silica aerogel content [%]
REF	40.00	60.00	0.00
AER2	40.00	58.82	1.18
AER5	40.00	57.14	2.86
AER10	40.00	54.55	5.45

Each composite laminate was made by vacuum bag technique from 510 mm (length) x 350 mm (width) x 2.5 mm (thickness) laminates, overlaying 7 layers of carbon fabric, using a flat glass plate as the bottom mould. A mesh of polymeric material was also placed on the upper carbon layer to improve resin distribution, which was otherwise made difficult by the low permeability of the bidirectional carbon fabric. The whole was then enclosed in a polymer material bag and epoxy resin infusion was then carried out.

Fig.°1 shows the final experimental set-up for the manufacture of the infused panels with the carbon fabric layers at the bottom, then the various film like the peel ply and the infusion mesh under the polymer bag.

Fig.°1 – Experimental set-up for the infusion process

For laminates named REF and AER2, the infusion process was successful in achieving full impregnation of the carbon fabrics in about 90 min, by applying a vacuum pressure of 0.9 bar. The complete cross-linking of the resin was achieved after 1 day at T=23 °C and RH=50% and 15 h at

Italian Manufacturing Association Conference - XVI AITeM   Materials Research Forum LLC
Materials Research Proceedings 35 (2023) 367-375   https://doi.org/10.21741/9781644902714-44

T=60 °C and RH=50%. The processing time was high due to the low permeability of the carbon fabric and the presence of the aerogel, which decreased the fluidity of the resin. However, for the laminates named AER5 and AER10, impregnation was not successful because after about 30 min, the introduced amount of aerogel caused the formation of clots that prevented the resin from advancing. These results were consistent with those obtained from researches by other authors, which have shown that high percentages of filler are better in the case of hydrophobic material. Conversely, in the case of hydrophilic fillers, such as the one used in this paper, the best results are obtained with low percentages[15].

Fig.°2 shows the clots formed in the silicone connector for the resin inlet. Note the resin feed stop line in green.

Fig.°2 – Clots in the connector circled in red

Experimental methods. From the only two laminates that could be produced by vacuum infusion (REF and AER2), ten specimens each were extracted: five for the tensile test and five for the three-point bending test. The specimens were mechanically characterized in order to assess whether and how much the addition of the filler affected the bonding properties between the fibre and the matrix and thus the mechanical performance. To do this, a universal testing machine (Instron 8802) equipped with a 50 kN load cell was used.

Specifically, tensile tests were performed according to ASTM D3039 on rectangular specimens without tabs on the ends measuring 250 mm (length) x 25 mm (width) x 2.5 mm (thickness). The speed of the moving crosshead was set at 1 mm/min.

The three-point bending test was performed according to ASTM D790 on rectangular specimens with dimensions of 100 mm (length) x 14 mm (width) x 2.5 mm (thickness). The distance between the support rollers was set at 40 mm with a moving crosshead speed set at 1.2 mm/min.

Each test was conducted under standard laboratory conditions (T=23 °C and RH=50%) and 5 replicates (N=5) were tested per case. Fig.°3 shows the shape of one of the tensile specimens cut from the laminate.

Fig.°3 – Example of specimen extracted from the composite panel

Italian Manufacturing Association Conference - XVI AITeM   Materials Research Forum LLC
Materials Research Proceedings 35 (2023) 367-375   https://doi.org/10.21741/9781644902714-44

In addition, specific parallelepiped samples were obtained from the REF and AER2 infused laminates for density evaluation. Each sample was weighed three times and, in this way, together with the dimensions of each sample, the average density of these laminates was obtained.

Result and discussion

Fig.°4 shows tensile stress–strain characteristic. The values of stress and strain were calculated according to the ASTM standard.

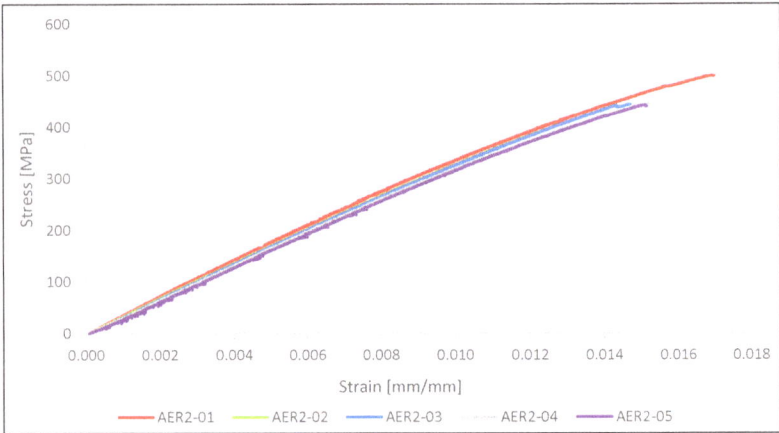

Fig.°4 – AER2 stress-strain graph for tensile test

Each test was considered valid because the failure of the specimens occurred in a section of the specimen within the gauge length and as can be seen from the graph, each test is practically superimposed on the other four.

Fig.°5 shows the fracture zone of one of the specimens after tensile testing.

Fig.°5 – Section of sample breakage AER2-01

The mechanical tests conducted on the materials have produced the results summarised in Table 2 and Table 3.

Italian Manufacturing Association Conference - XVI AITeM
Materials Research Proceedings 35 (2023) 367-375

Materials Research Forum LLC
https://doi.org/10.21741/9781644902714-44

Table 2 – Results obtained on the REF panel

	Tensile resistance [MPa]	Elongation at break [%]	Tensile modulus [GPa]
REF-01	413	1.6	33.7
REF-02	456	1.5	33.5
REF-03	374	1.5	32.9
REF-04	456	1.5	33.5
REF-05	439	1.4	35.4
Average Value	428	1.5	33.8
Standard Deviation	32	0.1	0.8

Table 3 - Results obtained on the AER2 panel

	Tensile resistance [MPa]	Elongation at break [%]	Tensile modulus [GPa]
AER2-01	499	1.7	35.3
AER2-02	400	1.3	34.9
AER2-03	444	1.5	35.7
AER2-04	448	1.5	34.2
AER2-05	443	1.5	32.3
Average Value	447	1.5	34.5
Standard Deviation	32	0.1	1.2

After the tensile tests, the fracture surfaces were observed under 10x to 40x magnification to assess which fracture mechanism affected the section.

Fig.°6 shows images of both fracture surfaces placed one on top of the other, of one of the reference specimens and one of the specimens with 2 wt% aerogel.

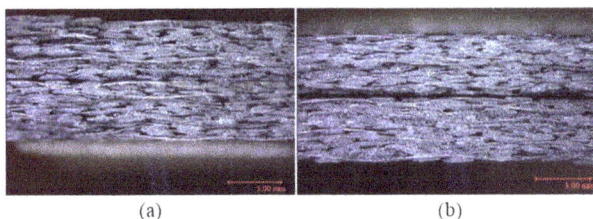

(a)   (b)

Fig.°6 – Sections of samples breakage at 10x magnification: REF-05 (a) and AER2-05 (b)

What can be seen from the images is that, on average, the fracture surface of the reference specimens is characterised by many inter-laminar cracks, which is less noticeable in the fracture surfaces of the specimens containing the aerogel; these are characterised by reinforcement failure or pull-out phenomena. This is indicative of a superior bond between the matrix and the reinforcement due to the presence of the additive.

Fig.°7 shows the stress-strain graph in the case of three-point bending tests on REF specimens.

Italian Manufacturing Association Conference - XVI AITeM

Materials Research Forum LLC

Materials Research Proceedings 35 (2023) 367-375

https://doi.org/10.21741/9781644902714-44

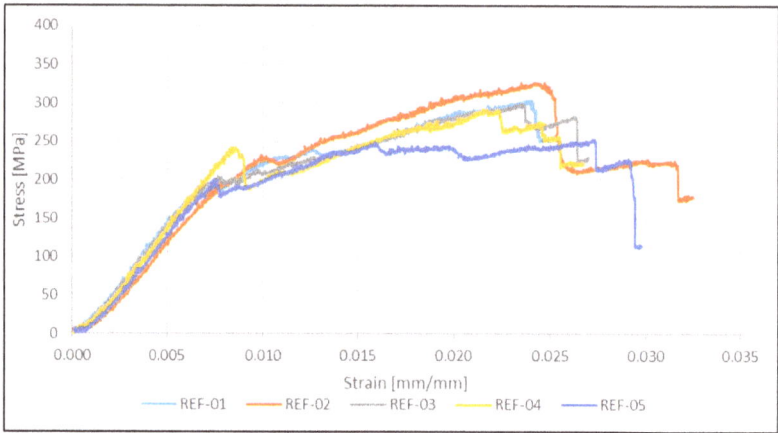

Fig.°7 – REF stress-strain graph for three-point bending test

The tension and strain values were derived from the force and displacement data output by the sensors using the formulae given in the standard.

Each test was considered valid since the failure occurred near the middle section of the specimen. A photo of test specimen AER-01 and its fracture section is shown as an example in Fig.°8.

Fig.°8 – Section of sample breakage AER2-01

Tables 4 and 5 summarise the results of the three-point bending tests conducted on the REF and AER2 laminates.

Table 4 – Results obtained on the REF panel

	Deflection at maximum load [mm]	Maximum load [N]	Strain at maximum tension [%]	Flexural strength [MPa]
REF-01	3.33	575.2	2.4	303
REF-02	3.37	612.8	2.4	327
REF-03	3.32	536.8	2.4	299
REF-04	2.99	533.7	2.1	292
REF-05	3.79	459.2	2.7	254
Average Value	3.36	543.5	2.4	295
Standard Deviation	0.28	57.1	0.2	26

Italian Manufacturing Association Conference - XVI AITeM   Materials Research Forum LLC
Materials Research Proceedings 35 (2023) 367-375   https://doi.org/10.21741/9781644902714-44

Table 5 - Results obtained on the AER2 panel

	Deflection at maximum load [mm]	Maximum load [N]	Strain at maximum tension [%]	Flexural strength [MPa]
AER2-01	3.61	676.6	2.6	324
AER2-02	3.81	591.4	2.7	337
AER2-03	3.53	545.5	2.5	294
AER2-04	3.36	609.3	2.4	357
AER2-05	3.38	561.5	2.4	308
Average Value	3.54	596.9	2.5	324
Standard Deviation	0.19	51.1	0.1	25

Furthermore, Tables 6 and 7 summarise the volume, average mass and density obtained for each sample of the REF and AER2 panels.

Table 6 - Volume, average mass and density of REF panel samples

	Volume [cm^3]	Average mass [g]	Density [g/cm^3]
REF-01	4.63	5.22	1.13
REF-02	4.58	5.17	1.13
REF-03	4.46	5.12	1.15
REF-04	4.44	5.07	1.14
REF-05	4.40	5.02	1.14
Average Value	4.50	5.12	1.14
Standard Deviation	0.10	0.08	0.01

Table 7 - Volume, average mass and density of AER2 panel samples

	Volume [cm^3]	Average mass [g]	Density [g/cm^3]
AER2-01	5.22	6.02	1.15
AER2-02	4.35	4.99	1.15
AER2-03	4.57	5.24	1.15
AER2-04	4.18	4.80	1.15
AER2-05	4.56	5.17	1.13
Average Value	4.58	5.24	1.14
Standard Deviation	0.40	0.47	0.01

Conclusion

The aim of this work was to evaluate the influence of silica aerogel in a carbon/epoxy composite produced via vacuum infusion.

The effect of the silica aerogel content on the tensile mechanical behaviour is not very relevant as the increases shown in the table are to be related to the relative sample standard deviation values. Even from the point of view of flexural behaviour, the performance increases obtained are in any case less than the uncertainty due to errors in the measurement chain.

What was certainly expected was a decrease in the density of the finished product, which was not achieved.

However, the results are nevertheless encouraging as it is important to emphasise that the tensile and flexural characteristics of the aerogel-added composite did not decrease, an event that could have occurred as the addition of an additive does not always maintain the adhesion between fibre and matrix. Furthermore, the addition of a quantity of aerogel makes it possible to obtain a product whose thermal characteristics could improve, an aspect that will be investigated after this work.

These results, therefore, confirm that the introduction of composite materials in the transport sector can be a viable alternative to traditional structural materials such as aluminium alloys as they have very similar tensile strengths despite having lower deformability and being less rigid, with Young's modulus values of approximately half. To the advantage of composite materials, however, due to their lower density compared to aluminium alloys, weight can be further reduced for the same size of the finished product. Furthermore, it has been shown how the introduction of additives can further increase the strength-to-weight ratio of composite materials.

It remains to be explored how the dynamic behaviour of composites is modified by the introduction of silica aerogels.

Since the results obtained with the addition of up to 2 wt% aerogel did not produce any performance gains in mechanical terms, it was decided to continue the research by further increasing the amount of aerogel within the epoxy matrix. However, an increase of more than 2 wt% created problems regarding the feasibility of impregnation by vacuum infusion.

To fully impregnate the reinforcements, it is necessary to adopt different production techniques, such as hand lay-up, with the knowledge that purely manual techniques such as the one mentioned above make it difficult to control the uniform distribution of filler that should be homogeneous.

References

[1]   K. C. Nagaraja, S. Rajanna, G. S. Prakash, and G. Rajeshkumar, "Improvement of mechanical and thermal properties of hybrid composites through addition of halloysite nanoclay for light weight structural applications," *Journal of Industrial Textiles*, vol. 51, no. 3, pp. 4880S-4898S, 2022. https://doi.org/10.1177/1528083720936624

[2]   N. J. Kanu *et al.*, "An Insight into Processing and Properties of Smart Carbon Nanotubes Reinforced Nanocomposites," *Smart Science*, vol. 10, no. 1, pp. 40–55, 2022. https://doi.org/10.1080/23080477.2021.1972913

[3]   V. Jain, S. Jaiswal, K. Dasgupta, and D. Lahiri, "Influence of carbon nanotube on interfacial and mechanical behavior of carbon fiber reinforced epoxy laminated composites," *Polym Compos*, vol. 43, no. 9, pp. 6344–6354, 2022. https://doi.org/10.1002/pc.26943.

[4]   M. Ozen, G. Demircan, M. Kisa, A. Acikgoz, G. Ceyhan, and Y. Işıker, "Thermal properties of surface-modified nano-Al$_2$O$_3$/Kevlar fiber/epoxy composites," *Mater Chem Phys*, vol. 278, 2022. https://doi.org/10.1016/j.matchemphys.2021.125689

[5]   H. B. Kaybal, H. Ulus, O. Demir, Ö. S. Şahin, and A. Avcı, "Effects of alumina nanoparticles on dynamic impact responses of carbon fiber reinforced epoxy matrix nanocomposites," *Engineering Science and Technology, an International Journal*, vol. 21, no. 3, pp. 399–407, 2018. https://doi.org/10.1016/j.jestch.2018.03.011

[6]   L. Boccarusso, L. Carrino, M. Durante, A. Formisano, A. Langella, and F. Memola Capece Minutolo, "Hemp fabric/epoxy composites manufactured by infusion process: Improvement of fire properties promoted by ammonium polyphosphate," *Compos B Eng*, vol. 89, pp. 117–126, Mar. 2016. https://doi.org/10.1016/j.compositesb.2015.10.045

[7]   A. Ślosarczyk, "Recent advances in research on the synthetic fiber based silica aerogel nanocomposites," *Nanomaterials*, vol. 7, no. 2. MDPI AG, Feb. 16, 2017. https://doi.org/10.3390/nano7020044

[8]   A. Du *et al.*, "Aerogel: A potential three-dimensional nanoporous filler for resins," *Journal of Reinforced Plastics and Composites*, vol. 30, no. 11, pp. 912–921, Jun. 2011. https://doi.org/10.1177/0731684411407948

[9]   Z. Mazrouei-Sebdani, H. Begum, S. Schoenwald, K. V. Horoshenkov, and W. J. Malfait, "A review on silica aerogel-based materials for acoustic applications," *Journal of Non-Crystalline Solids*, vol. 562. Elsevier B.V., Jun. 15, 2021. doi: 10.1016/j.jnoncrysol.2021.120770

[10]   D. Ge, L. Yang, Y. Li, and J. P. Zhao, "Hydrophobic and thermal insulation properties of silica aerogel/epoxy composite," *J Non Cryst Solids*, vol. 355, no. 52–54, pp. 2610–2615, Dec. 2009. https://doi.org/10.1016/j.jnoncrysol.2009.09.017

[11]   N. Mazlan, N. Termazi, S. Abdul Rashid, and S. Rahmanian, "Investigations on Composite Flexural Behaviour with Inclusion of CNT Enhanced Silica Aerogel in Epoxy Nanocomposites," *Applied Mechanics and Materials*, vol. 695, pp. 179–182, Nov. 2014. https://doi.org/10.4028/www.scientific.net/amm.695.179

[12]   R. Riahipour, M. S. Nemati, M. Zadehmohamad, M. reza Abadyan, M. Tehrani, and M. Baniassadi, "Mechanical properties of an epoxy-based coating reinforced with silica aerogel and ammonium polyphosphate additives," *Polymers and Polymer Composites*, vol. 30, Jan. 2022. https://doi.org/10.1177/09673911211069019

[13]   M. A. Ver Meer, B. Narasimhan, B. H. Shanks, and S. K. Mallapragada, "Effect of mesoporosity on thermal and mechanical properties of polystyrene/silica composites," *ACS Appl Mater Interfaces*, vol. 2, no. 1, pp. 41–47, Jan. 2010. https://doi.org/10.1021/am900540x

[14]   M. T. Run, S. Z. Wu, D. Y. Zhang, and G. Wu, "A polymer/mesoporous molecular sieve composite: Preparation, structure and properties," *Mater Chem Phys*, vol. 105, no. 2–3, pp. 341–347, Oct. 2007. https://doi.org/10.1016/J.MATCHEMPHYS.2007.04.070

[15]   S. Salimian *et al.*, "Silica Aerogel-Epoxy Nanocomposites: Understanding Epoxy Reinforcement in Terms of Aerogel Surface Chemistry and Epoxy-Silica Interface Compatibility," *ACS Appl Nano Mater*, vol. 1, no. 8, pp. 4179–4189, Aug. 2018. https://doi.org/10.1021/acsanm.8b00941

Italian Manufacturing Association Conference - XVI AITeM
Materials Research Proceedings 35 (2023) 376-384

Materials Research Forum LLC
https://doi.org/10.21741/9781644902714-45

PET foaming: development of a new class of rheological additives for improved processability

Clizia Aversa[1,a] *, Massimiliano Barletta[1,b], Annalisa Genovesi[1,c],
Annamaria Gisario[2,d]

[1] Università degli Studi Roma Tre, Dipartimento di Ingegneria Industriale, Elettronica e Meccanica, Via Vito Volterra 62, 00146 Roma (Italy)

[2] Sapienza Università di Roma, Dipartimento di Ingegneria Meccanica e Aerospaziale, Via Eudossiana 18, 00184 Roma (Italy)

[a]clizia.aversa@uniroma3.it, [b]massimiliano.barletta@uniroma3.it
[c]annalisa.genovesi@uniroma3.it, [d]annamaria.gisario@uniroma1.it

Keywords: Polymers, Extrusion, Foaming

Abstract. Polymer foaming is a process broadly used for manufacturing light weight packaging solutions. Polystyrene (PS) is the most widespread material for this application, as it combines easy processability, low cost and high performance of the resulting items. However, foamed PS is difficult to recycle and highly polluting for the oceans and aquatic environment. Polyethylene terephthalate (PET) is, instead, commonly recycled and R-PET is broadly used for several industrial applications. Yet, PET quickly loses viscosity during the foaming process, due to thermo-hydrolytic and oxidative degradation thus causing poor foaming. In this paper, an innovative combination of chain extenders, anti-oxidants and nucleating agents to modify PET rheology is studied. The additives were experimented both in off-line and in-line apparatus. The experimental results show PET rheology can be customized by appropriately modulating the content of the different additives, thus making PET suitable for foaming process of high-quality items.

Introduction

Expanded Polystyrene is widely used for containers and packing materials in fishing, agriculture, and household applications. Its characteristics of lightness, thermal/acoustic insulation and hygienic-food safety have favored its large scale development. EPS (expanded polystyrene, also called XPS), as well as PS is technically a potentially recyclable material through thermo-mechanical or chemical recycling. However, there are many factors that limit the effective recycling of this material: first of all, in most of the countries that carry out differentiated waste collection, including Italy, there is no supply chain dedicated to the collection of EPS, which is conferred to the plastics section and should hypothetically be sorted and separated from other waste to be sent to specific EPS recycling plants. The selection phase turns out to be extremely complicated because due to the mechanical characteristics of EPS products, they tend to crumble during the collection phase, making separation from other waste practically impossible. Moreover, unwanted dispersion of EPS products in the environments is very common thus causing formation of microplastics extremely polluting and harmful to the aquatic ecosystem and indirectly to human health. These reasons prompted the European Union to include disposable EPS products in the list of products banned by the SUP directive (dir. 904/2019). PET on the other hand is the most recyclable and recycled plastic in the world. The European average amount of recycled PET (r-PET) is more than 90% of virgin PET. The foam industry has therefore turned its attention towards the development of expanded PET in order to exploit the PET recycling chain already widely distributed in many countries, mainly driven by recycling of PET bottles. However, in the

Italian Manufacturing Association Conference - XVI AITeM   Materials Research Forum LLC
Materials Research Proceedings 35 (2023) 376-384   https://doi.org/10.21741/9781644902714-45

transition from EPS to EPET, producers of foam products have encountered some difficulties related to the rheological differences of the two polymers.

PET, like all polyesters, is sensitive to thermo-hydrolytic and thermo-oxidative degradation phenomena triggered by the presence of impurities and excessive humidity levels. Such phenomena cause breakage of the polymer chains, loss of molecular weight and therefore reduction of the viscosity. During the direct gas injection process used for the production of foams, a high viscosity is required to allow the expansion of the molten polymer and the formation of the typical cell structure.

A low molecular weight (MW) does not favor the foaming operation in a direct gas injection extrusion process of foamed sheets. During this process the polymeric melt undergoes intense elongational deformations which are not supported by the typically low MW of PET, characterized by low viscosity values, low strength and elasticity of the polymer melt. As a result, uncontrolled cell expansion and unstable growth of bubbles can occur. Higher viscosity (shear and intrinsic) and improvement of melt strength provides a higher resistance to bubble coalescence and a better stability of the cellular structure during the foaming process.

Improved rheology is generally associated with increases in MW and molecular weight distribution (MWD) as a result of polymer chain extension and branching. Degradation of PET during processing can be overcome by adding toughening reactive agents [1,3] and chain extenders in the presence of various catalysts [2,4–7] to obtain materials with controlled rheology and tailored mechanical performance. Chain extenders usually contain at least two functional groups capable of reacting with the carboxylic and/or alcoholic end groups of PET. According to the type, reactivity and number of functionalities present on the chain extender molecule, different reaction mechanisms of the functional groups and different macromolecular architectures can be obtained. Oxazolines [8,9], react very quickly with the carboxylic end groups of PET chains and are able to favour linear chain extension of the polymer by formation of stable bis-amide bridging segments. Pyromellitic dianhydride (PMDA) [10] and organic phosphites [11,12] have also been successfully used as PET chain extenders. Moreover, combinations of PMDA bis-oxazolines have also been studied [13]. Epoxides react with carboxyl moieties of PET [14], leading to the formation of esters and a secondary hydroxyl group which can eventually take part in transesterification reactions with other PET chains, resulting in chain branching and chain scission. Some studies [14, 15] describe the use of polyfunctional chain extenders such as branched low molecular weight or linear oligomeric polyepoxy compounds [14,15] and oligomeric polyisocyanates [16] to obtain branched chain extended PET with rheological properties appropriate for the production of PET foams.

In the present work the effect of three commercially available chain extenders and their combination on melt viscosity of PET was investigated. Inline test during extrusion trials allowed to measure torque and die pressure at various screw speeds. Offline tests on a measuring mixer allowed to observe the torque trend over an extend amount of time (20 minutes), comparable to process time during industrial direct gas injection.

Experimental
Materials
The PET grade selected for this study (Cleartuf P82, Gruppo Mossi & Ghisolfi, Italy) is a food grade characterized by intrinsic viscosity equal to 0.8 dl/g, melting point at 249 °C, with a content of carboxyl groups CC equal to about 27-28 equiv/10^6g. PET was dried, prior to all tests, at 130°C for 4 h and stored in oxygen barrier thermally sealed bags. Different additives and combinatios to promote linear chain extension and branching were selected.

Several chain extenders based on polyepoxides can be profitably used to modify the rheological properties of PET. Diepoxides (characterized by 2 functionalities), triepoxides (f=3) and tetraepoxides (f=4). In relation to the molecular weight of the individual chain extenders, a more or less marked effectiveness of the regradation process will be obtained. Poly-epoxides, i.e.

molecules characterized by a very high number of epoxy functionalities are widely available on the market. In this study Joncryl ADR 4468, a polyepoxy commercialized by Basf (Germany), containing more than 20 epoxy functionalities was considered. This kind of chain extender favours branching reactions which occur by combining the hydroxyl end groups of the PET molecule with the polyepoxide epoxies. The end result is a PET with a comb structure. By suitably controlling the chain extension reaction, it is possible to limit the formation of gels and obtain the desired increase in molecular weight and, therefore, in intrinsic viscosity required by the PET expansion process.

Oxazolines are, like polyepoxides, capable of combining with the carboxyl end groups present on PET. Differently from polyepoxides, oxazolines present a molecular configuration with functional groups placed at the end of the molecule which can favor purely linear chain extension mechanisms. This aspect is certainly important, as the use of oxazolines reduces the risk of gel formation during the extrusion process. Oxazolines have two functional groups and are relatively small molecules (low molecular weight), therefore they are potentially effective at relatively low concentrations in the finished product (0.2 - 0.3 %). In this study 1,3-phenylene-bis-oxazoline (PBO), by the commercial name of Nexamite A99 (Nexam Chemical, Sweden) was selected.

Rheology improvements of PET by using tetracarboxylic dianhydrides as chain extenders has been described as an efficient strategy by several patents throughout the years [17-22, 26-29]. For example [17-18] report how PET extrusion with the addition of 0.3 wt% pyromellitic dianhydride (PMDA) increased melt strength of PET. No parison sag was observed during extrusion blowing and intrinsic viscosity was increased to 0.86 dl/g in PET bottle walls.

Table 1 Reaction characteristics of selected chain extenders

Additive	F [eq./mol]	MW [g/mol]	F_{Eq} [g/eq.]	Reactive Group	Reaction	% Stoich.
PBO	2	140.14	70,1	Carboxylic	Linear CE	0.3
BTDA	4	322.23	80,6	Hydroxyl	Linear CE	0.72
PMDA	4	218.11	54,5	Hydroxyl	Linear CE	0.49
Polyepoxy	23.8	7250	304,6	Carboxylic	Branching	0.85

Known the number of carboxyl and hydroxyl end groups available in the specific PET commercial grade, the stoichiometric quantity of additive capable of reacting with PET may be calculated by applying the following formulas:

(1) Wt %= (MW x CC)/ f x 10^4

(2) Wt %= (MW x HC)/ f x 10^4

Where MW is the molecular weight of the additive, CC and HC are respectively the content of carboxyl and hydroxyl groups of PET in equiv/10^6g and F are the functionalities available in the additive. It is, therefore, possible to deduce the stoichiometric amount of each chain extender needed to saturate the carboxyl functionalities on the chosen PET grade (Table 2). Table 2 also reports characteristics of Benzophenone tetracarboxylic dianhydride (BTDA), to show how not only the number of functionalities, but also molecular weight influences the effective amount of chain extender needed to react with PET moieties.

Addition of small quantities of PE to PET leads to improved thermoformability of foams as described by [24]. In the former patent 2-4 wt% polyolefin (LLDPE) in combination with 0.6 wt% sterically hindered phenolic antioxidant was added to the PET recipe, used for foam extrusion of a thin cellular sheet. The foamed sheet was thermoformed into trays characterized by reduced density, 15% less than the cellular sheets. PET foam modified by polyolefin provides therefore a better flexural and impact properties according to [23,24].

Italian Manufacturing Association Conference - XVI AITeM   Materials Research Forum LLC
Materials Research Proceedings 35 (2023) 376-384   https://doi.org/10.21741/9781644902714-45

Anti-oxidants allow to compensate oxidative degradation, accelerated by the high temperatures during the extrusion process. Anti-oxidants are classified into primary antioxidants that terminate the chain and secondary antioxidants which act by decomposition of hydroperoxides. Mixtures of stabilizers with different mechanisms are today the state of the art.

Therefore, addition of High Density Polyethylene (HDPE) (DOW 410), and a primary/secondary antioxidant mixture (ADK Adeka Stab A611) was evaluated in this study.

Characterization methods

Offline and inline testing procedures were adopted in this study. Offline tests were conducted using a Brabender mixer (Brabender GmbH & Co., Germany) with a 50 ml internal chamber operating at 250°C and 50 rpm, with a mixing time of 20 min. Stoichiometric amounts of the given chain extender were added into the pre-heated mixing chamber 2 minutes after the introduction of PET. Table 2 summarizes the formulations tested on the Brabender measuring mixer. The evolution of the torque (Nm) during the mixing observation time was recorded.

Table 2 Recipes tested on the Brabender measuring mixer

Sample ID	#1a	#2a	#3a	#4a	#5a	#6a	#7a
Components	*wt. %*	*wt. %*	*wt. %*	*wt. %*	*wt. %*	*wt. %*	*wt. %*
PET	100	97,4	97,325	96,72	97,045	97,725	97,12
HDPE	0	2	2	2	2	2	2
PMDA	0	0,4	0,4	0,4	0	0	0
PBO	0	0	0,075	0	0,075	0,075	0
Joncryl	0	0	0	0,68	0,68	0	0,68
Adeka ADK Stab A-611	0	0,2	0,2	0,2	0,2	0,2	0,2
Totale	100	100	100	100	100	100	100

Inline tests consisted in reactive extrusion of PET with stoichiometric amounts of selected chain extender additives (Table 1), using a twin screw extruder (Leistritz 27ZSE iMAXX, Leistritz Gmbh, Germany). A constant temperature profile of 250°C was used along the barrel for all the extrusion tests. The feed rate was set constant at 5%, while the screw speed was varied from 100 to 300 rpm. At low feed rates the screw speed significantly influences the throughput time. The torque and melt pressure during the extrusion process was recorded. The resulting compounds were dried at 110°C for 4 hours and stocked in oxygen barrier thermally sealed bags. Subsequently the melt flow rate of each compound was measured at 260°C with a 2.16 kg load.

Results and Discussion

PET was processed as is as reference for processing of chain extended PET. MFR results show how at higher screw speed, meaning lower throughput time, processed PET features lower MFR, suggesting higher residence time caused major degradation of PET at lower speeds.

Table 3 MFR of as-is processed PET

Sample ID		FR [%]	Screw speed [rpm]	MFR (260°C, 2.16 kg)
		5	50	22.7
#1	PET	5	100	21.0
		5	300	16.0

Subsequently Joncryl, Oxazoline and PMDA were added to PET. Extrusion of PET with Joncry al 0.85% led to a stable process, torque values decrease from 49% to 23% by increasing the screw speed. Higher screw speeds allowed faster throughput, lower residence time and lower filling rate of the barrel, hence leading to a lower torque. The addition of Joncryl makes it possible to halve the MFR of PET, in all working conditions, confirming the regradation effectiveness of the additive. Even PET with PBO is easily processable. PET additivated by PBO features higher MFR values compared to PET+Joncryl samples, thus suggesting Joncryl might be more effective in chain extending PET compared to PBO. However, higher torque values were registered during the extrusion of PET+PBO compared to PET+Joncryl. Higher torque values are generally measured for more viscous polymers, hence suggesting PBO may indeed have modified rheology of PET.

Table 4 Results of inline tests

	Sample ID	Torque [%]	Die Pressure [bar]	FR [%]	Screws speed [rpm]	MFR (260 °C, 2.16 kg)
		58	43	5	50	4.6
#2	PET + PMDA	X	X	5	100	X
		X	X	5	300	X
		55	38	5	50	6.2
#3	PET + PMDA + PBO	X	X	5	100	X
		X	X	5	300	X
		55	55	5	50	2.9
#4	PET + Joncryl + PMDA	X	X	5	100	X
		X	X	5	300	X
		63	45	5	50	6.1
#5	PET + Joncryl + PBO	40	50	5	100	4.3
		30	40	5	300	3.8
		60	33	5	60	15.6
#6	PET + PBO	43	31	5	100	13.8
		30	24	5	300	21.7
		49	48	5	60	10.3
#7	PET + Joncryl	31	34	5	100	10.6
		23	24	5	300	13.4

Combination of the polyepoxy and PBO was also investigated. No processing issue was observed, the extrudate featured good melt strength and optimal pelletization. MFR progressively decreased with increasing screw speed, reaching the lowest value for the compound produced at 300 rpm, equal to 3.8 g/10 min. Therefore, the influence of screw speed is confirmed. The combination of PBO and Joncryl appears extremely effective in lowering MFR of PET. Higher viscosity is also confirmed by torque and die pressure values. Extrusion of PET with stoichiometric PMDA (0.49 wt.%) gave place to uncontrolled swelling of the extrudate, as shown in Figure 1. The amount of PMDA was therefore slightly decreased. The same swelling phenomena appears at 0.4 wt.%, while at 0.3 wt.% processing was possible. PMDA proved to be extremely efficient in reducing MFR of PET. Amounts lower than the stoichiometric quantity were able to reduce MFR of PET as low as 4.6 g/10 min. Issues caused by the die swelling phenomenon could be easily solved by adopting a different pelletizing system (i.e., under water cutting system). The screw speed had a significant effect on PET rheology and, subsequently, processing, for all scenarios where PMDA was involved. Screw speeds higher than 50 rpm led to viscosity increase in the molten polymer which caused a twisting effect on strands exiting the die. The strands were extremely unstable and strand transport across the cooling bath to the pelletizing system was impossible.

Italian Manufacturing Association Conference - XVI AITeM
Materials Research Proceedings 35 (2023) 376-384

Materials Research Forum LLC
https://doi.org/10.21741/9781644902714-45

Figure 1 Die swelling phenomenon for PET+PMDA

Blending PMDA with Joncryl and PBO gave place to the same die swelling phenomena, nonetheless the material was processable at 50 rpm with a 5% feed rate. Combination of PMDA and Joncryl determined a significant viscosity increase as shown by the measured MFR, equal to 2.6 g/10 min and die pressure equal to 55 bar. Combination of PMDA and PBO was also effective causing MFR reduction of PET to 6.2 g/10 min, comparable with MFR registered for the combination Joncryl+PBO under the same processing conditions.

Offline tests allowed to evaluate torque trends on a longer observation time. The extrusion residence time is 1-2 minutes depending on screw speed, whereas residence time in the measuring mixer is independent from other parameters. Torque values of PET and chain extended PET were measured after 5, 10, 14 and 20 minutes mixing inside the Brabender mixer (Table 5). Offline tests on the measuring mixer proved effectiveness of all the studied additives, coherently with inline tests.

Table 5 Torque values registered during measuring mixer tests

	Sample ID	5 min	10 min	14 min	20 min
#1a	PET	4.5 Nm	3.3 Nm	2.9 Nm	\
#2a	PET+PMDA	6.1 Nm	9.9 Nm	15 Nm	14.9 Nm
#3a	PET+PMDA+PBO	6.7 Nm	13.5 Nm	15.5 Nm	13.2 Nm
#4a	PET+PMDA+Joncryl	7.1 Nm	13.6 Nm	15.5 Nm	13.2 Nm
#5a	PET+Joncryl+PBO	9.2 Nm	8.3 Nm	7.1 Nm	6.3 Nm
#6a	PET+PBO	5.5 Nm	4.3 Nm	3.7 Nm	3.2 Nm
#7a	PET+Joncryl	8.8 Nm	6.8 Nm	5.5 Nm	4.6 Nm

PBO was the least effective in increasing torque, compared to as-is PET. PMDA and its combination with Joncryl and PBO gave place to the higher torque values. The combined additive solutions caused a faster torque increase, whereas PMDA alone was more persistent for a longer time.

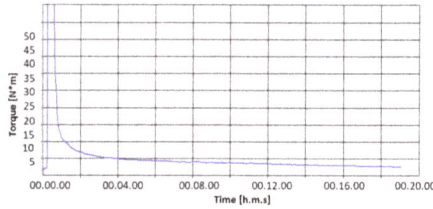

Figure 2 Torque trend of as-is PET

Figure 2 and Figure 3 show the torque trends for a 20 minute observation time. The figures clearly highlight the efficiency and persistency of PMDA in PET regradation. The combination of polyepoxy (Joncryl) and Oxazoline is rather persistent since a very small torque reduction was recorded throughout the observation time. Nevertheless absolute torque values are much lower compared to samples containing PMDA. This behaviour could be associated to the reaction mechanism of PMDA with PET. PMDA is indeed the only selected additive involving reaction with hydroxyl groups, whereas all other additives exploit carboxyl groups reactions.

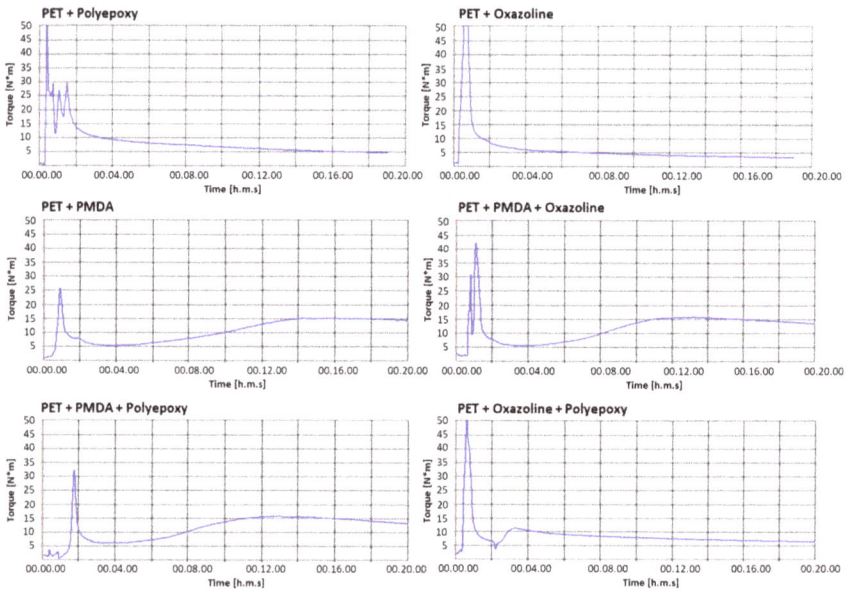

Figure 3 Torque trends of PET additivated with several chain extender additives

Conclusions

Inline and offline tests were performed on chain extend PET to verify the efficiency of PBO, PMDA, Polyepoxy (Joncryl) and their combinations in regradation and rheology modification of PET. Torque and die pressure was registered during reactive extrusion and MFR was measured on resulting pellets to verify rheology variations on chain extended PET. Torque was also measured at set intervals for compounds mixed for a longer period inside a heated chamber. Both tests showed the efficiency of the selected additives, with special reference to PMDA and its combination with other additives. Combination of PMDA with PBO and Joncryl gave place to faster reactions. The highest viscosity increase was registered for PET+PMDA in the inline tests and for PET+PMDA+Joncryl in the offline tests. Remarkable results were also observed for the Joncryl+Oxazoline combination. Inline tests showed good persistence of the combined additives and a nearly constant torque trend was observed during offline tests. In conclusion, the selected additives effectively increase viscosity of PET, thus allowing improved foaming of PET.

References

[1] W. Loyens, G. Groeninckx, Macromol. Chem. Phys. 203 (2002) 1702-1714. https://doi.org/10.1002/1521-3935(200207)203:10/11<1702::AID-MACP1702>3.0.CO;2-6

[2] M.-B. Coltelli, S. Bianchi, M. Aglietto, Polymer 48 (2007) 1276-1286. https://doi.org/10.1016/j.polymer.2006.12.043

[3] M. Aglietto, M.-B. Coltelli, S. Savi, F. Lochiatto, F. Ciardelli, M. Giani, J. Mater, Cycles Waste Manage. 6 (2004) 13-19. https://doi.org/10.1007/s10163-003-0100-z

[4] M.-B. Coltelli, Polym. Degrad. Stabil. 90 (2005) 211-223. https://doi.org/10.1016/j.polymdegradstab.2005.03.016

[5] M.-B. Coltelli, M. Aglietto, F. Ciardelli, Eur. Polym. J. 44 (2008) 1512-1524. https://doi.org/10.1016/j.eurpolymj.2008.02.007

[6] M.-B. Coltelli, S. Bianchi, S. Savi, V. Liuzzo, M. Aglietto, Macromol. Symp. 204 (2003) 227-236. https://doi.org/10.1002/masy.200351419

[7] M.-B. Coltelli, S. Savi, Macromol. Mater. Eng. 289 (2004) 400-412 https://doi.org/10.1002/mame.200300297

[8] N. Cardi, R. Pò, G. Giannotta, E. Occhiello, F. Garbassi, G. Messina, J. Appl. Polym. Sci. 50 (1993) 1501-1509. [9] G.P. Karayannidis, E.A. Psalida, J. Appl. Polym. Sci. 77 (2000) 2206-2211. https://doi.org/10.1002/1097-4628(20000906)77:10<2206::AID-APP14>3.0.CO;2-D

[10] L. Incarnato, P. Scarfato, L. Di Maio, D. Acierno, Polymer 41 (2000) 6825-6831. https://doi.org/10.1016/S0032-3861(00)00032-X

[11] B. Jacques, J. Devaux, R. Legras, E. Nield, Polymer 37 (1996) 1189-1200. https://doi.org/10.1016/0032-3861(96)80846-9

[12] F.N. Cavalcanti, E.T. Teòfilo, M.S. Rabello, S.M.L. Silva, Polym. Eng. Sci. 47 (2007) 2155-2163. https://doi.org/10.1002/pen.20912

[13] C.R. Nascimento, C. Azuma, R. Bretas, M. Farah, M.L. Dias, J. Appl. Polym. Sci. 115 (2010) 3177-3188. https://doi.org/10.1002/app.31400

[14] S. Japon, Y. Leterrier, J.-A.E. Manson, Polymer 41 (2000) 5809-5818. https://doi.org/10.1016/S0032-3861(99)00768-5

[15] M. Xanthos, M.-W. Young, G.P. Karayannidis, D.N. Bikiaris, Polym. Eng. Sci. 41 (2001) 643-655. https://doi.org/10.1002/pen.10760

[16] Y. Zhang, W. Guo, H. Zhang, C. Wu, Polym. Degrad. Stabil. 94 (2009) 1135- 1141. https://doi.org/10.1016/j.polymdegradstab.2009.03.010

[17] J. P. Leslie, C. A. Lane, R. P. Grant, Melt strength improvement of PET, (US Patent US4,145,466 (1979))

[18] Li, Jie, Gräter Horst, Chain-extenders and foamed thermoplastic cellular materials obtained by reactive extrusion process and with help of said chain-extenders, (EU Patent EP 2 163 577 B1 (2012))

[19] Phobos N.V., et al., Verfahren zur kontinuierlichen Herstellung von hochmolekularen Polyester-Harzen (EU Patent EP0422282 (1995))

[20] Al Ghatta, H., T. Severini, L.Astarita, Foamed cellular polyester resins and process for their preparation, (Worldwide patent WO9312164 (1992)

[21] Al Ghatta, H., T. Severini, L.Astarita, Geschäumte zellhaltige Polyesterharze und Verfahren zu ihrer Herstellung (European Patent, EP0866089 (1992))

[22] Sublett, B.J., Preparation of branched polyethylene terephthalate (Worldwide Patent WO9502623 (1993))

[23] K.C. Khemani, J. W. Mercer, Richard L.Mcconnell, Concentrates for improving polyester compositions and method of making same (Worldwide Patent WO9509884 (1994))

[24] T. M. Cheung, C. L. Dabis, J. E. Prince, Light weight polyester article (European Patent EP 0390 723 (1990))

[25] P. Raffa; M.-B. Coltelli; S. Savi; S. Bianchi; V. Castelvetro. Chain extension and branching of poly(ethylene terephthalate) (PET) with di- and multifunctional epoxy or isocyanate additives: An experimental and modelling study, Reactive & Functional Polymers, 72(1), (2012) 50-60. https://doi.org/10.1016/j.reactfunctpolym.2011.10.007

[26] EP3246349A12017-05-242017-11-22, Nexam Chemical AB, Process for increasing the melt strength of a thermoplastic polymer

[27] BE1027200B1 2019-04-192020-11-17, Nmc Sa, Composite profile pieces with a low density polyester foam core

[28] TWI705094B2019-04-252020-09-21, 南亞塑膠工業股份有限公司, Recycle pet foaming material and method for manufacturing the same

[29] CN113150256B *2021-04-212022-08-26, 浙江恒逸石化研究院有限公司, Branched copolyester for bead foaming and preparation method thereof

Italian Manufacturing Association Conference - XVI AITeM
Materials Research Proceedings 35 (2023) 385-392

Materials Research Forum LLC
https://doi.org/10.21741/9781644902714-46

Study of autoclave process to manufacture thermoplastic composites constituted by PP/flax fibers

Gianluca Parodo[1,a] *, Luca Sorrentino[1,b] and Sandro Turchetta[1,c]

[1]Department of Civil and Mechanical Engineering, University of Cassino and Southern Lazio, via G. di Biasio, 43, Cassino (FR), 03043, Italy

[a]gianluca.parodo@unicas.it, [b]luca.sorrentino@unicas.it, [c]sandro.turchetta@unicas.it

Keywords: Thermoplastic Composites, Natural Fibers, Thermoforming, Process Parameters, Experimental Tests

Abstract. Autoclave processes are widely used from industries that produce thermoset polymer composite parts. However, these materials show sustainability issues as they are non-recyclable and produced by energy-intensive processes. The use of thermoplastic matrices reinforced with natural fibers can solve these problems; however the optimal use of this material is linked to the knowledge of the forming parameters. In this work, starting from semipreg sheets, the autoclave forming process for parts in flax woven and polypropylene is studied and developed: it represents a fundamental step to develop the use of these new eco-friendly materials starting from the well-established industrial knowledge, not only in terms of environmental sustainability, but also economic and social sustainability. First, working temperatures were determined by DSC and TGA; while optimal forming pressure were determined by ILSS tests.

Introduction

The use of fiber-reinforced polymers for structural and semi-structural applications has seen a strong growth in recent decades, especially in the aeronautical and automotive sectors. This trend is essentially due to the lightness and high specific resistance that these materials have, characteristics that allow to reduce consumption and CO_2 emissions of aircraft, boats, railway and road vehicles, etc. [1–3].

Nowadays, most of structural and semi-structural components are manufactured using prepregs cured by autoclave [4]. This technique allows to obtain high volumetric fractions of reinforcement, excellent mechanical performance and a very low presence of defects inside the parts. For these reasons, the autoclave is one of the main and most important equipment present in all those industries that nowadays deal with the production of polymer composite parts.

One of the problems faced by traditional fiber-reinforced polymer composites is their non-recyclability and their highly energy-consuming production process. In addition, they are mostly derived from synthetic petroleum products, which nowadays present environmental sustainability issues [5]. For these reasons, recently there is a growing interest in vegetal reinforcements, which have shown to have specific resistances similar to glass reinforcement, so they can be a valid substitute for the latter in the case of semi-structural parts. Specifically, flax, hemp and jute fibers have been the subject of numerous studies due to their excellent performance, superior in terms of specific resistance to E-glass [6].

Plant fibers have a lower permeability than glass, which is why making parts with these types of reinforcement through infusion or injection processes is quite complicated [7]. Moreover, they would only be recyclable if coupled with thermoplastic matrices, which have high viscosities. A solution to this problem can be given by the production of the composite part through film compression molding, in which the dry reinforcement is layered alternately with thermoplastic polymer films [8]. In this way, the path that the resin must make inside the reinforcement is limited in the direction of the thickness. However, the bio composites obtained with this technique

Italian Manufacturing Association Conference - XVI AITeM   Materials Research Forum LLC
Materials Research Proceedings 35 (2023) 385-392   https://doi.org/10.21741/9781644902714-46

generally have a high volumetric percentage of resin. Furthermore, they may have any internal defects due to the possible absorption of environmental moisture in the reinforcement. A solution to this problem can be given by the use of semipregs [9]. As well as traditional prepregs, semipreg are pre-impregnated reinforcements with the optimal amount of resin, in this case no longer thermosetting but thermoplastic. This aspect allows to greatly simplify the management and logistics of the material because the semipreg is chemically stable at room temperature and pressure. In addition, being the reinforcement protected from the external environment by the same thermoplastic resin, it is not necessary to store it in a controlled humidity environment and dry it just before the production of the component.

Generally, parts made of polymer composite material should be made by thermoforming [10]. In the case of thermoforming of structural or semi-structural polymer composites, the material is first layered and formed in hot-plate press in order to obtain a consolidated laminate with the nominal thickness provided by the project specification, then it is cooled and analyzed by non-destructive testing. If there are no inner defects, it is tensioned by means of tensioners and heated again by IR furnaces at the process temperature. Once this is achieved, the consolidated laminate is placed in the press and formed [11]. Being a totally different process from the traditional one in autoclave, most of the industrial companies that have developed their know-how on thermosetting composites should completely convert their departments in order to be able to produce components in bio thermoplastic composite material. In the authors' best knowledge, nowadays there are no scientific publications that study the process of thermoforming of bio composite materials in autoclave. This intermediate step, which consists in the manufacturing of bio composite parts through autoclave processes, would be fundamental to allow the production of bio-recyclable composite laminates with processes consolidated by many industrial realities. The adoption of natural fiber composites with thermoplastic matrix as a replacement, for example, of fiberglass would greatly reduce the environmental impact in terms of emitted CO_2, as, in addition to requiring energy-intensive processes, the production of the reinforcement itself would result in a negative CO_2 balance obtained by the plant photosynthesis process [12].

The aim of this work is to study the range of processability of this new type of materials in autoclave and, subsequently, investigate the effect of the obtained process parameters on the interlaminar shear strength, representative of the adhesion between the layers of the formed laminates. This study represents an intermediate but indispensable step for the development of this type of materials not only in terms of environmental sustainability, but also of economic and social sustainability [13]. Specifically, the main process parameters in autoclave for the production of polypropylene and woven flax fiber laminates were identified and the mechanical performance were evaluated.

Materials and Methods
The semipreg used in this work is produced by Bcomp and it consists of a 2x2 twill fabric reinforcement in flax fibers known as Amplitex 5040 and impregnated in a polypropylene matrix. It is a semipreg with an area density of 300 g/m^2 and a thickness of about 0.8 mm. Its tension strength is 224 MPa and it has a Young modulus of about 20 GPa.

Identification of process parameters
Since the temperature of workability of the material was not known, the first step of the analysis consisted in its definition. This range was determined by DSC (Differential Scanning Calorimetry) tests on semipreg samples with a mass of about 10 mg and using the DSC module Q20 V24, supplied by TA Instruments. The analyses were carried out according to EN 6032. Specifically, the samples were equilibrated to -60°C, therefore subject to a ramp of 10 °C/min up to the target temperature of 250 °C, at which they remained in isothermal conditions for 1 min, then cooled

Italian Manufacturing Association Conference - XVI AITeM   Materials Research Forum LLC
Materials Research Proceedings 35 (2023) 385-392   https://doi.org/10.21741/9781644902714-46

with a cooling rate of -10 °C/min until the room temperature was reached. By means of DSC analysis, it was possible to define the temperature range for thermoforming.

In order to evaluate if thermal degradation of the semipreg occurs at the temperatures of interest, TGA (Thermo Gravimetric Analysis) analyses were carried out both on dry reinforcement and on semipreg, using the TGA Q500 V20 module, also supplied by TA Instruments. In this case, the analysis was carried out on samples of about 10 mg but with a heating ramp of 3 °C/min up to 250 °C. The TGA analysis made it possible to assess whether there was a degradation of the processed material in terms of mass loss in the range determined by the DSC tests. This analysis was carried out on both semipreg and dry woven in order to evaluate the barrier effect of the matrix against environmental moisture. Five replicas were made for both DSC and TGA analyses.

Once the working temperature range was determined, the range of pressures and optimal processing times were investigated by autoclave thermoforming of consolidated laminates. The consolidated laminates were made with 8 layers for a final thickness of about 5.4 mm. The layers were placed on the mold, so the bag was made and the vacuum applied. In this way the vacuum itself constrained the layers on the mold, preventing them from moving during the process. This approach, however, allows to stratify even more complex geometries, as the drapability of the individual ply at room temperature is always higher than that of the consolidated laminate, presenting the latter a much higher thickness (therefore a flexural stiffness which could cause the failure of the bag during the vacuum application). After compaction, the bag was placed in the autoclave and subject to the reference thermal cycle. The heating ramp adopted in this work was 3 C°/min, while the dwell time was obtained empirically. In particular, preliminary forming of flat laminates were carried out with type J thermocouples inside them in the position shown in Fig. 1. The dwell time was therefore defined as the time needed for the thermocouple to register a temperature within the formability range of the material obtained by DSC analysis.

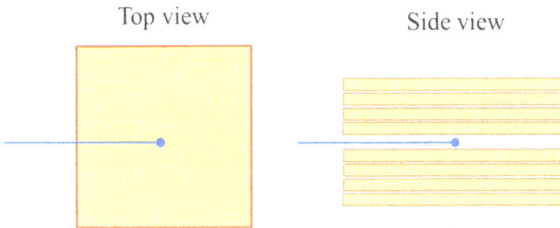

Fig. 1. Position of the control thermocouple.

For completeness, the values in terms of temperature and pressure used for the production of the consolidated in autoclave were also reported in Table 1.

Table 1. Experimental plan for the manufacturing of consolidated thermoplastic laminates.

Factors	# Level	Levels
Temperature [°C]	3	160 – 170 – 180
Pressure [bar]	3	2 – 4 – 6
Repetitions	3	

Italian Manufacturing Association Conference - XVI AITeM   Materials Research Forum LLC
Materials Research Proceedings 35 (2023) 385-392   https://doi.org/10.21741/9781644902714-46

Once consolidated, they were finished and cut to realize interlaminar shear strength specimens (ILSS, Fig. 2b), in order to evaluate their mechanical performance and adhesion between the layers. For each consolidated laminate, 5 specimens were made. ILSS tests were carried out according to ASTM D2344; in this case the dimensions were 40 mm x 12 mm x 5.4 mm, while the crosshead speed was 1 mm/min. All mechanical tests were carried out using a universal testing machine equipped with a 10 kN load cell.

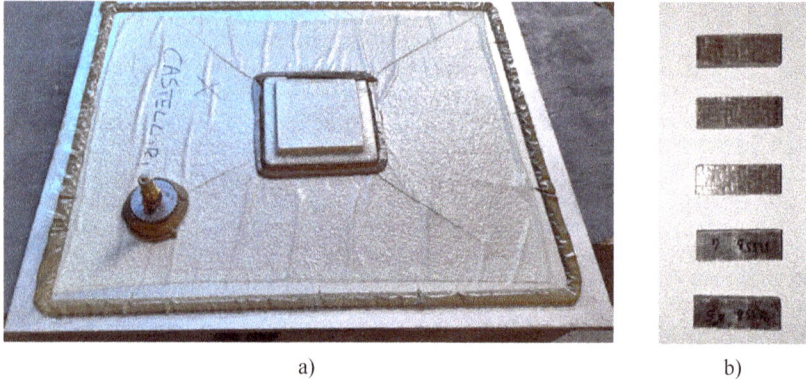

a)   b)

Fig. 2. Manufacturing of Flax/PP laminates: a) material compaction steps; b) ILSS specimens.

Results and Discussions

DSC results

The results of the DSC analyses showed a high repeatability and allowed the matrix melting temperature to be detected at 166.3 °C with a standard measurement dispersion of about 0.3 °C during heating. Moreover, recrystallization peaks between 130 and 120 °C were observed during the cooling phase. It is possible to state that the thermoforming process will have to consider a thermal cycle that should have a process temperature around 166.3 °C. An example of a DSC test result is reported in Fig. 3.

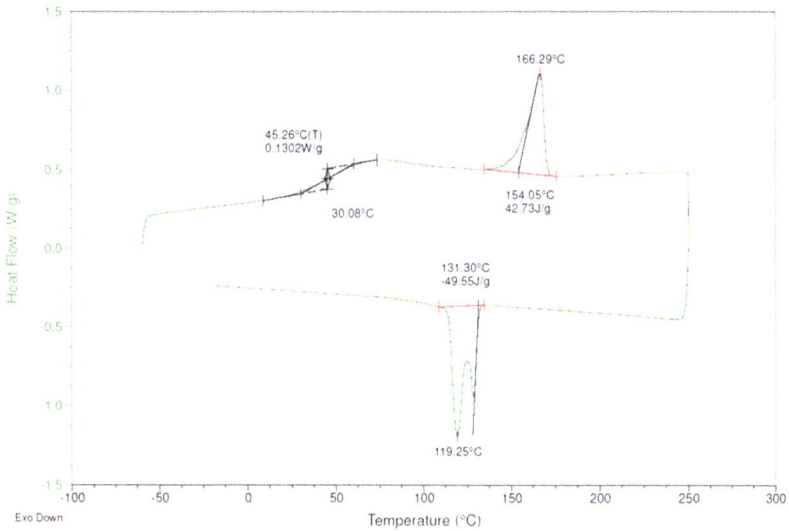

Fig. 3. Example of temperature /heat flow curve obtained from a DSC test on semipreg PP/flax fabric.

TGA results

While in the case of DSC tests it was possible to determine the range of the semipreg workability temperature, the TGA analyses allowed to evaluate the material degradation at the investigated process temperatures in terms of percentage of mass lost. Specifically, the specimens were subjected to a heating ramp of 3 °C/min up to the temperature of 250 °C. The results showed that the mass loss is always less than 3% on semipreg at the temperatures range obtained by DSC analysis, while the dry woven showed a strong variability, resulting in a mass variation between 7 and 9% near the processing temperature range (Fig. 4). This difference was mainly due to the presence of the PP matrix that provided a barrier effect on the reinforcement avoiding that the latter absorbs moisture from the environment. It is possible to state that the use of semipreg allowed an easier management of the forming process, as the storage conditions of the material will not lead to a moisture absorption that can deteriorate the reinforcement. Moreover, it will no longer be necessary to carry out the drying of the reinforcement itself as a processing step prior to stratification.

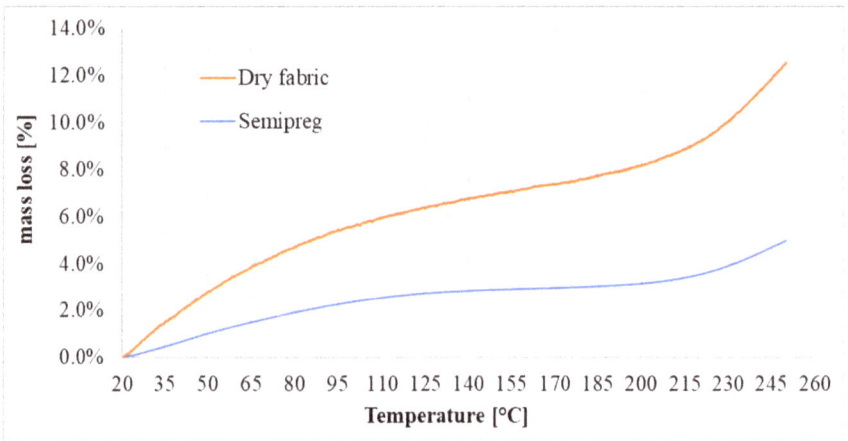

Fig. 4. Example of TGA results for dry flax fabric and PP/flax fabric semipreg.

ILSS test results

The results of the ILSS tests allowed the mechanical evaluation of adhesion between the layers as a function of the process parameters. The experimental results showed a strong variability in the range of the investigated process parameters, as shown in Fig. 3. In order to assess whether the effect of the individual process parameters on the interlaminar shear strength of the consolidated is significant or not, an ANOVA analysis of the results obtained was carried out. For this purpose, it has been hypothesized that the population investigated had a Gaussian distribution, was homoscedastic and that the observations made were independent. In addition, a significance level of 0.05 has been established for the analysis. The results obtained are reported in Table 2.

It is possible to say that the investigated process parameters, namely the forming temperature and pressure, had a not negligible effect on the interlaminar shear strength in the obtained laminates. Specifically, as shown in Fig. 5a, the most impacting factor was the processing temperature, with a contribution obtained of about 81%. This result was due to an increased energy available to allow the polymer to melt and subsequent solidification, resulting in an improvement in the adhesion between the layers. For this reason, the values obtained at 160 °C were found to be very low, while at 180 °C the maximum values were obtained. However, it has been observed that the pressure and the combination of temperature and pressure also had a not negligible effect, although with a lower contribution of about 8% and 10% respectively. In fact, observing Fig. 5 and Fig. 6b, it is possible to affirm that in the case of forming at 170 °C, the effect of the pressure on the interlaminar shear strength had led to a strong increase in performance as the compaction pressure increases, despite an increase in the dispersion of results.

Italian Manufacturing Association Conference - XVI AITeM   Materials Research Forum LLC
Materials Research Proceedings 35 (2023) 385-392   https://doi.org/10.21741/9781644902714-46

Table 2. ANOVA table for ILSS results.

Source	DoF	Seq SS	Contribution	Adj SS	Adj MS	F-Value	P-Value
Temperature [°C]	2	1092.80	81.40%	1092.8	546.4	754.51	0.000
Pressure [bar]	2	105.03	7.82%	105.03	52.517	72.52	0.000
Temperature*Pressure [°C] [bar]	4	131.57	9.80%	131.57	32.892	45.42	0.000
Error	18	13.04	0.97%	13.04	0.724		
Total	26	1342.44	100.00%				

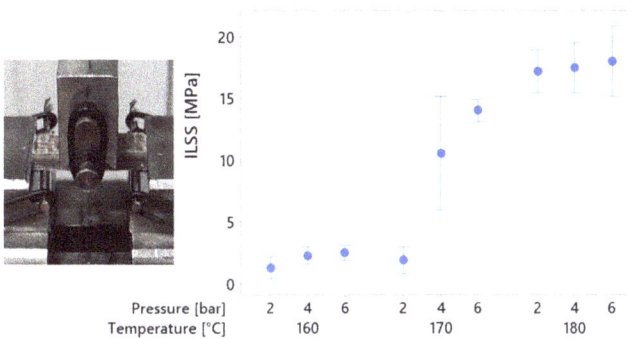

Fig. 5. Experimental results from ILSS tests.

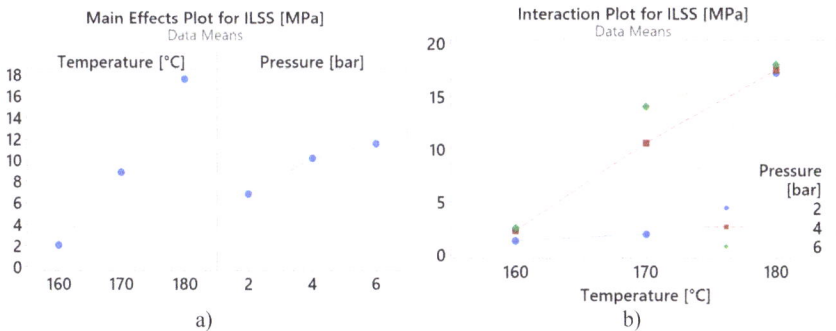

Fig. 6. ILSS experimental results: a) main effect plot; b) interaction plot.

Conclusions

In this work, starting from semipreg sheets, the study of autoclave process to manufacture thermoplastic composite laminates constituted by woven flax fibers and PP matrix was carried out. Specifically, process temperatures of the investigated material were identified between 160 and 180 °C by DSC and TGA tests. Subsequently, ILSS samples were produced with different pressures, in order to assess the good adhesion between the layers according to the process parameters. It was observed that the most impacting factor on performance was temperature, with a contribution of about 81%. In fact, average ILSS increased from about 2 MPa at 160 °C to about 18 Mpa at 180 °C. Finally, compaction pressure had an important effect especially at 170 °C, which was the temperature closest to the melting of investigated material.

References

[1]   A. Baker, S. Dutton, D. Kelly, Composite Materials for Aircraft Structures, AIAA; 2nd edition, 2004.

[2]   S.A. Pradeep, R.K. Iyer, H. Kazan, S. Pilla, Automotive Applications of Plastics: Past, Present, and Future, Second Edi, Elsevier Inc., 2017. https://doi.org/10.1016/B978-0-323-39040-8.00031-6

[3]   F. Rubino, A. Nisticò, F. Tucci, P. Carlone, Marine application of fiber reinforced composites: A review, J. Mar. Sci. Eng. 8 (2020). https://doi.org/10.3390/JMSE8010026

[4]   S. V Hoa, Principles of the Manufacturing of Composite Materials, DEStech Publications, Inc., 2009.

[5]   M. Akter, M.H. Uddin, I.S. Tania, Biocomposites based on natural fibers and polymers: A review on properties and potential applications, J. Reinf. Plast. Compos. 41 (2022) 705–742. https://doi.org/10.1177/07316844211070609

[6]   D.U. Shah, Developing plant fibre composites for structural applications by optimising composite parameters: a critical review, J. Mater. Sci. 48 (2013) 6083–6107. https://doi.org/10.1007/s10853-013-7458-7

[7]   L. Sorrentino, S. Turchetta, G. Parodo, R. Papa, E. Toto, M.G. Santonicola, S. Laurenzi, RIFT Process Analysis for the Production of Green Composites in Flax Fibers and Bio-Based Epoxy Resin, Materials (Basel). 15 (2022) 8173. https://doi.org/10.3390/ma15228173

[8]   L. Boccarusso, D. De Fazio, M. Durante, Production of PP Composites Reinforced with Flax and Hemp Woven Mesh Fabrics via Compression Molding, Inventions. 7 (2021) 5. https://doi.org/10.3390/inventions7010005

[9]   R.C. Adams, S. Advani, D.E. Alman, ASM Handbook, 2001. https://doi.org/10.1016/S0026-0576(03)90166-8

[10]   A. Stamopoulos, A. Di Ilio, Numerical and experimental analysis of the thermoforming process parameters of semi-spherical glass fibre thermoplastic parts, Procedia CIRP. 99 (2021) 420–425. https://doi.org/10.1016/j.procir.2021.03.060

[11]   S. Iwan, F. Althammer, S. Mueller, J. Troeltzsch, L. Kroll, Simulating the Forming of Thermoplastic, Fibre Reinforced Plastics - Demonstrated for a Side Impact Protection Beam, J. King Mongkut's Univ. Technol. North Bangkok. (2017). https://doi.org/10.14416/j.ijast.2017.05.006

[12]   L. Huang, B. Yan, L. Yan, Q. Xu, H. Tan, B. Kasal, Reinforced concrete beams strengthened with externally bonded natural flax FRP plates, Compos. Part B Eng. 91 (2016) 569–578. https://doi.org/10.1016/j.compositesb.2016.02.014

[13]   O. Iordache, Industry and Society, in: Stud. Syst. Decis. Control, Springer Science and Business Media Deutschland GmbH, 2023: pp. 159–176. https://doi.org/10.1007/978-3-031-07980-1_8

Italian Manufacturing Association Conference - XVI AITeM   Materials Research Forum LLC
Materials Research Proceedings 35 (2023) 393-401   https://doi.org/10.21741/9781644902714-47

A feasibility study to improve the processability of pure copper produced via laser powder bed fusion process

Abdollah Saboori[1,*], Marta Roccetti Campagnoli[2], Manuela Galati[1], Flaviana Calignano[1], Luca Iuliano[1]

[1] Integrated Additive Manufacturing Center, Department of Management and production Engineering, Politecnico di Torino, Corso duca Degli Abruzzi 24, 10129 Torino, Italy

[2] Department of Mechanical Engineering, Politecnico di Torino, Corso duca Degli Abruzzi 24, 10129 Torino, Italy

Keywords: Additive Manufacturing, Copper, Processability, Laser Powder Bed Fusion, Single Scan Tracks

Abstract. Additive Manufacturing (AM) refers to a family of layer-upon-layer building technologies capable of producing geometrically intricate parts in a single step. Today, the processability of many materials through AM is under development. One of the most interesting studies is the production of copper parts via laser-based technologies. Unluckily, mainly due to the high thermal conductivity and reflectivity of copper, its processability through AM processes is particularly challenging. Thus, in this research, a new material-based solution is proposed to improve the processability of copper through laser powder bed fusion. Therefore, a single scan track analysis is performed on pure copper and mixtures of copper/graphite. The outcomes show that adding graphite could increase copper's laser absorption and processability.

Introduction

These days, thanks to the new developments in 3D printing technology, as well as the improvements achieved in equipment and materials, metal Additive Manufacturing (AM) has become one of the most attractive research fields [1,2]. However, this technology is classified into two main classes; Powder Bed Fusion (PBF) and Directed Energy Deposition (DED) [3–5]. It is well reported that, even if their manufacturing concept is the same, their building mechanisms are rather different. In the case of the PBF process, the laser melts the powder particles, which are already spread like a powder bed and solidify afterwards [6,7]. One of the key advantages of the AM processes is freedom in design that brings the complexity for free in the part design and production. Therefore, the production of complex shape parts, including the lattice structures or internal channels, was facilitated using metal AM processes [8,9]. A heat exchanger made of copper and copper alloys is one of these complex components that can be the first candidate to be produced via metal AM technologies [10].

Copper (Cu) which is a malleable and ductile metallic material, presents a good corrosion resistance and low chemical reactivity [11,12]. In addition, copper is characterised by extraordinary machinability, formability, and high electrical and thermal conductivity [13]. Copper attracts much attention in applications like microelectronics, roofs and plumbing implants, radiators, charge air coolers, and heat exchangers [14,15]. This wide range of characteristics and applications makes copper a promising material in various industrial sectors like electro packaging, automotive, and construction [16]. On the other hand, copper is also frequently used as a base material for different alloys such as brass and bronze, that, in addition to copper, consist of zinc and tin, respectively [17]. Owing to its high formability, copper is commonly processed via Powder Metallurgy (PM) or conventional manufacturing processes (e.g. forging, machining, extrusion and casting) [18]. However, these manufacturing techniques suffer from various limitations mainly related to the difficulties in producing optimised finned heat exchangers and heat sinks [19]. As a matter of fact, traditional manufacturing processes for the complex components are characterised by high production costs as well as complex and time-consuming post-processing steps. The design of heat exchange components aims to minimise their size while using optimised thin fins to increase the surface area and the heat transfer rate between the heat exchanger surface and the surroundings [20]. In this context, AM processes results are attractive thanks to the possibility of producing topologically optimised geometries layer by layer, reducing the period of manufacturing and tooling requirements [2,18].

Despite the clear advantages of adopting AM technologies, it is well documented that the processability of copper and copper alloys via metal AM processes faces several challenges [18,21]. The high electrical and thermal conductivities of copper and its alloys increase the heat transfer rate from the melt pool to the surrounding area, generating high cooling rates and detrimental consequences for the process and the part quality [22]. In addition, for laser-based processes, the low laser absorption rate in the near-infrared region is another greatest issue. Both the rapid heat transfer and the high reflectivity that hinders the absorption of the laser power, result in high porosity and poor mechanical, thermal and electrical properties [18]. Moreover, the high sensitivity of copper to oxidation makes the powder handling of this alloy very difficult [21]. In fact, it requires an inert atmosphere during the process and special storage. The risk connected to the presence of copper oxides is the formation of gas bubbles inside the matrix of the final component, which reduces the density and electrical conductivity of the component.

Nevertheless, obtaining high-performance components following a layer-by-layer approach boosts wide scientific research to develop or enhance the processability of copper and its alloys through the AM processes [23]. Many works have been published in this context, particularly during the last years [24,25]. The adopted solutions can be grouped into technological-based or material based.

The technological-based solutions come back to modifying or optimising the design of machine or process parameters. For instance, Liu et al. succeeded to produced dense pure copper samples with a density of 99.6% using a blue laser [26]. Sciacca et al. also printed a dense pure copper heat sink through the LPBF process using the outcome of a process parameter optimization procedure [27].Instead, material-based solutions concern the addition of other elements to the pure copper and/or the surface modification of the copper powder particles [18].

So far, several studies have been conducted to investigate the problems faced and present new possible solutions to be adopted. However, each alloy modification for the sake of processability increased the cost of the powder and also deteriorated some characteristics of the final components. Therefore, this paper aims to contribute to this growing area of research by proposing a new material-based solution to enhance the processability of copper to produce copper parts via laser-based AM technologies. Moreover, one of the main goals is to develop a new solution without increasing the cost and sacrificing some features, such as the thermal and electrical conductivity of copper components.

Materials and Methods

A gas atomised spherical copper powder with a particle size range of 20-50 μm and a graphite powder with an average particle size range of 7-11 μm were used as the feedstock material.

In order to find the best graphite content from the flowability point of view, several powder mixtures containing different graphite contents were prepared in a low energy jar mill for 24 hours. Thereafter, the distribution of graphite powder within the copper particles was evaluated using a tabletop Scanning Electron Microscope (SEM, Phenom XL). The density, Hausner ratio (HR) and flowability of the powder mixtures were also assessed using a Hall flowmeter. The tapped density of powder mixtures was also evaluated using a container of 25 cm³ that filled up till the highest compaction that was achieved through a vibration. Thereafter, the relative tapped density was

Italian Manufacturing Association Conference - XVI AITeM
Materials Research Forum LLC
Materials Research Proceedings 35 (2023) 393-401
https://doi.org/10.21741/9781644902714-47

calculated using the theoretical density of the powder mixture. The same characterisations were also performed on the pure copper as a reference sample.

The spreadability of each composition was evaluated using a self-made spreading system that could simulate the powder spreading inside the LPBF machine. Then, the homogeneity of the powder layer was evaluated using a Leica Stereomicroscope.

Afterwards, two sets of single scan tracks (SSTs) with a length of 9 mm were produced for pure copper and Cu-0.5 wt.%C (Fig.1). For this reason, a design of experiment (DOE) consisting of 20 different combinations of process parameters was considered in this DOE, the laser power was between 85-95 W, and laser scanning speed was in the range of 100 to 250 mm/s, with a step size of 25 mm/s. The layer thickness was fixed at 0.02 mm. These combinations of process parameters resulted in a linear energy density in the range of 0.38 to 0.95 J/mm.The same DOE was used for pure copper to evaluate the role of graphite on the laser absorption of copper.

After the production, all the SSTs were analysed from the top using a Leica EZ4W stereomicroscope. Afterwards, two disks were cut, mounted and polished following the standard metallography reported for copper alloys. The as-polished surfaces were then etched chemically for further analysis by an optical microscope (OM). The geometry of the melt pools was measured from their cross-section using ImageJ software.

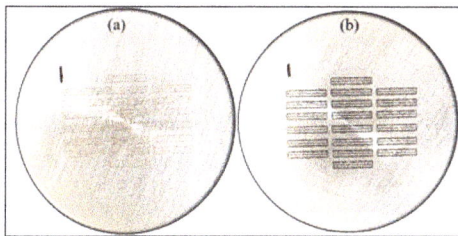

Fig. 1. (a) SSTs of pure coppers; (b) SSTs of Cu-0.5%C.

Results and Discussions

Powder characterisation

Fig. 2 shows the variation of relative tapped density of the Cu-C powder mixture as a function of graphite content. As can be seen, any addition of graphite to pure copper reduces the tapped density of the powder mixture and, consequently, the packing density of the powder layer.

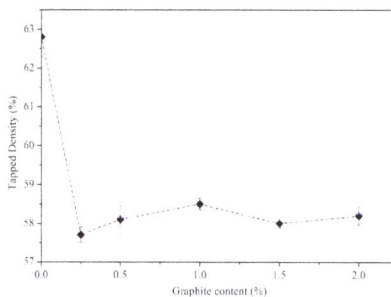

Fig. 2. Tapped density of Cu-C powder mixture as a function of graphite content.

Italian Manufacturing Association Conference - XVI AITeM   Materials Research Forum LLC
Materials Research Proceedings 35 (2023) 393-401   https://doi.org/10.21741/9781644902714-47

Table 1 compares the Hausner ratio of the powder mixtures with pure copper. It is shown that pure copper with a proper Hausner ratio (1.0<HR<1.1) exhibited excellent flowability. In contrast, the addition of graphite resulted in higher values (HR>1.19) and, as a consequence, poor flowability. The effect of graphite content on the flowability and spreadability of the copper powder is shown in Fig. 3, which compares a layer of copper powder with the powder mixtures.

Table 1. The hausner ratio of pure copper and copper-graphite powder mixture.

Composition (wt.%)	Hausner ratio
Pure Cu	1.10
Cu-0.25%C	1.14
Cu-0.5%C	1.20
Cu-1.0%C	1.22
Cu-1.5%C	1.25
Cu-2.0%C	1.26

This analysis confirmed the trends of the tapped density and Hausner ratio, decreasing the packing density and spreadability of the powder mixture as a function of the graphite content.

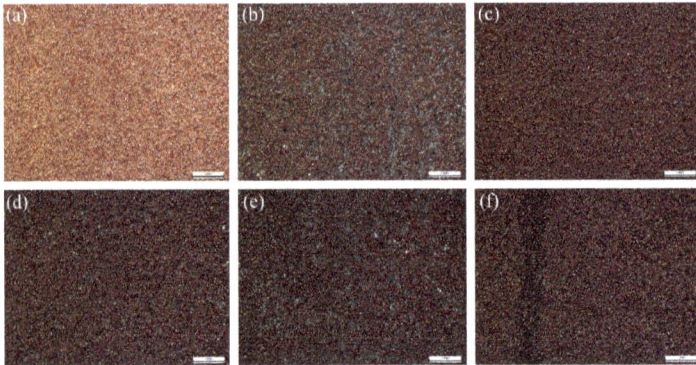

Fig. 3. OM micrograph of a layer of (a) pure Cu, (b) Cu-0.25%C, (c) Cu-0.5%C, (d) Cu-1.0%C, (e) Cu-1.5%Cu, (f) Cu-2.0%C.

As it is possible to see in Fig. 3, pure copper shows a perfect uniform distribution over the plate, thanks to the spherical shape of the particles. Adding a small quantity of graphite negatively affects the spreadability, and the mixture cannot fully cover the base plate. In the opposite case, if the amount of graphite surpasses 1.5 wt.%, it cannot mix homogeneously with the copper, forming agglomerates that deteriorate the uniformity of the powder layer. After that, the graphite distribution within the copper powder is evaluated through the SEM analysis, and the outcomes are reported in Fig. 4.

As can be seen in Fig. 4, the graphite plates adhered to the surface of copper particles, and they tended to form agglomerates as the graphite content increased and consequently deteriorated the flowability of the powder mixture.

Considering the outcomes of the powder characterisations, the composition of Cu-0.5%C was chosen as the most promising one for SSTs analysis. As mentioned earlier, the powder consisting of 0.5% graphite exhibited an acceptable tapped density and Hausner ratio than can guarantee the flowability of the powder mixture. Moreover, the spreadability test and SEM analysis confirmed that the spread powder layer is agglomerate free and uniform.

Italian Manufacturing Association Conference - XVI AITeM   Materials Research Forum LLC
Materials Research Proceedings 35 (2023) 393-401   https://doi.org/10.21741/9781644902714-47

After printing the SSTs following the DOE considered in this research, it was possible to group the SSTs into four different categories: samples with not enough Linear Energy Density (LED), melt pools with evident balling, thin and stable or irregular SSTs (As shown in Fig. 5).

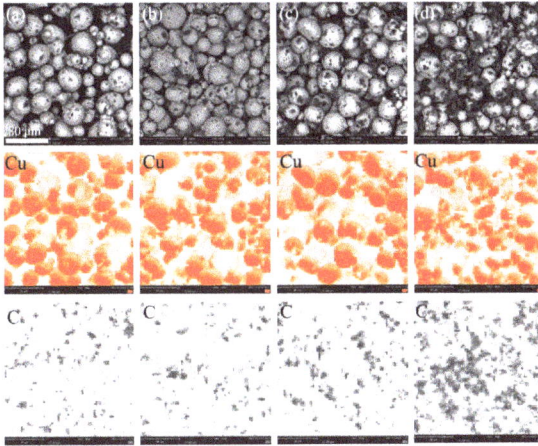

Fig. 4. SEM images of (a) Cu-0.5%C, (b) Cu-1.0%C, (c) Cu-1.5%C, (d) Cu-2.0%C.

No scan track was revealed in the "Not enough LED" case since the LED used was insufficient to melt the powder. In the second case, balling behaviour prevails: the scan track was discontinuous, and the melt pool is characterised by poor wetting. The "Thin and stable" category is individuated in a narrow range of LEDs, where the SSTs resulted in a uniform melt pool. Finally, the melt pools are strongly asymmetrical with higher LED values: these latter are defined as "Irregular".

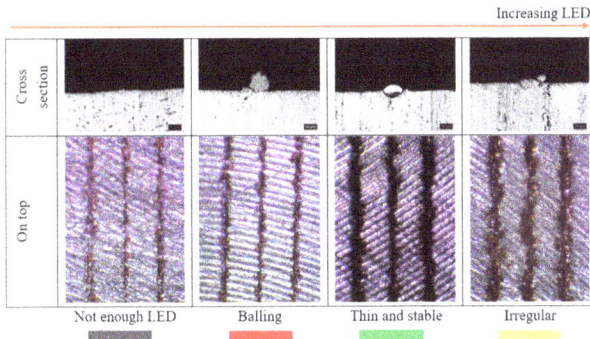

Fig. 5. Examples of different melt pools: not-enough LED, balling effect, thin and stable, irregular.

Fig. 6 demonstrates the cross-section of the SSTs of pure copper at different line energies. This figure shows that very high energy densities result in irregular melt pools, whereas very low line energies lead to unstable SSTs. However, as can be seen, thin and stable melt pools are formed at the medium level LED. On the other hand, it is clear that LED can not be considered a key factor

Italian Manufacturing Association Conference - XVI AITeM   Materials Research Forum LLC
Materials Research Proceedings 35 (2023) 393-401   https://doi.org/10.21741/9781644902714-47

in finding the optimum process parameters. This means two sets of processes with the same line energy do not essentially form the same melt pool size. For instance, the LED of 0.422 J/mm resulted in an unstable melt pool, while the melt pool generated using other sets of parameters with the LED of 0.425 was regular and stable.

Fig. 7 shows the variation of the melt pool width of pure copper as a function of LED. This graph also confirms that the melt pool width can be different at the constant energy density. Interestingly, the melt pools are quite small with a LED ranging between 0.4 and 0.5 J/mm, and the scan tracks are just fairly continuous. The increment in LED starts to be detrimental when it exceeds the value of 0.7 J/mm.

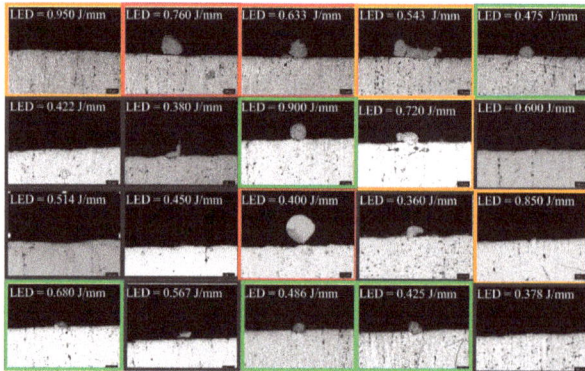

Fig. 6. Melt pool cross-section of pure copper SSTs.

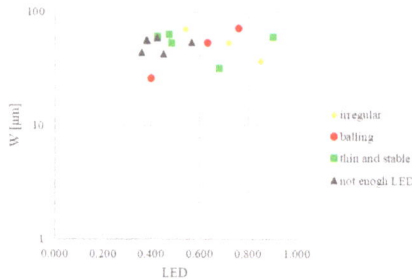

Fig. 7. Melt pool width of pure copper as a function of LED (J/mm).

Fig. 8 shows the cross-section of the Cu-0.5%C melt pools. As can be seen, after adding the graphite platelets, the undesired phenomena of balling reduced significantly, and the melt pools looked more stable.

Italian Manufacturing Association Conference - XVI AITeM   Materials Research Forum LLC
Materials Research Proceedings 35 (2023) 393-401   https://doi.org/10.21741/9781644902714-47

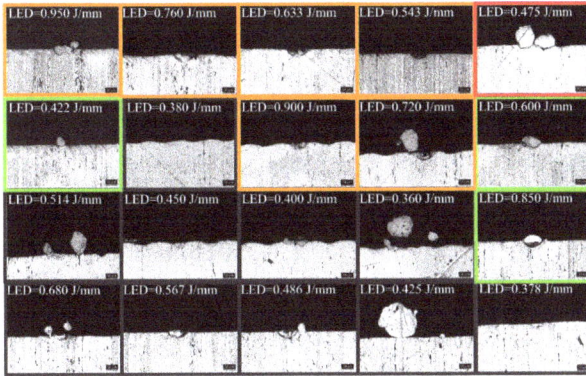

Fig. 8. Melt pool cross-section of Cu-0.5%C SSTs.

Fig. 9 compares the width of the Cu-0.5%C melt pool produced using different linear energy densities. This figure shows that as the melt pools result more stable with respect to the pure copper, the dimensions of the melt pools are larger, in particular the width of melt pools. This finding confirms that the addition of a small quantity of graphite could result in a higher LED absorption, consequently, the larger the melt pool formation.

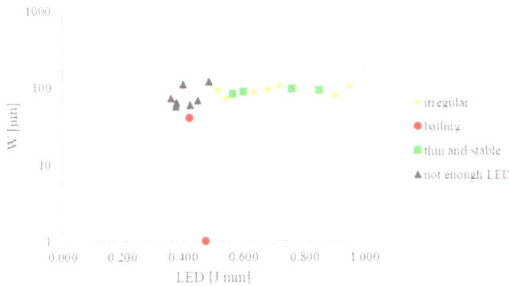

Fig. 9. Melt pool cross-section of Cu-0.5%C SSTs.

Conclusions

In this study, the effects of adding different amounts of graphite to copper powder were analyzed. With its higher laser absorption, graphite helps improve copper manufacturing through laser based AM processes. In contrast, the quantity and the dimensions of graphite flakes need to be carefully selected so as not to destroy the mixture's spreadability. The outcomes demonstrated that the best quantity of graphite to be added is 0.5 wt.%. A good density, flowability, and spreadability can be obtained with this value. Moreover, high coverage of copper particles is ensured, and no graphite agglomerations are detected. A comparison between the cross-section of the SSTs of pure copper and Cu-0.5%C confirmed that the melt pools of the Cu-0.5%C generally look more stable and wider with respect to the pure copper ones. This suggests that the mixture can absorb a larger amount of LED, increasing the dimensions of the melt pools. This study lays the groundwork for future research into the successful fabrication of complex shape components through the LPBF process, thanks to the addition of a small quantity of graphite.

References

[1] M. Dadkhah, M.H. Mosallanejad, L. Iuliano, A. Saboori, A Comprehensive Overview on the Latest Progress in the Additive Manufacturing of Metal Matrix Composites: Potential, Challenges, and Feasible Solutions, Acta Metall. Sin. (English Lett. 34 (2021) 1173–1200. https://doi.org/10.1007/s40195-021-01249-7

[2] M. Attaran, The rise of 3-D printing: The advantages of additive manufacturing over traditional manufacturing, Bus. Horiz. 60 (2017) 677–688. https://doi.org/https://doi.org/10.1016/j.bushor.2017.05.011

[3] A. Saboori, A. Aversa, G. Marchese, S. Biamino, M. Lombardi, P. Fino, Application of Directed Energy Deposition-Based Additive Manufacturing in Repair, Appl. Sci. 9 (2019). https://doi.org/10.3390/app9163316

[4] M.H. Mosallanejad, B. Niroumand, A. Aversa, D. Manfredi, A. Saboori, Laser Powder Bed Fusion in-situ alloying of Ti-5%Cu alloy: Process-structure relationships, J. Alloys Compd. 857 (2021) 157558. https://doi.org/10.1016/j.jallcom.2020.157558

[5] I.O. for S. ISO/ASTM, ASTM 52900: 2015 (ASTM F2792) Additive Manufacturing—General Principles—Terminology, ISO Geneva, Switz. (n.d.).

[6] M. Aristizabal, P. Jamshidi, A. Saboori, S.C. Cox, M.M. Attallah, Laser powder bed fusion of a Zr-alloy: Tensile properties and biocompatibility, Mater. Lett. 259 (2020) 126897. https://doi.org/https://doi.org/10.1016/j.matlet.2019.126897

[7] M.H. Mosallanejad, B. Niroumand, A. Aversa, A. Saboori, In-situ alloying in laser-based additive manufacturing processes: A critical review, J. Alloys Compd. 872 (2021) 159567. https://doi.org/https://doi.org/10.1016/j.jallcom.2021.159567

[8] G. Del Guercio, M. Galati, A. Saboori, Electron beam melting of Ti-6Al-4V lattice structures: correlation between post heat treatment and mechanical properties, Int. J. Adv. Manuf. Technol. 116 (2021) 3535–3547. https://doi.org/10.1007/s00170-021-07619-w

[9] G. Del Guercio, M. Galati, A. Saboori, P. Fino, L. Iuliano, Microstructure and Mechanical Performance of Ti–6Al–4V Lattice Structures Manufactured via Electron Beam Melting (EBM): A Review, Acta Metall. Sin. (English Lett. 33 (2020) 183–203. https://doi.org/10.1007/s40195-020-00998-1

[10] S.D. Jadhav, S. Dadbakhsh, J. Vleugels, J. Hofkens, P. Van Puyvelde, S. Yang, J.-P. Kruth, J. Van Humbeeck, K. Vanmeensel, Influence of Carbon Nanoparticle Addition (and Impurities) on Selective Laser Melting of Pure Copper, Materials (Basel). 12 (2019) 2469. https://doi.org/10.3390/ma12152469

[11] A. Saboori, M. Pavese, C. Badini, P. Fino, Development of Al- and Cu-based nanocomposites reinforced by graphene nanoplatelets: Fabrication and characterization, Front. Mater. Sci. 11 (2017). https://doi.org/10.1007/s11706-017-0377-9

[12] A. Saboori, M. Pavese, C. Badini, P. Fino, A Novel Cu–GNPs Nanocomposite with Improved Thermal and Mechanical Properties, Acta Metall. Sin. (English Lett. 31 (2018) 148–152. https://doi.org/10.1007/s40195-017-0643-y

[13] L. Kaden, G. Matthäus, T. Ullsperger, H. Engelhardt, M. Rettenmayr, A. Tünnermann, S. Nolte, Selective laser melting of copper using ultrashort laser pulses, Appl. Phys. A. 123 (2017) 596. https://doi.org/10.1007/s00339-017-1189-6

[14] A. Saboori, S.K. Moheimani, M. Pavese, C. Badini, P. Fino, New Nanocomposite Materials with Improved Mechanical Strength and Tailored Coefficient of Thermal Expansion for Electro-Packaging Applications, Met. (Basel). 7 (2017)

[15] V. Sufiiarov, E. Borisov, I. Polozov, SELECTIVE LASER MELTING OF COPPER ALLOY, Mater. Phys. Mech. 43 (2020) 65–71. https://doi.org/10.18720/MPM.4312020_8

[16] M.A. Lodes, R. Guschlbauer, C. Körner, Process development for the manufacturing of 99.94% pure copper via selective electron beam melting, Mater. Lett. 143 (2015) 298–301. https://doi.org/https://doi.org/10.1016/j.matlet.2014.12.105

[17] A. Popovich, V. Sufiiarov, I. Polozov, E. Borisov, D. Masaylo, A. Orlov, Microstructure and mechanical properties of additive manufactured copper alloy, Mater. Lett. 179 (2016) 38–41. https://doi.org/https://doi.org/10.1016/j.matlet.2016.05.064

[18] M. Roccetti Campagnoli, M. Galati, A. Saboori, On the processability of copper components via powder-based additive manufacturing processes: Potentials, challenges and feasible solutions, J. Manuf. Process. 72 (2021) 320–337. https://doi.org/https://doi.org/10.1016/j.jmapro.2021.10.038

[19] R. Neugebauer, B. Mueller, M. Gebauer, T. Töppel, Additive manufacturing boosts efficiency of heat transfer components, Assem. Autom. 31 (2011) 344–347. https://doi.org/10.1108/01445151111172925

[20] L. Benedetti, C. Comelli, C. Ahrens, Study on Selective Laser Melting of Copper, 2017. https://doi.org/10.26678/ABCM.COBEF2017.COF2017-0148

[21] T.I. El-Wardany, Y. She, V.N. Jagdale, J.K. Garofano, J.J. Liou, W.R. Schmidt, Challenges in Three-Dimensional Printing of High-Conductivity Copper, J. Electron. Packag. 140 (2018). https://doi.org/10.1115/1.4039974

[22] F. Singer, D.C. Deisenroth, D.M. Hymas, M.M. Ohadi, Additively manufactured copper components and composite structures for thermal management applications, in: 2017 16th IEEE Intersoc. Conf. Therm. Thermomechanical Phenom. Electron. Syst., 2017: pp. 174–183. https://doi.org/10.1109/ITHERM.2017.7992469

[23] T.Q. Tran, A. Chinnappan, J.K.Y. Lee, N.H. Loc, L.T. Tran, G. Wang, V. V Kumar, W.A.D.M. Jayathilaka, D. Ji, M. Doddamani, S. Ramakrishna, 3D printing of highly pure copper, Metals (Basel). 9 (2019). https://doi.org/10.3390/met9070756

[24] K. IMAI, T.-T. Ikeshoji, Y. SUGITANI, H. KYOGOKU, Densification of pure copper by selective laser melting process, Mech. Eng. J. (2020). https://doi.org/10.1299/mej.19-00272

[25] F. Sciammarella, M. Gonser, M. Styrcula, Laser Additive Manufacturing of Pure Copper, 2013.

[26] X. Liu, H. Wang, K. Kaufmann, K. Vecchio, Directed energy deposition of pure copper using blue laser, J. Manuf. Process. 85 (2023) 314–322. https://doi.org/https://doi.org/10.1016/j.jmapro.2022.11.064

[27] G. Sciacca, M. Sinico, G. Cogo, D. Bigolaro, A. Pepato, J. Esposito, Experimental and numerical characterization of pure copper heat sinks produced by laser powder bed fusion, Mater. Des. 214 (2022) 110415. https://doi.org/https://doi.org/10.1016/j.matdes.2022.110415

Italian Manufacturing Association Conference - XVI AITeM   Materials Research Forum LLC
Materials Research Proceedings 35 (2023) 402-410   https://doi.org/10.21741/9781644902714-48

Mechanical recycling of CFRPs: manufacturing and characterization of recycled laminates

Dario De Fazio[1,a] *, Luca Boccarusso[1,b], Antonio Formisano[1,c],
Antonio Langella[1,d], Fabrizio Memola Capece Minutolo[1,e] and Massimo Durante[1,f]

[1]Department of Chemical, Materials and Production Engineering, University of Naples
"Federico II", P.le Tecchio 80, 80125, Naples, Italy

[a]dario.defazio@unina.it, [b]luca.boccarusso@unina.it, [c]aformisa@unina.it, [d]antgella@unina.it,
[e]capece@unina.it, [f]mdurante@unina.it

Keywords: Fibre Reinforced Plastic, Composite Recycling, Compression Moulding

Abstract. Carbon fibre reinforced plastics (CFRPs) are a very attractive family of materials used in various application fields such as automotive, marine or aeronautic, due to their high specific mechanical properties. However, the large use of CFRPs dramatically increases the amount of waste materials that derives from the end-of-life products and the off-cuts generated during the manufacturing. In this contest, especially when thermosetting matrices are considered, the need to further study the recycling process of CFRPs is an open topic, both in academic and industrial research. Therefore, in this experimental campaign, CFRP materials deriving from the aeronautic field were recycled by using a milling process. The obtained chips were sieved and inspected with a confocal microscope aiming to evaluate the presence of residual matrix on the recovered fibre's surface. Then the sieved reinforcement was impregnated with new epoxy resin and three-point bending tests were performed to understand the mechanical properties of the recycled composite materials. To produce recycled composites, two manufacturing techniques, i.e. open moulding and compression moulding were considered.

Introduction

The increasing research for lightweight structural materials with improved mechanical properties makes CFRPs a very attractive alternative to their metal counterpart. The rising demand for this category of materials over the last decades can be attributed to their intrinsic high level of tailorability and design freedom, these aspects allow a large use of CFRPs in various application fields such as aerospace, marine, automotive or energy [1].

Indeed, it was revealed an increasing demand of CFRP in these industrial fields that was around 70 kTons in 2010 and reached a level of almost 170 kTons in 2020 [2]. However, a further increase is expected in the imminent future since it is estimated that the demand for carbon fibre composite materials will increase to around 190 kTons in 2050 [2, 3]. Therefore, the large amount of CFRP materials in all industrial sectors will inexorably increase the amount of waste materials that must be managed when they are decommissioned at the end of life. Then, in this context, it is evident that the need for a second life application of these materials is required, moving toward a circular economy that allows the reuse of the dismissed materials, reducing at the same time the production of additional wastes and toxic chemical agents [4]. Based on these considerations, the disposal and the waste management of the end-of-life CFRPs can be considered a rapid-developing challenge for the industrial sectors; therefore, in this perspective the European Union has implemented some directives such as the 2008/98/EC and the 2000/53/EC that are focused on principles of prevention and smart utilisation of composite materials, amount of recycled fibres in new industrial products and pollution payment [3, 5].

To date, a large part of the dismissed materials is usually disposed in landfill or burned to generate energy from the combustion; however, these methods are not in line with the principles

Italian Manufacturing Association Conference - XVI AITeM   Materials Research Forum LLC
Materials Research Proceedings 35 (2023) 402-410   https://doi.org/10.21741/9781644902714-48

of the circular economy and cannot be considered as an appropriate way to recover carbon fibres [6]. In comparison with landfill disposal and incineration, composite recycling is a more sustainable process that is in accordance with the philosophy of the circular economy. It is possible to distinguish three main recycling methods like thermal, chemical and mechanical that differ from each other by the fibre recovery process [7].

However, it is known from the literature that, although produce clean recovered fibres, chemical and thermal recycling processes require high specific energies and appropriate reactor vessels since aggressive and hazardous chemical substances are used [8–11]. All these aspects make these recycling methods unattractive in comparison with the mechanical process that can be considered suitable without limitations to recover the most used industrial fibre's typologies.

The mechanical recycling process usually consists of a reduction of CFRP materials in small pieces using shredding, milling or grinding techniques that on some occasions can be combined to obtain recycled material with the desired dimension. The shredding process is employed to reduce composite materials in small pieces, usually in form of flakes 50 – 100 mm in dimension. During this process, all fasteners and inserts embedded into the material are removed. Additional machining operations like milling and grinding are usually required to further reduce the dimensions of the recycling material and to obtain recovered fibres in form of bundles [12, 13]. Therefore, at the end of the machining operations, it is possible to obtain recovered composite material that can be classified by dimension in powders smaller than 0.2 mm, fine fibres with a length in a range of 0.2 – 20 mm and coarse fibres with a length of about 50 mm. The classification of the machined composite chips by dimension can be allowed through a cyclone or a shaking sieving machine.

Composite materials recycling is an open challenge in the research world since many works are focused on this issue; i.e. Palmer et al. [14] investigated the mechanical properties of recycled composite materials reinforced with recovered carbon fibres and the possibility to use these fibres in place of virgin glass fibres. At the end of the experimental campaign, the tests revealed a reduction of the flexural modulus and flexural strength of almost 3% and 9% in comparison with composite materials reinforced with virgin glass fibres. The research group attributed the reduced performances to the poor adhesion between the recovered fibres and the new resin. Thomas et al. [15] studied the influence of using recovered carbon fibres to increase the mechanical properties of epoxy resin. The mechanical tests revealed that the sample produced with 20% by weight of recycled carbon fibres is characterised by an increase in the flexural strength of almost 30% in comparison with the sample in pure resin.

An overview on the state of the art revealed that the researchers in their works find out that the use of mechanical recovered carbon fibres, although increase the mechanical properties of the pure resin, when used in recycled composite materials are characterised by reduced mechanical properties if compared with virgin CFRP materials [14–18]. This drawback that defines recycled composite materials can be attributed to geometrical characteristics and short dimensions of the recovered fibres and to the presence of residues of matrix that affect the adhesion efficiency at fibre-matrix interface. However, limitation in the mechanical properties can be further ascribed to the production technologies of the recycled composite materials, because in some cases several voids can be entrapped significantly, affecting the mechanical properties of this category of materials.

Therefore, based on these observations, in this research work CFRP materials have been machined with a milling recycling strategy and the recovered carbon fibres have been reused for the production of recycled composite materials adopting different production strategies in order to overcome the technological limitations and then increase the mechanical properties of composite materials reinforced with recycled fibres.

Italian Manufacturing Association Conference - XVI AITeM   Materials Research Forum LLC
Materials Research Proceedings 35 (2023) 402-410   https://doi.org/10.21741/9781644902714-48

Materials and methods

For the experimental campaign, a CFRP material was manufactured using a 200 g/m^2 twill wave carbon/epoxy prepreg supplied by Toray. A total number of 20 layers 200 x 200 mm^2 in dimensions were layered up producing $(0/90)_{20}$ laminates 4 mm in thickness. The composite materials were produced using the compression moulding technique and was cured at 130 °C for 8 hours with a pressure of about 0.8 MPa. At the end of the polymerization phase, CFRP laminates were mechanically recycled by using the milling process in order to obtain carbon fibre chips. To this purpose a CNC machine (C.B. Ferrari) equipped with a 20 mm in diameter three flute HSS end mill tool was used. A well-defined cutting strategy was applied for the mechanical recycling process of the CFRP laminates; in detail, a spindle speed s and a feed rate f of 1000 rpm and 1000 mm/min were respectively adopted and a depth of cut of 3 mm was used to produce a material removal rate MRR of 0.33 mm/tooth.

At the end of the machining operations, the recovered carbon fibres (rCFs) were sieved with a shaking table allowing to a chips dimension classification: (i) coarse fibre with a length >10 mm, (ii) fine fibres with a length in a range of 0.2 – 10 mm and (iii) powders with a length <0.2 mm. The rCFs were used to produce new composite laminates by using two manufacturing techniques, i.e. open mould and compression moulding techniques. In both cases, the classified chips were manually impregnated with epoxy resin and were placed into the mould at room temperature for 24 hours. Different recycled CFRP samples, that differ in the dimension of the reinforcement (powders, fine and coarse fibres) were produced using both the production strategies but fixing the same amount of recovered fibres, that is around 20% in weight of the overall fibre-matrix mixture. A further sample of pure epoxy resin was produced adopting the open mould technique and was used as reference. In the following figure are reported the main production steps adopted for the production of the recycled samples.

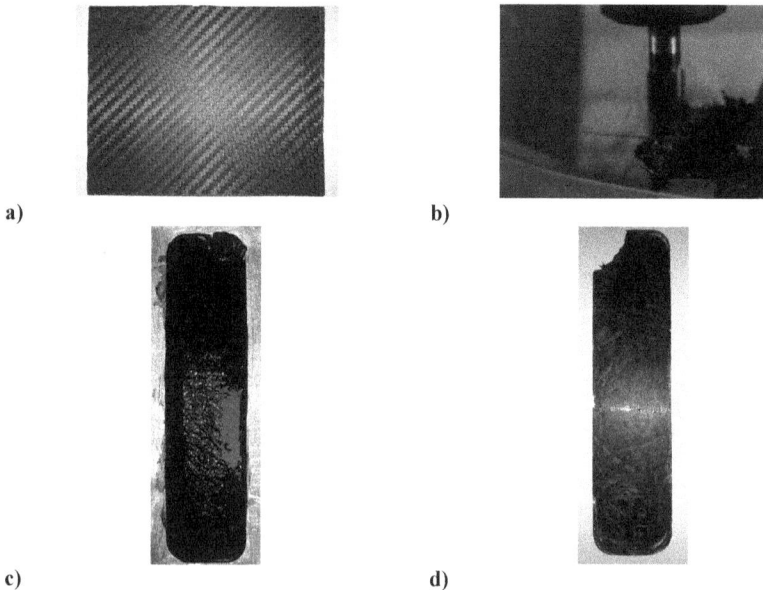

Figure 1: *Main production steps: laminate's manufacturing (a); mechanical recycling phase (b); recycled sample's production (c); recycled sample (d)*

The experimental campaign provided flexural tests in accordance with the ASTM D790 standard on each sample typology. For this purpose, a universal testing machine (Alliance RT/50) equipped with a 1kN piezoelectric load cell was used to evaluate the bending behaviour of the recycled samples. Then, a reference pure epoxy resin sample and a total number of three specimens for each family were tested by fixing the span-to-thickness ratio equal to 32. The bending behaviour, in terms of flexural strength was evaluated by means of the following equation:

$$\sigma_f = \frac{3Pl}{2bt^2} \tag{1}$$

where σ_f is the maximum stress in MPa, P is the maximum load in N, l is the support span in mm, b and t are respectively the width and the thickness in mm of the testing specimen.

Results and discussion
The mechanical recycling process of the CFRP laminates leads to rCFs in form of bundles that differ in dimensions. In detail, Figure 2 reports the weight distribution of the recovered fibres as function of the overall amount of recycling material.

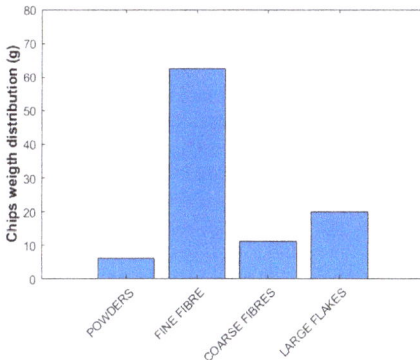

Figure 2: *Weight distribution of recovered carbon fibres as function of the total amount of machined material.*

Looking at Figure 2, it is possible to highlight that the process parameters adopted during the machining of the CFRP material lead to a large formation of fine fibres (more than 60% of the recovered material). This chip's composition can be attributed to delaminations and plies fragmentation phenomena that occur when the tooth of the cutting tool impacts on the recycling laminate. However, powders, fine fibres and coarse fibres were used to produce recycled composite samples, whose characteristics are listed in Table 1.

Table 1: Main properties of the recycled samples

Label	Production technology	Fibre type	Fibre volume percentage [%]
REF	Open mould	-	-
POWDER_OM	Open mould	Powder	12.00
POWDER_CM	Compression moulding	Powder	20.00
FINE_OM	Open mould	Fine fibres	12.00
FINE_CM	Compression moulding	Fine fibres	20.00
COARSE_OM	Open mould	Coarse	12.00
COARSE_CM	Compression moulding	Coarse	20.00

Looking at the Table 1, it is possible to highlight that when the open mould technique is adopted, the fibre volume fraction of the recycled samples does not exceed the 12%. On the other hand, when the compression moulding is used, although the fibre-matrix composition is the same, the volume fraction raised up to around 20%. The difference in the reinforcement amount is attributed to the strategy adopted during the sample's production, because the compression moulding technique is responsible of a thickness reduction and of the resin flow that pushes away the exceed of resin and the trapped porosities generated during the production phase of the recycled samples. All these aspects can be considered responsible of the increasing of the fibre's percentage that characterises all sample families produced with the compression moulding strategy.

In Figure 3 the main results of the three-point bending tests carried out on each sample typology in accordance with the ASTM standard are represented. In detail, the typical stress-strain curves (Figure 3a) and the flexural strength and modulus (figure 3b) are reported. This test can be very attractive since it can be used to have a better insight into the mechanical properties of the recycled composite materials in terms of interaction between the recovered fibres and the new resin. A global overview on the flexural curves (Figure 3a), lets to conclude that all samples under inspection reveal a linear trend of the flexural stress because of the imposed deformation. A further inspection of the flexural curves revealed that all samples produced with recovered material are characterised by an improved elastic modulus and a reduced strain at failure in comparison with the sample in pure epoxy resin used as reference. This behaviour can be related to the presence of the reinforcement that makes the sample stiffer than the reference.

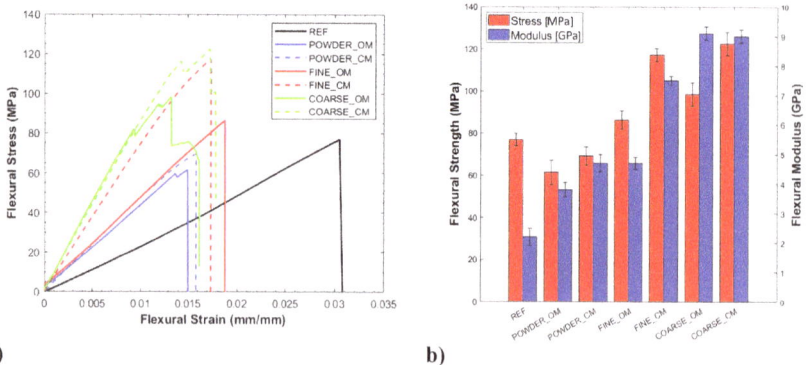

a)   b)

Figure 3: Typical stress-strain flexural curves (a); mean value of the flexural stress and modulus (b) of each sample typology

Italian Manufacturing Association Conference - XVI AITeM   Materials Research Forum LLC
Materials Research Proceedings 35 (2023) 402-410   https://doi.org/10.21741/9781644902714-48

Focusing the attention on the recycled sample reinforced with the shortest fibres (POWDER_OM), it is possible to highlight that even if it revealed a modulus increase of almost +72.7% (3.8 GPa) in comparison with the pure resin sample (2.2 GPa), it is characterised by a reduction of the flexural strength of around -20.1% if compared with the reference (76.9 MPa). The premature failure of this sample typology can be attributed to the geometry of the recovered material which is in form of particles that, in conjunction with physical porosities that generate into the fibre-matrix system during the production phase, acts as preferential surface where internal cracks propagate leading to the failure of the sample. Analogous conclusions can be drawn when the same fibre typology is used with the compression moulding technique (POWDER_CM sample). As highlighted in Table 1, this production technology leads to an increase in fibre volume fraction and then to an overall improvement of the elastic modulus and flexural strength respectively of almost +23.7% and +12.7% (respectively 4.7 GPa and 69.2 MPa) in comparison with the POWDER_OM sample. However, although the increase of the fibre's content, it is possible to assert that the flexural behaviour is still strongly influenced by the geometry of the recovered fibres, because the POWDER_CM sample revealed a reduction of the flexural strength of almost -10% in comparison with the pure resin sample. As well as for the POWDER_OM sample, the premature failure can be attributed to the dimension of the reinforcement that generates preferential cracks propagation surfaces.

On the other hand, when fine fibres are used for the production of recycled material (FINE_OM sample), it is possible to appreciate the effect of the reinforcement geometry over the mechanical properties since this sample typology revealed an increase of both the elastic modulus and flexural strength of almost +23.7% and +40.5% (4.7 GPa and 86.3 MPa) respectively if compared with the POWDER_OM type and around +113.6% and +12.2% respectively in comparison with the reference sample. However, the slight properties increase of FINE_OM sample respect to the reference can be attributed to the well-known poor adhesion efficiency between recovered fibres and new matrix that generates pull-out and debonding phenomena. A magnification of the mechanical recycled materials (Figure 4) shows the presence of residues of old matrix on the fibre's surface that is responsible for the poor adhesion efficiency then, for the reduced mechanical properties. However, the mechanical properties can be further affected by residual porosities that generate in proximity of fibre agglomerations (Figure 5), that as well as for the sample typologies reinforced with the shortest recycled fibres, act as cracks propagation ways. Interesting results can be observed when recovered fibres with the same dimensions reinforce a recycled composite material produced with the compression moulding technique. In this case, the increase in the fibre volumetric percentage leads to an improvement in the elastic modulus and flexural strength of respectively +59.6% and +35.8% (7.5 GPa and 117.2 MPa) in comparison with the FINE_OM sample. However, the FINE_CM sample revealed very interesting mechanical properties since if compared with the pure resin, it is characterised by an elastic modulus that is more than doubled (+240%) and a flexural strength that is almost 52.4% higher; therefore, it is possible to appreciate the contribution over the flexural properties of the recovered fibres.

Among the samples produced with the open mould process, the best results were achieved with the COARSE_OM sample since it revealed an improvement in the flexural properties of almost +313.6% (9.1 GPa) in modulus and +28.1% (98.5 MPa) in flexural strength if compared with the reference sample. However, an insight on the flexural curves revealed a clear drop of the flexural stress in correspondence of the failure of the FINE_OM sample; therefore, as well as that sample typology, it is possible to ascribe the failure of the COARSE_OM sample to the same mechanism that characterise the FINE_OM one.

All the issues about the presence of residual porosity into the composite material, poor adhesion at the fibre-matrix interface and the instauration of debonding and pull-out mechanisms are in part overcome with the compression moulding process. Indeed, even if the COARSE_CM sample

Italian Manufacturing Association Conference - XVI AITeM   Materials Research Forum LLC
Materials Research Proceedings 35 (2023) 402-410   https://doi.org/10.21741/9781644902714-48

revealed a flexural modulus that is comparable with that of the COARSE_OM one, it is characterised by an increased flexural strength of almost 24.4%. However, these results are not fully encouraging since this sample, although it is produced with the same production technology, revealed a flexural strength that is just +4.5% higher than the FINE_CM sample. A further focus on the bending curve of the COARSE_CM sample revealed the presence of a clear stress drop on the loading portion of the curve and a not sharp failure behaviour. However, despite the COARSE_OM sample, these singularities are typical of a progressive bending failure of the fibre bundles that reinforce the recycled composite material.

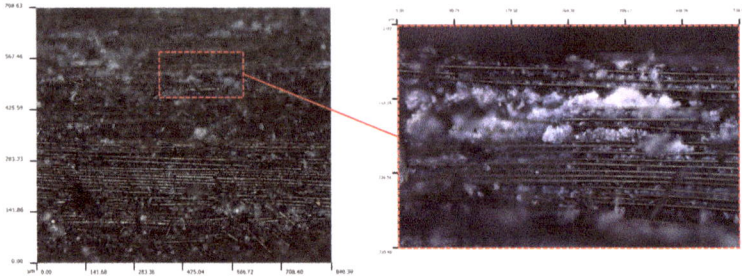

Figure 4: *Magnification of the milled fibres with residues of old matrix on their surface*

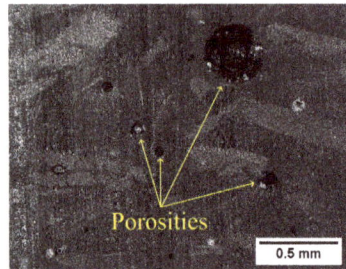

Figure 5: *Residual porosity localised in proximity of fibre agglomerations*

Conclusions

In the present research work it is done a preliminary study on the mechanical recycling process of CFRP material by using the milling method and on the application of different production strategies to obtain recycled CFRP materials. At the end of the experimental campaign, it was pointed out that by using the same production technique, the flexural properties are affected by the chip's geometry since an increase in dimension leads to an improvement of the mechanical properties of the recycled material. For both the production techniques analysed in the present study, the use of the reinforcement in form of powders leads to a recycled material with mechanical properties that are globally lower than the reference. In case of open mould production process, fine and coarse fibres reveal a flexural strength that is respectively +12.2% and +28.1% higher than the pure resin. However, it was showed that the open mould technique limits the mechanical properties of the recycled composite materials since it is responsible for the formation of residual porosity and the occurrence of debonding and pull-out mechanisms. These limitations can be overcome with the compression moulding strategy since it is able to emphasise the properties of the recovered fibres with an increase of the flexural strength of almost +52.4% and +59.3% respectively of FINE_CM and COARSE_CM samples in comparison with the reference in pure resin.

Reference

[1] Pinto F, Boccarusso L, De Fazio D, et al. Carbon/hemp bio-hybrid composites: Effects of the stacking sequence on flexural, damping and impact properties. Composite Structures 2020; 242: 112148. https://doi.org/10.1016/j.compstruct.2020.112148

[2] Yang Y, Boom R, Irion B, et al. Recycling of composite materials. Chemical Engineering and Processing: Process Intensification 2012; 51: 53-68. https://doi.org/10.1016/j.cep.2011.09.007

[3] Directive 2008/98/EC of the European Parliament and of the Council of 19 November 2008 on waste and repealing certain Directives., https://eur-lex.europa.eu/eli/dir/2008/98/oj (2008).

[4] MacArthur E. Towards the circular economy, economic and business rationale for an accelerated transition. Ellen MacArthur Foundation: Cowes, UK 2013; 21-34.

[5] Directive 2000/53/EC of the European Parliament and of the Council on end-of-life vehicles. 2000; 34-269.

[6] Gharde S, Kandasubramanian B. Mechanothermal and chemical recycling methodologies for the Fibre Reinforced Plastic (FRP). Environmental Technology & Innovation 2019; 14: 100311. https://doi.org/10.1016/j.eti.2019.01.005

[7] Pickering SJ. Recycling technologies for thermoset composite materials-current status. Composites Part A: Applied Science and Manufacturing 2006; 37: 1206-1215. https://doi.org/10.1016/j.compositesa.2005.05.030

[8] Nahil MA, Williams PT. Recycling of carbon fibre reinforced polymeric waste for the production of activated carbon fibres. Journal of Analytical and Applied Pyrolysis 2011; 91: 67-75. https://doi.org/10.1016/j.jaap.2011.01.005

[9] Mazzocchetti L, Benelli T, D'Angelo E, et al. Validation of carbon fibers recycling by pyro-gasification: The influence of oxidation conditions to obtain clean fibers and promote fiber/matrix adhesion in epoxy composites. Composites Part A: Applied Science and Manufacturing 2018; 112: 504-514. https://doi.org/10.1016/j.compositesa.2018.07.007

[10] Piñero-Hernanz R, García-Serna J, Dodds C, et al. Chemical recycling of carbon fibre composites using alcohols under subcritical and supercritical conditions. The Journal of Supercritical Fluids 2008; 46: 83-92. https://doi.org/10.1016/j.supflu.2008.02.008

[11] Zhu J-H, Chen P, Su M, et al. Recycling of carbon fibre reinforced plastics by electrically driven heterogeneous catalytic degradation of epoxy resin. Green Chemistry 2019; 21: 1635-1647. https://doi.org/10.1039/C8GC03672A

[12] Vincent G, Bruijn TA De, Iqbal M, et al. Fibre length distribution of shredded thermoplastic composite scrap. In: 21st International Conference on Composite Materials. 2017.

[13] Anane-Fenin K, Akinlabi ET. Recycling of Fibre Reinforced Composites: A Review of Current Technologies. In: Proceedings of the DII-2017 Conference on Infrastructure Development and Investment Strategies for Africa. 2017.

[14] Palmer J, Savage L, Ghita OR, et al. Sheet moulding compound (SMC) from carbon fibre recyclate. Composites Part A: Applied Science and Manufacturing 2010; 41: 1232-1237. https://doi.org/10.1016/j.compositesa.2010.05.005

Italian Manufacturing Association Conference - XVI AITeM   Materials Research Forum LLC
Materials Research Proceedings 35 (2023) 402-410   https://doi.org/10.21741/9781644902714-48

[15] Thomas C, Borges PHR, Panzera TH, et al. Epoxy composites containing CFRP powder wastes. Composites Part B: Engineering 2014; 59: 260-268. https://doi.org/10.1016/j.compositesb.2013.12.013

[16] Durante M, Boccarusso L, De Fazio D, et al. Investigation on the Mechanical Recycling of Carbon Fiber-Reinforced Polymers by Peripheral Down-Milling. Polymers; 15. Epub ahead of print 2023. DOI: https://doi.org/10.3390/polym15040854. https://doi.org/10.3390/polym15040854

[17] De Fazio D, Boccarusso L, Formisano A, et al. A Review on the Recycling Technologies of Fibre-Reinforced Plastic (FRP) Materials Used in Industrial Fields.

[18] Quadrini F, Bellisario D, Santo L. Molding articles made of 100 % recycled fiberglass. Epub ahead of print 2016. DOI: 10.1177/0021998315615199. https://doi.org/10.1177/0021998315615199

Miscellaneous

Italian Manufacturing Association Conference - XVI AITeM   Materials Research Forum LLC
Materials Research Proceedings 35 (2023) 412-419   https://doi.org/10.21741/9781644902714-49

A novel quality map for monitoring human well-being and overall defectiveness in product variants manufacturing

Elisa Verna[1,a] *, Stefano Puttero[1,b], Gianfranco Genta[1,c] and Maurizio Galetto[1,d]

[1]Politecnico di Torino, Department of Management and Production Engineering, Corso Duca degli Abruzzi 24, 10129 Torino, Italy

[a]elisa.verna@polito.it, [b]stefano.puttero@polito.it, [c]gianfranco.genta@polito.it, [d]maurizio.galetto@polito.it

Keywords: Quality, Performance Indicators, Industry 5.0

Abstract. Nowadays, companies are faced with demands for increasingly customised products, shifting from mass production to mass customisation. Thus, operators typically have to produce multiple product variants, often characterised by different complexity levels, while meeting quality standards. Companies, however, cannot only be concerned with production quality, but also with the quality and well-being of workers, as demanded by the human-centred paradigm of Industry 5.0. Therefore, this paper proposes a combined analysis of (i) production quality in terms of overall defects generated during product variants manufacturing and (ii) human well-being in terms of stress response. The combination of the two indicators results in a novel tool called "Quality Map", which enables the evaluation and monitoring of quality systems during the production of product variants from a broad standpoint. To demonstrate the viability of the method, a collaborative human-robot assembly is used as a case study.

Introduction

In recent years, the traditional approach to mass production is shifting towards mass customisation driven by technological advances, increased consumer demand for customisation and growing awareness of the environmental and social impact of mass production. However, this shift requires a flexible production system to adapt to product type and volume variations. Human-Robot Collaboration (HRC) seems to be an effective approach to achieve such mass customisation, combining the flexibility and versatility of operators with the precision of collaborative robots, i.e. cobots [1]. Interest in HRC has grown with the development of Industry 5.0, in which human well-being is placed at the centre of production systems to provide a more sustainable manufacturing sector that enables mass customisation [2].

However, many existing approaches to HRC prioritise task completion over realising its full potential. To achieve a more human-centred society and industry, HRC researchers need to broaden their focus [3,4]. To address this issue, the paper proposes the new "Quality Map" tool, which combines performance-centred and human-centred perspectives to assess the quality of HRC systems and enable more effective human-robot collaboration. The "Quality Map" evaluates and monitors the quality of a production system, combining two indicators, the process quality indicator, and the human operator stress indicator. The Quality Map provides a comprehensive view of the system's overall quality during the production of different product variants and allows for in-progress monitoring and diagnosis. The paper presents a real-life case study of the assembly of electronic board variants using an HRC system, showing the potential of the Quality Map for identifying critical production scenarios and implementing necessary corrective actions to maintain the desired quality level while considering the well-being of human operators.

Italian Manufacturing Association Conference - XVI AITeM   Materials Research Forum LLC
Materials Research Proceedings 35 (2023) 412-419   https://doi.org/10.21741/9781644902714-49

Case study

An experimental campaign is conducted to assemble six different customised variants of electronic boards (from variant A to variant F) using the ARDUINO UNO starter kit (ARDUINO®), as shown in Fig. 1(a). This starter kit consists of three main elements: (i) the components, i.e. the parts that are assembled to produce the different boards, which are listed in Table 1; (ii) the microcontroller, i.e. a small computer that allows the circuits to function; (iii) the Breadboard, i.e. a board on which the actual circuit can be built. Each electronic board has a different level of complexity and allows real-time verification of their proper functioning, i.e., the correct assembly of the products.

The experimental campaign to assemble the six selected electronic boards involved six skilled operators. The boards were assembled with the support of the UR3e cobot from Universal Robots™, equipped with an OnRobot RG6 gripper (OnRobot™), as shown in Fig. 1(b). The six operators assembled the electronic board variants randomly to avoid learning effects. The experimental campaign included an assembly phase and a quality control phase. In the former phase, the cobot passed the parts to the operator, who assembled the electronic boards following a strategy defined a priori according to the circuit theory [5]. The operator was in control of the process and activated the cobot through a pushbutton. In the quality control phase performed offline, an experienced external operator (different from assembly operators) checked the quality of the assembly, identifying any product defect which was left in the final assembly. Data about overall defectiveness and human stress response was collected during the trials.

(a)   (b)

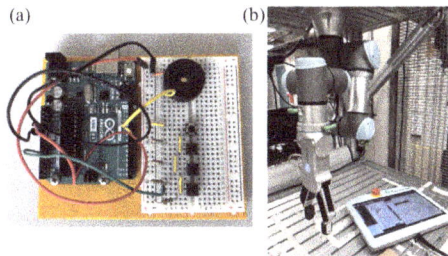

Fig. 1. (a) Example of an assembled electronic board (variant C) and (b) HRC workstation showing the single-armed UR3e cobot equipped with the OnRobot RG6 gripper.

Table 1. Components of the six electronic board variants (A-F).

	A	B	C	D	E	F
Long wires	-	1	2	8	9	13
Short wires	1	3	5	3	6	4
Resistors	1	1	4	6	2	2
Pushbuttons	-	2	4	-	2	1
LED	1	1	-	1	-	-
Phototransistor	-	-	-	3	-	-
Potentiometer	-	-	-	-	1	1
Piezo	-	-	1	-	-	-
LCD	-	-	-	-	-	1
Battery snap	-	-	-	-	1	-
DC Motor	-	-	-	-	1	-
H-bridge	-	-	-	-	1	-
Total parts number	3	8	16	21	23	22

Italian Manufacturing Association Conference - XVI AITeM   Materials Research Forum LLC
Materials Research Proceedings 35 (2023) 412-419   https://doi.org/10.21741/9781644902714-49

Complexity analysis

The scientific literature typically employs complexity as a metric to predict production performances, including production times and defects. In fact, a decrease in complexity is often found to correspond with a significant improvement in performance [6,7]. The structural complexity model, first introduced by Sinha et al. [8] and later adapted by Alkan and Harrison [9], and Verna et al. [7], is used in this study to assess the complexity of the assembly of selected ARDUINO products. This model defines the structural complexity of any network-based engineering system as a function of the complexity of individual parts (C_1), the pair-wise interaction between connected parts (C_2), and the effects of the system's overall topology (C_3). The structural complexity, represented as C, is a combination of these factors and can be expressed as:

$$C = C_1 + C_2 \cdot C_3. \tag{1}$$

In Eq. (1), C_1 represents the handling complexity of the product, i.e. the complexity of managing the individual components of a product when they are considered separately. One of the most accredited models to calculate a handling complexity index of individual parts is the Lucas method [9] based on Design For Assembly (DFA). C_2 is the complexity of connections and liaisons between parts, calculated as the sum of the complexities of pair-wise connections existing in the product structure. It may be estimated by the Lucas method [9], using the symmetrical binary adjacency matrix of the product. Each entry in the matrix denotes an assembly link between two components. Finally, C_3 represents the topological complexity related to the product's architectural pattern. It is calculated as the average of singular values of the adjacency matrix of the product [7]. It increases as the system topology shifts from centralised to more distributed architectures [8].

According to increasing total assembly complexity C, Table 2 lists the complexities C_1, C_2 and C_3 of the selected product variants. Notably, an increase in complexity does not always imply an increase in the number of parts (see Table 1 for comparison).

Table 2. Complexities of the six variants of electronic boards (A-F).

	A	B	C	D	E	F
C_1	1.39	2.87	5.10	6.35	7.25	6.72
C_2	2.98	5.44	13.84	14.58	21.79	26.02
C_3	0.94	0.90	0.90	0.93	0.83	0.84
C	4.20	7.77	17.51	19.95	25.35	28.61

HRC system quality analysis

During the manufacturing process, quality data on the overall defectiveness of product and process were collected to assess the quality of the HRC system. Experimental data were then statistically analysed to identify and exclude any possible outliers [10]. Then, the relationship between the total number of defects recorded by the 6 operators for each of the 6 variants of electronic boards and the complexity of assembly (calculated as described in the previous section) was analysed. The "operator factor" was not considered in the analysis after checking its non-significance at 95% confidence level using a two-way ANOVA (p-value of 0.290). The Poisson regression model was adopted for the analysis, as total defects are count data [11]. The selection of the most appropriate Poisson model and link function (log, square root and identity link functions) was made based on Akaike's Corrected Information Criterion (AICc) and Bayesian Information Criterion (BIC), goodness-of-fit tests (Deviance and Pearson tests), and deviance residual plots [11].

According to the results, the most appropriate Poisson model describing the relationship between defects and complexity was the one using the square root link function [11], defined as:

$$D = (k_1 \cdot C)^2, \tag{2}$$

where D is the total number of defects, C is the assembly complexity evaluated according to Eq. (1), and k_1 is the regression coefficient. The results of the Poisson regression analysis, reported in Table 3 and Fig. 2(a), showed that the relationship between D and C was statistically significant. Additionally, the analysis of deviance residuals and the goodness-of-fit tests of Deviance and Pearson (in which p-values are higher than the significance level of 0.05) indicated that the model fit the data well. Furthermore, a very high value of the deviance R^2 was obtained. Results obtained for product and process quality show that the increase in assembly complexity of the variants leads to an increase in total defects following a nonlinear trend.

Table 3. Poisson regression output for total defects (D) vs assembly complexity (C). Model is in the form $D = (k_1 \cdot C)^2$.

k_1	$SE(k_1)$	Coefficient p-value	Deviance R^2	Goodness-of-Fit Tests	
0.079	0.004	<0.0005	99.32%	Deviance Test p-value	0.619
				Pearson Test p-value	0.649

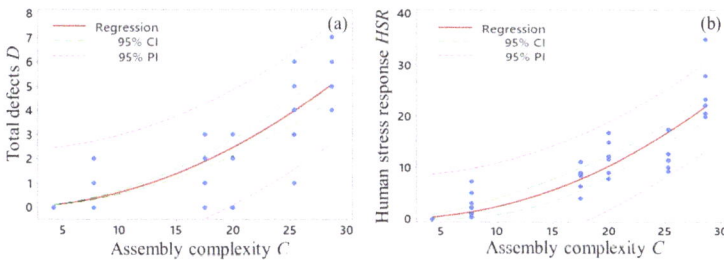

Fig. 2. (a) Poisson regression model of total defects vs assembly complexity, and (b) nonlinear regression model of human stress response vs assembly complexity.

On the other hand, physiological measures can be used to objectively assess the state of human well-being during production. Electrodermal activity (EDA) data are used in this study to measure human well-being, as they are commonly used as an indicator of human stress response [12]. The Empatica E4 wristband, a non-invasive biosensor that records EDA information at 4 Hz, was used to collect the EDA data.

For each test performed by the operators, the raw signal was recorded and then analysed using the EDA Explorer software, which removes external noise and separates the EDA signal into tonic signals related to Skin Conductance Level (SCL) and phasic signals related to Skin Conductance Response (SCR) [12,13]. According to its widespread use [12], this study used the average value of SCR peaks amplitude as a stress indicator for each assembly operator. The peak amplitude values were standardised to compute the final stress indicator to remove individual differences between individuals. As a result, for each operator, the human stress response (HSR) indicator results:

$$HSR = \left| \frac{\frac{\sum_{i=1}^{N_P} p_i}{N_P} - p_{min}}{p_{max} - p_{min}} \right| \cdot 100, \tag{3}$$

Italian Manufacturing Association Conference - XVI AITeM   Materials Research Forum LLC
Materials Research Proceedings 35 (2023) 412-419   https://doi.org/10.21741/9781644902714-49

Where p_i is the amplitude of the i-th SCR peak, N_p is the total number of SCR peaks during the assembly of a certain product variant, p_{min} is the minimum amplitude of SRC peaks and p_{max} is the maximum amplitude of SRC peaks (both referring to each operator).

Human stress response data obtained during the 36 assembly processes (i.e., the 6 product variants assembly performed by each of the 6 operators) are related to the assembly complexity. The "operator factor" was not considered in the analysis after checking its non-significance at 95% confidence level using a two-way ANOVA (p-value of 0.999). Fig. 2(b) represents the two-term power curve fitting relating human stress response and product variants assembly complexity in the form:

$$HSR = k_2 \cdot C^{k_3}, \tag{4}$$

where HSR is the human stress response, C is assembly complexity evaluated according to Eq. (1), and k_2 and k_3 are the regression coefficients. This model was the best-fitting model compared to various linear and nonlinear models, considering the goodness-of-fit statistics and residual analysis [14]. The statistical significance of the parameter estimate is confirmed by verifying that the 95% confidence intervals for the parameters, calculated from the corresponding Standard Errors (SE) reported in Table 4, exclude the zero [14]. Note that nonlinear regression is preferable to linear quadratic regression, as using a logarithmic transformation can lead to bias in the predictions [15].

Table 4. Nonlinear regression output for human stress response (HSR) vs assembly complexity (C). Model is in the form $HSR = k_2 \cdot C^{k_3}$.

k_2	SE(k_2)	95% CI for k_2	k_3	SE(k_3)	95% CI for k_3	S
0.019	0.020	(0.001, 0.158)	2.109	0.336	(1.444, 2.998)	4.067

This result, which is one of the first attempts at studying the relationship between assembly complexity and human stress response, shows that as the complexity of the product assembly increases, the assembly process becomes more challenging and requires a higher degree of cognitive effort, leading to an increase more than proportional in human stress response.

Quality Map

This section introduces the "Quality Map", a tool designed to synthesise previous HRC system quality analyses by directly relating HSR and D, regardless of the complexity of the product assembled. Two types of Quality Maps are proposed: one for single variant production, where each product is produced separately, even if it is produced several times in the HRC system, and the other for small-batch variant productions, where each variant is produced in small batches. Both types use the same tool in the use phase, but they differ in the realisation phase of the Quality Map.

To construct the Quality Map, the following operational steps should be followed. Firstly, a set of historical experimental data representative of production must be collected. In the case of the Quality Map for single variant production, a reasonable number of products (at least about thirty, for robust regression parameter estimates) should be produced, and quality and human stress responses should be measured (as described in previous sections). On the other hand, regarding the Quality Map for small-batch variant production, an adequate number of production units should be collected for each batch (at least about fifteen units for each product type, if possible [14]), and the average performance measures should be obtained for each batch. As mentioned above, it is advisable to perform a preliminary data analysis using traditional statistical techniques to detect and filter outliers [10].

Materials Research Forum LLC
https://doi.org/10.21741/9781644902714-49

Secondly, the model relating the two performance measures should be developed, depicting the system's overall quality in terms of product/process quality and human well-being. In the case study, when considering single variant production, the combination of the models in Eq. (2) and (4) leads to a linear model, by considering the goodness-of-fit statistics and residual analysis [14]. Fig. 3(a) depicts the prediction model relating human stress response HSR with total defects D. On the other hand, when considering small batches of products from the same variant, average values of HSR and D should be obtained for each variant, and the prediction model using these averages should be derived. In the case study, six small batches are considered, one for each product variant, each consisting of six products. Fig. 3(b) illustrates the best fitting model, i.e. a linear regression model. Regression outputs are shown in Table 5.

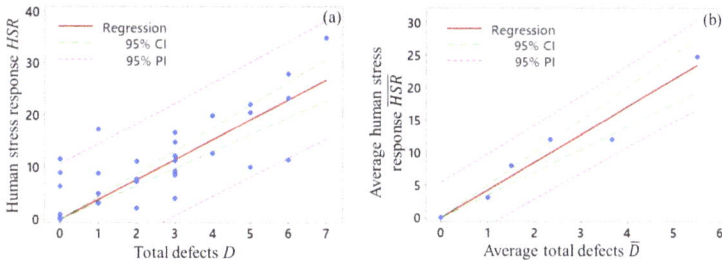

Fig. 3. Linear regression model of (a) human stress response (HSR) vs total defects (D) for single variant production, and (b) average human stress response ($\overline{HSR}$) vs average total defects ($\overline{D}$) for small-batch variant production.

Table 5. Linear regression output for human stress response (HSR) vs total defects (D). Model is in the form $HSR = a \cdot D$.

	a	SE(a)	Coefficient p-value	R^2	R^2 pred.	S
Single variant production	3.821	0.278	<0.0005	84.38%	82.99%	5.243
Small-batch variant production	4.257	0.294	<0.0005	97.67%	95.64%	2.127

The diagnostic tool Quality Map (see Fig. 4) employs the model as a reference for prediction and considers the associated uncertainty range. Specifically, the two prediction limits (Lower Prediction Limit LPL and Upper Prediction Limit UPL) derived from the regression models, illustrated in Fig. 3, are used as thresholds for identifying critical products or small batches, respectively. Products and small batches are classified as critical if a special source of variation i.e., sources not inherent to the process, occurs [14]. It should be noted that negative values of LPL are set equal to zero being physically not possible. The two prediction limits can be calculated as follows:

$$LPL = \widehat{HSR} - t_{1-\frac{\alpha}{2},N-1}\sqrt{[SE(Fit)]^2 + S^2}, \; UPL = \widehat{HSR} + t_{1-\frac{\alpha}{2},N-1}\sqrt{[SE(Fit)]^2 + S^2}, \quad (5)$$

where $\widehat{HSR}$ is the predicted value of the regression curve, $t_{1-\frac{\alpha}{2},N-1}$ is the point of Student's t distribution with level of significance α and $N-1$ degrees of freedom (where N is the total number of observations), $SE(Fit)$ is the standard error of the fit, and S is the standard error of the regression [14].

Italian Manufacturing Association Conference - XVI AITeM   Materials Research Forum LLC
Materials Research Proceedings 35 (2023) 412-419   https://doi.org/10.21741/9781644902714-49

During the utilisation phase, when new single products or batches are produced, the observed (HSR, D) value is compared with the corresponding prediction limits from the Quality Map for single variant or small-batch variant production, respectively. Accordingly, (i) if the observed (HSR, D) value falls within the prediction range (LPL, UPL), the product or batch is not deemed critical; (ii) if the observed (HSR, D) value is higher than UPL (region A in Fig. 4) or lower than LPL (region B in Fig. 4), it indicates a mismatch between HSR and D and an abnormal situation is present, resulting in the product or batch being signalled as critical.

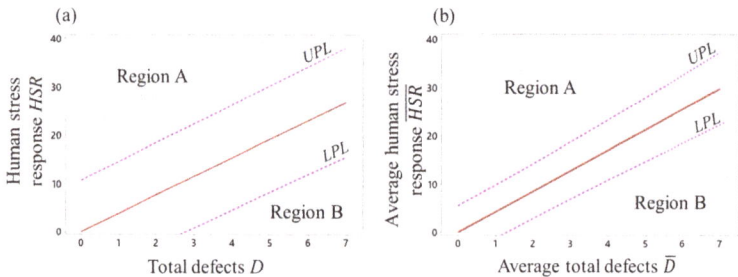

Fig. 4. Quality Map for (a) single variant production and (b) small-batch variant production.

The diagnostic tool has a dual purpose: to position products accurately in the Quality Map, aiding in quality control decisions and identifying areas for improvement, and to detect critical and out-of-control situations for prompt corrective action. This enables high-quality production and serves as an in-progress control approach. Overall, the diagnostic tool is a significant step forward in quality control and monitoring, providing real-time quality correction and consistent system quality.

Conclusions

This research aimed to propose a novel tool called the Quality Map, which combines two indicators to evaluate and monitor the quality of a production system: the overall defects generated during manufacturing product variants and the human stress response. The research was conducted using a collaborative human-robot system to assemble electronic boards as a case study to show the feasibility of the Quality Map approach. The Quality Map is implemented by collecting historical experimental data and developing a model relating the two performance measures that depict the system's overall quality. This tool can be utilised for both single variant production and small-batch variant production. It is worth noting that the proposed approach is general and can be applied to different case studies after refining and tailoring the model parameters used to build the Quality Map.

The study demonstrates that the Quality Map offers a comprehensive assessment of quality systems, encompassing both production/process quality and human well-being, in line with the human-centred approach of Industry 5.0. This highlights the significance of considering both technical and human factors in the quality assessment of production systems.

The proposed approach has some limitations, such as the use of a structural complexity model originally designed for manual and fully automated processes, and the comparison in the Quality Map being based only on two indicators. Future research efforts will be aimed at overcoming these limitations by refining the complexity model and performing a validation using different products and including other environmental/economic sustainability indicators.

References

[1] J. Krüger, T.K. Lien, A. Verl, Cooperation of human and machines in assembly lines, CIRP Ann. 58 (2009) 628–646. https://doi.org/10.1016/J.CIRP.2009.09.009

[2] H. ElMaraghy, G. Schuh, W. ElMaraghy, F. Piller, P. Schönsleben, M. Tseng, A. Bernard, Product variety management, Cirp Ann. 62 (2013) 629–652. https://doi.org/10.1016/j.cirp.2013.05.007

[3] E. Coronado, T. Kiyokawa, G.A.G. Ricardez, I.G. Ramirez-Alpizar, G. Venture, N. Yamanobe, Evaluating quality in human-robot interaction: A systematic search and classification of performance and human-centered factors, measures and metrics towards an industry 5.0, J. Manuf. Syst. 63 (2022) 392–410. https://doi.org/10.1016/j.jmsy.2022.04.007

[4] P. Damacharla, A.Y. Javaid, J.J. Gallimore, V.K. Devabhaktuni, Common metrics to benchmark human-machine teams (HMT): A review, IEEE Access. 6 (2018) 38637–38655. https://doi.org/10.1109/ACCESS.2018.2853560

[5] L. Zadeh, From Circuit Theory to System Theory, Proc. IRE. 50 (1962) 856–865. https://doi.org/10.1109/JRPROC.1962.288302.

[6] W. ElMaraghy, H. ElMaraghy, T. Tomiyama, L. Monostori, Complexity in engineering design and manufacturing, CIRP Ann. 61 (2012) 793–814. https://doi.org/10.1016/j.cirp.2012.05.001

[7] E. Verna, G. Genta, M. Galetto, F. Franceschini, Defect prediction for assembled products: a novel model based on the structural complexity paradigm, Int. J. Adv. Manuf. Technol. 120 (2022) 3405–3426. https://doi.org/10.1007/s00170-022-08942-6

[8] K. Sinha, Structural complexity and its implications for design of cyber-physical systems, PhD dissertation, Engineering Systems Division, Massachusetts Institute of Technology, 2014.

[9] B. Alkan, R. Harrison, A virtual engineering based approach to verify structural complexity of component-based automation systems in early design phase, J. Manuf. Syst. 53 (2019) 18–31. https://doi.org/10.1016/j.jmsy.2019.09.001

[10] G. Barbato, E.M. Barini, G. Genta, R. Levi, Features and performance of some outlier detection methods, Http://Dx.Doi.Org/10.1080/02664763.2010.545119. 38 (2011) 2133–2149. https://doi.org/10.1080/02664763.2010.545119

[11] R.H. Myers, D.C. Montgomery, G.G. Vining, T.J. Robinson, Generalized linear models: with applications in engineering and the sciences, John Wiley & Sons, Hoboken, NJ, USA, 2012.

[12] R. Gervasi, K. Aliev, L. Mastrogiacomo, F. Franceschini, User Experience and Physiological Response in Human-Robot Collaboration: A Preliminary Investigation, J. Intell. Robot. Syst. 106 (2022) 36. https://doi.org/10.1007/s10846-022-01744-8

[13] S. Taylor, N. Jaques, W. Chen, S. Fedor, A. Sano, R. Picard, Automatic identification of artifacts in electrodermal activity data, Proc. Annu. Int. Conf. IEEE Eng. Med. Biol. Soc. EMBS. 2015-Novem (2015) 1934–1937. https://doi.org/10.1109/EMBC.2015.7318762

[14] D.C. Montgomery, Introduction to statistical quality control, 8th ed., Wiley Global Education, 2019.

[15] M. Galetto, E. Verna, G. Genta, Accurate estimation of prediction models for operator-induced defects in assembly manufacturing processes, Qual. Eng. 32 (2020) 595–613. https://doi.org/10.1080/08982112.2019.1700274

Italian Manufacturing Association Conference - XVI AITeM | Materials Research Forum LLC
Materials Research Proceedings 35 (2023) 420-427 | https://doi.org/10.21741/9781644902714-50

Feasibility study and stress analysis of friction stir extruded rods and pipes: a simulative model

Sara Bocchi[1,a] *, Cristian Cappellini[1,b], Gianluca D'Urso[1,c] and Claudio Giardini[1,d]

[1] University of Bergamo - Department of Management, Information and Production Engineering, via Pasubio 7b, Dalmine (BG), 24044, Italy

[a]sara.bocchi@unibg.it, [b]cristian.cappellini@unibg.it, [c]gianluca.d-urso@unibg.it, [d]claudio.giardini@unibg.it

Keywords: Sustainable Processes, Aluminium Alloys, Friction Stir Extrusion

Abstract. The traditional aluminium recycling process consumes a lot of energy and adversely affects the metallurgical quality of the secondary alloys produced. With the increasing need for resolving this problem, Friction Stir Extrusion (FSE) has been patented. FSE is a new solid-state recycling process through which parts can be extruded directly from waste. In this research, the analysis was focused on different process parameters, process set ups and geometries of the extruded parts. The traditional setup, where the tool rotates and advances while the chamber remains stationary, was considered, and a new one was introduced. In this configuration, the tool has only an advance feed, while rotation is performed through the chamber. Moreover, for each combination of process parameters the bonding phenomena occurrence, considering both the thermal and the stress conditions generated by the parameters, was analysed. For this purpose, the Piwnik and Plata criterion was chosen.

Introduction

The process of recycling aluminium traditionally involves re-melting scraps, forming new ingots, and reworking them into new billets for extrusion. Unfortunately, this process is highly energy-intensive due to the necessary combustion process, which also negatively affects the metallurgical quality of the secondary aluminium alloys obtained. Moreover, it is very difficult to completely eliminate all impurities from the aluminium bulk, which further compounds the problem [1].

Given the needs of a modern and sustainable industry, traditional recycling technologies have so far proved insufficient. The Friction Stir Extrusion (FSE) is a new solid-state recycling process that enables the direct extrusion of pieces from scraps, bypassing the most energy-intensive phases of traditional recycling. During the FSE process, a rotating tool is plunged into a hollow chamber to compact, stir, and back-extrude the scraps into a full, dense rod.

The FSE process is not limited to aluminium, either; it has been demonstrated in literature that it is also possible to use the FSE process to process magnesium and its alloys, including biodegradable ones [2], as well as dissimilar metals like aluminium and steel [3].

The innovation behind FSE lies in the use of heat developed exclusively through friction between the scraps and the tool: this leads to a reduced energy requirement of approximately 15% compared to traditional recycling technologies [4].

Despite its many benefits, the use of FSE in the industrial field is still limited due to the challenging correlation between process parameters and the final characteristics of the piece obtained, as well as the limitation of extrudable geometries. Currently, only axisymmetric pieces with diameters around ten millimetres can be obtained, because of to the complexity of predicting the behaviour of the chips in contact with each other, which leads to mechanical characteristics that limit the possibility of extruding large pieces.

Italian Manufacturing Association Conference - XVI AITeM   Materials Research Forum LLC
Materials Research Proceedings 35 (2023) 420-427   https://doi.org/10.21741/9781644902714-50

As FSE is a relatively new technology, the development of simulative models is very useful for studying the relationship between the parameters and the physics of the process. For this reason, several researchers have developed simulative models using different software.

The thermal, mechanical, and microstructural behaviour of FSEed magnesium chips were analysed using ABAQUS software. The authors demonstrated that the rotational speed of the tool influenced the quantity of heat exchanged more than the descent feed of the tool [5].

A more focused thermal model was built using ANSYS FLUENT, in which only a linear heat flux was considered. This model proved to be able to efficiently simulate the temperature trend reached during an FSE process of a AA6061 cylinder [6].

It is worth noting that almost no model is actually able to predict the integrity of the Friction Stir Extruded products using a unique FEM model considering both the thermal and stress conditions reached during the FSE process [7]. Due to this lack of knowledge, it is important to consider other numerical models to be embedded in the simulation models. One such model is the Piwnik and Plata criterion [8]. This model, traditionally considered suitable for traditional extrusion processes but compatible with the FSE process [9], enables the evaluation of the effectiveness of the FSE process by considering the internal stress generated during the extrusion process.

The purpose of this study is to assess the feasibility of using FSE technology not only for the extrusion of solid rods but also for pipes, taking advantage of the dual configuration of direct and inverse frictional extrusion. To do that, the focus will be on two different process setups and on the extrusion of both rods and pipes. The traditional setup will be considered, in which the tool rotates and advances while the chamber remains stationary, as well as a new setup in which the tool only has an advance feed while the rotation is performed by the bottom of the chamber. Additionally, for each combination of process parameters and setups, the analysis will consider if and how the bonding phenomena occur, considering both the thermal and stress conditions generated by the parameters. For this purpose, the Piwnik and Plata criterion was chosen.

Materials and methods
In order to achieve the investigation of the thermo-mechanical properties of the simulated processes, the implicit Lagrangian 3D simulation software DEFORM 3D was used. To create the simulation model, four distinct objects were taken into account: a tool, a hollow chamber, a bottom, and the material to be extruded.

The first three components were simulated as rigid objects with meshes consisting of 20,000, 27,000, and 8,000 elements, respectively. Meshes were also assigned to the rigid bodies to allow for thermal analysis of the components during the process. With regard to the material to be extruded, even though it was composed of metal shavings, it was included in the simulation model as a single porous object consisting of 61,000 tetrahedral elements. A complete representation of the model setup can be seen in Fig. 1.

Italian Manufacturing Association Conference - XVI AITeM   Materials Research Forum LLC
Materials Research Proceedings 35 (2023) 420-427   https://doi.org/10.21741/9781644902714-50

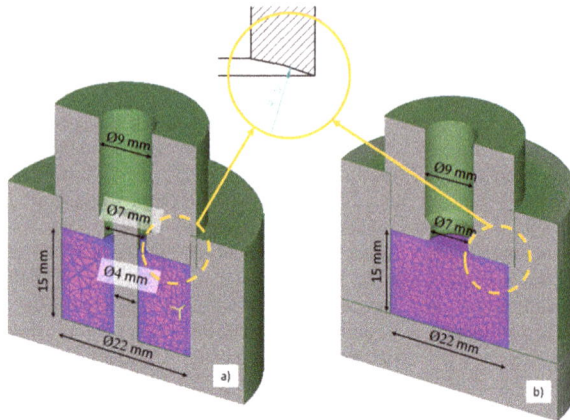

Fig. 1. Models setup for pipes (a) and for rods (b).

This choice was justified by the possibility to pre-compact the metal shavings, demonstrated in previous experimental tests, resulting in a single cylinder with a density of 2.11 g/cm^3, which is 78% of the density of the base aluminium [9]. Additionally, the AISI 1043 material was assigned to the rigid bodies, while the porous workpiece was assigned AA6061 Machining-Johnson aluminium, whose mechanical properties are defined between 20°C and 550°C.

For both the direct and inverse extrusion processes, the setups reported in Fig. 1 were used. In the case of inverse extrusion, the chamber and the bottom remained static while the tool had a rotational and translational movement. Conversely, in direct extrusion modelling, the camera and the tool were kept fixed, and the bottom underwent both a translational movement along Z and a rotational motion.

The simulations were carried out by varying two parameters, namely the rotational speed (S) and vertical feed (F) of the tool/bottom, considering ranges already optimized in a previous work [9] and further analysis conducted by the authors. These parameters were both tested at two levels, as illustrated in Table 1.

Table 1. *Process parameters combinations.*

Combination	S [rpm]	F [mm/s]
1	400	1
2	400	3
3	800	1
4	800	3

To accurately capture the steady state condition of the processed material, the stop criterion for displacement in the Z direction was set at 5 mm for the primary die (either the tool for the inverse extrusion or the bottom for the direct one).

The thermal behaviour of aluminium and steel was held constant throughout all simulations, using the values presented in Table 2, which had been optimized in a prior study [9].

Table 2. *Boundaries parameters for aluminum and steel.*

Parameter	Value
Heat transfer coefficient aluminum-tool [N/s/mm/°C]	11.00
Heat exchange with the environment [N/s/mm/°C]	0.02
Thermal conductivity [N/(s·°C)]	450.00
Steel emissivity	0.70
Aluminum emissivity	0.25
Mechanical conversion to heat	0.80
Friction coefficient aluminum-tool	0.60

Previous researches in literature have demonstrated that relying solely on thermal analysis or density verification at the end of the FSE process is insufficient to ensure the successful extrusion of large pieces [9]. Therefore, it has been chosen to consider necessary to assess the internal stress state of the workpiece using the Piwnik and Plata criterion. According to this criterion, material bonding occurs when the parameter w exceeds a limit value (w_{lim}), which is determined based on temperature. The parameter w is defined as the time integral of the ratio of pressure (p) and the actual stress acting on the material (σ_{eff}), as shown in Eq. 1:

$$w = \int_0^t \frac{p}{\sigma_{eff}} \cdot dt \tag{1}$$

To account for each step of the simulations, Eq. 1 needs to be revised as the total sum presented in Eq. 2:

$$w_{i,n} = \sum_{j=1}^n \left(\frac{p}{\sigma_{eff}} \right)_{i,j} \cdot \Delta t_j \tag{2}$$

With n = total number of steps, j = generic j-th step, i = generic i-th node and Δt_j = time per step of the j-th step.

For the Friction Stir Extrusion simulations, the pressure (p) in Eq. 1 and 2 was substituted with σ_{mean}, which represents the average stress acting in the porous material. This choice was made because the workpiece was modelled as a single porous material, making it impossible to calculate the local pressure resulting from interactions between individual chips.

E. Ceretti et al. developed an empirical relationship for determining w_{lim} as a function of the steady-state temperature achieved during the extrusion process [10]. However, the experimental interpolation curve presented in Eq. 3 was only validated for temperatures greater than 320°C.

$$w_{lim} = 4.9063 e^{-0.0017\,T} \tag{3}$$

To calculate w and w_{lim} for each node at every step, a specialized Fortran routine was developed and integrated into the simulation model, enabling their automated computation and the relationship between these two parameters.

The data on bonding conditions, torque, energy from axial thrust, and energy from rotational movement were extracted and analysed after the simulations. In addition, the total energy expenditure for machining was calculated by adding the energy from axial thrust and the energy required for rotational movement, calculated as reported in Eq. 4.

Italian Manufacturing Association Conference - XVI AITeM
Materials Research Proceedings 35 (2023) 420-427

Materials Research Forum LLC
https://doi.org/10.21741/9781644902714-50

$$E_r = \omega \cdot \int_0^t C\, dt \tag{4}$$

With E_r = energy required for rotational movement, ω = rotational speed [rad/s], C = torque and t = time per step.

Results and discussion

The implementation of a dedicated Fortran routine in the simulation models facilitated the automatic calculation of w and w_{lim} for each node and enabled the relationship between these two quantities to be embedded. The Piwnik and Plata criterion was then graphically represented by introducing a new user-variable called *welding*. This variable color-coded the elements in which $w>w_{lim}$ as red and the regions in which $w<w_{lim}$ as blue, allowing the effective proof of the Piwnik and Plata criterion to be easily identified.

By employing this method, it was possible to demonstrate the feasibility of using both direct and inverse FSE methods to produce sound and completely massive products, including rods and pipes. As already mentioned, it has been already demonstrated that the exclusive thermal analysis of the FSE process are not sufficient to ensure the extrusion of massive pieces [9]. In addition to this, based on the data collected through simulations, it was also possible to demonstrate that not even the verification of the final density of the workpiece can ensure the effective extrusion of massive pieces. In fact, in Fig.2 it is possible to observe that, considering the inverse extrusion with S=400 rpm and F=3 mm/s, although the temperature reached was on average higher than 350°C (Fig. 2a), the temperature at which chip welding begins, and that the density relative reached in the extruded part is equal to 1 (Fig. 2b), the bonding conditions have not been verified (Fig. 2c).

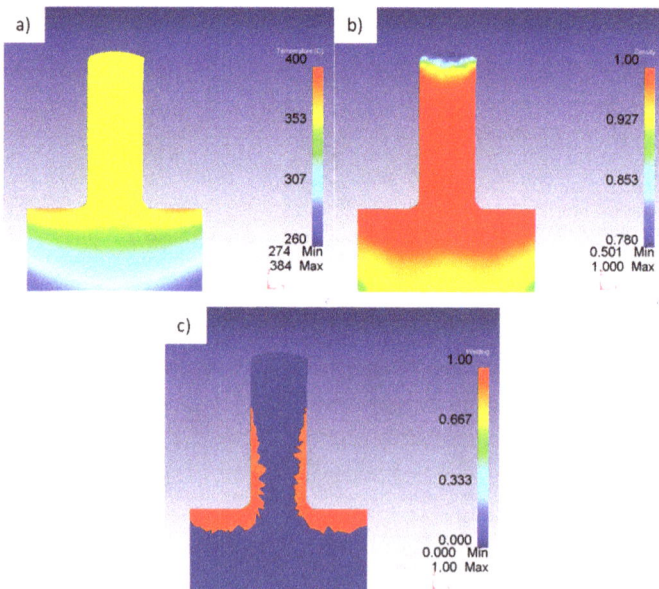

Fig. 2. FEM results: a) Temperature, b) density and c) welding parameter.

By way of example, Fig. 3 illustrates the outcome obtained by considering the direct extrusion of a pipe with S=800 rpm and F=1 mm/s, which enabled both extruded geometries to be produced.

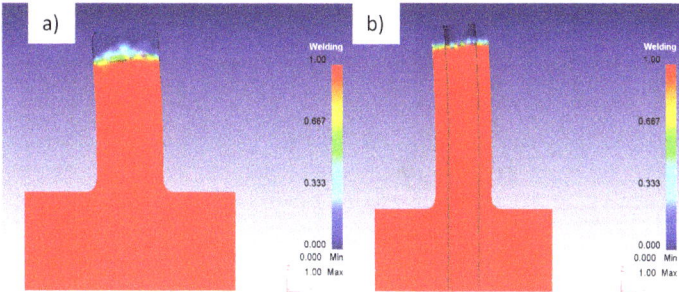

Fig. 3. FEM results obtained: a) rod with direct extrusion and b) pipe with inverse FSE process.

The torque resulting from the different simulations are reported in Fig. 4 and Fig. 5.

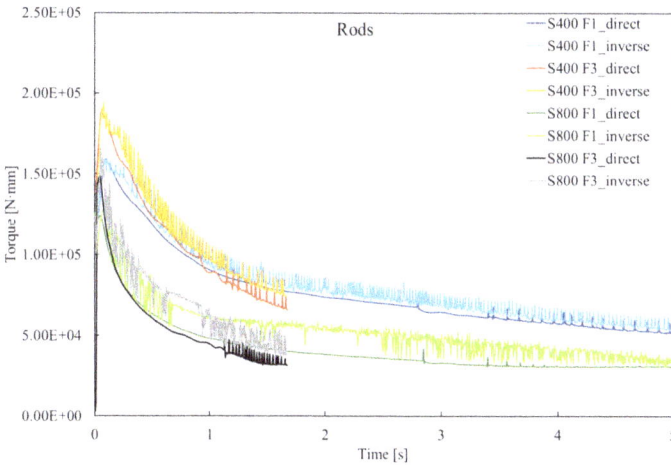

Fig. 4. Torque results from the direct and inverse FSE simulation of rods.

Italian Manufacturing Association Conference - XVI AITeM   Materials Research Forum LLC
Materials Research Proceedings 35 (2023) 420-427   https://doi.org/10.21741/9781644902714-50

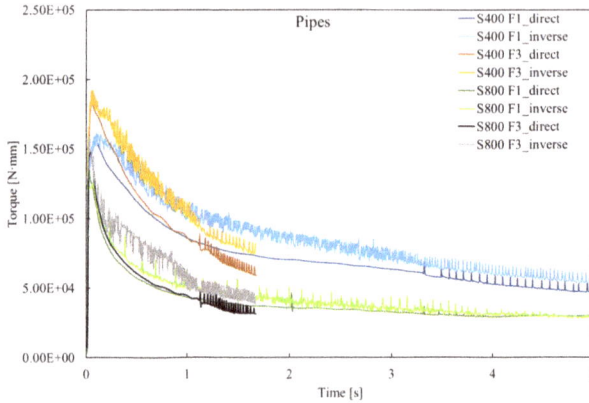

Fig. 5. Torque results from the direct and inverse FSE simulation of pipes.

Starting from these results, by using the Eq. 4, the E_r was calculated for each combination of process parameters. This energy component was added to the energy demand resulted directly from the Deform post-processor, obtaining the total energy demand (E_{tot}).

In Table 3 all the results of the bonding occurrence as a function of both the process parameters and the calculated energy demands are reported. The cells corresponding to the conditions in which perfect bonding was found are highlighted in green, in red where there was no bonding.

Table 3. Simulation results of the energy demands.

		Direct extrusion		Inverse extrusion	
		Rods	**Pipes**	**Rods**	**Pipes**
S [rpm]	**F [mm/s]**	**E$_{tot}$ [KJ]**	**E$_{tot}$ [KJ]**	**E$_{tot}$ [KJ]**	**E$_{tot}$ [KJ]**
400	1	16.82	16.15	17.61	18.45
400	3	8.56	8.27	8.81	9.03
800	1	18.2	17.25	22.94	18.92
800	3	8.39	8.37	10.73	10.06

The differences between the direct and inverse extrusion configurations in terms of energy, with the same process parameters, are almost negligible.

However, it immediately became clear that, with the same energy demand -or even lower-, with direct extrusion it was possible to obtain massive pieces also considering the pairs of parameters which did not lead to bonding in the inverse extrusion.

This is due to the benefit of the direct FSE which lies in complete compaction of the chips before the start of extrusion inside the tool cavity. This led to a more thermally uniform starting workpiece, which was ready for fully compacted material extrusion. As a result, the forces required at the interface between the tool and the workpiece were reduced, allowing for bonding even at lower temperatures and creating a state of stress that facilitated the production of completely massive pieces.

Conclusions

The objective of this study is to assess the feasibility of using FSE technology for direct and inverse extrusion of fully dense rods and pipes, by analysing bonding conditions through the Piwnik and

Plata criterion. A robust FEM model was developed to consider the rotational and descent speeds of the tool/bottom, and determine a reliable technology window for FSE. The study demonstrates that both direct and inverse extrusion configurations can be employed, and full dense rods and pipes can be produced with both configurations. Furthermore, it seems that the direct FSE provides a better starting point for the active extrusion phase, allowing the reduction of the energy consumption and limiting, at the same time, the non-bonding technological window.

References

[1] D. Baffari, G. Buffa, D. Campanella, L. Fratini, A.P. Reynolds, Process mechanics in Friction Stir Extrusion of magnesium alloys chips through experiments and numerical simulation, J. Manuf. Process. 29 (2017) 41–49.

[2] V.C. Shunmugasamy, E. Khalid, B. Mansoor, Friction stir extrusion of ultra-thin wall biodegradable magnesium alloy tubes — Microstructure and corrosion response, Mater. Today Commun. 26 (2021) 102129. https://doi.org/10.1016/j.mtcomm.2021.102129

[3] W.T. Evans, B.T. Gibson, J.T. Reynolds, A.M. Strauss, G.E. Cook, Friction Stir Extrusion: A new process for joining dissimilar materials, Manuf. Lett. 5 (2015) 25–28. https://doi.org/10.1016/j.mfglet.2015.07.001

[4] R.M. Izatt, Metal Sustainability: Global Challenges, Consequences, and Prospects, Wiley, 2016. https://doi.org/10.1002/9781119009115

[5] R.A. Behnagh, N. Shen, M.A. Ansari, M. Narvan, M. Kazem, B. Givi, H. Ding, Experimental analysis and microstructure modeling of friction stir extrusion of magnesium chips, J. Manuf. Sci. Eng. Trans. ASME. 138 (2016). https://doi.org/10.1115/1.4031281

[6] H. Zhang, X. Li, W. Tang, X. Deng, A.P. Reynolds, M.A. Sutton, Heat transfer modeling of the friction extrusion process, J. Mater. Process. Technol. 221 (2015) 21–30. https://doi.org/10.1016/j.jmatprotec.2015.01.032

[7] D. Baffari, G. Buffa, L. Fratini, A numerical model for Wire integrity prediction in Friction Stir Extrusion of magnesium alloys, J. Mater. Process. Technol. 247 (2017) 1–10. https://doi.org/10.1016/j.jmatprotec.2017.04.007

[8] M. Plata, J. Piwnik, Theoretical and experimental analysis of seam weld formation in hot extrusion of aluminum alloys, in: 7th Int. Alum. Extrus. Technol., 2000: pp. 205–211.

[9] S. Bocchi, G. D'urso, C. Giardini, G. Maccarini, A Simulative Method for Studying the Bonding Condition of Friction Stir Extrusion, Key Eng. Mater. 926 KEM (2022) 2333–2341. https://doi.org/10.4028/P-FT5355

[10] E. Ceretti, L. Fratini, F. Gagliardi, C. Giardini, A new approach to study material bonding in extrusion porthole dies, CIRP Ann. - Manuf. Technol. 58 (2009) 259–262. https://doi.org/10.1016/j.cirp.2009.03.010

Italian Manufacturing Association Conference - XVI AITeM   Materials Research Forum LLC
Materials Research Proceedings 35 (2023) 428-436   https://doi.org/10.21741/9781644902714-51

Reinforcement learning for energy-efficient control of multi-stage production lines with parallel machine workstations

Alberto Loffredo[1] *, Marvin Carl May[2] and Andrea Matta[1]

[1] Politecnico di Milano, Via G. la Masa 1, Milan, 20156, Italy

[2] Karlsruhe Institute of Technology, Kaiserstraße 12, Karlsruhe, 76131, Germany

Keywords: Artificial Intelligence, Sustainability, Manufacturing Systems

Abstract. An effective approach to enhancing the sustainability of production systems is to use energy-efficient control (EEC) policies for optimal balancing of production rate and energy demand. Reinforcement learning (RL) algorithms can be employed to successfully control production systems, even when there is a lack of prior knowledge about system parameters. Furthermore, recent research demonstrated that RL can be also applied for the optimal EEC of a single manufacturing workstation with parallel machines. The purpose of this study is to apply an RL for EEC approach to more workstations belonging to the same industrial production system from the automotive sector, without relying on full knowledge of system dynamics. This work aims to show how the RL for EEC of more workstations affects the overall production system in terms of throughput and energy consumption. Numerical results demonstrate the benefits of the proposed model.

1. Introduction

Energy efficiency has become a crucial focus of research in production systems, alongside productivity and quality improvements. Manufacturing is responsible for nearly 40% of global energy consumption [1], making it a significant area for eco-friendliness and sustainability advancements. Using energy-efficient control (EEC) policies can effectively minimize machines' environmental impact by implementing real-time actions during production to reduce energy consumption during idle periods. EEC aims to switch off machines during idle periods and turn them back on when required for production, achieving the optimal balance between production rate and energy demand. However, the "Always-On" (AOn) policy keeps machines running during idle periods and wastes energy. AOn is still a common practice despite its drawbacks. Existing EEC methods assume complete knowledge of system dynamics and parameters, which can limit the accuracy and generality of EEC models and results.

Recent research shows the potential of Reinforcement Learning (RL) to perform optimal control for complex systems. RL is a Machine Learning approach that enables agents to learn by interacting with their environment, even in the presence of incomplete or uncertain information. RL algorithms are indeed adaptive: they are designed to learn how to deal with the system dynamics and adjust their strategies accordingly. Recent research also demonstrated the effectiveness of RL for the optimal EEC of a manufacturing workstations composed of parallel machines, a widely used layout to obtain a balanced production system in terms of workstations workload. This work is focused on this type of configuration. A literature RL-based model for EEC is applied to more parallel machine workstations that are part of the same production line. The literature model is able to reduce the workstation system energy consumption while maintaining its throughput, even without full knowledge of the system dynamics. The focus is then moved on the effect that these energy-efficient actions have on the overall production system in terms of throughput and energy consumption.

The paper is structured as follows. Section 2 presents a literature review of EEC and RL for production control, and highlights the main contribution of this work. Section 3 includes a

Italian Manufacturing Association Conference - XVI AITeM   Materials Research Forum LLC
Materials Research Proceedings 35 (2023) 428-436   https://doi.org/10.21741/9781644902714-51

description of the industrial case studied in this work. Section 4 presents an overview on the used RL-based algorithm from literature. Section 5 presents the results of the numerical experiments carried out, demonstrating the benefits of the algorithm when applied to more workstations in the same industrial case. Finally, Section 6 concludes the paper and discusses possible further developments.

2. Literature Review

2.1 Energy-Efficient Control of Manufacturing Systems

EEC problem for manufacturing equipment is addressed in two ways: (i) controlling systems with a single buffer before a single machine and (ii) controlling systems with a single buffer followed by parallel machines in a workstation. The literature on EEC for manufacturing systems has been growing, and a review can be found in [2].

Recent examples of EEC for the single-buffer-single-machine layout can be found in the following. Mouzon et al. [3] were the first to address the topic, proposing various switch off rules for a non-bottleneck workstation in a production system. Duque et al. [4] contributed to the development of a fuzzy controller that can be used to turn on/off a single workstation. Later, in Frigerio and Matta [5], the authors conducted an analytical study of an EEC policy for a single machine. In Jia et al. [6] an alternative approach was introduced where the authors used Work-In-Process (WIP) data to formulate efficient EEC policies for a multi-stage production line that included single machine workstations. Finally, in a recent work by Cui et al. [7], an optimal EEC technique was suggested for the entire production line using buffer level information.

Research stream for EEC of workstations with parallel machines is less developed. Loffredo et al. [8] introduced a model using a Markov Decision Process (MDP) that generates effective EEC policies for a single workstation with parallel machines. Furthermore, they extended this approach with an MDP-based model for controlling multi-stage production lines with parallel machine workstations [9]. However, these studies were constrained by their reliance on MDP that necessitates complete knowledge of the system dynamics. This approach resulted in a solution that did not utilize any machine learning methods. To fill this gap, in [10] it is possible to find an RL-based model for the EEC of a single parallel machine workstation, but in this model they considered the stand-alone workstation without any interaction with the shop floor.

2.2 Reinforcement Learning in Production Systems Control

RL algorithms consist of two elements: an agent and an environment. The agent continuously interacts with its environment to optimize its behavior and finally achieve a specific goal. This involves recognizing the best action in each state to optimize an objective function, such as the total discounted reward over a given time horizon. A basic overview of the RL framework is explained in the following and is extracted from [11]. Everything begins with the agent observing the environment state $s_t \in \mathbb{S}$, and selecting an action at $a_t \in \mathbb{A}$. The state space, denoted by $\mathbb{S}$, includes all the possible observations an agent can make of the environment, providing information on the current production state to allow the agent to choose actions that optimally solve the control problem. On the other hand, the action space, represented by $\mathbb{A}$, includes all the possible actions the agent can take. After a_t is performed, the environment responds with the resulting state s_{t+1} while the agent is rewarded with a reward r_{t+1} and the next iteration can start (see Fig. 1).

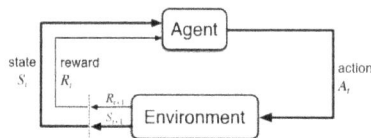

Fig. 1 Overview of RL-Framework [11].

Italian Manufacturing Association Conference - XVI AITeM   Materials Research Forum LLC
Materials Research Proceedings 35 (2023) 428-436   https://doi.org/10.21741/9781644902714-51

The primary objective of the agent is to maximize the long term cumulative reward by optimizing its action-selection policy. This involves learning the best control policy, or EEC policy for this work scope, to apply to the environment.

Literature offers plenty of successful RL applications in production systems (complete and recent literature review in [12]). For instance, RL proved to be strongly effective in production scheduling, dispatching and plant-internal logistic of applications (examples in [13, 14, 15]). However, despite RL's proven effectiveness for production planning and control problems, only one work dealing with RL for EEC of manufacturing equipment can be identified [10].

2.3 Contribution

Energy efficiency is a key factor for achieving sustainability in manufacturing. The EEC approach offers solutions to minimize the environmental impact of manufacturing equipment, but it faces limitations due to assumptions of complete knowledge of system dynamics and parameters. Reinforcement Learning can address this challenge by handling uncertain information. Nevertheless, in literature there is only work addressing the RL for EEC approach in manufacturing but, even in this case, the study focuses on a single workstation without considering its interactions with the shop floor. To fill this gap, this work analyzes the impact that using a literature RL for EEC model to more parallel machines workstations has on the overall production system, in terms of throughput and energy consumption. The study analyzes a real-world example in the automotive industry, using various RL agents to identify the optimal one for achieving the best performance. The algorithm parameters are fine-tuned to find the optimal trade-off between throughput and energy consumption.

3. System Description

3.1 System Layout

A real industrial system is used as reference case study where the proposed model can be applied and its effects can be analyzed. The production line under investigation is a manufacturing system producing cylinder heads in the automotive sector (see layout in Fig. 2).

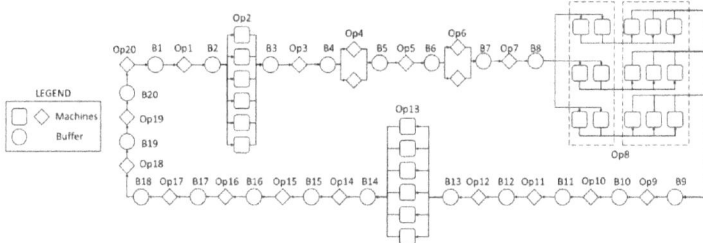

Fig. 2 Layout of the real industrial system under investigation.

The production process consists of 20 stages (m = 20). Each stage $S_i | i \in \{1, \ldots, m\}$ is characterized by an upstream buffer $B_i | i \in \{1, \ldots, m\}$ with a finite capacity K_i, followed by c_i parallel machines $M_{ij} | i \in \{1, \ldots, m\}, j \in \{1, \ldots, c_i\}$. Machines M_{ij} are considered starved if they have the ability to process parts, but there are none available in B_i. Conversely, they are blocked if they can process parts, but B_{i+1} is already full. It is worth noting that machines M_{mj} in S_m cannot be blocked, as there is a downstream infinite capacity buffer that allows processed parts to exit the system immediately after they are completed. All the machines in the line operate on a single type of part, blocking after service rule and first-come-first-served rule are enforced. Also, machines cannot be switched off while they are operating on items: part processing cannot be interrupted by the control. Also five stages are characterized by parallel machines ($c_i >$ 1) while the remaining have a single-buffer-single-machine layout ($c_i = 1$). The line is composed by 12 controllable stages that are fully automated while eight are not controllable due to the presence of human operators involved in the operation performed. In detail, the subset of controllable stages is $S_i | i = [2, 3, 4, 7, 8, 9, 10, 12, 13, 14, 15, 19]$. The production system parameters are not reported because of a confidentiality agreement with the company owning the system.

The system is characterized by different stochastic processes, including the arrival rate of parts to the first stage S_1, machines processing and startup times, and, also time between failures (TBF), and time to repair (TTR) of the machines. All of these are assumed to be independent of each other and stationary. The arrival of parts to S_1 follows a stochastic process with an expected value of λ. Each machine M_{ij} has startup and processing times that follow a stochastic process, with expected values equal to δ_{ij} and μ_{ij}, respectively. Additionally, the machines are unreliable and can be subject to operation-dependent failures. Each machine M_{ij} is characterized by stochastic TBF and TTR, with expected values equal to ψ_{ij} and ξ_{ij}, respectively. All the mentioned expected values vary for each stage in the production line. Lastly, all the line machines are consistent with the energetic state model detailed in Section 3.2.

3.2 Machine States and Associated Power Consumptions
All the machines are characterized by the following states: working (w), standby (sb), startup (su), and failed (f); the working state is then divided into two sub-states: idle (id) and busy (b) (see Fig. 3).

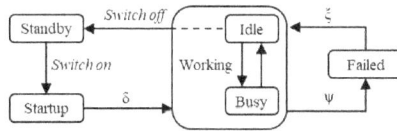

Fig. 3 Machine state model.

When the machine is turned off, it is in the standby state. When it is switched on to transition to the working state, the machine goes first through the startup state, which has a stochastic startup time with an expected value of δ. While working, the machine is either busy processing parts with stochastic processing times with an expected value of μ, or it is idle and ready for processing. However, the machine is unreliable and can fail, which causes it to transition to the failed state, determined by the stochastic TBF with an expected value of ψ. If the machine is fully repaired, including being switched off and back on, and tested, it transitions back to the working state, which is ruled by the machine's stochastic TTR with an expected value of ξ. Finally, the machine can be switched off immediately from the idle sub-state sending it back to the standby state, i.e. it is not possible to switch off the machine while it is busy. Each machine state necessitates a power requirement $w_s | s = \{w, sb, su, f, id, b\}$, which is non-negative and dependent on the active components in the respective state or sub-state. It is assumed that $w_b > w_{su} > w_{id} > w_{sb} \simeq w_f \simeq 0$. Additionally, it is worth noticing that w_w depends on w_{id} and w_b.

4. Literature RL for EEC Model
The elements of the literature RL-based model [10] are summarized in this section as follows: the agent state space in section 4.1, action space in section 4.2, and reward function in section 4.3. The environment is the single workstation to be controlled. The model goal is to achieve the optimum

Italian Manufacturing Association Conference - XVI AITeM   Materials Research Forum LLC
Materials Research Proceedings 35 (2023) 428-436   https://doi.org/10.21741/9781644902714-51

trade-off between system productivity and energy demand for the single workstation without relying on full knowledge of the system dynamics.

4.1 State Space

In the RL framework, the state is the representation of the environment at a particular time, which includes all the relevant information necessary for the agent to make decisions about what action to take next. Consequently, the state space $\mathbb{S}$ represents the set of possible agent observations of the environment, i.e. the single workstation S_i. In the literature model the state $\boldsymbol{s} \in \mathbb{S}$ is represented with the ordered vector of size $\boldsymbol{s} = [\frac{n_i}{K_i}, x_1, \ldots, x_{c_i}]$. For the generic stage S_i, in $\boldsymbol{s}$, $n_i \in \{0, \ldots, K_i\}$ is the number of parts in buffer B_i. Furthermore, each binary variable $x_{ij} \in \{0,1\}$ provides information to the agent on the working state of each stage machine: $x_{ij} = 1$ means machine M_{ij} is in working state while if $x_{ij} = 0$ machine M_{ij} is not in working state.

4.2 Action Space

In the literature model, the action the agent has to perform is quite straightforward: it has to select, at any time, how many machines should be in working state in the workstation. In this way, starting from the actual number of working machines in the system, some machines might then be switched on or switched off, according to the agent's action. The action space $\mathbb{A}$ is the set of all possible actions that an agent can take. Therefore, the action $\boldsymbol{a} \in \mathbb{A}$ is applied to control the number of machines to be in working state in the controlled stage S_i. Therefore, in S_i, $\mathbb{A} = \{0, \ldots, c_i\}$, since the control action $\boldsymbol{a}$ can only assume integer values ranging from 0 (all machines in the stage must be switched off and not working) to c_i (all machines in the stage must be switched on and working). Note that system assumptions dictate the allowable actions, such as not interrupting part processing or startup procedures. Therefore, the agent cannot immediately set a machine undergoing startup to a working state.

4.3 Reward Function

The reward function is an essential component of RL, as it provides the feedback that guides the agent's learning process. Therefore, it must be designed to reflect the goals of the task being learned. The EEC problem is characterized by a multi-objective nature since there are two goals, or Key Performance Indicators (KPIs), to be considered: reducing the energy consumption and maintaining the system throughput.

The literature model reward function is based on two elements: the *Throughput Component* R_t (see Eq.1) and the *Consumption Component* R_{cons} (see Eq.2):

$$R_t = \frac{\theta}{e} \quad with \quad \theta = e^{\frac{TH_t}{TH_{t,max}}} \tag{1}$$

$$R_{cons} = e^{-z\Delta cons} \quad with \quad \Delta cons = cons_t - cons_{t-1} \tag{2}$$

In Eq. 1, TH_t is the workstation throughput from at the actual time t, i.e. number of produced parts until time t divided by t itself. $TH_{t,max}$ is the maximum throughput the stage S_i can reach in the same time-period by maintaining all the line machines always switched on. Consequently, $0 \leq \frac{TH_t}{TH_{t,max}} \leq 1$, $1 \leq \theta \leq e$, and $\frac{1}{e} \leq R_t \leq 1$, where R_t grows as TH_t approaches $TH_{t,max}$. Through R_t, the agent is directed to increase productivity by maintaining the stage machines in working state to produce a larger quantity of parts to increase R_t. In Eq. 2, $\Delta cons$ represents the increasing consumption. This is the difference between (i) $cons_t$, i.e. the stage S_i energy consumption in the time-period at actual time t, and (ii) $cons_{t-1}$, i.e. the stage S_i energy consumption until time t-1. The latter is the time when the previous reward was given to the agent. Considering the use of a scale factor z, it is possible to state that $0 \leq R_{cons} = e^{-z\Delta cons} \leq 1$, where R_{cons} grows as $\Delta cons$ approaches zero. This means that, through R_{cons}, the agent is directed to decrease productivity and save energy by maintaining the stage machines not in working state to produce fewer parts.

R_t and R_{cons} compose the reward R through the following reward function:

$$R = \phi R_t + (1 - \phi)R_{cons} \tag{3}$$

Where $0 \leq \phi \leq 1$ is a key element in the literature model that balances the multi-objective targets of the problem: if $\phi = 0$ then $R = R_{cons}$ and the agent only aims at reducing the consumption, while, if $\phi = 1$ then $R = R_t$ and the agent only aims at increasing the production. ϕ determines the weight of energy consumption and throughput in the action-selection process and must be calibrated. However, it is possible to affirm that the optimal ϕ will tend to 1, since this leads to a null or almost-null productivity drop. Finally, since $0 \leq \phi \leq 1, 0 \leq R_{cons} \leq 1$, and $0 \leq R_t \leq 1$, then $0 \leq R \leq 1$.

5. Numerical Experiments

The objective of the experimental study is to assess the impact that applying the literature RL for EEC model on more parallel machines workstations has on the overall production system in the real-world industrial system (Section 3). In particular, 12 stages are controlled in the system, $S_i| i = [2, 3, 4, 7, 8, 9, 10, 12, 13, 14, 15, 19]$. The literature model is applied independently to all the 12 controllable stages in the line: there are 12 single RL agents controlling only the respective workstation observing an environment that is only a part of the overall production system. Different agent-types are compared to identify the most suitable for the use case and then, with the selected agent, ϕ is optimally calibrated for the line under study.

In the experiments, a discrete-event simulator of the system, as described in Section 3, has been implemented using the SimPy library, and the agent, as described in Section 4, has been built using the TensorForce library [16]. Both interact through Python code. Every experiment involves the assessment of two KPIs: throughput loss and energy saving. The former is determined by calculating the difference in system throughput when the AOn policy is implemented versus when the RL-based model is applied to the controlled stages; the latter is determined by measuring the difference in total system energy consumption between the AOn policy and the RL-based model. The KPIs are extracted by comparing the case when the system is managed with AOn policy. 10 replications were carried out for each case, with a simulation length of 30 days. The experiments were characterized by random number generation.

To evaluate the most effective approach for applying the EEC in the real-world case, 4 commonly used types of RL agents are implemented in each stage and compared (results in Fig. 4):

1. The Tensorforce-General agent, an agent included in the Tensorforce library [16].
2. The TRPO agent exploiting the Trust Region Policy Optimization algorithm [17].
3. The PPO agent which exploits the Proximal Policy Optimization method [18].
4. The DQN agent which uses Deep Q-Network [19]

To not jeopardize the throughput, only high values of ϕ ($\phi \geq 0.90$) are tested. Among all agents, DQN is best performing since at the same it is characterized by lower throughput loss and higher savings. For all the agents, the default TensorForce library NN hyperparameters are used.

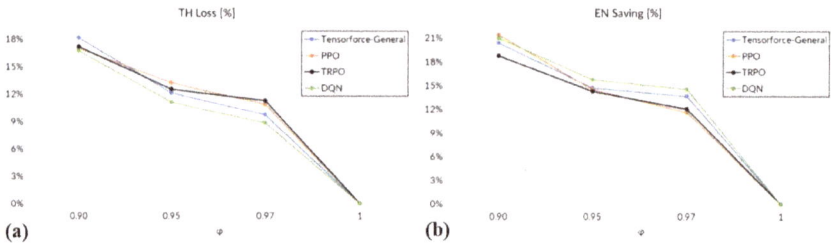

Fig. 4 Comparison of different agents when applied to the industrial case: throughput loss (a) and energy saving (b) in respect to the AOn Policy. Values are shown considering a confidence level of 90% on the mean value. However, being all the confidence intervals strict, i.e. with a width in all the cases lower than 2%, confidence intervals are not visible with the selected figure scale.

The subsequent step regarded the optimal calibration of ϕ when the DQN agent is applied. ϕ has been calibrated only for high values, varying it from 0.95 to 1 with a step of 0.01. Results of this analysis are shown in Fig. 5 .

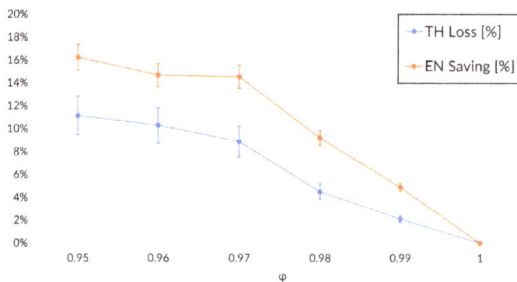

Fig. 5 Energy saving and throughput loss when ϕ varies for the industrial case. A PPO agent is used. Values are shown considering a confidence level of 90% on the mean value.

The optimal value appears to be $\phi = 0.99$, leading to a slight throughput loss ($2.11 \pm 0.07\%$) and a corresponding significant energy saving equal to $4.91 \pm 0.17\%$. Considering the annual system performance, this would lead, on average, to a throughput decrease of about 8000 units over a total production volume of more than 500,000 parts, while saving more than 105 barrels of oil equivalent. It must be noted that, if the company owning the manufacturing line would prefer to avoid also the slight throughput loss and produce that $2.11 \pm 0.07\%$ of lost parts during extra working hours, the system would be subject to additional energy consumption. In this case, the corresponding average savings will be decreased to $3.11 \pm 0.14\%$ but with a corresponding null throughput loss. This managerial choice will improve the system sustainability by saving, on average, more than 70 barrels of oil equivalent without any productivity drop. This confirms that the RL-based algorithm significantly enhances the sustainability of the industrial use-case, while maintaining its productivity, even when applied independently to more line stages.

6. Conclusions and Further Developments
In this work, an RL for EEC model has been applied to more parallel machines workstation used in a manufacturing system. Afterwards, it has been studied the impact that this energy-efficiency action has on the entire production line. Numerical results are presented, showing the

Italian Manufacturing Association Conference - XVI AITeM   Materials Research Forum LLC
Materials Research Proceedings 35 (2023) 428-436   https://doi.org/10.21741/9781644902714-51

corresponding benefits when model is implemented. The latter uses Reinforcement Learning strategies to implement energy-efficient control actions efficiently without relying on complete knowledge of system dynamics and parameters. The model adapts and evolves over time during the training process, making it promising for direct application in industry with minimal effort. The potential of the proposed approach, combined with the increasing ease of use and knowledge of reinforcement learning techniques, presents a promising opportunity for direct and successful application in industry. This application can be achieved in a short time and with minimal effort, making it appealing from a managerial standpoint.

However, a challenging topic might be the creation of a novel RL-based model leading to where the control is executed jointly in all the workstations, considering the overall system as the environment to be controlled. Future research will focus on developing this method.

References

[1] Bipartisan Policy Center. Annual energy outlook 2022. EIA, Washington, DC, 2020.

[2] Renna, P. and Materi, S. (2021) A literature review of energy efficiency and sustainability in manufacturing systems. Applied Sciences, 11(16), 7366. https://doi.org/10.3390/app11167366

[3] Mouzon, G., Yildirim, M.B. and Twomey, J. (2007) Operational methods for minimization of energy consumption of manufacturing equipment. International Journal of Production Research, 45(18-19), 4247-4271. https://doi.org/10.1080/00207540701450013

[4] Duque, E.T., Fei, Z., Wang, J., Li, S., Li, Y., 2018. Energy consumption control of one machine manufacturing system with stochastic arrivals based on fuzzy logic, in: 2018 IEEE International Conference on Industrial Engineering and Engineering Management (IEEM), IEEE. pp. 1503-1507. https://doi.org/10.1109/IEEM.2018.8607749

[5] Frigerio, N., Cornaggia, C.F. and Matta, A. (2021) An adaptive policy for on-line energy-efficient control of machine tools under throughput constraint. Journal of Cleaner Production, 287, 125367. https://doi.org/10.1016/j.jclepro.2020.125367

[6] Jia, Z., Zhang, L., Arinez, J. and Xiao, G. (2016) Performance analysis for serial production lines with Bernoulli machines and real-time WIP-based machine switch-on/off control. International Journal of Production Research, 54(21), 6285-6301. https://doi.org/10.1080/00207543.2016.1197438

[7] Cui, P.-H., Wang, J.-Q., Li, Y. and Yan, F.-Y. (2021) Energy-efficient control in serial production lines: Modeling, analysis and improvement. Journal of Manufacturing Systems, 60, 11-21. https://doi.org/10.1016/j.jmsy.2021.04.002

[8] Loffredo, A., Frigerio, N., Lanzarone, E., Matta, A., 2021. Energy-efficient control policy for parallel and identical machines with availability constraint. IEEE Robotics and Automation Letters. Vol. 6(3), pp 5713-5719 https://doi.org/10.1109/LRA.2021.3085169

[9] Loffredo, A., Frigerio, N., Lanzarone, E., Matta, A., 2023. Energy-efficient control in multi-stage production lines with parallel machine workstations and production constraints. IISE Transactions. https://doi.org/10.1080/24725854.2023.2168321

[10] Loffredo, A., May, M.C., Schäfer L., Matta, A., & Lanza G. Reinforcement Learning for Energy-Efficient Control of Parallel and Identical Machines. CIRP Journal of Manufacturing Science and Technology. Under Review.

[11] Sutton, R.S., and Barto, A.G., 2018. Reinforcement learning: An introduction. MIT press,

[12] Panzer, M., Bender, B., 2022. Deep reinforcement learning in production systems: a systematic literature review. International Journal of Production Research 60, 4316-4341. https://doi.org/10.1080/00207543.2021.1973138

[13] Baer, S., Turner, D., Mohanty, P., Samsonov, V., Bakakeu, R., Meisen, T., 2020. Multi agent deep q-network approach for online job shop scheduling in flexible manufacturing, in: Proceedings of the 16th International Joint Conference on Artificial Intelligence, pp. 1-9.

[14] Stricker, N., Kuhnle, A., Sturm, R., Friess, S., 2018. Reinforcement learning for adaptive order dispatching in the semiconductor industry. CIRP Annals 67, 511-514. https://doi.org/10.1016/j.cirp.2018.04.041

[15] Malus, A., Kozjek, D., et al., 2020. Real-time order dispatching for a fleet of autonomous mobile robots using multi-agent reinforcement learning. CIRP annals 69, 397-400. https://doi.org/10.1016/j.cirp.2020.04.001

[16] Kuhnle, A., Schaarschmidt, M., Fricke, K., 2017. Tensorforce: a tensorflow library for applied reinforcement learning.

[17] Schulman, J., Levine, S., Abbeel, P., Jordan, M., Moritz, P., 2015. Trust region policy optimization, in: International conference on machine learning, PMLR. pp. 1889-1897.

[18] Schulman, J., Wolski, F., Dhariwal, P., Radford, A., Klimov, O., 2017. Proximal policy optimization algorithms.

[19] Mnih, V., Kavukcuoglu, K., Silver, D., Graves, A., Antonoglou, I., Wierstra, D., Riedmiller, M., 2013. Playing atari with deep reinforcement learning.

Italian Manufacturing Association Conference - XVI AITeM   Materials Research Forum LLC
Materials Research Proceedings 35 (2023) 437-443   https://doi.org/10.21741/9781644902714-52

Tool path strategies for a new Agile incremental bending system

Enrico Simonetto[1,a] *, Andrea Ghiotti[1,b], Enrico Savio[1,c] and Stefania Bruschi[1,d]

[1]Department of Industrial Engineering, University of Padova, Via Venezia 1, Padova, 35131, Italy

[a]enrico.simonetto.1@unipd.it, [b]andrea.ghiotti@unipd.it, [c]enrico.savio@unipd.it, [d]stefania.bruschi@unipd.it

Keywords: Sheet Forming, Incremental Sheet Forming, Tool Path

Abstract. In the last years, driven by the advent of industry 4.0 and more recently by the Covid19 pandemic, the paradigm of Agile manufacturing has become an important issue. To address these needs, manufacturing companies are looking for new processes to reduce production rap-ups times and costs related to traditional bending and stamping technologies. This paper proposes a new system, to produce bent elongated profiles of different thickness and shape by incremental forming. The novel machine uses two opposite rollers, to locally deform a workpiece by repeatedly moving back and forth along the profile to be bent. However, to fully exploit the new system potentialities ensuring the quality of the final products, the correct design of the tools paths are of utmost importance. An analysis of different tool path strategies to bend profiles up to 90° is presented. Finally, the proposed paths were validated on AISI304 sheets of 1 and 3 mm thickness.

Introduction

Bending and profiling operations are sheet metal forming operations commonly employed in a variety of industrial sectors [1], including shipbuilding, automotive, power plants and shell construction, to realize structural parts and increase their stiffness [2]. However, such operations typically require specialized dies or skilled personnel capable of performing complex tasks with manual processes. Large-size parts present an additional challenge, as traditional metal forming presses often lack the necessary force or table size to handle bending operations. In recent years, there has also been a growing demand for small-batch and single-part production runs, leading to decreased profitability for these operations [3]. As a result, manufacturers are seeking ways to remain competitive by offering greater flexibility, high quality, shorter time-to-market, and lower production costs [4].

Incremental Forming (IF) processes emerged as a promising alternative, as they do not require costly dies and can accommodate a wide range of part shapes and batch sizes [5]. IF processes involve limited contact between the tools and the metal sheet, enabling controlled plastic deformation with relatively low applied forces [6].

Numerous studies have explored the use of IF processes for axis-symmetric and elongated parts. In the former case metal spinning [7] represent one of the most promising processes to manufacture large and thick parts. In this case the tool path design [8] represent a key issue for the workpiece quality as the circumferential stress generated by the reduction of the part average diameter often lead to wrinkles [9] or other defects. Regarding the latter case, Voswinckel et al. investigated stretch and shrink flanging [10] and assessed the impact of tool path on part accuracy [11]. Other research has explored pre-forming flanging operations for stretch, shrink, and hole-shaped flanges [12]. While many investigations have relied on pin-like tools, some have proposed the use of a single roller-shaped tool for hemming operations [13] combined with robotic arms. However, this combination to move the tools along complex 3D trajectories [14], while offering greater flexibility, may be limited by the low stiffness of the equipment and may not accommodate thicker workpieces [15].

Italian Manufacturing Association Conference - XVI AITeM   Materials Research Forum LLC
Materials Research Proceedings 35 (2023) 437-443   https://doi.org/10.21741/9781644902714-52

This paper proposes a novel IF machine concept for flexible sheet forming operations on straight components, using two working rollers [16]. The use of simple and cost-effective tools is compensated by an increase in the complexity of the toolpath necessary to obtain the finished component. Therefore, the aim of this paper is to analyze the effects of two different possible toolpaths. The paper focuses on the effects of such parameters on the process loads, with a validation on two bent stainless-steel profiles with two different thicknesses respectively of 1 and 3 mm.

Equipment and material

The primary aim of the proposed technology is to profile long metal sheets of various geometries by utilizing simple roller-shaped tools that are highly flexible and require short set-up time, in adherence with the principles of Agile manufacturing. In Fig. 1a, the prototype of the novel incremental forming machine is shown with a detailed description of its main features. The machine has a gantry configuration consisting of two side uprights connected at the basement that support the guides for the profiling head. The upper crown, which serves as a blank holder for the sheet metal blank during the process, can be moved vertically by four servo-hydraulic actuators. It can apply a maximum vertical load of 160 kN and mounts the guides for the upper profiling head. The prototype's overall structure is 2650 mm tall, 4540 mm long, and 1250 mm deep, while the length of the blank-holder is 2100 mm.

The profiling head is a carriage that moves repeatedly back and forth along the bending line (y-direction, Fig. 1a) using fully electric engines on two ball bearings guides, for a maximum speed of 1000 mm/s. It carries the roller and actuators to adjust the tool position during the process. Each roller can independently move along the vertical (x) and horizontal (z) directions, as shown in Fig. 1b, in these cases with a maximum speed of 3 mm/s. As a result, the machine has six electric-driven axes and one hydraulic-driven axis controlled by a PLC system, allowing the interpolated motion of the electric axes. The rollers are made of 18NiCrMo5 steel, have a diameter of 175 mm, a width of 35 mm, and a 5 mm rounded edge.

Three three-axial Kistler 9067C load cells were embedded in the lower roller support, to measure the forming forces. Each load cell has a range of ± 60 kN along the vertical direction, and of ± 30 kN along the other. To test the effects of the tool path on the process and on the final component, the machine was tested by bending 1 ± 0.05 mm and 3 ± 0.05 mm thick sheets of AISI304 stainless steel.

Tool path and Forces

The newly proposed forming concept offers a shift towards simpler and cost-effective tools, but it comes at the expense of increased tool path complexity. To achieve the desired final shape based on the available tools, various tool paths strategies have been developed. Initially, the blank is secured by the blank-holder, which is operated by four hydraulic actuators, bringing the upper crown into contact with the machine bead. The in the simpler configuration the sheet is formed by the rollers, with the distance between them maintained at a minimum equal to the thickness of the metal sheet to prevent a workpiece thinning. As a consequence, no significant variations in thickness were measured on the bent workpieces. In the presented case, the upper roller is kept in a fixed xz-position and is only moved along the y-axis at each forming step to bend the entire workpiece length. In this case the target bending radius is equal to the upper roller edge radius.

Italian Manufacturing Association Conference - XVI AITeM   Materials Research Forum LLC
Materials Research Proceedings 35 (2023) 437-443   https://doi.org/10.21741/9781644902714-52

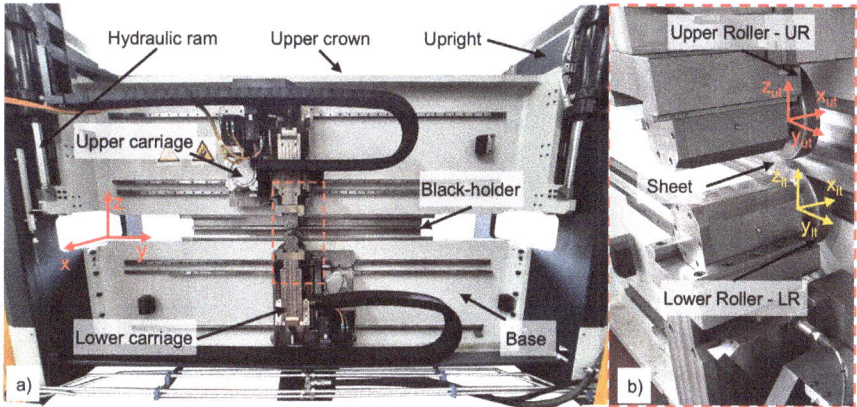

Figure 1. (a) Incremental Forming prototype; (b) detail of the upper and lower tools.

As shown in Fig. 2, several tool paths can be designed to achieve the same final configuration of the workpiece. Two opposite forming paths, called internal and external, are presented in Fig. 2a and 2b, respectively. In the internal path, the bottom roller moves, remaining tangent to the target geometry, forming it in n-steps. After each ith step, the lower roller moves sequentially along the x and z directions, minimizing the distance traveled between two successive steps. On the other hand, in the external path, the lower roller is first brought outside the forming zone, and then the subsequent i^{th} steps are performed, maximizing the distance between the two rollers. At each step, once the target xz position is reached, the rollers move along the y-axis back and forth to form the entire profile. While the number and increment of each step influence the final geometrical and surface quality of the bending zone, as well as the total production cycle time, the optimization procedure for this value is currently under investigation. Compared to the internal path, the external path increases the Δx_i and Δz_i distances traveled by the roller at each new step.

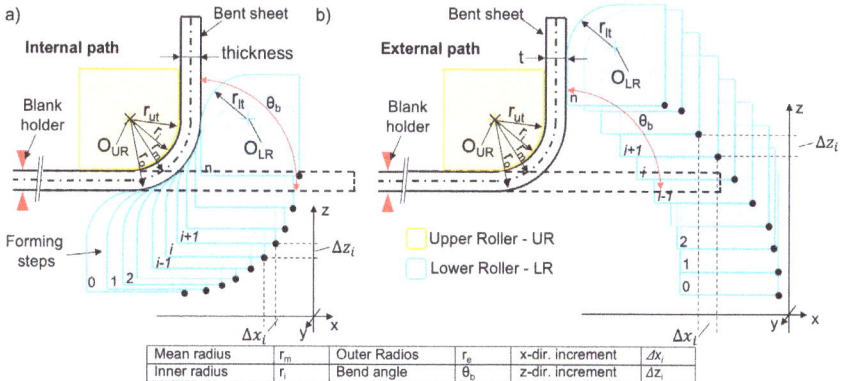

Figure 2. a) internal and b) external tool paths

439

Fig. 3a and b show that during each forming step, the two rollers experience opposing forces in the x, y, and z directions, which generate the necessary bending moment M to bend the component. The internal path strategy shown in Fig. 3a minimizes the distance between the rollers and thus the available moment arm l_i, while the external path strategy presented in Fig. 3b maximizes the distance l_i, reducing the forces required to exert the same bending moment M. Importantly, the force along the y-direction is negligible compared to the force required along the xz bending plane in both cases.

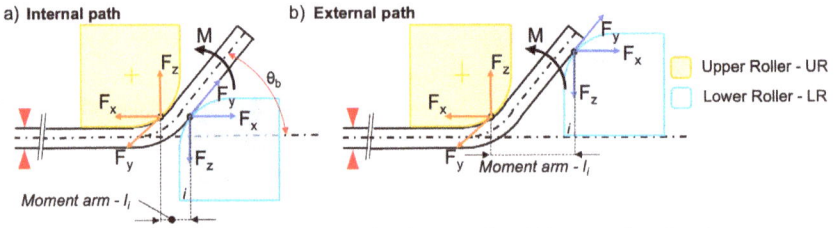

Figure 3. Forces and moment arm for a) internal and b) external tool paths

Results

The two strategies were tested on AISI304 sheets of 1 mm and 3 mm thickness, up to a nominal bending angle of 90 °, over a sheet length of 400 mm. In the case of 1 mm thickness, 8 forming steps were used to shape the profile, whereas 14 steps were employed for the 3 mm AISI304 sheets. The number of steps remained the same regardless of the applied strategy. In Fig. 4a, the mean absolute force F_x is presented for a 1 mm sheet following both tested paths, as the bending angle θ_b is increased. Likewise, Fig. 4b displays the force measured along the z-direction for the same cases. As the bending angle approaches 90°, the force along the x-direction increases while F_z decreases due to the tool process kinematics changing the resulting force orientation. However, both cases show that following the external path can reduce the required forces in both directions. Specifically, for the considered case, the peak force along the x-direction was reduced from 5300 N to 3000 N, and there was a similar reduction from 1680 N to 815 N in the z-direction. The internal path strategy required 155 s, while the external one required 182 s.

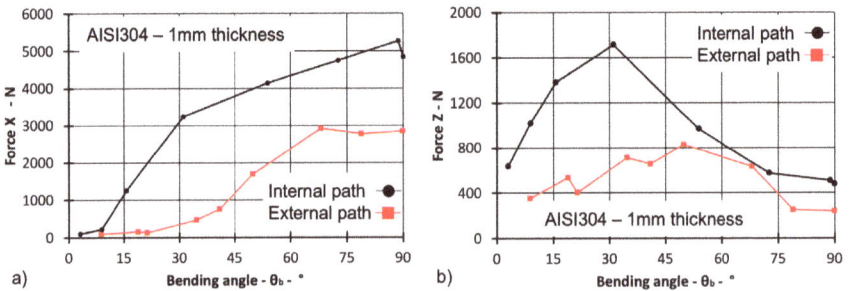

Figure 4. Bending loads for 1 mm of AISI304 steel along the: a) x and b) z directions.

Fig. 5 presents similar results observed in the forming process of a 3 mm AISI304 sheet. The peak F_x values in this case rise to 21580 N and 14350 N for the internal and external paths, respectively. However, the most significant difference was observed in F_z, with the peak force of

18930 N measured for the internal path case, that decreased by 60 % to 7875 N following the external path. As before in the two cases the production time was respectively of 188 s and 212 s.

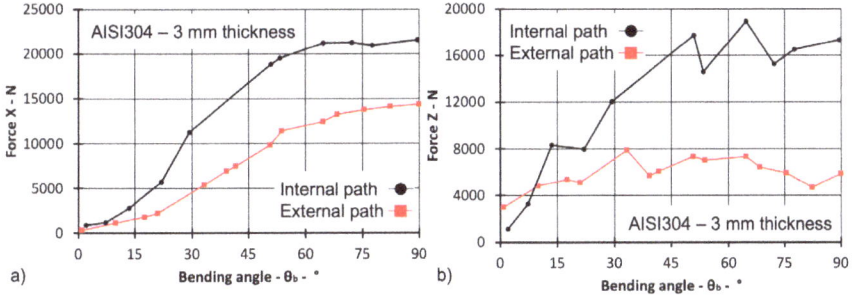

Figure 5. Bending loads for 3 mm of AISI304 steel along the: a) x and b) z directions.

The results demonstrate that maximizing the distance between the rollers can significantly reduce the required forming loads, which, in turn, reduces machine distortions resulting from applied loads. This approach also allows for better control of tool positions, leading to a more accurate process control. Fig. 6a presents an image of the formed profile during an intermediate stage and the final formed part, following the internal path strategy. While the material in between the two rollers bends with a curvature consistent with the upper roller edge radius, along the overall workpiece section a double curvature shape was obtained. This defect in the formed part was due to the presence of unbent material along the y-direction constraining the material over the rollers side to not bend according to the desired shape. On the contrary, by maximizing the distance between the rollers and limiting the amount of material over the rollers side, this effect can be significantly reduced. Fig. 6b further highlights this point by presenting an image of an intermediate forming step and the final part formed according to the external path strategy. While the profiles were bent with a nominal angle θ_b of 90 °, at the end of the process the difference Δ between the angle at the outer edge and at the inner radius (See Fig. 6) was measured along the workpiece middle section. Following the internal path strategy, for 1 mm and 3 mm sheets, the measured Δ were respectively of 42 ± 2.1 ° and 47 ± 1.6 °. On the contrary, the external path strategy allowed to decrease these values respectively at 0.8 ± 0.3 ° and 1.3 ± 0.4 °.

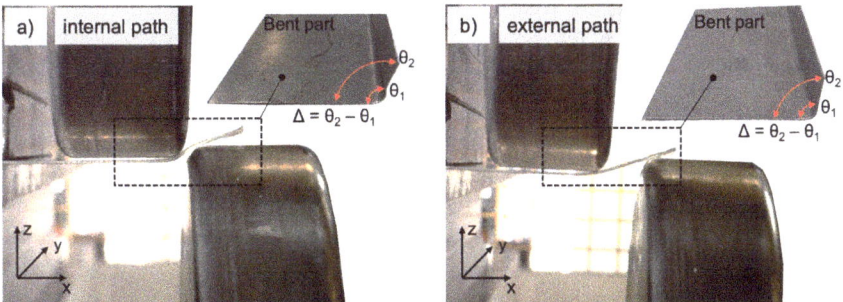

Figure 6. Lateral view of the forming process and final bent part following a) the internal and b) external paths.

Italian Manufacturing Association Conference - XVI AITeM   Materials Research Forum LLC
Materials Research Proceedings 35 (2023) 437-443   https://doi.org/10.21741/9781644902714-52

Conclusions

A new machine, developed to manufacture complex shape bend profile was presented. The effects of different tool paths, one minimizing and one maximizing the tools relative distance during the forming process were analyzed. While the former case allows to reduce the tools necessary displacement, the latter one allows to reduce the forming loads required to achieve the same final components. The two paths were tested on AISI304 sheets with a thickness of 1 and 3 mm, and by maximizing the rollers distance a significant load reduction was obtained. Especially in the case of a 3 mm stainless steel sheet, a 53 % difference in the necessary peak load has been measured between the two cases. At the same time, by maximizing the tools distance it was possible to minimize the influence of the unbent material leading to the appearance of geometrical defects such as a double curvature radius bend. For the 3 mm case, the latter strategy allowed a reduction of the average difference between the angle measured at the outer edge and next to the bending zone from 46.9 ° to 1.3 °. In conclusion, the proposed system offers a cost-effective solution for bending freeform elongated profiles, especially for small batch sizes. The system has potential applications in various sectors, including construction, power plants, and industrial machinery.

References

[1] G. Ambrogio, L. Filice, F. Gagliardi. Formability of lightweight alloys by hot incremental sheet forming. Materials and Design (2012) 34, 501-508. https://doi.org/10.1016/j.matdes.2011.08.024

[2] X. Dang, K. He, F. Zhang, Q. Zuo, R. Du. Multi-stage incremental bending to form doubly curbed plates based on bending limit diagram. International Journal of Mechanical Sciences 155 (2019) 19-30. https://doi.org/10.1016/j.ijmecsci.2019.02.001

[3] G. Ambrogio, L. De Napoli, L. Filice, F. Gagliardi, M. Mazzupappa. Application of incremental forming process for high customized medical product manufacturing. Journal of Material Processing Technology 162-163 (2005) 156-162. https://doi.org/10.1016/j.jmatprotec.2005.02.148

[4] A. Attanasio, E. Ceretti, C. Giardini, L. Mazzoni. Asymmetric two points incremental forming: Improving surface quality and geometric accuracy by tool path optimization. Journal of material process technology 197 (2008) 59-67. https://doi.org/10.1016/j.jmatprotec.2007.05.053

[5] J. Jeswiet, F. Micari, G. Hirt, A. Bramley, J. Duflou, J. Allwood. Asymmetric single point incremental forming of sheet metal. CIRP Annals Volume 54, Issue 2 (2005) 88-114. https://doi.org/10.1016/S0007-8506(07)60021-3

[6] D. Patel, A. Gandhi, A. A review article on process parameters affecting Incremental Sheet Forming (ISF). Materials Today: Proceedings (2022). https://doi.org/10.1016/j.matpr.2022.03.208

[7] K. Jackson, J. Allwood. The mechanics of incremental sheet forming. Journal of materials processing technology (2009) 209 1158-1174. https://doi.org/10.1016/j.jmatprotec.2008.03.025

[8] O. Music, J. Allwood, J., K. Kawai. A review of the mechanics of metal spinning. Journal of materials processing technology (2010) 210, 3-23. https://doi.org/10.1016/j.jmatprotec.2009.08.021

[9] I.M. Russo, J.C. Christopher, J. Allwood. Seven principles of toolpath design in conventional metal spinning. Journal of Materials Processing Technology (2021) 117131. https://doi.org/10.1016/j.jmatprotec.2021.117131

[10] H. Voswinckel, M. Bambach, G. Hirt. Process limits of stretch and shrink flanging by incremental sheet metal forming. Key engineering materials vol. 549 (2013) 45-52. https://doi.org/10.4028/www.scientific.net/KEM.549.45

[11] H. Voswinckel, M. Bambach, G. Hirt. Improving geometrical accuracy for flanging by incremental sheet metal forming. Int J Mater form (2015) 8 391-399. https://doi.org/10.1007/s12289-014-1182-y

[12] T. Buranathiti, J. Cao, W. Chen, L. Baghdasaryan, Z. Cedric Xia. Approaches for Model Validation: Methodology and Illustration on a Sheet Metal Flanging Process. J. Manuf. Sci. Eng. (2006) 128(2) 588-597. https://doi.org/10.1115/1.1807852

[13] N. Le Maoût, S. Thuillier, P.Y. Manach. Classical and Roll-hemming Processes of Pre-strained Metallic Sheets. Experimental Mechanics (2010) 50 1087-1097. https://doi.org/10.1007/s11340-009-9297-7

[14] J.Q. Wang, Y.W. Huang, G.R. Zhou, Z.L Niu, Z.W. Cheng. Process Parameters Study on Robot Rope Hemming of the Hood. Applied Mechanics and Materials (2014) 950-953. https://doi.org/10.4028/www.scientific.net/AMM.602-605.950

[15] X. Hu, Y.X. Zhao, S.H. Li, C. Liu. Numerical Simulation of Dimensional Variations for Roller Hemming. Advanced Materials Research (2010),160-162 1601-1605. https://doi.org/10.4028/www.scientific.net/AMR.160-162.1601

[16] E. Simonetto, A. Ghiotti, S. Bruschi, S. Filippi. Flexible Incremental Roller Flanging process for metal sheets profiles. Procedia CIRP (2021) 103 219-224. https://doi.org/10.1016/j.procir.2021.10.035

Italian Manufacturing Association Conference - XVI AITeM   Materials Research Forum LLC
Materials Research Proceedings 35 (2023) 444-451   https://doi.org/10.21741/9781644902714-53

The effects of dry grinding processing parameters on the electromagnetic and geometrical properties of Nd2Fe14B permanent magnets

Lorenzo Cestone[a,*], Erica Liverani[b], Alessandro Ascari[c], Alessandro Fortunato[d]

Department of Industrial Engineering (DIN), University of Bologna, viale Risorgimento 2, Bologna, Italy

[a]lorenzo.cestone2@unibo.it, [b]erica.liverani2@unibo.it, [c]a.ascari@unibo.it, [d]alessandro.fortunato@unibo.it

Keywords: Grinding, Cubic Boron Nitride (CBN)

Abstract. The automotive industry has grown increasingly interested in electric propulsion over the past few years, which has led to an increase in engine and battery efficiency improvements. The study of rare earth permanent magnets has recently become essential for the development and improvement of electric engines because rotors are constructed of permanent magnets that interact with the stator windings. Due to its strong remanence and coercive field, neodymium magnet is the most frequent rare earth employed in electric motors. Nd2Fe14B magnets are made by sintering in the simple geometries of prismatic, cubic, and cylindrical, and they typically require machining to achieve the final shape necessary for the construction and assembly of the rotor. However, prismatic Nd2Fe14B row materials that have just been sintered are extremely brittle and challenging to produce; as a result, they are typically finished through grinding with a CBN grinding wheel. This study's objective is to evaluate the effects of a dry grinding process with a wet traditional one through an experimental campaign. Process parameters such as cutting speed and feed rate were varied and surface roughness and morphology were compared, together with the magnetic field loss due to the increment of the temperature occurring during the processes. Due to the large number of electric motors that are anticipated to be manufactured in the upcoming years, dry grinding could represent the turning point in terms of eco-sustainability of the process.

Introduction

Rare earth elements (REEs), or "rare earth metals", are a group of seventeen elements which consists of fifteen lanthanides, as well as scandium and yttrium. These REEs have similar properties and are often found together in geologic deposits. Most of them are used in wind turbines, electric vehicles, and solar panels, the main function for which these materials are used is that of permanent magnets [2]. Due to the electrification of the automotive industry, rare earth magnet applications surged by more than 56% between 2010 and 2015; this trend seems destined to continue in the upcoming years [3]. In this study we will focus on Neodymium magnets that are characterized by very high flows and the highest energy products among all state-of-the-art magnets. We will focus on the effects induced on magnetism by the grinding process after having studied and fixed the chemical composition of examined magnets. Initially the Neodymium could resist up to 120 ° C, after which the material tended to demagnetize. This greatly reduced the fields of application, but today new technologies have raised this limit up to 230 °C. This temperature limit can be reached in two main moments: during the on-site application in the machine and in the vehicles or in the production cycle of the components that include magnets like rotors in our case. The main goal of this work is to understand the consequences that grinding induces on the magnetic field generated from the rotors. In the contact area between the grinding wheel and the workpiece heat is generated but it was hard in our situation to detect the trend of the temperature during the process. We therefore decided to investigate the degradation through the measurement

Italian Manufacturing Association Conference - XVI AITeM   Materials Research Forum LLC
Materials Research Proceedings 35 (2023) 444-451   https://doi.org/10.21741/9781644902714-53

of the magnetism pre and post process correlating it to the parameters of the process itself. Another aim is to do this process in a sustainable way and to reach this result, one of the main factors is to eliminate the lubro-refrigerant fluid. A single grinding machine normally requires 2000–4000 L of oil to fill it, of which 100–200 L must be changed each month [4], having an especially negative impact on the environment and the workers. Oil is used to reduce grinding forces, absorb heat created during the process, and make it easier to remove grinding swarf and wheel debris from the cutting zone [5]. The tool used to carry out the test plan is a CBN grinding wheel that we describe in the following chapter. The work done by Lin et al. [6] enables to state that using CBN wheels is better to keep temperature low than using aluminum oxide wheels, in this study they also demonstrate that the differences between the temperatures for dry and wet grinding with CBN wheel appeared to be small [7]. Considering these statements, it should be possible to work magnetic material without the use of coolant liquids, provided that the component's magnetic properties are not compromised. In this work, authors find the optimized parameters to carry out the process by reducing the environmental impact while maintaining the performance of the material.

Materials and methods
Before starting with the removal process on the magnet and with the definition of the processing parameters, a study of the raw material was carried out. This study includes the analysis of the material composition, the surface morphology and the distribution of the punctual phases. After understanding the characteristics of the material, will be described the type of tools used and the parameters on which attention will be paid in the test plan.

Microstructural analysis
A standard metallographic preparation was used to examine the material's structure: the sample was prepared by hot mounting in a phenolic resin and then polished with a metallographic machine by using SiC abrasive papers with decreasing grain size from 80 to 2500. The last two polishing stages instead were performed with the use of two different diamond suspensions with respective granulometry of six and one micron. The ultrasonic washing concludes each polishing stage of the sample to avoid contamination of the process with the subsequent decreasing granulometry. Vilella's reagent (5 ml HCl, 1 g picric acid, and 100 ml ethanol) was finally used to etch the polished surface for 10 to 15 seconds before cleaning it with ethanol and drying it with hot air [8]. A Zeiss Axio optical microscope (OM) was used to observe the sample thus prepared and chemically etched. Following OM, the samples were analyzed with scanning electron microscopy (SEM) coupled with energy dispersive spectroscopy (EDS, X-Act/INCA,10mm^2 SDD Oxford Instruments) to identify the distribution of different elements and phases in the center of the grains and in their borders. This first analysis of the material was conducted to learn more about the chemistry inside the magnets and understand how it affects magnetism. The dissolution of the magnetic material in acids (nitric and hydrochloric) was used as part of the analysis to study the material; using this method, it is possible to precisely determine the compositional percentages by weight of the elements inside the magnet. The analysis took place through an analytical instrument capable of measuring the light (optical emission) produced by a liquid sample when introduced into an inductively coupled argon gas plasma. Through this mechanism it is possible to quantify the metals contained in the sample by measuring, for each one, the intensity of the light emitted with a specific optical bench. The intensity of the light emitted allowed us to quantify the presence of each element. The instrument used to perform this type of analysis is the MP-AES following the standard procedures of the ICP-OES analysis. (UNI EN ISO 17072-1).

Italian Manufacturing Association Conference - XVI AITeM   Materials Research Forum LLC
Materials Research Proceedings 35 (2023) 444-451   https://doi.org/10.21741/9781644902714-53

Characterization of the magnet: roughness, hardness, and magnetism
The base material was observed with a non-contact optical profiler (Taylor Hobson CCI MP-L) after the sintering process and prior the grinding test plan. With the same instrument the roughness of the samples after machining was measured to appreciate the differences induced by the variation of the machining parameters. To find out the hardness of the material that composes the permanent magnets, a Rockwell hardness tester ERGOTEST COMP 25 from the Officine Galileo was used. The magnetism was tested before and after the grinding process to check if it was influenced by the grinding process. The 18 samples were measured with a gaussmeter (PCE Instruments, PCE-MFM 3000) prior to machining to identify the magnetic field generated in the unprocessed state for each component in order to comprehend the entity of the magnetic degradation brought on by the grinding. The gaussmeter probe was installed on a centesimal motion system with three degrees of freedom (x,y,z); the precision of the handling system is essential to maintain the constant air gap between the probe and the sample being measured. This air gap has been set at 0.1 mm from the material and it must be respected when measuring the magnetic field of the raw and machined material.

Magnets grinding
In the traditional production process of the rotors, the parallelepiped-shaped magnets (Fig. 1(a)) are glued together and grounded simultaneously on a cylindrical grinding machine. In this case the samples will be subjected to a tangential grinding machine on a single magnetic element of the rotor at a time to simplify the test plan and to have a greater number of samples. The post-process surface integrity and the corresponding magnetic degradation were evaluated by working on a single magnet stick [9]. The CBN grinding wheel is composed of cubic-shaped grains with a CBN concentration of 40% and a porosity of 37%. The hardness of the binder is low and the porosity is chemically induced. The grinding machine (Rettifica Delta, model TP 1200 x 500) on which the 18 samples have been dry ground is made up of a grey cast iron frame which supports a workpiece table equipped with a permanent magnet table for anchoring the components to be machined. The table runs along "V" guides with recirculating ball bearings and oil scraper lubricated by an intermittent control. Longitudinal and transversal motions of the table, as well as the vertical motion of the wheel spindle, are manually controlled by handwheels. As shown in Table 1, the cutting speed and feed rate were varied on three and two levels, respectively, with three repetitions being carried out for each parameter combination.

Table 1 Grinding test plan.

N °samples	Wheel speed [RPM]	Cutting speed [m/s]	Depth [mm]	Feed rate [mm/s]
1-3	1400	14,6		90
4-6				132
7-9	1900	19,9	3 x 0.01	90
10-12				132
13-15	2400	25,1		90
16-18				132

Result and discussion
Chip formation depends on the grit size of the grinding wheel and the peripheral speed with which it works. With the same wheel selected for a process, chip size can be changed by changing the grinding parameters. The cutting speed greatly affects the size of the chip; at high peripheral speeds, the chip is smaller, generating less force on the grain and applying less stress on the wheel. At the same time, small chips and high cutting speeds increase the amount of heat dissipated, increasing the risks of the wheel wearing. When grains and binder break, other new sharp elements

Italian Manufacturing Association Conference - XVI AITeM   Materials Research Forum LLC
Materials Research Proceedings 35 (2023) 444-451   https://doi.org/10.21741/9781644902714-53

will be made available, and the removal capacity will increase [10]. Theoretically, wheels with soft behavior generate less heat because they always have new cutting edges which facilitate removal. Wheels with hard behavior have a longer life but if the chips are too small, they can cause clogging and overheating of the piece. The most influential parameter on the volume of the chip formed is the feed rate, the substantial difference between these two parameters is that the cutting speed does not affect the cycle time. Increasing the feed rate is possible to reduce the production time but it must be taken into consideration that the mechanical stresses on the grinding wheel will increase and the precision of the machining will decrease.

Metallurgical characterization

Starting from metallographic images captured by OM and following ISO 4499-2:2020 standard, the mean grain dimension was measured and the final results is 8 microns. Another aspect of such results shows the presence of a sintering binder. The binder is used by the manufacturers of rare earth magnets to perform the sintering operations of the components, usually this process takes place by placing powders and binders in the matrices subjecting them to high temperatures and high pressures [10]. The Fig. 1(a) shows the sample before grinding and Fig. 1(b) represents the microscopic characteristics of the base material.

Fig. 1 (a) stick magnet (b) Metallographic structure

Chemical composition

The results obtained with EDS analysis are shown in Table 2.

Table 2 Result from the dissolution analysis

	Fe	Nd	B	Pr	Dy
Wt %	65	27	0.77	<LdR	3.3
S. Dev.	3	1	0.04		0.1

The Nd2Fe14B material's stoichiometric composition is confirmed by this test, along with one additional finding: the presence of dysprosium. According to Brown's et al. [4] research, the presence of dysprosium alters magnetic properties of permanents components. Dysprosium could increase the modulus of the coercivity which in practice corresponds to the increase in resistance to demagnetization, on the other hand this alloying causes a reduction in the magnetic remanence [11]. When using these types of magnets, temperatures tend to rise, and the dysprosium helps maintain a significant resistance. In Brown's et al. [4] study dysprosium and praseodymium were analyzed and an amount of 0% and 10% of dysprosium was measured. In this condition, the magnetic coercivity decrease with the grow of the temperature in both materials but in the material where more dysprosium was measured, the temperature starts to decrease after reaching a higher value [4]. It can be concluded that materials rich in dysprosium have higher magnetic coercivity at the same operating temperature and, as a result, are more resistant to demagnetization.

Additionally, it has been observed that the magnetism of the same material obtained using different manufacturing processes can vary significantly. Sintering processes produce magnets that have a higher magnetic remanence at the expense of coercivity. The starting magnet conditions studied in the paper, with 3.3% of dysprosium and the use of sintering process, have probably mitigated the effects on magnetism that ultimately result.

SEM characterization

Analysing the composition at the centre of the grain, the stoichiometric constitution of the material occurs (spectrum 7) in Fig. 2 (a). At the grain boundaries it was noticed instead that the elements in these areas differ widely from the standard ones (spectrum 6) Fig. 2 (b), mainly because praseodymium and neodymium are there concentrated. The presence of praseodymium was particularly studied by H. Sepehri-Amin and explains the growing of the magnetic coercivity from these intergranular phases [12]. Traces of C,O,Al,Si,Cl were identified by SEM analysis, as shown in Fig. 2(b). This could be due to the sample preparation steps that require the use of SiC abrasive papers and a lot of water to perform the cold polishing. The presence of copper could indicate that some of the elements used to form the sintering powder may have come from recycled sources, usually these magnets work near the stator windings which are made from copper which is likely to contaminate the finished product. Sometimes zirconium is used as a binder in the constriction of powders, so it is not difficult to find it at the grain boundaries.

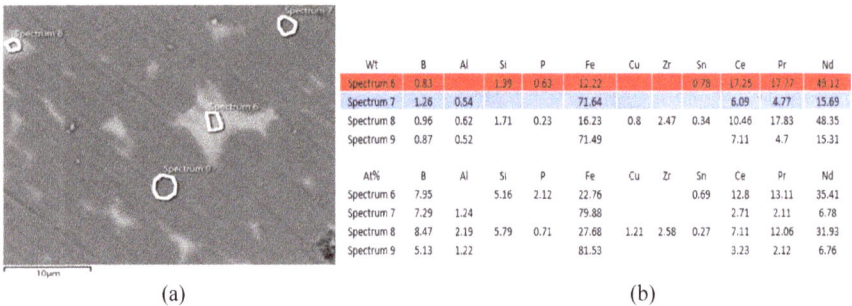

Wt	B	Al	Si	P	Fe	Cu	Zr	Sn	Ce	Pr	Nd
Spectrum 6	0.83	1.39	0.63		12.22			0.78	17.25	17.77	49.12
Spectrum 7	1.26	0.54			71.64				6.09	4.77	15.69
Spectrum 8	0.96	0.62	1.71	0.23	16.23	0.8	2.47	0.34	10.46	17.83	48.35
Spectrum 9	0.87	0.52			71.49				7.11	4.7	15.31
At%	B	Al	Si	P	Fe	Cu	Zr	Sn	Ce	Pr	Nd
Spectrum 6	7.95		5.16	2.12	22.76			0.69	12.8	13.11	35.41
Spectrum 7	7.29	1.24			79.88				2.71	2.11	6.78
Spectrum 8	8.47	2.19	5.79	0.71	27.68	1.21	2.58	0.27	7.11	12.06	31.93
Spectrum 9	5.13	1.22			81.53				3.23	2.12	6.76

(a)   (b)

Fig. 2 SEM image (a) and element concentration (wt% and at%) measured on the selected areas (b)

Hardness and roughness characterization

The average hardness measured in each of the 18 samples is about 50 ±2 HRC, this high value combined with the high brittleness of the sintered material, making the magnets difficult to machine. The roughness measured with non-contact optical profiler (Taylor Hobson CCI MP-L) on the sintered component before grinding is very low (Ra = 0.27 µm) and subsequent cutting operation negatively affects this value. In Fig. 3 the correlation between grinding parameters and roughness can be seen. The roughness decreases as the cutting speed increases and therefore the rotation speed of the grinding wheel increases.

It is interesting to observe that rare earth-based materials follow the same trend that iron alloys and steel would have if subjected to the same process. Although feed rate has a negative impact on surface finish, which should remain below 0.42 µm to allow for proper assembly of rotor parts, increasing the feed rate enables to shorten production cycle times. Trough non-contact optical profiler (Taylor Hobson CCI MP-L) the surface of the worked magnets was scanned and collected; here the worst and the best results (respectively sample number 5 and 16). Fig. 4 (a) shows the 2-D of the surface with the highest roughness value among all (0,51 µm). On the other hand, Fig. 4

(d) shows proper cutting that allowed for the best roughness value (0,33 μm), highlighting that the tool effectively cut the material's surface while leaving clean lines on the same layer. Fig. 4 (b) and Fig. 4 (e) are useful to understand the profile that the instrument reads to produce the average value of respective roughness.

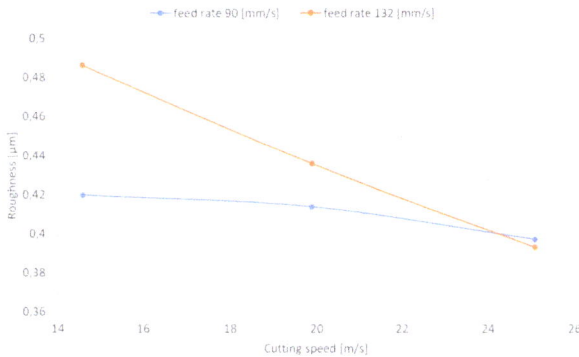

Fig 3: Effect of grinding process parameters on the roughness

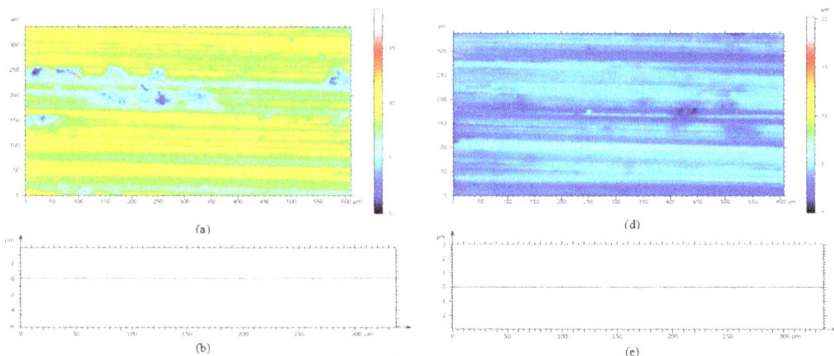

Fig. 4 Ra roughness measured on samples 5 (a,b) and 16 (d,e): (a,d) show 2-D reconstructions of samples surface and (b,e) represent a linear roughness acquisition through the grind marks.

Sample characterization after grinding
Several studies have already analyzed the distribution of temperatures reached in the grinding area, in this work authors focus the experiments on process parameters that have relevant impact on heat generation. As we know increasing the cutting speed helps the removal of material and the obtaining of a better final surface, as reported before in the paragraph about the roughness, but at the same time the heat generated increases as well. According to the graph in Fig. 5 and Table 3, when feed rate around 90 mm/s was selected, both low (14,6 m/s) and high cutting speeds (25,1 m/s) do not lead to significant magnetism losses which are in the range of 10-15 mT. On the other hand, when the feed rate increases to 132 mm/s, the highest cutting speed (25,1 m/s) generates a substantial loss of magnetism of 25 mT. Until this last case the magnetism degradation is simply

Italian Manufacturing Association Conference - XVI AITeM   Materials Research Forum LLC
Materials Research Proceedings 35 (2023) 444-451   https://doi.org/10.21741/9781644902714-53

imputable to the volume reduction of the active material since a linear correlation between the volume of the magnet and the intensity of the magnetic field produced was highlighted. When the decrease in magnetism reaches more significant values, like twice the previous ones (20-25 mT) the heat becomes the main cause. The increase of the cutting speed, as demonstrated by the work of Lin et al. [6], causes the increase of the temperature and consequently is responsible for the magnetic degradation. With the introduction of lubro-refrigerants, it is possible to increase the feed rate and cutting speed but thereby compromising the environmental considerations. Before analyzing the correlation between the loss of magnetism and both the cutting speed and the feed rate it is necessary to keep in mind that the principal antagonist of this property is the high temperature. In the grinding process it is easy to generate heat in the contact zone between the piece and the grinding wheel especially if the parameters are not optimized.

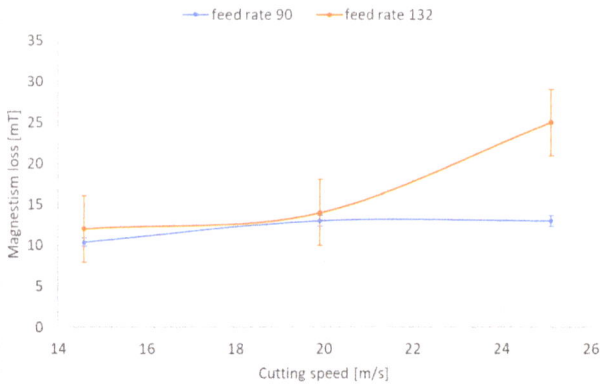

Fig. 5 Correlation between magnetism degradation and grinding parameters.

Conclusions

As the numerical results show, it is possible to grind permanent magnets without using any fluids, maintaining the magnetic field within an acceptable range. This can be considered a major achievement as these hard magnetic materials will surely be processed more frequently in the evolution of the manufacturing in the automotive sector. The rare earth permanent magnets machined with CBN wheels exhibit trends like those of conventional steels in terms of surface finish. Except for samples that are machined simultaneously at a high feed rate and high cutting speed, most of the time, the magnetic degradation brought on by the grinding process is negligible. The presence of 3.3% of Dysprosium increases the coercivity and consequently the resistance to the demagnetization at the same operating temperature making the application range of this material wider. The better roughness (Ra) obtained after grinding is around 0.4 μm and it has been reached with a cutting speed corresponding to 25,1 m/s and a feed rate of 132 mm/s. This corresponds to an overall productivity improvement of compared to the current practice. The only combination of parameters that involves a non-negligible magnetic degradation, which is around 5%, has a cutting speed of 25,1 m/s and a feed rate of 132 mm/s.

References

[1] Zhou, X., Huang, A., Cui, B., Sutherland, J.W., 2021. Techno-economic Assessment of a Novel SmCo Permanent Magnet Manufacturing Method. Procedia CIRP 98, 127–132. https://doi.org/10.1016/j.procir.2021.01.017

[2] Baldi, L., Peri, M., Vandone, D., 2014. Clean energy industries and rare earth materials: Economic and financial issues. Energy Policy 66,53–61. https://doi.org/10.1016/j.enpol.2013.10.067

[3] Constantinides, S., n.d. The demand for rare earth materials in permanent magnets

[4] Brown, D.N., Smith, B., Ma, B.M., Campbell, P., 2004a. The Dependence of Magnetic Properties and Hot Workability of Rare Earth-Iron-Boride Magnets Upon Composition. IEEE Trans. Magn. 40, 2895–2897. https://doi.org/10.1109/TMAG.2004.832240

[5] Mallipeddi, D., Norell, M., Sosa, M., Nyborg, L., 2019. The effect of manufacturing method and running-in load on the surface integrity of efficiency tested ground, honed and superfinished gears. Tribology International 131, 277–287. https://doi.org/10.1016/j.triboint.2018.10.051

[6] Influence of grinding parameters on surface temperature and burn behaviours of grinding rail; B. Lin, K. Zhou, J. Guo, Q.Y. Liu, W.J. Wang

[7] Britton, R.D., Elcoate, C.D., Alanou, M.P., Evans, H.P., Snidle, R.W., 2000. Effect of Surface Finish on Gear Tooth Friction. Journal of Tribology 122, 354–360. https://doi.org/10.1115/1.555367

[8] Pavel, R., Srivastava, A., 2007. An experimental investigation of temperatures during conventional and CBN grinding. Int J Adv Manuf Technol 33, 412–418. https://doi.org/10.1007/s00170-006-0771-4

[9] Sarriegui, G., Martín, J.M., Burgos, N., Ipatov, M., Zhukov, A.P., Gonzalez, J., 2021. Effect of neodymium content and niobium addition on grain growth of Nd-Fe-B powders produced by gas atomization. Materials Characterization 172,110844.https://doi.org/10.1016/j.matchar.2020.110844

[10] Lerra, F., Grippo, F., Landi, E., Fortunato, A., 2022. Surface integrity evaluation within dry grinding process on automotive gears. Cleaner Engineering and Technology 9, 00522. https://doi.org/10.1016/j.clet.2022.100522

[11] Sharma, P., Verma, A., Sidhu, R.K., Pandey, O.P., 2005. Process parameter selection for strontium ferrite sintered magnets using Taguchi L9 orthogonal design. Journal of Materials Processing Technology 168, 147–151. https://doi.org/10.1016/j.jmatprotec.2004.12.003

[12] Sepehri-Amin, H., Liu, L., Ohkubo, T., Yano, M., Shoji, T., Kato, A., Schrefl, T., Hono, K., 2015a. Microstructure and temperature dependent of coercivity of hot-deformed Nd–Fe–B magnets diffusion processed with Pr–Cu alloy. Acta Materialia 99, 297–306. https://doi.org/10.1016/j.actamat.2015.08.013

Work in progress

Italian Manufacturing Association Conference - XVI AITeM   Materials Research Forum LLC
Materials Research Proceedings 35 (2023) 453-466   https://doi.org/10.21741/9781644902714-54

Laser welding in e-mobility: process characterization and monitoring

Caterina Angeloni[1,a] *, Michele Francioso[1] ,Erica Liverani[1] ,
Alessandro Ascari[1], Alessandro Fortunato[1], Luca Tomesani[1]

[1] University of Bologna, Department of Industrial Engineering (DIN), Viale del Risorgimento 2,
Bologna 40136, Italy

[a] caterina.angeloni2@unibo.it

Keywords Laser Welding, E-Mobility, Quality Control

Abstract Nowadays car manufacturers have to face the challenge of electric mobility transition and it is therefore becoming an industrial priority to accomplish a large-scale battery manufacturing based on high production rates and zero-defect processes that must be obtained over a very high variety of products. The above challenges are currently examined, at the Department of Industrial Engineering of the University of Bologna, by studying and optimizing laser processes (manly cutting and welding) in case of battery fabrication and assembly. Materials investigated are manly pure copper and aluminum, processed by means of different sources, wavelengths and scanning heads. In order to reach the quality targets requested by the market, an in-process monitoring system is implemented as well as physical and data driven simulation models are under development. Finally, comparisons between the monitoring output, the keyhole simulation results, and experimental data will be carried out.

Introduction

Nowadays it is well known that the future of Mobility is Electric, ranking first among the solutions in the automotive field to address the reduction of emissions into the environment. European Parliament states that by 2035, all new cars manufactured for the EU market should produce *zero-emissions*[1]. However, switching to electric mobility represents for many companies a fundamental revamp of production technology. The electrical energy storage system is the most critical feature since the battery is the most expensive and the heaviest component inside the electric vehicle. Many companies operating in this field are adopting lithium-ion batteries because of their advantages in terms of high energy density, making them easier and faster to charge and long-lasting. There are three battery geometries for automotive applications: cylindrical, prismatic and pouch. These geometries are assembled into modules which are put together into a battery pack. The number of modules and, consequently, batteries are chosen depending on the power/energy that must be supplied by the vehicle[2].The pouch cells that contain the battery are equipped with two contact tabs made of copper and aluminum for anode and cathode. When more cells are mounted in a battery pack, electrical connections are needed between these contact tabs in order to get a serial connection of the cells[3]. Welding processes are suitable to create such contacts characterized by high strength and low electrical resistance.

In this context, laser is becoming a fundamental tool thanks to its flexibility in terms of automation and control, therefore can be easily inserted into industrial production. In particular laser-welding is widespread thanks to its production speeds and accuracy, which are the highest in the entire panorama of technologies. Automated production itself calls for a high number of welds and a wide range of differences in terms of materials, thickness and welding units involved in the process.

Lap- or butt-joint welding of dissimilar highly reflective metals, such as aluminum and copper, has been widely investigated for electrodes, as shown by Katayama et al.[4]. The latter discern thoroughly more than one-hundred papers on laser welding of thin metal sheets. It is therefore well known how to weld a proper seam in hybrid configurations. The challenge today is when it comes to high production rates and high demand of quality required by a zero-defect process. Industrial production speeds can reach up to 5-10 batteries/sec and the assembly of a single battery pack can contain up to 20000 welds [5]. It is estimated that each Gigafactory produces 6% circa of defected cells and battery modules[6]. Since most of the materials (electrodes, cell separators, electrolytes) are not fully recyclable and the whole disposal process is pricy, the quality target of the process has been set to 99.7%[6]. This highlights the central role of a monitoring system that gathers information from the process, improving the understanding of the detecting phenomena. It uses the collected data to create quality control methods and adaptive, closed loop control of the process[7]. Hence, as the number of destructive samples inspections are minimized, the implementation of a monitoring system can be seen as a product certification.

A lot of studies of online quality control have been carried out aiming at reducing or eliminating product quality defects and process errors. Nowadays monitoring systems range from simple systems using single sensors to more sophisticated systems which utilize a great deal of sensors and detection methods, as shown by Katayama et al.[8] and W. Cai et al.[9]. The latter reviewed three-hundred-ish papers describing the typical sensors used for laser welding and adaptive control: it ranges from photodiodes, visual sensor, spectrometer, acoustical sensor, pyrometer, plasma charge sensor to the application of artificial intelligence algorithms. It states that, to date, it is consolidated that the geometry of the melt pool exerts a major influence on the weld. That's why researchers are now assessing imaging as a tool in monitoring and predicting weld's quality. Moreover, machine learning models have been exploited since they are able to learn the error between the predicted value and the real value. Especially, deep learning is the present research highlight [9].

A. Bluga et al. implemented a closed-loop system based on the observation of the penetration depth of keyhole welding processes on aluminum foils. A stochastic approach was implemented for the relation between laser power and the probability to detect full penetration hole. A CNN based system called Q-eye developed by Anafocus was used to execute the algorithms for the detection of the full penetration hole. The optics units were mounted to the coaxial process window of the welding head in order to create the thermal image in the spectral range of 820 to 980 nm. A 20m cable connects the camera with the CNN control unit, which contains the interface to the laser control unit on the one hand and the PC on the other hand. The latter runs the control software which functions as the user interface. The results show that the standard deviation of the laser power was in the order of 2% [10].

Ascari et al. [11] proposed a method to ascertain the laser weld depth of battery connector tabs (Al e Cu thin foils) using *optical coherence tomography* (OCT) equipment. An adjustable ring mode (ARM) laser integrating OCT technology with two beams was exploited: one pointing at the bottom of the keyhole and the other one referring at the sample's surface. They used the "Keyhole Mapping" approach, which identifies the optimum positioning of the OCT measuring ray. Considerations on both the measurement's accuracy and the keyhole stability were made. It was demonstrated that ARM laser returns a more stable process as it reduced the fluctuations of the opening of the keyhole and it improved the measurement accuracy by the 50%.

Recently, part of the research focuses on the implementation of *deep learning* e *digital twin algorithms*. Good results were achieved by Franciosa et al. studies [12]. It presents a digital twin framework for assembly systems combining sensors with deep learning and CAE simulations. This study developed a closed-loop in-process control using a fully digital developed remote welding process for aluminum doors for e-mobility applications. It can identify the main causes of quality defects and suggest corrective actions for automatic defects mitigation.

The above studies are based mostly on deep-learning methods that can achieve only one prediction task. As laser welding is a complex phenomenon that is influenced by multiple parameters, multiple tasks have to be accomplished, as the work of Kim et al. shows [13]. Another study carried out by Franciosa et al. [14] developed a method for closed-loop in-process quality control in battery assembly lines. It is based on a holistic approach exploiting the potential use of *light-based technology*. The authors exploit both in and off-process methodologies to monitor the process: SEM scanning (off-process) to correlate grain structure with the material strength status, CT scanning (off-process), Optical measure (off-process) to evaluate weld penetration depth and weld interface width, Laser radiation (in-process) to measure weld depth penetration through OCT technology, thermal radiation (in-process) inspected with a IR camera. The closed-loop system is based on the development of a low fidelity artificial intelligence model, which identifies critical patterns in data flows that are then improved by implementing Multiphysics simulations.

Marc Seibold et al. [15] realized a process control by real-time pulse shaping where the power is adjusted in each pulse. Al-Cu thin foils were welded, and it was found out that the steps of the welding process can be captured and recognized by using photodiodes with band-pass filters.

Currently, there is little research on effective closed-loop industrial solution that can be implemented in an industrial environment: most of the technologies cited previously involved an equipment that is complicated, bulky and pricy, making them suitable just for experimental studies. Only a simple and flexible monitoring system can cope with rapid product variety in terms of materials and geometries that can be processed.

In this direction, this paper considers a low-level system which includes photodiodes sensors (Laser Welding Monitor LWM 4.0). A *low-level* system is a *ready to use* industrial system: a PD can be defined as a commercial sensor that can perform direct measurements and be easily integrated into different setups and detect different wavelengths ranges. These can be easily integrated into different setups and detect different wavelengths ranges. This kind of monitoring systems for mass production integrated in the production line nowadays do not allow a real-time modification of the process parameters, they only state, at the end of the process, if any signal values have gone out of range, which has been established previously during high-quality welding.

In literature it is described which parameters have the major influence on the process quality in laser welding but there is a consistent lack of information on the correlation between physical phenomena and sensor response. Consistently, the aim of this paper is to investigate the in-process signals measured by photodiodes and the level of correlation between the three signals in order to improve our understanding of the feedback they provide, both individually and together. The three photodiodes detect the luminous intensity of three radiations emitted over three wavelength ranges. These detectors collect plasma signal (PS), temperature signal (TS) and back-reflection signal (BS). Further details are reported in 'Process monitoring' Section. To accomplish this, experimental tests on Cu and Al thin plate were performed by varying laser power, welding speed and the length of the trace while monitoring the signals of the three photodiodes. The experimental data was evaluated offline and analyzed to evaluate the relationships between technological and signals outputs.

Materials and Method

The welding experiments were carried out using pure copper (Cu>99.6%, 0.3 mm thick), coated with a thin nickel layer in order to improve optical absorptivity of the laser radiation [16] and to avoid surface oxidation, and commercially pure aluminum AA1060 (99.4% Al, 0.25% Si e 0.35% Fe, 0.45 mm thick). The physical properties of both materials are shown in Table 1. Welds were performed on samples in a lap-joint configuration. For all the tested process parameters, both Al-Cu (aluminum on top) and Cu-Al (copper on top) configurations were examined in order to understand the importance of laser absorption and melt pool dynamics during re-solidification. For the purpose of keeping the adhesion between the two sheets during the process, a clamping device was used.

Table 1 Physical properties of aluminum and copper

	Copper	Aluminum
Density [g/cm^3]	8.9	2.71
Melting temperature [°C]	1083	660
Thermal conductivity [W/ (m ·K)]	390	226
Thermal expansion coefficient [°C^{-1}]	$17 \cdot 10^{-6}$	$24 \cdot 10^{-6}$
Specific heat capacity [J/ (kg K)]	385	880

Methodology
Laser welding equipment and Optical Setup
Two near-infrared fiber laser sources were selected for the experiments. A single mode and a multimode laser source delivered by feeding fiber with core diameter respectively of 49 and 68 μm were used in a continuous (CW) mode for weld tests. The multimode source was a nLight Alta 3KW, while the single mode source an nLight CFL-1200. The characteristics of both laser sources and optical systems are shown in Table 2. The motions of the laser beam were achieved by a galvo scanning head for both configurations to maximize the laser speed, while the initial displacement of each weld was carried out by (ROBOT Motoman HP-20 six axes). The Precitec monitoring system, fully described in the following 'Process Monitoring' section, was mounted on the scanning head, as Fig.1 shows.

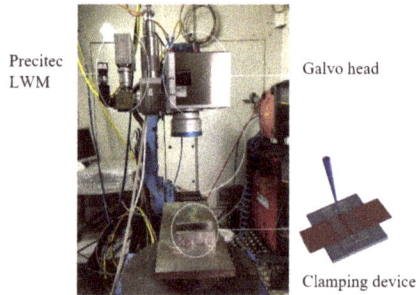

Fig. 1 Laser system set-up. Galvo head, Precitec monitoring system and clamping device position are highlight with white circles

The full experimental campaign is described in Table 1. After defining a feasibility window, the effects of power and welding speed on the weld bead geometry were assessed. Only the first test was conducted with the multimode laser source, while the single mode laser source was implemented in the first, the second and the third trial, with a resulting campaign of 27 weld beads for each configuration. All experiments were performed without shielding gas and without filler wire for reducing external influences on monitoring signals. In order to avoid surface contamination, samples' surface was cleaned before welding.

The two different F-Theta lenses installed in the galvo scan head lead to different spot dimensions. A 19% larger spot results from multimode laser configuration, therefore higher laser power and slower speeds are needed to get roughly the same energy density, defined as the product between power density and interaction time [J/cm3]. See Table 3.

Table 2 *Multimode and Single mode Laser source properties*

Laser source			Optical Path		
Specific	Value		Specific	Value	
	Multi mode	*Single mode*		*Multi mode*	*Single mode*
Beam's wavelength [nm]	1064	1064	Collimation length [mm]	120	120
Operative Method	CW	CW	Focal length galvo [mm]	163	420
Max Laser Power [W]	3000	1200	Fiber diameter [μm]	50	14
BPP [mm·mrad]	2	0.4	spot [μm]	68	49

In the third test, only the length of the linear weld track was increased from 48 mm to 100 mm to discern the influence of weld-path on weld-bead geometry and measured signals. A comparison was then made with the previous tests.

At the end of the experimental campaign, the welds were sectioned in the middle of the bead and were prepared for metallurgical analysis by using standard criteria for sample preparation. Cross sections of the weld joints were mounted in resin and then polished with SiC paper grit from 800 to 2500 followed by 1 μm alumina suspension. The metallurgical specimens were etched with Keller's reagent (1 ml HF, 1.5 ml HCL, 2.5 ml HNO3 and 95 ml H2O) with an etching time of 20 s for preliminary observation. Micrographs and weld bead measurements were taken with a ZEISS Axio Vert.AlM for a visual correspondence of the results obtained.

Table 3 *Process parameters investigated*

			Power [W]	Speed [mm/s]	Weld length [mm]
Multi mode	Test 1	Al-Cu	900-1200-1500	100-200-300	48
		Cu-Al	1200-1500-1800	100-200-300	48
Single mode	Test 2	Al-Cu	800-1000-1200	120-240-360-480	48
		Cu-Al	800-1000-1200	200-300-400-500	48
	Test 3	Al-Cu	800-1000-1200	120-240-360-480	100
		Cu-Al	800-1000-1200	200-300-400-500	100

Process monitoring

The Precitec monitoring system used (LWM 4.0; Precitec GmbH, Germany) assures non-destructive online control in real time. As Fig. 1 shows, the system is mounted on the camera flange located on the laser head. Hence, the internal optical path in the welding head is used, meaning that the sensors are always coaxially aligned to the welding spot and preserved from contamination. The system contains 3 photodiodes which detect the luminous intensity of three radiations emitted over three wavelength ranges. These detectors collect plasma signal (PS), temperature signal (TS) and back-reflection signal (BS) at a maximum sampling rate of 1kHz.

- The plasma sensor documents the UV light from the plasma plume by recording and analyzing the amplitude. The plasma control variable itself indicates the amount of metal vapor ionized during the keyhole formation process [17].

Italian Manufacturing Association Conference - XVI AITeM   Materials Research Forum LLC
Materials Research Proceedings 35 (2023) 453-466   https://doi.org/10.21741/9781644902714-54

- Back-reflection is that fraction of the laser beam which is not absorbed by the material and therefore emits at the same wavelength as the laser (which emits infrared). It is consolidated that the BS is directly correlated with the keyhole geometry, thus providing information about the penetration depth [18].
- The temperature detector captures radiation in the near infrared and gives information on the thermal condition of the irradiated surface. Hence, the TS enables the identification of lack of fusion.

The system stores all the process signals in the SQL-Database, which is used to manage the Measurements and Configurations. The Control Module happens to be the core connection between the database, one or more View modules and possible external units connected by the customer. The View Module in the software guarantees a visualization of the processed signals. All the measurements were then selected and exported as .txt files, which were then given as an input for a short MATLAB script whose purpose was to graph the data and perform statistical analyses. The script was then readapted to all tests carried out in the experimental campaign, the results of which were then compared.

Results and Discussion
With the aim of developing better in-process monitoring, this paper focuses on finding correlations between the data detected by the monitoring system with the typical welding process characteristics.

Plasma and Temperature
In the first place, the welds performed with the multimode fiber source were initially analyzed and the results of test 1 were compared with each other. Al-Cu configuration was analyzed in the first place. PS and TS monitored during the laser welding process are reported in Fig. 2. If speed is fixed (v=100 mm/s), there is a linear correspondence between the two signals and power: if power steps up from 900W (blue line) to 1500W (yellow line), PS and TS increase also. It can be noticed that the trend of PS along the trace over time mirrors that of the temperature: a progressively increasing value can be noticed, which gets steeper as the laser power applied grows. For example, the increase of PS along the trace changes from 3.24% for lower power to 28% for 1500W.

This similarity between the two signals can be related to the fact that the temperature measured on the surface directly affects aluminum's vaporization and its alloy elements leading to the opening or closing of the keyhole. This observation agrees with the findings of Franciosa [6] and I Eriksson et al. [19] who state that temperature and plasma signals are strongly correlated with Pearson's correlation coefficient above 94%. This suggests that the plasma plume emits not only in the UV/visible spectrum but also contributes to thermal radiation in the IR, it is therefore necessary to remove the PS from the TS.

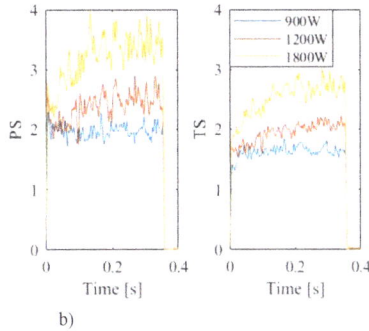

Fig. 2 *a)Plasma and b)Temperature vs time as power ranges from 900W to 1500W for fixed velocity of 100 mm/s. Current data refers to Al-Cu configuration.*

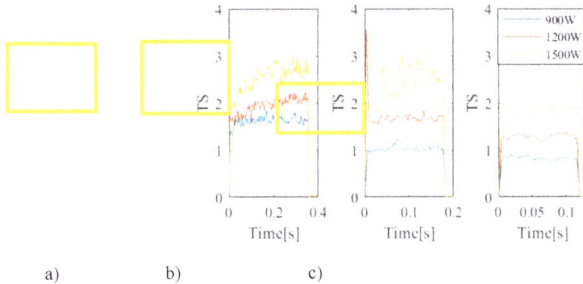

Fig. 3 *Influence of speed on the welding process in test 1: for higher values of speed which ranges from a)100mm/s b)200 mm/s to c)300 mm/s, the standard deviation of the signal decreases*

The graph in Fig. 3 describes the effect of *speed* on the process: within the process parameters window analyzed, the trend of the signals becomes more stable (less oscillating) for high-speed values up to assuming an almost constant trend (like for v=300 mm/s). From a process point of view, this data behavior could be explained as follows: for low speeds, during the laser-aluminum interaction, the component increases its temperature moment by moment since not all the heat can be dissipated by conduction. Therefore, it is logical to think that, by heating continuously, the characteristics of the joint vary from the beginning to the end of the weld bead. This problem of heat accumulation during welding is emphasized for low speeds that lead to a longer interaction time and the amount of heat absorbed by the material increases with it. In the graphs in Fig. 3 this behavior is very evident: as the speed increases from 100 mm/s (Fig. 3 a) to 300 mm/s (Fig.3 c) the signals are much less oscillating and more centered on the average value, which, presumably, is equivalent to a more homogeneous, sound weld bead along the track. To sum up, BS and TS standard deviation (T_{std}, P_{std}) increases as laser power increases and decreases for higher speeds, which comes to a higher energy density on the material. This feature is found to be more relevant for the Cu-Al configuration, hence the Fig.4 is reported. This means that the more energy density is applied, the more the signal oscillates, hence the more unstable the keyhole gets.

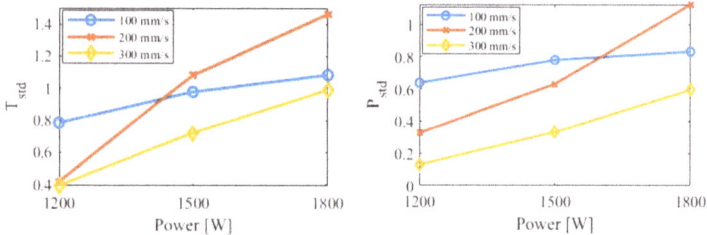

Fig. 4: *Temperature and Plasma signals standard deviation for test 1 Cu-Al. As laser power increases and speed is diminished, an increasing trend can be noticed, leading to keyhole instability.*

The increasing linear relationship is consistent not only for T_{std} and P_{std}, but also for PS. As previously anticipated, the mean value of PS (P_m) and TS (T_m) was calculated for every test over the whole weld bead when a multimode laser source was exploited. The resulting values are plotted in Fig 5, both for Al-Cu and Cu-Al configuration. There is clear evidence of a linear relationship between power and monitoring signals. Similar trend was identified also for weld-bead width. The micrographs reported in Fig.6 point out the different characteristics of the weld bead morphology of Al-Cu and Cu-Al configurations. In Al-Cu the laser beam interacts with aluminum, and its weld bead tends to be larger at the top and narrower at the bottom, while if the first interaction is with the copper, the situation is the opposite. This behavior is due to the different physical properties of the two metals involved in the process, in terms of heat conduction coefficient and melting temperature (see Table 1). When the aluminum is at the top side, the laser beam melts, at first, the material that has the lower temperature melting point and then the copper that is situated at the interface between the two sheets. Copper has a much higher melting point and a higher thermal conductivity than aluminum, that's why it melts to a much lower extent.

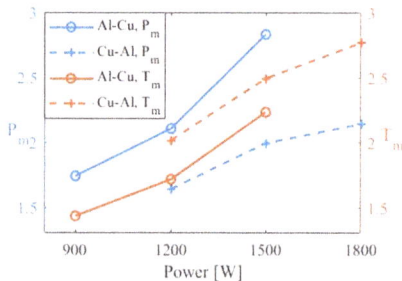

Fig. 5 *Mean values of PS when a multimode source is exploited and speed is fixed at 200mm/s for both Al-Cu and Cu-Al configurations.*

Italian Manufacturing Association Conference - XVI AITeM   Materials Research Forum LLC
Materials Research Proceedings 35 (2023) 453-466   https://doi.org/10.21741/9781644902714-54

P[W]	900	1200	1500	1800
Al-Cu	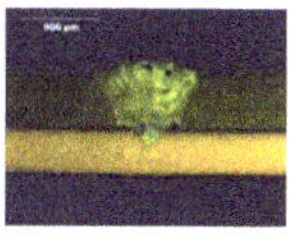	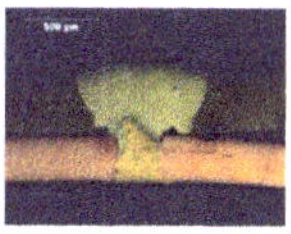	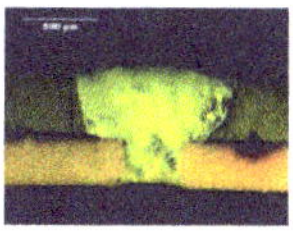	-
Weld Width [μm]	687	868	1054	-
Cu-Al	-	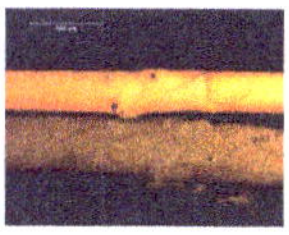	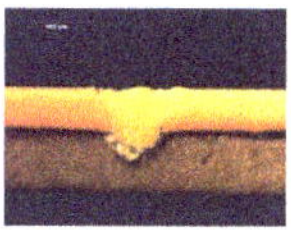	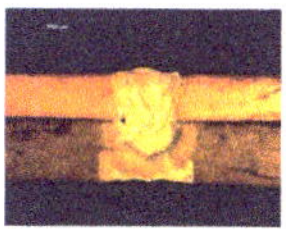
Weld Width [μm]	-	170	495	590

Fig. 6 *Corresponding micrographs and weld bead width taken at the middle of trace.*

When copper is at the top side, heat is transferred very quickly at the interface between the sheets because of the higher melting temperature, if compared to aluminum. The latter melts abruptly and, if the energy is excessive, tends to favor a complete penetration of the specimen [20].

Another thing that can be noticed is that the average amount of metallic vapor produced during the interaction between the laser beam and the aluminum is way higher than that found for welding the copper first: in

Fig. 5 Al-Cu P_m curve assumes plasma values ranging in an interval that is higher than the Cu-Al P_m. This happens because aluminum has a lower phase transition temperature than copper and therefore vaporizes before. The width variation can then be monitored by the signal values of either plasma or thermal radiation during welding.

It is consequently confirmed that the control of the laser power based on thermal radiation values can minimize the variation in the bead width.

Back-Reflection
In the following paragraph BS is analyzed in test 1 Al-Cu. A comparison is then made with the data obtained in test 3 to comment the influence of the welding length. For tests 2 and 3 similar results were obtained, therefore the discussion can be extended for all Al-Cu tests. For the Cu-Al configuration such a marked trend was not evident: as the power and speed change, the signals mean values were almost steady. In general, it is well known that material absorptivity decreases with an increase in the welding speed [4]. As the welding speed increases, the molten pool becomes smaller, and the melt zone in front of the keyhole is also narrower (or thinner). That is, at higher welding speeds, a part of the laser beam is irradiated on the thin melt zone and solid metal surface; consequently, the laser back reflection increases, and the ratio of a laser beam absorbed into the keyhole decreases. The histogram in Fig. 7 testimony the latter consideration: increasing speed, the back reflection signal increases.

Italian Manufacturing Association Conference - XVI AITeM Materials Research Forum LLC
Materials Research Proceedings 35 (2023) 453-466 https://doi.org/10.21741/9781644902714-54

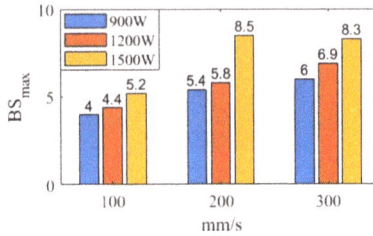

Fig. 7 Peak values of the Back reflection-time signal at varying power and laser speed for test 1 Al-Cu. It can be noticed that as the speed increases also BS increases as a smaller amount of laser beam is absorbed because of the narrower weld bead formed.

The BS itself is characterized by a fast peak near the central area followed by a fast drop, decreasing towards the end of the track. The bell-shaped pattern is probably due to the fact that the sensor, remaining integral with the laser head during the whole welding process (as a galvo head is expected to work), is able to collect all the radiations only in the configuration in which the laser head is positioned at the center of the trace and the beam is perpendicular to the surface. In all other welding positions, the inclination of the laser beam towards the surface increases and makes a good percentage of the light radiation going out from the sensor's reading range. For this reason, to make comparison analysis, the maximum value is considered, which is the peak point of the hilly trend. In support of this assumption, test 3 was carried out, where only the length of the track was changed. It was found that the bell-shaped trend is more marked: the sensor will receive very weak light radiation from the points furthest from the center of the weld bead because of the greater inclination. By making the weld bead longer, the surface temperature of the aluminum and the plasma detected decreases, if compared with the previous tests.

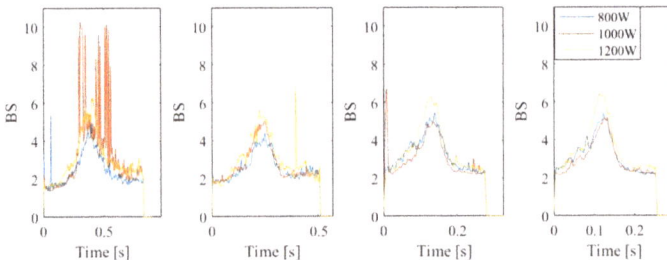

Fig. 8 Hilly trend of the BS-time at varying power and speed that ranges from 120 mm/s (to the left) to 480 mm/s (on the right) for test 3. The bell-shaped trend in longer weld beads is more marked: the sensor will receive very weak light radiation from the points furthest from the center of the weld bead because of the greater inclination of the laser beam.

Looking at test 1, a comparison between BS collected during Al-Cu and Cu-Al configuration was carried out. It is clear that BS on copper is always greater because it has a lower absorption coefficient than aluminum (for the percentage increase look at Table 4).

Table 4 *Percentage increase values from Al-Cu to Cu-Al configuration in the operating window considered*

	Speed [mm/s] →		
	106%	165%	170%
Power [W] ↓	65%	154%	133%
	133%	82%	96%

Statistical Analysis

In order to hold together the results from the whole experimental campaign, the spot dimension has to be taken into account as single and multimode sources were exploited. For every set of parameters, the mean value and the standard deviation were calculated for each weld bead and plotted against energy density.

As Fig. 9 states, TS can be modelled with a linear regression with a good accuracy. In fact, the statistical analysis obtained with MATLAB shows a very low p-value and an R^2 very close to unity ($R^2 = 90\%$), suggesting that the linear model used for temperature variable interpolates the data very well. However, this correlation is not true for back reflection and plasma variables, whose intensity depends on the laser source exploited for the experiment, as Fig. 10 points out. Hence, two different models must be implemented in order to describe the welding process carried out with two different sources. In order to monitor the stability of the process, the standard deviation of plasma (PS_{std}), temperature (TS_{std}) and back-reflection (B_{std}) versus Energy density Es were plotted. In this account, some considerations were raised: the TS_{std} graph in Fig.11a) shows clearly three patterns for both multi and single mode sources. These linear patterns were analyzed, especially for the multimode tests where it was more evident that the separate trends, with three points each, were characterized by equal speed. The histogram in Fig. 11b) shows how the TS_{std}, which is directly correlated to process stability, depends more on power than speed: keeping the process at a constant speed, it becomes more instable as the power increases, and not vice versa. Consequently, it is possible to suggest changing the *power* as the variable input in the closed-loop system if any signal exceeds the threshold.

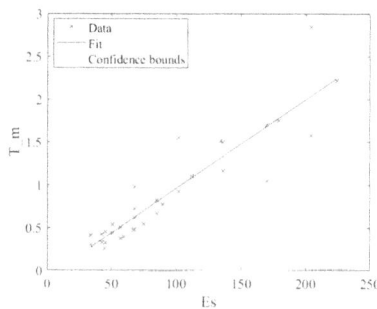

Fig. 9 *Linear regression for TS with a $R^2 = 90\%$*

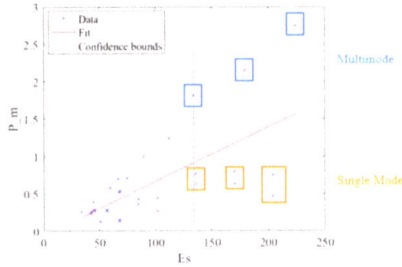

Fig. 10 *Plasma values for both single mode and multimode sources. If a multimode source is exploited, the intensity of plasma radiation recorded grows with those of the power density (E_s). On the other hand, if a single source is exploited, the variation of PS from the mean value is not significant*

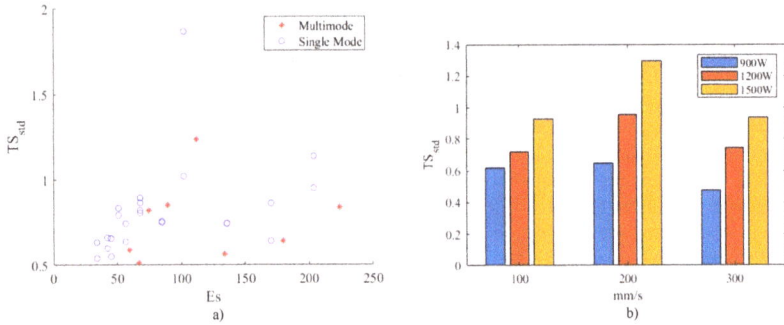

Fig. 11 *a) TS_{std} against energy density Es for single and multimode sources. Three linear trends can be noticed with three scatter points each in every of which the speed is kept constant and the Power is increased. This consideration is validated by the histography in b) where for every speed, for an increased value of power, the process becomes more instable (standard deviation values increase).*

The same studies were made on Cu-Al configurations and the same results were observed.

Conclusions and future work

To sum up, the following results were obtained:

- Signal standard deviation or signal variance data can be used to improve the accuracy of laser welding monitoring devices, as Olsson states in his work[21].
- The plasma trend mirrors that of the temperature: the temperature directly influences the vaporization of the aluminum and its alloying elements leading to the opening or closing of the keyhole. This observation agrees with the findings of Franciosa [6] and I Eriksson et al. [19] who state that temperature and plasma signals are strongly correlated with Pearson's correlation coefficient above 94%, and thus a similar signal is delivered. This suggests that the plasma plume emits not only in the UV/visible spectrum but also contributes to thermal radiation in the IR.

Italian Manufacturing Association Conference - XVI AITeM   Materials Research Forum LLC
Materials Research Proceedings 35 (2023) 453-466   https://doi.org/10.21741/9781644902714-54

- There is an increasing linear relationship between plasma and power for the multimode source tests, a decreasing linear trend for the single mode ones. Observing the micrographs, the variation of the weld seam's width seems to follow the same tendency varying the power with the other parameters in play being equal.
- With the same power density as E_s the plasma values obtained in the multimode tests are greater than those obtained for the single mode tests for the Al-Cu configuration. This increased value is probably due to the greater width of the bead (the sensor detects more radiation).
- Within the window of process parameters analyzed, if high speeds were set, the trend of the signals stabilizes (less oscillating) until it assumes an almost constant trend. Hence, the standard deviation of the signal disclosed an explicit correlation with the stability of the process. This consideration is valid within the considered operative window. It is possible that the signal turns to be unstable if the speed is increased more.
- The back-reflection varies with a bell-shaped trend: the sensor is able to detect all radiations only in the configuration in which the laser head is located in the center of the trace and the light beam is perpendicular to the surface. In all other positions, a good percentage of the light radiation goes out of the reading range of the sensor. This behavior only affects the signals but not the weld quality itself [6].
- The average amount of metal vapor produced during the interaction between the laser beam and aluminum is much higher than that detected for copper welding because aluminum has lower phase transition temperatures than copper and therefore vaporizes first.
- The back reflection on copper is always greater because it has a lower absorption coefficient than aluminum.
- The power is expected to be the most influent variable for the closed-loop monitoring system.

Based on these observations, future technological implementations might follow:

- It is conceptually possible to exploit plasma as an input to regulate the power with a feedback control to keep the temperatures constant during the process.
- The analysis of the plasma vs energy density trend confirms that any threshold value you want to define for the control of the seam strictly depends on the laser source used.
- If the consideration that was made on the back-reflection signal is true, it therefore makes this signal of little use for the analysis of the single trace but can only be evaluated in a comparative way with a pre-acquired signal on a bead, whose properties are considered acceptable.
- The takeaway from the comparison between Al-Cu and Cu-Al configurations is that any threshold value for the bead control strictly depends on the first material irradiated.

During the analysis of the dataset, the width of the bead emerged to be a relevant parameter for the control of the opening/closing of the keyhole. This observation led us to consider the implementation of a camera system on the laser head capable of measuring the variations in shape and size of the keyhole as a possible evolution of this research. In this way, the correlation between the data detected by the sensors and the welding behavior observed by the high-speed imaging shall provide greater precision in welding control, guaranteeing the required quality. It is essential to have a larger dataset available in order to apply Machine Learning and Artificial Intelligence techniques to be able to link the behavior of one or more sensors and match it to a specific real defect and ultimately, prevent it through feedback control. That is to say, humanizing the whole laser welding monitoring process (acquiring signals, analyze them and making monitoring targets) [9].

References

[1] "Deal confirms zero-emissions target for new cars and vans in 2035,".

[2] A. Ascari, A. Fortunato, in Joining Processes for Dissimilar and Advanced Materials, Elsevier, 2022, pp. 579-645. https://doi.org/10.1016/B978-0-323-85399-6.00006-0

[3] C. Wunderling, C. Bernauer, C. Geiger, K. Goetz, S. Grabmann, L. Hille, A. Hofer, M. K. Kick, J. Kriegler, L. Mayr, M. Schmoeller, C. Stadter, L. Tomcic, T. Weiss, A. Zapata, M. F. Zaeh.

[4] S. Katayama, Fundamentals and Details of Laser Welding, Springer Singapore, Singapore, 2020. https://doi.org/10.1007/978-981-15-7933-2

[5] M. Kogel-Hollacher, The Laser User Magazine 2020.

[6] G. Chianese, P. Franciosa, J. Nolte, D. Ceglarek, S. Patalano, Journal of Manufacturing Science and Engineering 2022, 144, 071004. https://doi.org/10.1115/1.4052725

[7] T. Purtonen, A. Kalliosaari, A. Salminen, Physics Procedia 2014, 56, 1218. https://doi.org/10.1016/j.phpro.2014.08.038

[8] D. Y. You, X. D. Gao, S. Katayama, Science and Technology of Welding and Joining 2014, 19, 181. https://doi.org/10.1179/1362171813Y.0000000180

[9] W. Cai, J. Wang, P. Jiang, L. Cao, G. Mi, Q. Zhou, Journal of Manufacturing Systems 2020, 57, 1. https://doi.org/10.1016/j.jmsy.2020.07.021

[10] A. Blug, D. Carl, H. Höfler, F. Abt, A. Heider, R. Weber, L. Nicolosi, R. Tetzlaff, Physics Procedia 2011, 12, 720. https://doi.org/10.1016/j.phpro.2011.03.090

[11] M. Sokolov, P. Franciosa, T. Sun, D. Ceglarek, V. Dimatteo, A. Ascari, A. Fortunato, F. Nagel, Journal of Laser Applications 2021, 33, 012028. https://doi.org/10.2351/7.0000336

[12] P. Franciosa, M. Sokolov, S. Sinha, T. Sun, D. Ceglarek, CIRP Annals 2020, 69, 369. https://doi.org/10.1016/j.cirp.2020.04.110

[13] H. Kim, K. Nam, S. Oh, H. Ki, Journal of Manufacturing Processes 2021, 68, 1018. https://doi.org/10.1016/j.jmapro.2021.06.029

[14] P. Franciosa, T. Sun, D. Ceglarek, S. Gerbino, A. Lanzotti, in Multimodal Sensing: Technologies and Applications (Eds.: S. Negahdaripour, E. Stella, D. Ceglarek, C. Möller), SPIE, Munich, Germany, 2019, p. 9.

[15] M. Seibold, H. Friedmann, K. Schricker, J. P. Bergmann, Procedia CIRP 2020, 94, 769. https://doi.org/10.1016/j.procir.2020.09.137

[16] V. Dimatteo, A. Ascari, A. Fortunato, Journal of Manufacturing Processes 2019, 44, 158. https://doi.org/10.1016/j.jmapro.2019.06.002

[17] Precitec, LWM Expert Training Book.

[18] A. Andreev, LTJ 2009, 6, 20. https://doi.org/10.1002/latj.200990068

[19] I. Eriksson, J. Powell, A. F. H. Kaplan, Meas. Sci. Technol. 2010, 21, 105705. https://doi.org/10.1088/0957-0233/21/10/105705

[20] A. Fortunato, A. Ascari, Lasers Manuf. Mater. Process. 2019, 6, 136. https://doi.org/10.1007/s40516-019-00085-z

[21] R. Olsson, I. Eriksson, J. Powell, A. F. H. Kaplan, Optics and Lasers in Engineering 2011, 49, 1352. https://doi.org/10.1016/j.optlaseng.2011.05.010

Italian Manufacturing Association Conference - XVI AITeM   Materials Research Forum LLC
Materials Research Proceedings 35 (2023) 467-475   https://doi.org/10.21741/9781644902714-55

Development of depositions strategies for edge repair using a WAAM process

Francesco Baffa[1,a] *, Giuseppe Venturini[1,b], Gianni Campatelli[1,c]

[1] Università degli Studi di Firenze – DIEF, Via Santa Marta, 3 - 50139 – Firenze, Italy

[a] francesco.baffa@unifi.it, [b] giuseppe.venturini@unifi.it, [c] gianni.campatelli@unifi.it

Keywords: Wire Arc Additive Manufacturing (WAAM), Circular Economy, Remanufacturing

Abstract. Remanufacturing is an industrial process able to restore a component to at least its original performance, and it is considered one of the key processes to support the transition to circular economy. For restoring a metal component, additive manufacturing processes based on Direct Energy Deposition (DED) techniques are the most widely used, since they can process a damaged part with a complex geometry. Among these, Wire Arc Additive Manufacturing (WAAM) has several advantages including a high deposition rate, lower operative, material, and equipment costs. Nevertheless, it is also characterized by low accuracy and a high risk of defects if the process is not tuned correctly. It is therefore crucial to develop smart deposition strategies to ensure defect-free deposition and high efficiency. This study focuses on the repair of steel edges, and specific toolpaths has been designed and tested for repairing this geometrical feature, both concave and convex, coupled with the selection of proper welding parameters and torch tilting angle.

Introduction

Circular Economy appears to be one of the main approaches adopted to address the issue of sustainability [1]. In addition, Covid-19 pandemic has also highlighted supply chains fragility, especially in Europe, thus the necessity of more resilient supply system [2]. In industry, the issues of sustainability and resilience have found a meeting point in the circular economy, as a possible strategy to simultaneously reduce waste (and therefore reduce the environmental impact) and dependence on external suppliers (increasing resilience of supply chains). Aim of circular economy is to decouple growth and resource consumption and sometimes it is summarized as "doing more with less". In the manufacturing field, it means to keep the product – or at least its material - as much as possible within the boundaries of the manufacturing system Russell and Nasr defined five processes, called Value Retention Processes, which aim to preserve the value that the component has acquired from the extraction phase to the use phase: Direct Reuse, Repair, Refurbishment, Comprehensive refurbishment, and Remanufacturing [3].

Transition to circular economy therefore involves many aspects, from an industrial engineering point of view, these concern mainly supply chain management, product design, and implementation of recovery operations.

This study focuses on remanufacturing, an operation whose objective is to restore (at least) the initial performance of the product. The remanufacturing process involves many stages: disassembly, cleaning, inspection, repair, replacement, reassembly [4]. From a technological point of view, the repair phase presents the greater technical difficulties. This project focuses on the development of repair strategies able to achieve a high-quality repaired part for a metal component through a hybrid process based on Wire Arc Additive Manufacturing (WAAM) and machining. For remanufacturing operations, Additive Manufacturing (AM) technologies are particularly suitable. Among these, the most suitable ones are Direct Energy Deposition (DED) techniques, as they have the possibility of working on the pre-existing part, that it is precluded to powder-bed

Italian Manufacturing Association Conference - XVI AITeM   Materials Research Forum LLC
Materials Research Proceedings 35 (2023) 467-475   https://doi.org/10.21741/9781644902714-55

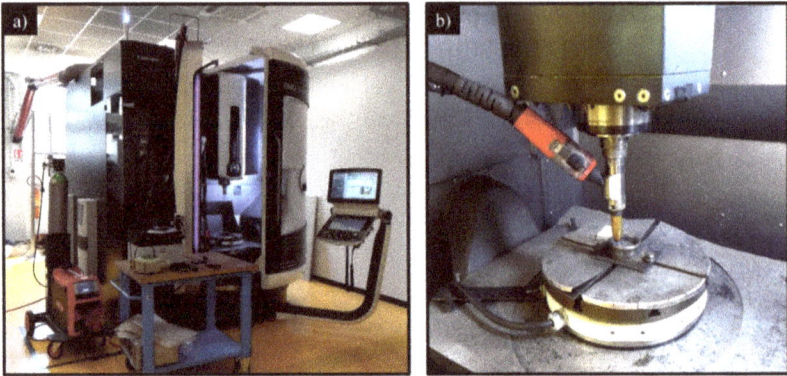

Figure 1 DMU 75-Monoblock machining center (a) and (b) the welding torch mounted on.

solutions. WAAM has the advantages of a high deposition rate, a wide choice of materials, an easy management of the filler material, a lower cost of the material, and a low cost of equipment. Moreover, it presents fewer health issues since it uses metal wire instead than metallic powder [5].

In the last years, also another feature of WAAM has increased its interest for the users: its reduced environmental footprint respect to its main competitor, i.e. AM based on laser as source and metal powder as feedstock [6], since both the production of the filler material and the process itself are less energy-intensive. Nonetheless, WAAM has the disadvantage of a lower accuracy and it is characterize by typical defects: porosity, high residual stress levels, and cracking [7]. Since remanufacturing often deals with non-standard components, it is necessary to develop process strategies able to guarantee the correct deposition of the beads of new material. Industrial sectors where repairing process are significantly adopted are turbomachinery [8], dies and molds [9], and marine transportation [10]. However, the most adopted technologies are based on laser and metal powder.

The remanufacturing process of a metal component by means of AM essentially consists of three stages [9]:

i.   First machining phase: worn and damaged material is removed, and a geometrically known shape of the surface is obtained.

ii. Deposition phase: new material is added through AM technique to restore the worn features.

iii. Second machining phase: exceeding material is removed, and the final geometry is achieved.

Therefore, additive process is just one of the three main phases of the remanufacturing operation. The significant presence of subtractive phases justifies the adoption of a hybrid process, that integrates in the same machine both machining operations and additive manufacturing. This choice has several advantages, among them the streamlining of setup operations.

Aim of the study

This study aims to develop a remanufacturing strategy for metal components through a hybrid process based on WAAM, including toolpath, torch inclination and deposition parameters.

Material and method

The equipment for repairing operations is a hybrid machine, realized by retrofitting a 5-axis machining center DMG Mori DMU 75-Monoblock (Figure 1a). A special tool that supports a GMAW (Gas metal Arc Welding) torch has been designed and implemented in the automatic tool

468

changer of the machine in order to alternate quickly and smoothly milling and WAAM operations (Figure 1b). The welding machine is a Fronius TPS 320i able to perform GMAW in different modes, e.g. Cold Metal Transfer (CMT) and Low Spatter Control (LSC) [11]. Mild steel has been adopted both for substrate (made by S235JR) and wire feedstock (ER70S-6). Shielding gas mixture is made of 82% Ar and 18% CO_2. Substrate and wire composition are listed in Table 1.

Table 1 *Chemical composition of the wire and the substrate used for the tests; reported data are from producers' datasheets.*

	C%	**Mn%**	**Si%**	**P%**	**S%**	**Cu%**	**Fe%**
Wire	0.08	1.5	0.9	≤0.025	≤0.025	-	Bal.
As deposited	0.08	1.1	0.6	≤0.025	≤0.025	-	Bal.
Substrate	0.19	1.5	-	0.045	0.045	0.45	Bal.

Repairing operations usually involve reconstruction of a solid feature (e.g. a gear tooth) and surface cladding (e.g. in case of a worn mold). WAAM cladding processes are much less studied in literature, and it has a potential risk for some quality issues. Two of the most common defects when cladding a surface are the lack of material in correspondence of the undercut zone of two adjacent layers and the material stacking due to a too low stepover between two layers. To solve this issue, bead modeling has been studied extensively in literature for defining the optimal value to obtain a planar surface [12]. However, most of the results are specific for traditional pulsed GMAW, that is characterized by a high heat input, that usually must be avoided since it is responsible for residual stresses on the component. The optimal solutions for new solutions with lower heat input, like CMT [13], must still be exploited. The presented study wants to take advantage also of the effect of inclination angle of the torch, namely tilting angle, on cladding a surface, since it allows quality improvement [14], and to integrate it in the design of a smart strategy for repairing a surface. The results are summarized in Figure 2. The inclination overcomes the difficulties related to undercut areas of adjacent beads, which also allows easier access to the torch, for filling milled pockets. Current activity is focused on repairing of edges, both concave and convex, since it is a very common feature that could be found whenever repairing a metallic part, i.e., a concave edge can be found on the base of a milled pocket, and a convex edge can be found on a metal sheet punch. Moreover, it is a feature that could be commonly found also after the first preparation (machining) phase to remove the worn material. The edges that have been included in this study are both sharp and chamfered.

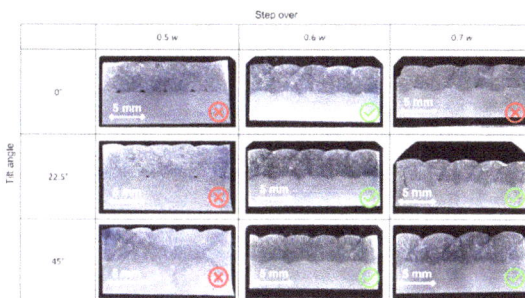

Figure 2 *Overview of material voids assessment. The parameter w is the bead width.*

Italian Manufacturing Association Conference - XVI AITeM   Materials Research Forum LLC
Materials Research Proceedings 35 (2023) 467-475   https://doi.org/10.21741/9781644902714-55

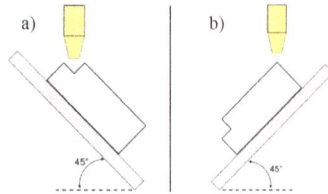

Figure 3 Experimental setup for (a) concave and (b) convex edges.

First, different set of parameters have been tested on a flat surface, and their coverage of the worn surface has been measured to correlate this value when depositing on a plane and on an edge. Then, twelve sets of parameters were selected grouped by similar geometrical dimensions. They are listed in Table 2, reporting the electrical parameters, welding parameters, i.e. Wire Feed Speed (WFS) and Welding Speed (WS), and energy input, i.e. Arc Power (AP) and Heat Input (HI). AP and HI are calculated as follow:

$$AP = \frac{V \cdot I}{1000} \tag{1}$$

$$HI = 60 \cdot \frac{AP}{WS} \tag{2}$$

Numerical coefficients are needed to give AP in kW and HI in kJ/mm.

As shown in Figure 3, deposition on the edges is performed rotating – by mean of the tilting table – the workpiece by 45 degrees and putting the welding torch on the bisector of the edge angle. For analyzing concave edges has been measured height (h), bead sides (L_1 and L_2), the bead area (A_b), the area added to the preexisting geometry (A^+), and dilution (D). Similarly, for convex edges has been measured A_b, A^+, and D. A geometric representation of the measures is shown in Figure 4, while the dilution has been calculated according to Eq. 3:

$$D = \frac{A_b}{B} \cdot 100 \tag{3}$$

Where B is the part of the bead penetrated in the substrate (Figure 4b and 4c).

For observing the tests after the deposition, standard metallurgical techniques have been followed. Test specimens were prepared by cutting, mounting in epoxy resin, grinding, and polishing with abrasive paper up to 1200 grit. Nital (1% nitric acid and ethanol mixture) was used as etching solution.

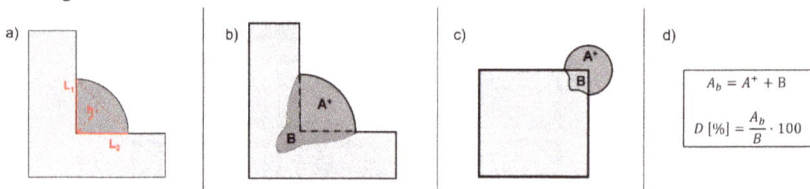

Figure 3 *Measures geometrical representation for (a, b) concave and (c) convex edges and (d) used relations.*

Italian Manufacturing Association Conference - XVI AITeM

Materials Research Proceedings 35 (2023) 467-475

Materials Research Forum LLC

https://doi.org/10.21741/9781644902714-55

Table 2 *Parameters adopted for tests grouped by the similar bead width.*

Test	Reference width [mm]	Current [A]	Voltage [V]	WFS [m/min]	WS [mm/min]	Heat Input [kJ/mm]	Arc Power [kW]
1	6.5	130	21.1	12	600	0.274	2.74
2	6.7	91	18.9	8	400	0.258	1.72
3	6.7	111	20.2	10	500	0.269	2.24
4	6.8	179	27.4	20	1000	0.294	4.91
5	7.8	114	20.0	10	300	0.456	2.28
6	7.9	119	20.1	10	400	0.359	2.39
7	8.0	150	22.1	14	500	0.398	3.31
8	8.1	92	20.8	8	200	0.574	1.91
9	8.2	138	21.2	12	400	0.439	2.93
10	9.1	152	21.7	14	400	0.495	3.30
11	9.2	117	19.7	10	200	0.691	2.30
12	9.3	124	22.5	12	300	0.558	2.79

Table 3 *Data from deposition on concave edges*

Test	L_1 [mm]	L_2 [mm]	h [mm]	A_b [mm^2]	A^+ [mm^2]	D [%]
1	4.7	4.7	3.5	15.1	11.1	26.1
2	3.7	5.0	2.7	11.5	9.0	21.5
3	4.4	4.1	2.9	10.8	9.4	13.7
4	5.0	4.1	3.0	15.9	10.1	36.4
5	6.1	5.4	4.0	19.8	17.5	11.5
6	4.9	4.5	3.3	17.4	16.3	6.5
7	5.1	5.3	3.6	20.0	17.8	10.9
8	6.1	5.9	4.1	21.0	19.8	5.6
9	5.2	5.3	3.6	17.7	14.2	20.1
10	5.3	5.9	3.9	21.4	17.0	20.6
11	8.2	6.3	4.9	27.7	26.2	5.5
12	5.3	7.8	4.2	26.8	20.6	23.1

Preliminary results

Results of deposition are listed in Table 3 for concave edges and in Table 4 for convex edges. It could be notice that there is not a clear correlation between the value of the reference width of the beads deposited on a plane and the mean value of L_1-L_2, this is probably due to the different heat input that is responsible for a variation in the melting of the substrate and the aspect ratio (width-height) of the bead and different thermal conductivity for convex and concave geometries, as shown in Figure 5. Specimens in Figure 6 show this issue. Nonetheless, this correlation must be further investigated. It could be noted also a relevant variability in the value of L_1 and L_2 both for convex and concave edges. The variability could be assessed by calculating the residuals respect to the mean value, and this assume a value of 0.56 for concave and 0.97 for convex, that is nearly twice the value for concave.

Indeed, the deposition on a sharp convex edge could be source of uncertainty since the bead easily tends to a side or to the other. This would affect the subsequent cladding of the vertical and the horizontal side since the resulting geometry is undefined and this will create a lack or stacking of material with an automatic repairing cycle. This issue could be hardly solved since the instability of the arc on a sharp edge it is difficult to control and the measure of the deposited bead and the following adjustment of the toolpath for the deposition of the next bead is time consuming and avoid the implementation of an automatic repairing cycle. The solution that we are pursuing is to create a geometry before deposition that will reduce the variability of the bead position and

Italian Manufacturing Association Conference - XVI AITeM
Materials Research Proceedings 35 (2023) 467-475

Materials Research Forum LLC
https://doi.org/10.21741/9781644902714-55

geometry. This could be solved by introducing a 45° chamfer of different dimensions on the sharp edge where it is possible to deposit the corner bead on a flat surface. The initial tests had a single parameter set, and the width of chamfers is related to the bead width, as reported in Table 5.

Deposition on chamfers requires one or two beads accordingly with their dimensions. A_b, A^+, and D have been measured for these tests as well.

When two beads are needed for chamfer cladding the stepover was fixed at $0.6w$. Therefore, set the chamfer width at $1.6w$ means to set chamfer width equal to the overlapped beads width. This seems to partially solve the issue, in fact, the deposition with two beads on a chamfer narrower than the overlapped beads, shows quite promising results, with consistent deposition geometry. On a chamfer characterized by a width of $1.2w$, deposition is nearly symmetric, and beads cover the whole chamfer (Figure 7).

Table 4 *Data from deposition on convex edges.*

Test	L_1 [mm]	L_2 [mm]	A_b [mm^2]	A^+ [mm^2]	D [%]
1	4.3	4.5	15.6	9.0	42.3
2	2.4	4.8	14.3	9.3	35.0
3	4.8	4.1	16.3	8.7	46.6
4	4.2	4.0	17.6	8.1	54.3
5	4.3	6.0	25.1	14.4	42.7
6	3.2	5.4	20.1	11.2	44.4
7	4.3	5.9	25.0	13.1	47.7
8	5.1	5.3	26.9	17.9	33.4
9	2.4	6.1	19.9	10.9	45.2
10	7.0	4.2	28.7	16.9	40.9
11	5.6	6.7	36.5	23.7	35.1
12	6.8	5.1	32.2	19.6	39.1

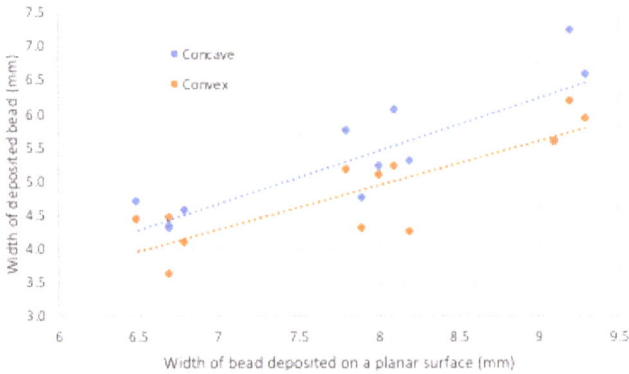

Figure 4 *Correlation between planar and edge bead width.*

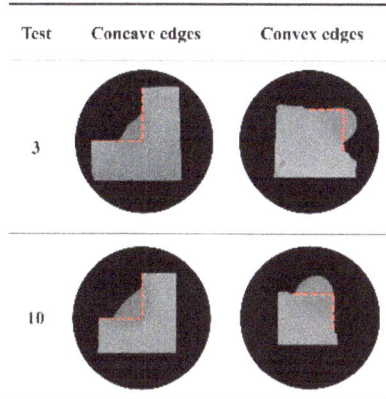

Figure 5 *Uncertainty is more significant in case of convex edges.*

Table 5 *Chamfers cladding tests, d is the chamfer width, w is the bead width.*

Test	d	d [mm]	N of beads
1	$0.8 \cdot w$	4	1
2	w	5	1
3	$1.2 \cdot w$	6	2
4	$1.6 \cdot w$	8	2

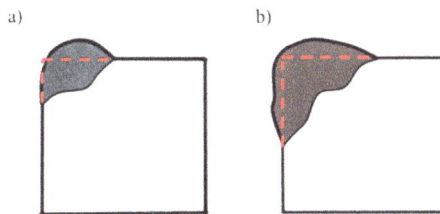

Figure 6 *Chamfers of test 1 (a) and 3 (b).*

Conclusion and future perspectives

The restoring of convex edge is characterized by a high uncertainty of the deposition geometry that is often asymmetric, thus affecting the reliability of the process. A solution is the preparation of the surface by machining a chamfer that enable a more stable deposition process. Starting from these preliminary results, some future actions could be planned, like fillets repair, where also the effect of the corner radius must be considered to predict the coverage of the surface and plan the position of the next beads accordingly. A model for fillets could be extended to repair free-form surfaces, that can be found, for instance, in molds and dies, which are a typical component subject to remanufacturing. Once the approaches to repair the most common geometrical features will be developed, it will be required to solve the issue of the connection of different toolpaths strategies to achieve a single and consistent remanufacturing cycle. In fact, a complex component could

Italian Manufacturing Association Conference - XVI AITeM   Materials Research Forum LLC
Materials Research Proceedings 35 (2023) 467-475   https://doi.org/10.21741/9781644902714-55

require different strategies for each of its features and these must be consistently and smoothly connected to avoid lack or stacking of material. In the meantime, also material tests need to be performed to ensure mechanical properties of the remanufactured parts. These tests could involve hardness test, grain analysis as well as tests for specific application, e.g. corrosion resistance assessment and thermo-mechanical properties. This final result will support a wider adoption of remanufacturing both for standard and non-standard components, ensuring at least the same performance of the original component and lower environmental impact and supporting the introduction of a greener manufacturing cycle.

References

[1] European Commission, "A new Circular Economy Action Plan," 2020.

[2] T. Ibn-Mohammed, K. B. Mustapha, J. Godsell, Z. Adamu, K. A. Babatunde, D. D. Akintade, A. Acquaye, H. Fujii, M. M. Ndiaye, F. A. Yamoah, and S. C. L. Koh, "A critical review of the impacts of COVID-19 on the global economy and ecosystems and opportunities for circular economy strategies," Resour. Conserv. Recycl., vol.164, 2021. https://doi.org/10.1016/j.resconrec.2020.105169

[3] J. Russell and N. Nasr, "Value-Retention Processes within the Circular Economy," in Remanufacturing in the Circular Economy, 1st ed., N. Nasr, ed. Wiley, 2019, pp.1-29. https://doi.org/10.1002/9781119664383.ch1

[4] C.-M. Lee, W.-S. Woo, and Y.-H. Roh, "Remanufacturing: Trends and issues," Int. J. Precis. Eng. Manuf.-Green Technol., vol.4, no. 1, pp.113-125, Jan. 2017. https://doi.org/10.1007/s40684-017-0015-0

[5] R. Chen, H. Yin, I. S. Cole, S. Shen, X. Zhou, Y. Wang, and S. Tang, "Exposure, assessment and health hazards of particulate matter in metal additive manufacturing: A review," Chemosphere, vol.259, p.127452, Nov. 2020. https://doi.org/10.1016/j.chemosphere.2020.127452

[6] A. Wippermann, T. G. Gutowski, B. Denkena, M.-A. Dittrich, and Y. Wessarges, "Electrical energy and material efficiency analysis of machining, additive and hybrid manufacturing," J. Clean. Prod., vol.251, p.119731, Apr. 2020. https://doi.org/10.1016/j.jclepro.2019.119731

[7] B. Wu, Z. Pan, D. Ding, D. Cuiuri, H. Li, J. Xu, and J. Norrish, "A review of the wire arc additive manufacturing of metals: properties, defects and quality improvement," J. Manuf. Process., vol.35, pp.127-139, Oct. 2018. https://doi.org/10.1016/j.jmapro.2018.08.001

[8] J. M. Wilson, C. Piya, Y. C. Shin, F. Zhao, and K. Ramani, "Remanufacturing of turbine blades by laser direct deposition with its energy and environmental impact analysis," J. Clean. Prod., vol.80, pp.170-178, Oct. 2014. https://doi.org/10.1016/j.jclepro.2014.05.084

[9] C. Chen, Y. Wang, H. Ou, Y. He, and X. Tang, "A review on remanufacture of dies and moulds," J. Clean. Prod., vol.64, pp.13-23, Feb. 2014. https://doi.org/10.1016/j.jclepro.2013.09.014

[10] M. Vishnukumar, R. Pramod, and A. Rajesh Kannan, "Wire arc additive manufacturing for repairing aluminium structures in marine applications," Mater. Lett., vol.299, p.130112, Sep. 2021. https://doi.org/10.1016/j.matlet.2021.130112

[11] G. Campatelli, G. Venturini, N. Grossi, F. Baffa, A. Scippa, and K. Yamazaki, "Design and testing of a waam retrofit kit for repairing operations on a milling machine," Machines, vol.9, no. 12, 2021. https://doi.org/10.3390/machines9120322

[12] D. Ding, Z. Pan, D. Cuiuri, and H. Li, "A multi-bead overlapping model for robotic wire and arc additive manufacturing (WAAM)," Robot. Comput.-Integr. Manuf., vol.31, pp.101-110, Feb. 2015. https://doi.org/10.1016/j.rcim.2014.08.008

[13] C. G. Pickin, S. W. Williams, and M. Lunt, "Characterisation of the cold metal transfer (CMT) process and its application for low dilution cladding," J. Mater. Process. Technol., vol.211, no. 3, pp.496-502, Mar. 2011. https://doi.org/10.1016/j.jmatprotec.2010.11.005

[14] F. Baffa, G. Venturini, G. Campatelli, and E. Galvanetto, "Effect of stepover and torch tilting angle on a repair process using WAAM," Adv. Manuf., vol.10, no. 4, pp.541-555, Dec. 2022. https://doi.org/10.1007/s40436-022-00393-2

Italian Manufacturing Association Conference - XVI AITeM
Materials Research Proceedings 35 (2023) 476-485

Materials Research Forum LLC
https://doi.org/10.21741/9781644902714-56

Measurement and analysis of tooth movements during orthodontic treatment with clear aligners

Marino Calefati[1,a], Francesco De Palo[1,b], Maria Derosa[1,c],
Ugo Marco Ferrulli[1,d], Andrea Verani[1,e], Giorgio Giustizieri[1,f*],
Eliana Di Gioia[1,g], and Luigi Maria Galantucci[1,h]

[1]Dept. of Mechanics, Mathematics and Management, Politecnico di Bari, Via Orabona 4 - 70125 Bari, Italy

[2]Studio Odontoiatrico Associato Di Gioia, via Dante. 97, -70122 Bari, Italy

[a]m.calefati@studenti.poliba.it,[b]f.depalo3@studenti.poliba.it, [c]m.derosa@studenti.poliba.it, [d]u.ferrulli@studenti.poliba.it, [e]a.verani@studenti.poliba.it, [f]giorgio.giustizieri@poliba.it, [g]eliana@studiodigioia.com, [h]luigimaria.galantucci@poliba.it

Keywords: Metrology & Tolerancing, Health Care, Medical Devices, Orthodontics

Abstract. This article aims to study, measure and verify through Reverse Engineering techniques and 3D analysis, the results obtained by an orthodontic treatment for the alignment of the teeth, using clear aligners (CA). A case study of a patient who followed orthodontic treatment using a set of 19 CA was analyzed: dental impressions by intraoral scanner were made before and after treatment; the study also analyzed the project planned by Invisalign ClinCheck® software, comparing it to the results clinically obtained, using specific software and hardware. The analysis of the obtained results shows that the measurements carried out are affected by a residual error respect to the final situation designed for the patient, caused by several factors here investigated; moreover, it is very determinant to do a correct diagnosis, to formulate an adequate treatment plan and to correct wear the chosen orthodontic appliances.

Introduction

Clear Aligners (CA) allow to perform an almost invisible orthodontic treatment that leads the teeth into the desired position through a set of clear thermo-molded plastic dental masks to be worn all day in sequence during the treatment, for a total duration that could go from few months to one or two years. The final alignment is digitally designed in the preliminary phase of the treatment: at the end, teeth will reach the pre-established positions, giving to the patient the desired aesthetic appearance and function [1] [2].

Generally, the impressions of the dental arches are acquired by an intraoral 3D scanner. The digital project defines the final shape of the arches and divides the total movements into treatment steps, distributing them over a variable number of aligners. The application of each individual aligner induces specific forces and torques to each tooth, which should cause the required movements to reach the right final position of the teeth; but this results are not always predictable in size and accuracy [2] [3]. CA are an alternative to classic metal brackets, but it is always necessary to rely on the judgment of a specialist to undertake the most suitable therapeutic path for each medical condition and for the type of coronal and root movements required [4]. Some studies have used the forces needed to obtain teeth movements with CA using Finite Elements Method for the prediction [5] [6] [7].

This work aims to measure and verify through Reverse Engineering techniques and 3D analysis, the results obtained in a case study using CA.

Italian Manufacturing Association Conference - XVI AITeM   Materials Research Forum LLC
Materials Research Proceedings 35 (2023) 476-485   https://doi.org/10.21741/9781644902714-56

Materials and Methods

A case study of a patient treated with CA was analyzed. A completely digital data flow was adopted: from the dental impressions with intraoral scanner, to the planning of the treatment and definition of the pairs and sequences of aligners, to the simulation of the treatment result with Invisalign™ ClinCheck® proprietary software [8], up to the scanning of each pair of aligners (19) and comparison with the intraoral scans taken before and after treatment. The aim was to verify the real tooth movements of each affected tooth. The study also analyzed the project provided by Invisalign, complete with *.stl format files and related documentation such as the values of the single displacements from the initial situation to the final designed one.

Special attention was given to the first and last couples of aligners acquired by the initial and final intraoral scans. provided by a dentist specialist in orthodontics.

The software and hardware used in are:
- Revo Scan for the acquisition of the aligners using the Revopoint POP 2 scanner [9].
- Meshmixer [10]for cleaning the aligners and for joining the roots to the teeth.
- GOM Inspect [11] for all alignment operations and three-dimensional analysis [12].
- Matlab [13] for data analysis and drawing up graphs.

Results and discussion

Once the files in *.stl format of the intraoral scans at the end of the treatment and of the final situation designed on Invisalign have been obtained, it is possible to check at first glance whether the results actually obtained from the treatment are in line with the design initially realised.

Using the GOM Inspect software, the two models were imported and once the alignment was performed, it was possible to carry out the dimensional analysis. The software allows to measure deviations in terms of distances between the two models using various color maps such as those shown in Figure 1.

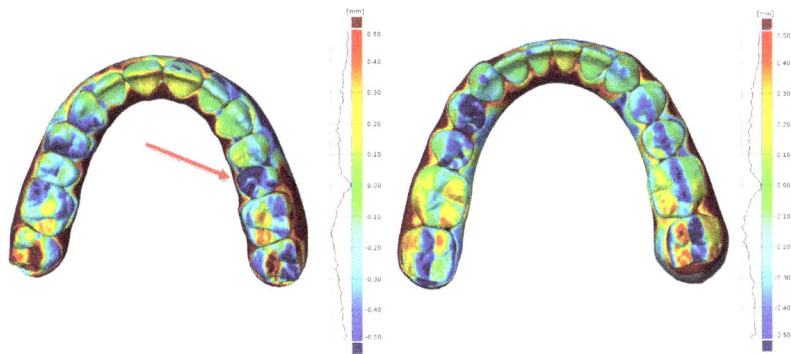

Figure 1 - Distance comparison of final intraoral scans and Invisalign final design, upper arch (left) and lower arch (right)- scale [0.5; -0.5] mm.

The intraoral scanner used was a TRIOS from 3shape, which has a long-span measurement accuracy of approximately 80 microns [14]. As can be seen from the figures, there is no perfect alignment between the two models. In fact, as the sensitivity of the scale increases to [0.5; -0.5], the alignment defects become more evident.

The teeth subject to the greatest error are the second molars and the four premolars of each arch, while the frontal and lateral incisors are the teeth whose final result is closest to the situation designed by ClinCheck®. The dental element that deviated the most from the project is 2.5

Italian Manufacturing Association Conference - XVI AITeM   Materials Research Forum LLC
Materials Research Proceedings 35 (2023) 476-485   https://doi.org/10.21741/9781644902714-56

(marked in the figure), whose correction entity is significantly lower than expected, so much so that it was necessary to proceed with a brief phase of alignment of this element.

The attachment made by the aligner supplier was not able to produce all the expected tooth movements for the most critical tooth, maybe due to a design error by the same company. In fact, post-treatment the tooth needed an additional set of 18 aligners to achieve the set objectives, for which the dentist opted for a traditional overcorrection treatment to reduce therapy times and costs.

However, it is possible to state that the error is less than a millimeter, as can be obtained from the previous maps.

Acquisition of Aligners

Parallel to the result obtained with the treatment, to better understand the movements that the teeth make inside the aligners, it is necessary to align the individual teeth inside the aligners. This operation is made possible by using the GOM software [11], which also allows you to compare the result of the alignment with the initial ClinCheck® project and with the final intraoral scans.

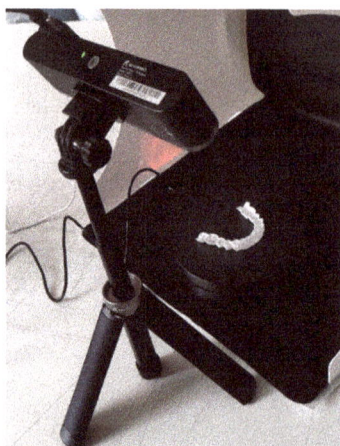

For this purpose, the 19 aligners were scanned using the structured light "Revopoint POP 2" scanner [9] and related software which provided the various point clouds and then the specific meshes. The transparency of the aligners made it necessary to coat them with a generic opacifying spray so that the scanner could detect the individual points (Figure 2).

According to data provided by the manufacturer of the scanner, it has a dimensional accuracy in the single frame acquired of 50 microns.

Once the aligners were acquired, a revision and cleaning operation was carried out on the Meshmixer software, by Autodesk [10]. Finally, as the last step of acquiring and cleaning the meshes, the latter were oriented automatically when some triangles had their inverted normal.

Figure 2 - Setup for acquisition of aligner Point cloud captured by Revopoint POP 2 scanner

Teeth Alignment

Analysis and Preliminary Operations

To align the teeth to the final situation, it is necessary to insert the model of each tooth inside the last tray of the treatment and using local best-fit, translate and rotate the affected tooth to have the most suitable alignment to the tray.

Before proceeding with the alignment, it is necessary to carry out some preliminary steps to obtain an appropriate starting point that allows comparisons to be made with the intraoral scans and with the situations designed by ClinCheck®:

- each dental element was segmented and renamed according to medical convention.
- to obtain a greater understanding of the overall movement of the tooth and its root, the respective 3D models of the tooth roots were added to the 3D models of the teeth crown provided by ClinCheck® [8] and this operation was performed using 3D models of roots not belonging to the patient, joined to the teeth on the Meshmixer software [10].

Italian Manufacturing Association Conference - XVI AITeM
Materials Research Proceedings 35 (2023) 476-485

Materials Research Forum LLC
https://doi.org/10.21741/9781644902714-56

Alignment of the final couple of Aligners

Before inserting the individual teeth, aligning them inside the final tray, the latter was aligned to the respective arch (lower or upper) of the pre-treatment ClinCheck® [8] project. This activity is necessary because by loading the arch and the template in the same GOM Inspect project [11], the two models are arranged in different points in space.

Therefore, it was necessary to identify which teeth by design would undergo the least displacements, since they already had good general positioning over the entire dental arch, and thus proceed to align the final tray according to the position of these teeth considered "fixed." The teeth in question are teeth 1.6, 2.6, 3.6 and 4.6, respectively, as shown in Figure 3 according to traditional nomenclature.

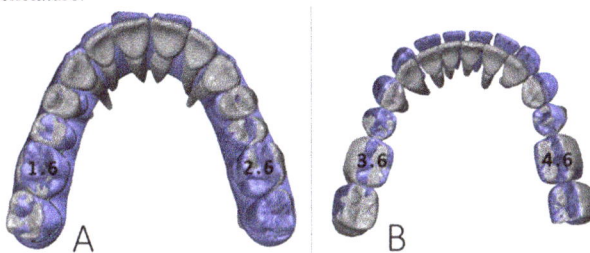

Figure 3 - A) Comparison of final aligned upper arch and initial upper ClinCheck® design; B) Comparison of final aligned lower arch and initial lower ClinCheck® design.

Using these alignments, it is already possible to deduce what the movement of the teeth will be, in particular, that of the front and lateral incisors which will have to move in the lingual direction.

The main advantage of this alignment consists in the possibility of proceeding with the direct comparison between the dental arches, without carrying out further alignment operations, once the individual teeth have been aligned as described below.

Alignment of Individual Teeth

The operator who carries out the analysis is responsible for choosing the alignment that best translates the set objective. It is useful to report the criteria pursued for the realization of the alignment:

- The colour map for alignment evaluation has the same setup for all teeth, precisely it has a tolerance range between [-0.2; 0.2] mm able to achieve maximum precision.
- The tooth must lie perfectly in its special space inside the mask and its tip must coincide with the edge of this. The local best-fit output does not always result in proper tooth placement and to overcome this problem, it is sufficient to carry out a "best-fit" on the tip of the crown and finally on the frontal part of the tooth;
- The best result was evaluated on the buccal part of the tooth for the incisors and canines, while on the entire dental crown for the molars.

Attention is drawn to the fact that the mask model does not match exactly the dental model. In fact, due to the elastic properties of the aligner material, during use, the elastic deformation of each individual aligner is able to apply the forces necessary for the required movements to the teeth. The material patented by the company is a particular polycarbonate, created specifically for medical use, multi-layer, highly elastic and transparent [4].

Once the desired alignment has been achieved, the tooth is exported in ".stl" format.

The output of the alignment between the teeth and the aligners (upper and lower aligner no. 19) is shown in Figure 4 and 5.

Italian Manufacturing Association Conference - XVI AITeM   Materials Research Forum LLC
Materials Research Proceedings 35 (2023) 476-485   https://doi.org/10.21741/9781644902714-56

The results obtained are considered by the team to be the best compromise between the above criteria. On the teeth where the attachments have been removed and the holes automatically closed, the colour map shows blue areas (-0.2 deviation between elements), these errors are therefore accepted.

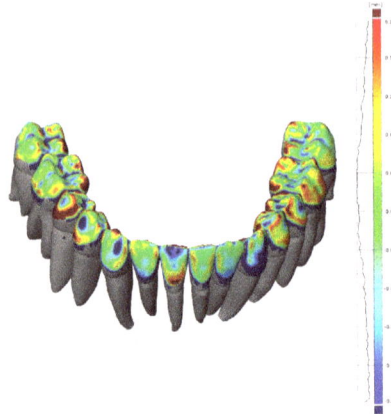

Figure 4 Lower arch: comparison between aligned teeth and aligner 19, scale [0.2; -0.2] mm

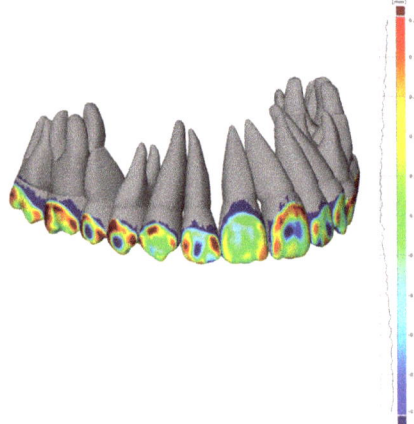

Figure 5 Upper arch: comparison between aligned teeth and aligner 19, scale [0.2; -0.2] mm

Assessment of Movements

Tooth Movements (Figure 6)

From a clinical point of view, tooth movements can be divided into [15]:

- Horizontal, when the tooth moves in a mesio-distal or buccal-lingual direction.
- Vertical, when the tooth moves in an occlusal or gingival direction.
- Circular, when the tooth moves around the long axis of the tooth.
- Torque, when the tooth changes its position within the bone structure (move the tooth buccally-lingually around its centre so that the crown and the root move in opposite ways).

Materials Research Forum LLC

https://doi.org/10.21741/9781644902714-56

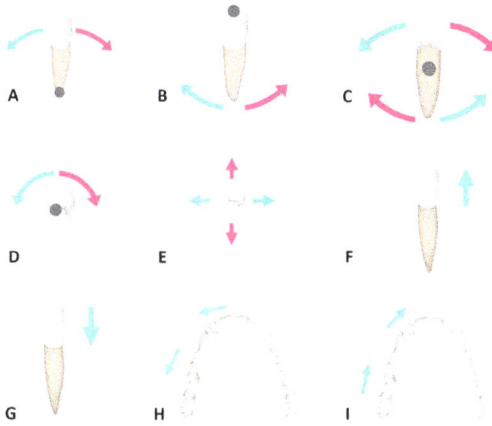

Figure 6 Main tooth movements: A) Crown tipping B) Root tipping C) Torquing D) Rotation E) Translation F) Extrusion G) Intrusion H) Distalization I) Mesialization [15] [16]

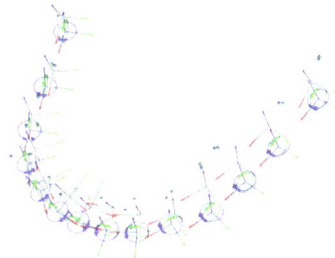

Figure 6 Reference systems on mesh and CAD models on the 'lower' arch for the evaluation of rotations

Having obtained the displacements between the initial situation (after the first phase of the treatment) and the final intraoral scans, using the Matlab software it was possible to calculate the displacement vector by performing the "norm two" of the three components of the displacement. The vector was calculated for the movements designed by ClinCheck® and those detected by measurements using GOM. The difference between the two displacements can then be calculated (Figure 7,8,9 and 10).

Italian Manufacturing Association Conference - XVI AITeM
Materials Research Forum LLC
Materials Research Proceedings 35 (2023) 476-485
https://doi.org/10.21741/9781644902714-56

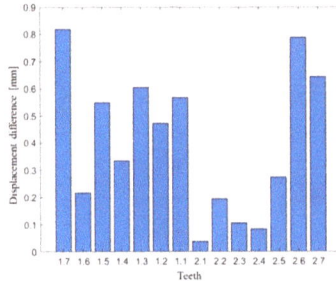

Figure 7 - Upper arch: difference in displacements between the ClinCheck® design and teeth movements detected by GOM Inspect

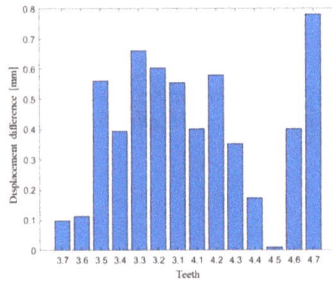

Figure 8 - Lower arch: difference in displacements between the ClinCheck® design and teeth movements detected by GOM Inspect

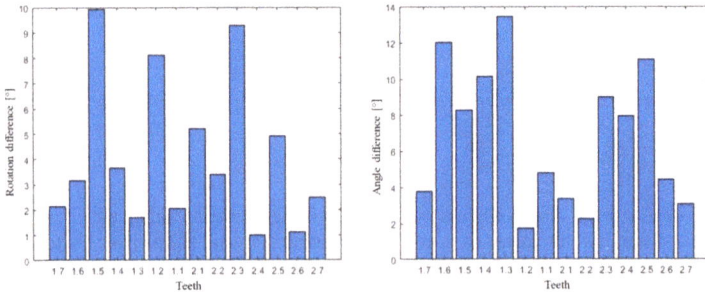

Figure 9 - Absolute error of rotation and angulation of the teeth of the upper arch

Error Considerations

The GOM displacement evaluation procedure is not immune to errors for the following reasons:

The reference systems considered may not coincide with the reference systems considered by the Invisalign ClinCheck® software for the evaluation of displacements and rotations, which may also be non-Cartesian (see [16]).

Italian Manufacturing Association Conference - XVI AITeM   Materials Research Forum LLC
Materials Research Proceedings 35 (2023) 476-485   https://doi.org/10.21741/9781644902714-56

The alignment procedure was carried out on elements that were not fixed in space: the so-called "fixed teeth" were actually subject to movement, especially rotation. All these factors could affect the measurement as the error is of the same order of magnitude as the variable to be measured.

By calculating the average value of the difference between the final displacements designed by ClinCheck® and those obtained from the GOM analysis, we obtain a value equal to $\Delta_{s\_sup}=$ 0.4061 mm for the upper arch and $\Delta_{s\_inf}=0.4058$ mm for the lower arch.

The values obtained are compatible with the colour maps. The average error for the rotation and angulation of the lower arch is equal to $\Delta_{rot\_inf}=4.4°$ and $\Delta_{ang\_inf}=4.9°$. For the upper arch instead $\Delta_{rot\_sup}=4.2°$ and $\Delta_{ang\_sup}=6.8°$.

Conclusions

The analysis of the results obtained shows that the measurements carried out in the project work are affected by a residual error with respect to the final situation designed for the patient. The error is caused by several factors:

- Inaccuracies maybe due to the scanner used for the acquisition of the aligners (Revopoint POP 2) not specific to the application.
- The operations carried out by the study group do not have an automatic and unambiguous procedure by the GOM Inspect software. In fact, to obtain greater precision in the work performed, it would have been advisable to use dedicated software capable of reducing the discretionarily of some operations such as the construction of the plans, the definition of the reference systems, the choice of significant points and the alignment operations. For example, the plane construction operation is based on the manual choice of points by the operator, which is why the accuracy and precision of the result do not depend on objective factors.
- The difficulty of the measurement operations is due to a degree of precision required for the case study of the order of tenths of a millimeter and to the intrinsic complexity of the tooth surfaces. In fact, the absence of standardization (dental elements that are different from each other and without significant geometric points) results in a very complex engineering situation.

A possible solution to minimize all the error components, in particular that of alignment, is to keep part of the palate present in the intraoral scan from the ClinCheck® project, in order to carry out the alignment on a fixed surface.

At the end of the following analysis work it was possible to notice how the end point of the orthodontic treatment does not coincide perfectly with the ClinCheck® project, as reported in [17], with an estimated deviation of less than one millimetre.

By minimizing the error and in the presence of data on the progressive movements of the teeth, a more in-depth analysis could be carried out by repeating the operating method illustrated for all subsequent pairs of masks.

But considering that each aligner can produce a maximum of 0.25 mm of dental movement, 2 degrees of rotation and 1 degree of torque [1], not respecting the instructions for use of a single aligner also means compromising the movements of subsequent aligners. To all this must be added the unpredictability of the movements of the teeth, which, being non-inert organs and considering the reactions of their support tissues, oppose the movements.

Therefore, explaining the reason for this difference, albeit small, it is very complex as the success of an orthodontic treatment depends on the combination of two factors:

- a correct diagnosis and consequent formulation of an adequate treatment plan.
- a correct clinical application of the chosen orthodontic technique.

The professionalism of the specialist must therefore also be accompanied by the determination of the patient, but it is essential to motivate the latter to wear the aligners correctly for 20-22 hours a day for the entire treatment period.

If from an engineering point of view the treatment has errors that are not negligible, from a clinical point of view it can be said that the patient, thanks to the Orthodontist practice, obtained a result very close to the desired one, i.e. a good balance between the dental arches and a progressive improvement of the aesthetic appearance.

References

[1] R. Cozza, P., Pavoni, C., & Lione, Approccio sistematico alla terapia ortodontica con allineatori. Milano: Edra, 2020.

[2] V. D'Antò, R. Valletta, R. Ferretti, R. Bucci, R. Kirlis, and R. Rongo, "Predictability of Maxillary Molar Distalization and Derotation with Clear Aligners: A Prospective Study," *International journal of environmental research and public health*, vol. 20, no. 4, 2023. https://doi.org/10.3390/ijerph20042941

[3] J. M. Smith, T. Weir, A. Kaang, and M. Farella, "Predictability of lower incisor tip using clear aligner therapy," *Progress in Orthodontics*, vol. 23, no. 1, pp. 1–12, 2022. https://doi.org/10.1186/s40510-022-00433-4

[4] G. Rossini, S. Parrini, T. Castroflorio, A. Deregibus, and C. L. Debernardi, "Efficacy of clear aligners in controlling orthodontic tooth movement: A systematic review," *Angle Orthodontist*, vol. 85, no. 5, pp. 881–889, 2015. https://doi.org/10.2319/061614-436.1

[5] R. Savignano, R. Valentino, A. V. Razionale, A. Michelotti, S. Barone, and V. D'Antò, "Biomechanical Effects of Different Auxiliary-Aligner Designs for the Extrusion of an Upper Central Incisor: A Finite Element Analysis," *Journal of Healthcare Engineering*, vol. 2019, 2019. https://doi.org/10.1155/2019/9687127

[6] J. H. Seo *et al.*, "Biomechanical Efficacy and Effectiveness of Orthodontic Treatment with Transparent Aligners in Mild Crowding Dentition—A Finite Element Analysis," *Materials*, vol. 15, no. 9, 2022. https://doi.org/10.3390/ma15093118

[7] Y. Li *et al.*, "Stress and movement trend of lower incisors with different IMPA intruded by clear aligner: a three-dimensional finite element analysis," *Progress in orthodontics*, vol. 24, no. 1, p. 5, Dec. 2023. https://doi.org/10.1186/S40510-023-00454-7/FIGURES/9

[8] "Your digital Invisalign® experience." https://www.invisalign.ca/invisalign-digital-experience (accessed Feb. 14, 2023).

[9] "POP 2 3D Scanner（Infrared Light｜Precision 0.05mm." https://shop.revopoint3d.com/products/pop2-3d-scanner?variant=42265546653931#shopify-section-template--15966931878088__2f3f314d-a371-4465-a2d9-d2107bff12c2 (accessed Feb. 14, 2023).

[10] "Autodesk Meshmixer free software for making awesome stuff." https://meshmixer.com/ (accessed Feb. 14, 2023).

[11] "GOM Inspect Pro Industry standard for 3D inspections and evaluations." https://www.gom.com/en/products/zeiss-quality-suite/gom-inspect-pro (accessed Feb. 14, 2023).

[12] L. Lo russo, C. Ercoli, L. Guida, M. Merli, and L. Laino, "Surgical guides for dental implants: measurement of the accuracy using a freeware metrology software program," *Journal of Prosthodontic Research*, 2022, doi: 10.2186/jpr.jpr_d_22_00069.

[13] "MATLAB - Matematica. Grafica. Programmazione." https://it.mathworks.com/products/matlab.html (accessed Feb. 14, 2023).

[14] F. Kernen *et al.*, "Accuracy of intraoral scans: An in vivo study of different scanning devices," *J Prosthet Dent*, vol. 128, no. 6, pp. 1303–1309, 2022. https://doi.org/10.1016/j.prosdent.2021.03.007

[15] Jamie L. Somers, "Spostamenti dei denti." https://support.clearcorrect.com/hc/it/articles/4402323236247-Spostamenti-dei-denti (accessed Feb. 14, 2023).

[16] "Crown Coordinate System." https://www.onyxwiki.net/doku.php?id=en:crowncoordinatesystem (accessed Feb. 14, 2023).

[17] R. Tien *et al.*, "The predictability of expansion with Invisalign: A retrospective cohort study," *American Journal of Orthodontics and Dentofacial Orthopedics*, vol. 163, no. 1, pp. 47–53, 2023. https://doi.org/10.1016/j.ajodo.2021.07.032

Italian Manufacturing Association Conference - XVI AITeM
Materials Research Proceedings 35 (2023) 485-494

Materials Research Forum LLC
https://doi.org/10.21741/9781644902714-57

A digital solution for slender workpiece turning: the DRITTO project

Niccolò Grossi[1,a*], Antonio Scippa[1,b], Lorenzo Sallese[2,c],
Lorenzo Morelli[1,d], Gianni Campatelli[1,e]

[1]Department of Industrial Engineering, University of Florence, Via di Santa Marta 3, 50139, Firenze, Italy

[2]Meccanica Ceccarelli e Rossi, Loc. Vallone Zona P.I.P. 35/C, 52044, Camucia, Cortona (AR), Italy

[a]niccolo.grossi@unifi.it, [b]antonio.scippa@unifi.it, [c]lorenzo.sallese@meccanicaceccarellierossi.it, [c]lorenzo.morelli@unifi.it, [c]gianni.campatelli@unifi.it

Keywords: Turning, Tool Path, Stiffness

Abstract. Turning slender components is a critical task since workpiece flexibility entails relevant deformations during the process, leading to potential loss of accuracy, lower machining efficiency and higher manufacturing costs. The DRITTO project aims at developing an easy-to-use digital solution to support manufacturing of flexible axisymmetric components. The proposed support system, starting from the not-optimized toolpath, stock geometry and tool parameters, it will compute the optimized toolpath by integrating three different modules: a) workpiece FE modelling, b) turning process modelling, c) toolpath optimization. The project is ongoing, but, at the current stage, preliminary validation of the proposed solution has been carried out. DRITTO is funded as an experiment of DIH-World Horizon2020 project, and the consortium is composed by the machining services SME Meccanica Ceccarelli & Rossi and the University of Florence as part of the Digital Innovation Hub ARTES4.0.

Introduction

Manufacturing slender axisymmetric components is still a challenging task even with modern machining processes [1]. The turning process represents the main technology for the realization of such components because of its versatility and the high-quality standards achievable (i.e., surface roughness and geometrical/dimensional accuracy). However, demanding requirements in terms of quality usually conflict with the achievable productivity rates. Therefore, defining a proper machining cycle represents a crucial task in attaining the suitable trade-off between those two aspects. While surface roughness mainly depends on cutting parameters (i.e., feed rate) and tool geometry, the geometrical errors are influenced by the workpiece compliance: the deflection induced by the cutting forces, indeed, impacts on the actual depth of cut, introducing form errors, potentially leading to scraps or unacceptable defects [2]. This issue is critical for flexible components (e.g., slender shafts), since significant workpiece deflection could occur during machining. Therefore, the minimization of geometrical errors while maintaining high productivity entails generating a machining cycle based on both the component stiffness and the cutting forces (i.e., the cutting parameters and workpiece material). The simplest approach that could be pursued to achieve such goal is based on trial-and-error procedures, that often reflect in uncertain manufacturing lead times. Moreover, this method gets less acceptable as the batch dimension decreases and the material cost increases, and it does not ensure the selection of an optimal solution, feasible only by getting a deeper understanding of the process behavior.

Digital Twin (DT) of machining processes can be exploited to reach such a goal [3]. DTs are virtual replica of a physical entity that could be used to analyze the process and make decisions through interaction between physical and virtual world. In the specific case a DT that includes cutting mechanism is required [3]. In this context, Zhu et al. developed a DT for machining process

Italian Manufacturing Association Conference - XVI AITeM   Materials Research Forum LLC
Materials Research Proceedings 35 (2023) 485-494   https://doi.org/10.21741/9781644902714-57

of thin-walled parts [4], while Afazov and Scrimieri focused their work on chatter vibrations [5]. This work presents a mechanism model for turning of slender workpiece that allows deflection compensation and could potentially enable the development of a Digital Twin when connected with the physical world (e.g., machine tool sensors).

On one side, the cutting forces can be estimated by means of simplified models based on cutting conditions, tool geometry and material proprieties [1]. The most adopted approach is taking advantage of mechanistic force models, tuned using experimentally identified cutting force coefficients [6]. In turning such an approach is generally used to compute the cutting force (in the cutting speed direction) and the rake face force. If decomposition of rake face force on feed and depth directions is required, as in the case of deflection estimation, chip flow angle needs to be computed. The simplest and most used approximation of such angle can be obtained by using the formulation proposed by Colwell [7].

On the other hand predictive models of workpiece deflection have been proposed for turning of slender shaft [8–10] The most effective methods are based on numerical analysis [2,11] (Finite Element Method, FEM), that is nowadays a commonly used tool, but requires specific high level knowledge and expertise.

This work presents the DRITTO (Deflection Reduction In Turning by Toolpath Optimization) project that aims at developing a digital solution for turning of flexible components with the purpose of generating optimized toolpaths to minimize geometrical errors, compensating the workpiece deflection. First the paper presents the proposed digital solution, describing the different blocks in which is composed. The numerical analyses involved in the toolpath optimization process are simplified to make their automation feasible and time effective. Cutting forces are estimated using a mechanistic force model and using Colwell formulation for chip flow. Workpiece behavior is modeled using Timoshenko beam model, its generation and update during the machining process are automatic, only toolpath and stock geometry are needed. Second, experimental validation is presented, specific tests were carried out of simplified case studies focusing on roughing operations, where geometrical errors are relevant, and the machined geometry could affect the subsequent phases (i.e., semi-finishing and finishing). Finally, conclusions are drawn, and future activities described.

Proposed digital solution
The proposed digital solution in schematized in Fig. 1. At the background level, the digital solution will include a toolpath generation model that will be interfaced with a simplified FEM environment to simulate the workpiece behavior under the effect of cutting forces. The system is composed by three modules: a) workpiece FE modelling, b) turning process modelling, c) toolpath optimization.

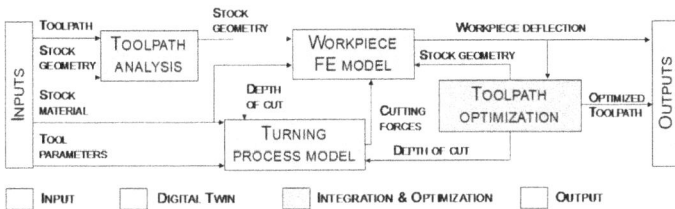

Fig. 1 General overview of the DRITTO digital solution.

The modules will be configured as an integrated solution: only the stock geometry and material, toolpath and tool geometry will be needed. The innovative idea underpinning the DRITTO solution is to fully integrate the workpiece deflection predictive model, so that the toolpath computation can be performed considering workpiece compliance changing during the turning process, as effect of material removal, and the instantaneous cutting conditions.

Input. The proposed approach requires the toolpath and the stock to compute and update the actual geometry of the workpiece and estimate the actual depth of cut. Toolpath is input as a standard ISO code (i.e., G Code), from which the system extracts the actual toolpath and the cutting parameters (i.e., cutting velocity and feed). The toolpath is then discretized to analyze the process with the desired resolution.

Stock geometry is included as a text file, written in a specific format: starting from tailstock (or free end) of the workpiece the segments with continuous radius variation along the axis are identified. Each segment is characterized by outer and inner radius at both its ends and by its length, hence every line of the text file represents one segment. Text file is reporting five different values for each line: initial outer radius, initial inner radius, final outer radius, final inner radius, length of the segment. This approach allows to represent any axisymmetric workpiece geometry. Portion inside the chuck should not be included in this representation.

In addition to these two inputs, tool geometry and workpiece material data are required. For the tool, lead angle and corner radius are needed, while for the workpiece material both elastic material proprieties (i.e., Elastic Modulus and Poisson Ratio) and cutting force coefficients should be input.

Turning process model. The cutting force model implemented in this work is provided below:

$$F_t = K_{tc}bh + K_{te}b \qquad F_{rf} = K_{rfc}bh + K_{rfe}b \tag{1}$$

$$F_f = F_{rf}\cos(\Omega) \qquad F_{ap} = F_{rf}\sin(\Omega) \tag{2}$$

where F_t is the cutting force in the cutting speed direction, while F_{rf} on the rake face plane, decomposed in feed force (F_f) and depth of cut force (F_{ap}), K_{ic} are the cutting force coefficients and K_{ie} the edge coefficients, b is the contact length and h is the chip thickness, Ω is the chip flow angle. In this work the Colwell approximation for such angle was used [7].

Workpiece FE model. Workpiece deflection is estimated by applying predicted cutting forces on a FE model of the component. Since slender workpieces are the target of the proposed approach Timoshenko beam 1D model [12] was selected as modeling strategy. A dedicated algorithm was implemented starting from the workpiece geometry to create nodes distribution (i.e., mesh) and element stiffness matrices, then assembled in the unconstrained component stiffness matrix K (Fig. 2). At each machining step the geometry is updated, and stiffness matrix reconstructed.

Constrained stiffness matrix is obtained by considering boundary conditions of chuck and tailstock (if present). In this work constraints are not considered rigid, therefore a 6x6 diagonal stiffness matrices are adopted as follows:

$$K_{chuck} = diag(K_{xc}, K_{yc}, K_{zc}, K_{rotxc}, K_{rotyc}, K_{rotzc}) \tag{1}$$

$$K_{tail} = diag(K_{xt}, K_{yt}, K_{zt}, K_{rotxt}, K_{rotyt}, K_{rotzt}) \tag{2}$$

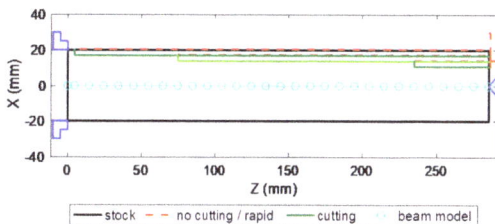

Fig. 2 Stock and toolpath example.

where diag() is the diagonal matrix that is characterized on its diagonal by the values provided in the bracket and K_{ij} are the stiffness value on the i degree of freedom for the j constraint. K_{chuck} is then assembled to the unconstrained matrix K by adding such matrix to the last node, while K_{tail}, if present, is assembled to the first node (i.e., end of the workpiece).

Toolpath optimization. Using the predicted cutting forces and the proposed modeling strategy it is possible to estimate workpiece deflection during the process by performing static analysis at each step. An iterative approach was used to consider the actual workpiece geometry and depth of cut: first the deflection was estimated using the commanded depth of cut, such first-attempt deflection was used to update both workpiece geometry and depth of cut, and a new deflection was evaluated, such cycle was repeated until convergence (minimization of the error on predicted deflections). Workpiece deflections are then used to compute the effective machined geometry (i.e., the effective workpiece radius, R_{eff}) as follows:

$$R_{eff} = \sqrt{(X - dx)^2 + dy^2} \tag{3}$$

where X is the commanded motion of the tool (i.e., desired radius), dx is the deflection on depth of cut direction and dy is the deflection on the cutting direction. Starting from such values, compensated toolpath is derived and written in a new file using ISO standard.

Integration. The different modules are integrated by exchanging data as highlighted in Fig. 1. Toolpath analysis computes depth of cut and workpiece geometry at the different steps of the machining operations, the first is input to the process modules to predict cutting forces, while workpiece geometry is essential for the beam model generation. Cutting forces are applied to such model to predict deflection. The first prototype of the digital solution was developed in MATLAB.

Experimental results

An experimental validation of the proposed approach was carried out at Meccanica Ceccarelli & Rossi facility. Turning operations were performed on a CNC lathe Mori Seiki SL-2500Y, equipped with a dynamometer (Kistler 9257A) to acquire cutting forces (Fig. 3).

Case studies. The proposed approach was tested on different geometries, using the same tool and material (C45 Steel). A Sandvik Coromant CNMG 120408-PM 4425 insert was used (corner radius 0.8 mm), mounted on a T-Max toolholder P DCLNL 2525M 12 (lead angle -5°). Four case studies were machined starting from a 40 mm bar: three simple single diameter cylinders (analyzing a single pass) and one shaft with three different diameters (analyzing three subsequent passes), their geometries are shown in the figures (Fig. 4, Fig. 5, Fig. 6, Fig. 7). Tailstock was used for all the case studies.

Italian Manufacturing Association Conference - XVI AITeM   Materials Research Forum LLC
Materials Research Proceedings 35 (2023) 485-494   https://doi.org/10.21741/9781644902714-57

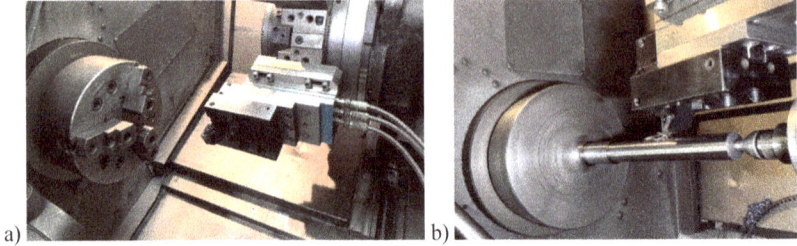

a)   b)

Fig. 3 a) experimental set-up b) turning of a case study.

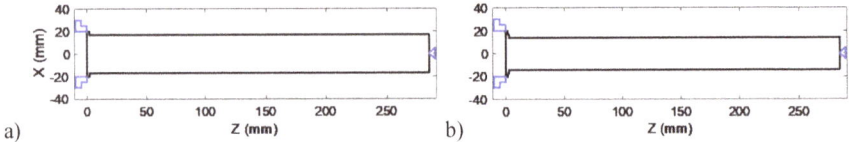

a)   b)

Fig. 4 Case A (overhang 287.5 mm) a) stock D: 34 mm b) final D: 28 mm.

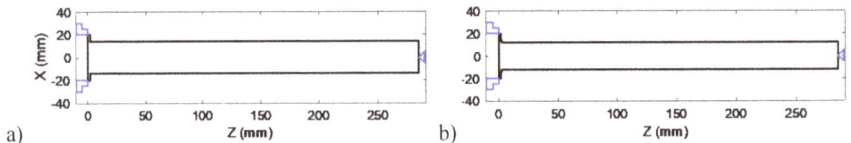

a)   b)

Fig. 5 Case B (overhang 287.5 mm) a) stock D: 28 mm b) final D: 24 mm.

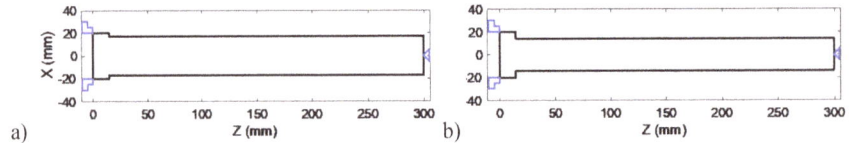

a)   b)

Fig. 6 Case C (overhang 299.5 mm) a) stock D: 34 mm b) final D: 28 mm.

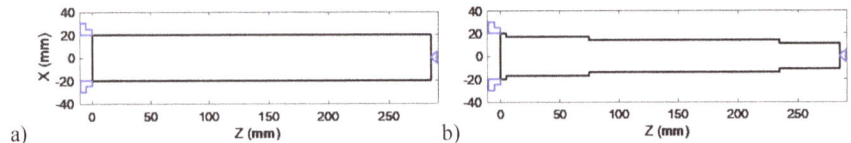

a)   b)

Fig. 7 Shaft (overhang 287.5 mm) a) stock D: 40 mm b) after 1st pass D: 34 mm L: 280 mm, 2nd pass D: 28 mm L: 210 mm and final pass D: 22 mm L: 50 mm.

Roughing operations were investigated using 200 m/min cutting velocity, feed 0.2 mm/r and radial depth of cut 2 mm (case study B) and 3 mm (all the other case studies).

Cutting forces. Cutting force coefficients were identified for the specific tool-material couple performing preliminary tests, acquiring cutting forces and using the procedure reported by Altintas [1], results are summarized in Table 1.

Italian Manufacturing Association Conference - XVI AITeM   Materials Research Forum LLC
Materials Research Proceedings 35 (2023) 485-494   https://doi.org/10.21741/9781644902714-57

Table 1 Cutting force coefficients

K_{tc} [MPa]	K_{te} [N/mm]	K_{rfc} [MPa]	K_{rfe} [N/mm]
1748.5	99.2	703.0	92.5

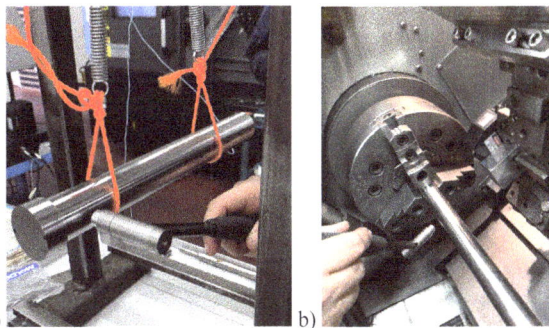

a)   b)

Fig. 8 Impact testing a) free-free boundary condition b) constrained.

Workpiece. To test the proposed approach without the uncertainties of material proprieties and constraints stiffnesses, some preliminary tests were performed to tune such values. Free-free modal analysis on bar specimens was carried out to identify material properties (Fig. 8a) through impact testing. Typical steel values were identified: Young Modulus: 210150 MPa, Poisson Ration: 0.28.

In addition, experimental modal analysis in the constrained configurations was used to estimate chuck and tailstock stiffnesses (Fig. 8b). Results are presented in Table 2. The chuck was modeled as a fixed end, while tailstock as a pinned end (i.e., free rotations).

Prediction results. To investigate the effectiveness of the proposed solution in estimating the machined workpiece geometry, a comparison between predicted and measured diametral errors was carried out and shown in Fig. 10. Results show good agreement between measured and predicted values, especially in terms of overall error difference between tailstock and chuck (average deviation on predicting the errors of 6.0, 5.4, 4.7, 4.1 μm in case studies a, b, c and shaft respectively). However, some discrepancies are found in the error shape, probably due to the tailstock modeling. Indeed, it is worth to point out that in the tested scenarios, the tailstock stiffness plays a crucial role in determining the error.

Toolpath compensation. The proposed approach was then applied to compute the compensated toolpath for all the case studies and machined geometries were measured for both compensated and non-compensated toolpath to evaluate its effectiveness. Results are shown in Fig. 10. As clearly emerges from the results, the compensated toolpath has proven to be effective in drastically reducing the errors, by at least halving the maximum error and by smoothing the shape. The average reduction achieved was about 62%, 74%, 68%, 72%. in case studies a, b, c and shaft respectively.

Table 2 Constraints stiffnesses

	Kx / Ky [N/mm]	Kz [N/mm]	K rotx / K roty [N mm/rad]	K rotz [N mm/rad]
Chuck	3.30e4	1e15	6.87e7	3e7
Tail	5.50e3	1e15	0	0

Italian Manufacturing Association Conference - XVI AITeM   Materials Research Forum LLC
Materials Research Proceedings 35 (2023) 485-494   https://doi.org/10.21741/9781644902714-57

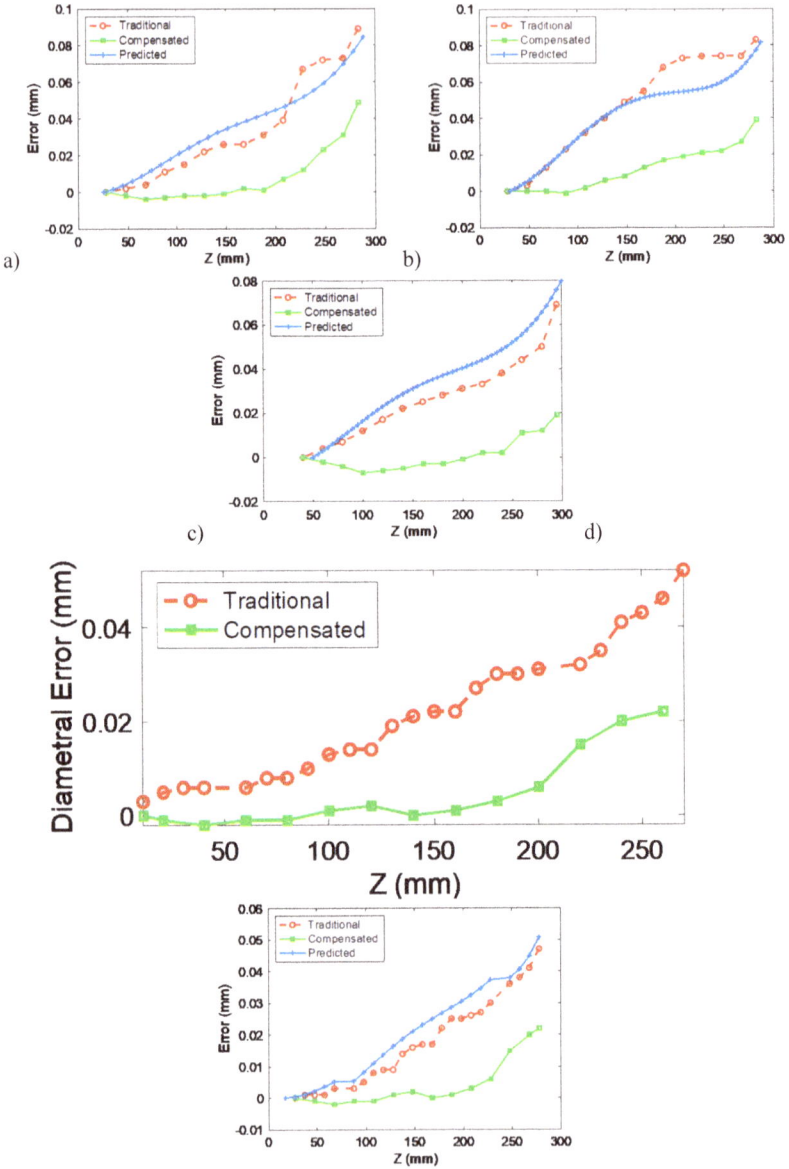

Fig. 9 Comparison between predicted and measured diametral errors and measured values with compensation strategy on: a) case study A, b) case study B, c) case study C, d) shaft.

The overall error appears to be well-compensated until the area close to the tailstock. This is probably due to local effects of tailstock constraints and the non-modelled impacts of the cut entry. These aspects need to be further investigated to improve the solution. However, even in this part the reduction achieved was significant: about 45%, 53%, 50%, 55% in case studies a, b, c and shaft respectively.

Conclusions

The DRITTO project aims at developing a tailored solution for the computation of optimized toolpaths for turning of slender workpiece. The project is still ongoing, and an intermediate validation phase has just concluded. The solution developed is composed of different modules that have been, at the current stage, individually validated. The prediction of workpiece deflection, based on simplified FE models and cutting forces estimation, has shown to be adequate accurately in estimating the overall shape errors. Further activities will be focused on investigating alternative constraints modeling strategy for the tailstock to considering local effects and improve the prediction accuracy. Such predictive module was exploited to compute the compensated toolpath through a dedicated approach and results confirm that providing an accurate prediction of the shape error could represent an effective approach for its reduction. The experimental validation shows that the error is globally reduced by at least half compared to the non-compensated tests. Such results were consistent for all the case studies investigated.

Although some residual errors could be highlighted approaching the tailstock, even at the current stage the solution seems promising in drastically reducing the shape error in roughing operations which could be exploited to avoid the need of semi-finishing phases, in line with the goal of the DRITTO project.

Further developments will be focused on:

- Investigating the effects of the tailstock constraints.
- Extending the validation to finishing operations.
- Studying the potential synergies with machine tool sensors to build an actual DT.
- Developing a Graphical User Interface for the implementation of the solution in the SME manufacturing environment.

Acknowledgements

This research was developed within the DRITTO project, funded as an experiment of DIH-World, an Horizon2020 project (grant agreement 952176). The authors wish to thank all the project partners.

References

[1] Y. Altintas, Manufacturing Automation: Metal Cutting Mechanics, Machine Tool Vibrations, and CNC Design, (2012).

[2] Y. Altintas, O. Tuysuz, M. Habibi, Z.L. Li, Virtual compensation of deflection errors in ball end milling of flexible blades, CIRP Ann. 67 (2018) 365–368. https://doi.org/10.1016/j.cirp.2018.03.001

[3] S. Liu, J. Bao, P. Zheng, A review of digital twin-driven machining: From digitization to intellectualization, J. Manuf. Syst. 67 (2023) 361–378. https://doi.org/10.1016/j.jmsy.2023.02.010

[4] Z. Zhu, X. Xi, X. Xu, Y. Cai, Digital Twin-driven machining process for thin-walled part manufacturing, J. Manuf. Syst. 59 (2021) 453–466. https://doi.org/10.1016/j.jmsy.2021.03.015

[5] S. Afazov, D. Scrimieri, Chatter model for enabling a digital twin in machining, Int. J. Adv. Manuf. Technol. 110 (2020) 2439–2444. https://doi.org/10.1007/s00170-020-06028-9

[6] N. Grossi, L. Sallese, A. Scippa, G. Campatelli, Speed-varying cutting force coefficient identification in milling, Precis. Eng. 42 (2015) 321–334. https://doi.org/10.1016/j.precisioneng.2015.04.006

[7] L. V Colwell, Predicting the Angle of Chip Flow for Single-Point Cutting Tools, Trans. Am. Soc. Mech. Eng. 76 (2022) 199–203. v10.1115/1.4014795

[8] P.G. Benardos, G.C. Vosniakos, Prediction of surface roughness in CNC face milling using neural networks and Taguchi's design of experiments, Robot. Comput. Integr. Manuf. 18 (2002) 343–354. https://doi.org/10.1016/S0736-5845(02)00005-4

[9] G. Jianliang, H. Rongdi, A united model of diametral error in slender bar turning with a follower rest, Int. J. Mach. Tools Manuf. 46 (2006) 1002–1012. https://doi.org/10.1016/j.ijmachtools.2005.07.042

[10] J.R.R. Mayer, A.-V. Phan, G. Cloutier, Prediction of diameter errors in bar turning: a computationally effective model, Appl. Math. Model. 24 (2000) 943–956. https://doi.org/10.1016/S0307-904X(00)00027-5

[11] M. Soori, B. Arezoo, M. Habibi, Tool Deflection Error of Three-Axis Computer Numerical Control Milling Machines, Monitoring and Minimizing by a Virtual Machining System, J. Manuf. Sci. Eng. 138 (2016). https://doi.org/10.1115/1.4032393

[12] A. Ertürk, H.N. Özgüven, E. Budak, Analytical modeling of spindle–tool dynamics on machine tools using Timoshenko beam model and receptance coupling for the prediction of tool point FRF, Int. J. Mach. Tools Manuf. 46 (2006) 1901–1912. https://doi.org/10.1016/j.ijmachtools.2006.01.032

Italian Manufacturing Association Conference - XVI AITeM
Materials Research Proceedings 35 (2023) 495-

Materials Research Forum LLC
https://doi.org/10.21741/9781644902714-58

Influence of long and short glass fiber on the mechanical behaviour of a single cell metamaterial

Luca Giorleo[1,a*], Antonio Fiorentino[1,b], Stefano Pandini[1,c] and Elisabetta Ceretti[1,d]

[1]Università degli Studi di Brescia, Brescia, Via Branze 38, 25123, Italy

[a]luca.giorleo@unibs.it, [b]antonio.fiorentino@unibs.it, [c]stefano.pandini@unibs.it, [d]elisabetta.ceretti@unibs.it

Keywords: Material Extrusion, FFF, Fibre Reinforced Plastic, Metamaterial, Tensile Strength, Poisson Ratio

Abstract. Additive manufacturing is presenting new challenges in various aspects of part production. Among these, the potential benefits derived from material complexity have been growing in recent years, especially when using polymeric materials. In fact, mixing polymers with long/short fibres lead to moderate to significant improvements in the mechanical properties of the parts. The degree of improvement strongly depends on the part geometry and can become critical in the case of a workpiece with a repeating pattern, such as metamaterials. In this preliminary research, the authors investigate the mechanical performance of a single- hourglass cell which is a common auxetic geometry used to achieve a negative Poisson ratio in metamaterials. Nylon was used as the matrix, and glass as the fibre. FFF additive process was used to produce samples with different cell designs (in width, size, inclination) and the nature of the fibres (long and short). The results were analysed using statistical methods.

Introduction

Auxetic materials are a unique class of materials that exhibit a counterintuitive property: they expand in all directions when stretched. Unlike conventional materials, which contract when stretched, auxetic materials have a negative Poisson's ratio, which means that they become wider and thicker when subjected to tensile forces [1,2].

One of the most interesting properties of auxetic materials is their ability to absorb impact and dissipate energy more effectively than conventional materials. This makes them ideal for use in a wide range of applications, including protective gear, shock absorbers and insulation [3].

Another advantage of auxetic materials is their ability to conform to complex shapes and surfaces, making them useful in applications where conventional materials would be difficult to use. For example, auxetic foam can be used in the design of customized medical implants, while auxetic fabrics can be used in the production of high-performance sports clothing [4, 5].

In Figure 1a there is an example of auxetic hourglass geometry while in figure 1b their behavior bulk behavior under compression load is represented.

(a) Auxetic cell design   (b) Behaviour scheme under tensile strength

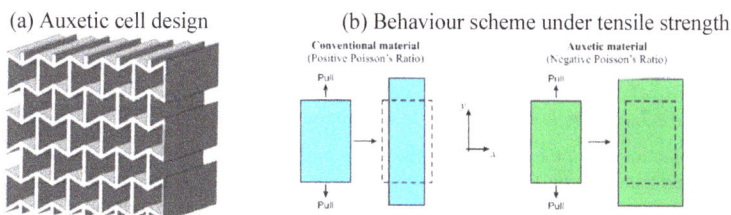

Figure 1. Design and characteristic of auxetic materials.

Italian Manufacturing Association Conference - XVI AITeM   Materials Research Forum LLC
Materials Research Proceedings 35 (2023) 485-494   https://doi.org/10.21741/9781644902714-57

Due to their complex shapes, additive manufacturing proves to be a more advantageous geometry to produce auxetic materials. Various studies can be found in literature regarding the characterization of auxetic cells produced through the Fused Filament Fabrication (FFF) process [6-10]. Mainly, materials such as PLA and Nylon. However, FFF can produce composite materials with high mechanical properties by extruding carbon, glass, or Kevlar fibers into the component. Fibers can be supplied as long or short, significantly changing the material properties. From the literature, it is known that long fiber composites have considerably better performance [11]. It should be noted that such an improvement has been verified on relatively simple geometries (rectangular specimens, dog-bone specimens, cylindrical specimens), and it is not yet clear if the gap between long and short fibers still holds true for complex geometries such as those of auxetic structures.

The aim of this research was to compare the mechanical properties of auxetic cells with a specific geometry created through additive manufacturing using long and short glass fiber composite materials. Through the FFF 3D printing technique, samples with a hourglass geometry were fabricated to study their tensile behavior, focusing on the analysis of the forces required to deform the sample, stiffness, and measured Poisson's ratio.

Materials and Methods
In this research, an hourglass geometry was used as an auxetic cell to analyse the performance of short and long fiber composites. Three levels of cell angle α (60°, 70°, and 80°) and three cell widths B (20 mm, 30 mm, and 40 mm) were designed, resulting in nine different combinations; thickness of sample was equal to 2 mm. Each of these combinations was fabricated using fused filament fabrication technology. The S3 printer provided by Ultimaker was used to print samples using nylon reinforced with short glass fiber (SGF), while the Mark2 printer provided by Markforged was used to produce samples using nylon reinforced with long glass fiber (LGF). The quantity of fiber designed for LGF was equal in weight with SGF. Figure 2a shows the auxetic cell design, while figures 2b and 2c show the different path strategies used for SGF-Nylon and LGF-Nylon, respectively

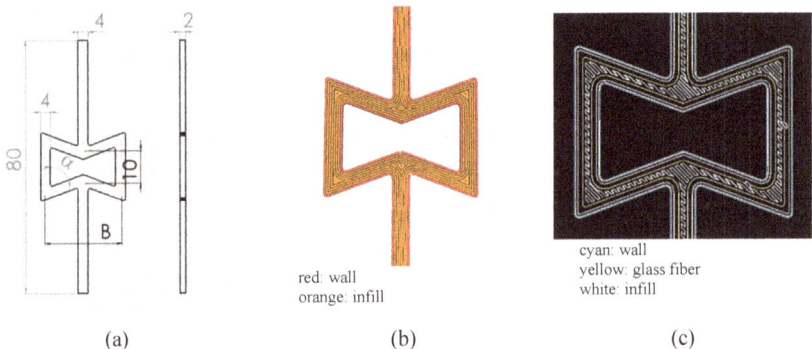

red: wall
orange: infill

cyan: wall
yellow: glass fiber
white: infill

(a)   (b)   (c)

Figure 2. a) Sample geometry, b) SGF-Nylon and c) LGF-Nylon infill strategies.

After production, tensile tests were conducted to measure the performance of the samples. The tests were performed using the Instron 3360 machine, with a gauge speed set at 1 mm/min. A reference pattern was applied on the surface of each sample before the tests and a digital camera was used to observe the tensile tests and acquire digital images. GOM correlate software was then used to evaluate the sample deformations. Finally, the maximum force and Poisson's ratio values were estimated.

Results and Preliminary Conclusion

Figure 3 plots preliminary results in terms of main effective plots and Tukey range test.
Regarding the Poisson ratio of the sample:

- is directly proportional to the cell dimension B and the α value.
- the type of fibre is not influent.

Regarding the Maximum force:

- is directly proportional to the cell dimension B, while α angle is not influent.
- it is higher in short fibre samples.

Overall, the preliminary results suggest that the type of fibres (short or long) does not influence the auxetic behavior in hourglass cells. On the contrary, the geometry of the cell itself determines the value of the Poisson ratio. The influence of the fiber type is more significant on the stability of the cell. In fact, the point of maximum force corresponds to the moment at which geometrical instability occurs and the sample is no more planar. Therefore, samples with short fibers are able to withstand higher loads before instability.

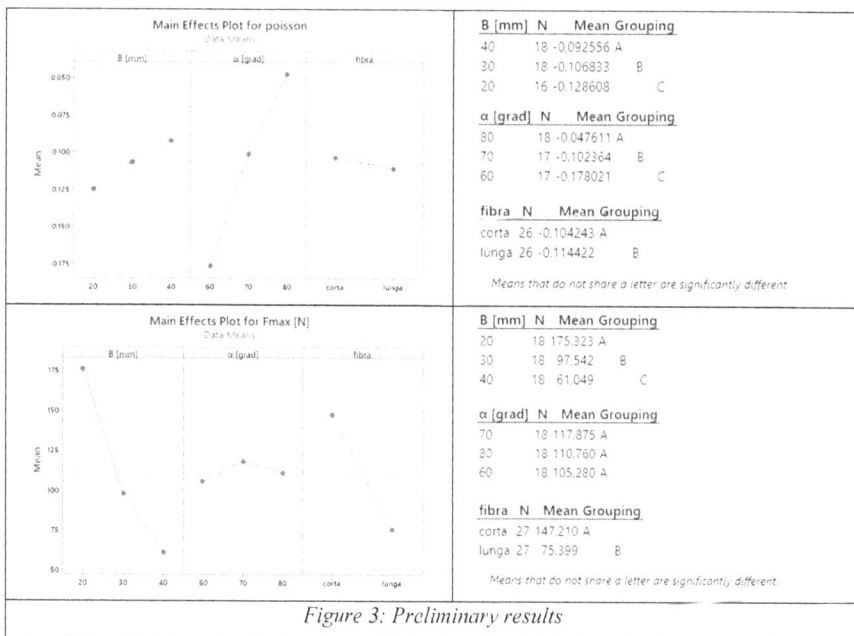

Figure 3: Preliminary results

References

[1] Essassi K., Rebiere J.-L., mahi A.E., Souf M.A.B., Bouguecha A., Haddar M., Indentation Characteristics of Auxetic Composite Material (2023) Lecture Notes in Mechanical Engineering, pp. 520 – 526. https://doi.org/10.1007/978-3-031-14615-2_58

[2] Li Z.-Y., Wang X.-T., Ma L., Wu L.-Z., Wang L., Auxetic and failure characteristics of composite stacked origami cellular materials under compression (2023) Thin-Walled Structures, 184, art. no. 110453. https://doi.org/10.1016/j.tws.2022.110453

[3] Carneiro V.H., Meireles J., Puga H., Auxetic materials - A review (2013) Materials Science-Poland, 31 (4), pp. 561 – 571. https://doi.org/10.2478/s13536-013-0140-6

[4] Yang W., Li Z.-M., Shi W., Xie B.-H., Yang M.-B., On auxetic materials (2004) Journal of Materials Science, 39 (10), pp. 3269 – 3279. https://doi.org/10.1023/B:JMSC.0000026928.93231.e0

[5] Duncan O., Shepherd T., Moroney C., Foster L., Venkatraman P.D., Winwood K., Allen T., Alderson A., Review of auxetic materials for sports applications: Expanding options in comfort and protection (2018) Applied Sciences (Switzerland), 8 (6), art. no. 941. https://doi.org/10.3390/app8060941

[6] Chen, Y., He, Q. 3D-printed short carbon fibre reinforced perforated structures with negative Poisson's ratios: Mechanisms and design (2020) Composite Structures, 236, art. no. 111859. https://doi.org/10.1016/j.compstruct.2020.111859

[7] Ingrole, A., Hao, A., Liang, R. Design, and modeling of auxetic and hybrid honeycomb structures for in-plane property enhancement (2017) Materials and Design, 117, pp. 72-83. https://doi.org/10.1016/j.matdes.2016.12.067

[8] Quan, C., Han, B., Hou, Z., Zhang, Q., Tian, X., Lu, T.J. 3d printed continuous fiber reinforced composite auxetic honeycomb structures (2020) Composites Part B: Engineering, 187, art. no. 107858. https://doi.org/10.1016/j.compositesb.2020.107858

[9] Najafi, M., Ahmadi, H., Liaghat, G. Experimental investigation on energy absorption of auxetic structures (2019) Materials Today: Proceedings, 34, pp. 350-355. https://doi.org/10.1016/j.matpr.2020.06.075

[10] Lvov, V.A., Senatov, F.S., Korsunsky, A.M., Salimon, A.I. Design and mechanical properties of 3D-printed auxetic honeycomb structure (2020) Materials Today Communications, 24, art. no. 101173. v10.1016/j.mtcomm.2020.101173

[11] Pappas J.M., Thakur A.R., Leu M.C., Dong X. A comparative study of pellet-based extrusion deposition of short, long, and continuous carbon fiber-reinforced polymer composites for large-scale additive manufacturing (2021) Journal of Manufacturing Science and Engineering, Transactions of the ASME, 143 (7), art. no. 071012. https://doi.org/10.1115/1.4049646

Keyword Index

About the Editors

Prof. Luigi Maria Galantucci
Brief Biography

Politecnico di Bari, Dipartimento di Meccanica, Matematica e Management, Via Orabona 4 - 70125 Bari, ITALIA
+390805963796

luigimaria.galantucci@poliba.it
https://www.dmmm.poliba.it/index.php/it/profile/109-lgalantucci
- ORCID https://orcid.org/0000-0003-2892-7433
- Google scholar profile Luigi Maria Galantucci - Google Scholar
- Scopus 6701626861

Full Professor in Technologies and Production Systems Dept. of Mechanics, Mathematics and Management, Politecnico di Bari (Italy) since 2000.

2019-2021 Deputy Rector for the Strategic Planning, Head of several laboratories of the DMMM – Politecnico di Bari: *Rapid Prototyping and Reverse, Engineering* and the *MICROTRONIC Micromachining and Micro-measurement.*

Fellow for life since 2006 of CIRP – Collège International pour l'étude scientifique des techniques de Production mécanique, now International Academy for Production Engineering.

"Doctor Honoris Causa" conferred by the Academic Senate of the Politechnic University of Tirana – PTU (Albania - October 16th, 2009).

CEO and President of Polishape 3D srl, a spin-off company of Politecnico di Bari (2011-2022).

Since the year 1981, he is involved in several research projects funded by the European Union, the Italian Minister of Public Education, the Italian Minister of the Scientific and Technological Research, the National Council of the Research, on: 3D scanning and measurement of micro components, Reverse Engineering, Rapid Prototyping and Additive Manufacturing, Laser Material Processing, Thermomechanical simulation of manufacturing processes (Welding, Heat Treatment, Forming), Manufacturing Processes, Computer Aided Manufacturing, Process Planning, Feature Technology, Manufacturing System Analysis and simulation, Biomechanics, Anthropometry.

Recent research projects coordination:

• 2013-2016 Coordinator for the Politecnico di Bari of the European project ADRIATInn - An Adriatic Network for Advancing Research Development and Innovation towards the Creation of new Policies for Sustainable Competiveness and Technological Capacity of SMEs, Consortium with di 20 European partners.

• 2016- now Responsible of the research activity on "Reverse Engineering of aeronautical components aimed at their repair by means laser deposition (DL) and cold spray (CS) processes" in the "Apulia Development Center for Additive Repair" laboratory done by the GE Avio and the Politecnico di Bari, installed at the facilities of the Politecnico di Bari.

• 2018-2020 and 2022-2023 Project Leader of the INTERREG IPA CBC ITALY–ALBANIA–MONTENEGRO PROGRAMME 3D-IMP-ACT (Virtual reality and 3D experiences to IMProve territorial Attractiveness, Cultural heritage, smart management and Touristic development), that involves 5 partners from Italia, Albania and Montenegro.

- 2019-2022 Scientific Responsible for the Politecnico di Bari of the project PON MIUR ARS01_00806 "Innovative solutions for quality and sustainability of ADDitive manufacturing processes (SIADD)", that involves 14 partners among Industries, Research Centres and Universities.
- 2022-2026 Scientific Responsible for the Politecnico di Bari of Spoke 6 "Additive Manufacturing" of the PNRR Mission 4 "Education and Research" (PNRR Extended Partnership 11 entitled " MICS - Made in Italy Circolare e Sostenibile", funded by the European Union under the NextGenerationEU.

Prof. Luca Settineri, PhD
Brief Biography

Politecnico di Torino, Department of Management and Production Engineering, Corso Duca degli Abruzzi 24 - 10129 Torino, ITALY
+390110907230

luca.settineri@polito.it
https://www.polito.it
- ORCID https://orcid.org/0000-0002-5816-170X
- Scopus ID 6602463988

2011 – to date: Full Professor in Manufacturing Technologies and Systems in the Dept. of Management and Production Engineering of Politecnico di Torino (Italy) since 2011.

2018 – to date: ViceRector for Planning and Infrastructures of Politecnico di Torino

2022 – to date: President of the Italian Manufacturing Association AITeM

2011 – to date: Fellow Member of the International Academy for Production Engineering (CIRP)

2015 – 2018: Head of the Department of Management and Production Engineering

2010 – 2015: Dean of the PhD Course in Management, Production and Design

Teaching activities:

Since A.Y. 2012/2013, he is teaching the courses "Analysis and Management of Production Systems" and "Sustainable Manufacturing" for the Management Engineering program of Politecnico di Torino, as well as lectures on the topic of Sustainability in Manufacturing in several doctoral courses.

Furthermore, over the last thirty years, Prof. Settineri has delivered lectures at undergraduate, graduate, doctoral level as well as for the Master School of Politecnico, on all subjects related to Manufacturing Technologies and Systems, in Italian, in English and in French, in Italy, Belgium and China.

Prof. Settineri has coordinated several regional, national and international research projects, among which:

- 2017: Project "Sustainable integration of additive/subtractive manufacturing processes for automotive applications," funded by FCA Group.
- 2018-2020: "Sustainable integration of HYbrid additive/subtractive MANufacturing for difficult-to-cut materials (Acronym: HY-MAN)", funded under the 'POR-FESR 2014/2020 - Axis I.

• 2021-2022: "Production of complex geometries through additive manufacturing using wire deposition - WireAdd," funded within the POR-FESR 2014-2020, call PRISM-E.

Prof. Settineri is a member of the Editorial Review Board of the CIRP Journal of Manufacturing Science and Technology and of the Green Manufacturing Open journal.

Prof. Settineri has authored or co-authored over 200 publications, on machining, joining, coating, rapid prototyping and additive manufacturing, sustainability in manufacturing processes.

www.ingramcontent.com/pod-product-compliance
Lightning Source LLC
Chambersburg PA
CBHW061202220326
41597CB00015BA/1224